Motives and Algebraic Cycles

FIELDS INSTITUTE
COMMUNICATIONS

THE FIELDS INSTITUTE FOR RESEARCH IN MATHEMATICAL SCIENCES

Motives and Algebraic Cycles

A Celebration in Honour of
Spencer J. Bloch

Rob de Jeu
James D. Lewis
Editors

American Mathematical Society
Providence, Rhode Island

The Fields Institute
for Research in Mathematical Sciences
Toronto, Ontario

FIELDS

The Fields Institute
for Research in Mathematical Sciences

The Fields Institute is a center for mathematical research, located in Toronto, Canada. Our mission is to provide a supportive and stimulating environment for mathematics research, innovation and education. The Institute is supported by the Ontario Ministry of Training, Colleges and Universities, the Natural Sciences and Engineering Research Council of Canada, and seven Ontario universities (Carleton, McMaster, Ottawa, Toronto, Waterloo, Western Ontario, and York). In addition there are several affiliated universities and corporate sponsors in both Canada and the United States.

Fields Institute Editorial Board: Carl R. Riehm (Managing Editor), Juris Steprans (Acting Director of the Institute), Matthias Neufang (Interim Deputy Director), James G. Arthur (Toronto), Kenneth R. Davidson (Waterloo), Lisa Jeffrey (Toronto), Barbara Lee Keyfitz (Ohio State), Thomas S. Salisbury (York), Noriko Yui (Queen's).

2000 *Mathematics Subject Classification.* Primary 11-XX, 14-XX, 16-XX, 19-XX, 55-XX.

Library of Congress Cataloging-in-Publication Data

Motives and algebraic cycles : a celebration in honour of Spencer J. Bloch / Rob de Jeu, James D. Lewis, editors.
    p. cm. — (Fields Institute Communications ; v. 56)
    Includes bibliographical references.
    ISBN 978-0-8218-4494-6 (alk. paper)
    1. Algebraic cycles. 2. Motives (Mathematics). I. Bloch, Spencer. II. Jeu, Rob de, 1964– III. Lewis, James Dominic, 1953–

QA564.M683 2009
512′.66–dc22

2009023440

Contents

Introduction

Spencer J. Bloch, one of the world's leading mathematicians, has had a substantial impact on, in particular, algebraic K-theory, algebraic cycles and motives. This conference and subsequent proceedings served as a tribute to and a celebration of his work and a dedication to his mathematical heritage. Among those who participated in this conference were his collaborators, former students, a majority of those who have benefited from his trail blazing work, as well as recent post-docs and students. The atmosphere of this conference could be accurately described as "electric". The feeling one had among participants and speakers alike, was that of being identified as "adopted" students of Spencer Bloch in that they had read a rather large proportion of his work and were strongly influenced by him. His ability to connect physics with the subject of motives and degenerating mixed Hodge structures is very typical of his style of doing mathematics. One instance of this, earlier on, was his insight into how algebraic K-theory could be useful to algebraic geometry. This in turn revolutionized the subject of algebraic cycles. Virtually all the talks would begin with a few words on how Spencer influenced their research (sometimes rather humourously), followed by a presentation of recent developments in the field of motives and algebraic cycles.

The agreeable environment at Fields, specifically the excellent support and facilities of the Fields Institute, together with the backdrop of a bustling major city, added to the upbeat mood of this conference. Many of the participants voiced their appreciation of this fantastic conference, one which they are unlikely to ever forget!

Summary of works in this volume. This is a volume comprised of a number of independent research articles. In particular, these papers give a snapshot on the evolving nature of the subject of motives and algebraic cycles, written by leading experts in the field. A breezy summary of the flavour of these articles goes as follows. Motivated by the celebrated Hodge conjecture, D. Arapura's paper concerns a nonnegative integral invariant of a complex smooth projective variety X, which is zero precisely when the cohomology of X is generated by algebraic cycles. In general it gives a measure of the amount of transcendental cohomology of X, or alternatively can be rather loosely thought of as measuring the complexity of the motive of X. A. Beilinson's article, which addresses an earlier question of P. Deligne, provides an ϵ-factorization of the determinant of the period isomorphism associated to a holonomic D-module on a compact complex curve. The paper of H. Esnault and A. Ogus deals with a conjecture of Pink and Roessler on the nature of Hodge cohomology for a smooth projective scheme over an algebraically closed field, twisted by an invertible sheaf. H. Gillet's paper deals with extending arithmetic intersection theory to the case of Deligne-Mumford stacks. Such an extension is important since many computations involving arithmetic intersection theory are carried over modular varieties that are best viewed as stacks. S. Gorchinskiy's article discusses four approaches to the biextension of Chow groups and their equivalences, including an explicit construction given by S. Bloch. B. Kahn provides a new proof of the classical finiteness theorem of Lang and Néron on abelian varieties. The proof is different from the traditional one which uses heights. Rather, it has a cycle-theoretic flavour employing a famous correspondence trick used by S. Bloch. The paper of Shun-ichi Kimura concerns the following. For a smooth projective variety X over $\mathbb{C}$, Uwe Jannsen proved that if the cycle map cl : $CH^*(X; \mathbb{Q}) \to H^*(X, \mathbb{Q})$ is surjective, then it is actually bijective. Kimura generalizes this result to Chow motives. The work of N. Mohan Kumar, A. P. Rao, G. V. Ravindra revisits an earlier question of P. Griffiths and J. Harris of a Noether-Lefschetz nature, by providing the existence of a large class of counterexamples, which subsumes C. Voisinõs earlier counterexamples. M. Levine's paper concerns the construction of DG categories of smooth motives over a smooth k-scheme S (k a field) essentially of finite type. This in turn gives a well-behaved triangulated category of mixed motives over S generated by smooth and projective S-schemes. J. D. Lewis' contribution involves constructing infinite rank subspaces of cycles with prescribed transcendence degree (over $\mathbb{Q}$) in graded pieces of a certain candidate Bloch-Beilinson filtration for smooth projective varieties defined over subfields of $\mathbb{C}$, complementing some earlier works of C. Schoen and P. Griffiths, M. Green, and K. Paranjape. S. Lichtenbaum's article provides some evidence for the philosophy that all special values of arithmetic zeta and L-functions are given by Euler characteristics. Regarding the work of J. P. Murre and D. Ramakrishnan, the Galois symbol on elliptic curves is closely related to the cycle map of a product of elliptic curves. They study the Galois symbol for an elliptic curve E over a local field F with good ordinary reduction. In particular, they provide a class of examples of surfaces (self products of such elliptic curves), for which the cycle map is injective on the the Albanese kernel modulo p, where p is the characteristic of the residue field, even when the p-torsion of E is not F-rational. K. Murty's article deals with the semiregularity map defined by Bloch, and semiregularity in the context of abelian varieties. The paper by N. Naumann, M. Spitzweck, and P. A. Østvær uses the motivic Landweber exact functor theorem to deduce that the Bott inverted infinite projective space is homotopy algebraic K-theory. V. Snaith's paper draws attention to a phenomena in motivic stable homotopy in which Adams operations should control reduced power operations in motivic cohomology. J. Stienstra's paper recasts the correspondence between 3-dimensional toric Calabi-Yau singularities and quivers with superpotential in the setting of an $(N-2)$-dimensional abelian algebraic group acting on a linear space $\mathbb{C}^N$. He shows how the quiver with superpotential gives a simple explicit description of the Chow forms of the closures of the orbits in the projective space $\mathbb{P}^{N-1}$.

Acknowledgments

The editors are extremely grateful to the Fields Institute, the Clay Mathematics Institute, the National Science Foundation (NSF), as well as the organizers connected with the thematic program on geometric aspects of homotopy theory of schemes, for including this conference in their program, and for providing some NSF funds for participants from the USA. A special thanks goes to Rick Jardine, both logistically and financially, for his unwavering support in making this conference possible. We also wish to thank the contributors to this volume, and the referees, for doing a splendid job, particularly under pressing time constraints. Finally, it is a pleasure to acknowledge the excellent support provided by the staff of the Fields Institute both before and during the conference, as well as the efforts beyond the call of duty by Debbie Iscoe for the preparation of these proceedings.

The Editors,
Rob de Jeu
James D. Lewis

Speakers and Talks

Alexander A. Beilinson (Chicago),   *Towards a motivic descent*

Spencer Bloch (Chicago),   *Mixed Hodge structurers and motives in physics*

Hélène Esnault (Essen),   *Remarks and questions on coniveau*

Eric M. Friedlander (Northwestern),   *Musings about algebraic cycles modulo algebraic equivalence*

Henri Gillet (Univ. Illinois, Chicago),   *Arithmetic intersection theory on stacks*

Alexander Goncharov (Brown),   *Motivic fundamental groups of curves and Feynman integrals*

Luc Illusie (Paris-Sud),   *On Gabber's recent work in étale cohomology*

Uwe Jannsen (Regensburg),   *Finiteness of motivic cohomology and resolution of singularities*

Bruno Kahn (Paris 7),   *Questions on weights and Albanese varieties*

Kazuya Kato (Kyoto),   *Non-commutative Iwasawa theory and Hilbert modular forms*

Marc Levine (Northeastern),   *Motivic Postnikov towers*

Stephen Lichtenbaum (Brown),   *The conjecture of Birch and Swinnerton-Dyer is misleading*

Arthur Ogus (Berkeley),   *Functoriality of the Cartier transform*

Wayne Raskind (USC),   *p-adic intermediate Jacobians*

Takashi Saito (Tokyo),   *Wild ramification and the characteristic cycle of an ℓ-adic étale sheaf*

Chad Schoen (Duke),   *Some surfaces of general type in abelian varieties*

Tony Scholl (Cambridge),   *Some Eisenstein series*

Vasudevan Srinivas (TIFR),   *Oriented intersection multiplicities*

Fields Institute Communications
Volume **56**, 2009

Varieties with very Little Transcendental Cohomology

Donu Arapura
Department of Mathematics
Purdue University
West Lafayette, IN 47907 USA
arapura@math.purdue.edu

To Spencer Bloch

Given a complex smooth projective algebraic variety X, we define a natural number called the motivic dimension $\mu(X)$ which is zero precisely when all the cohomology of X is generated by algebraic cycles. In general, it gives a measure of the amount of transcendental cohomology of X. Alternatively, $\mu(X)$ may be rather loosely thought of as measuring the complexity of the motive of X, with Tate motives having $\mu = 0$, motives of curves having $\mu \leq 1$ and so on. Our interest in this notion stems from the relation to the Hodge conjecture: it is easy to see that it holds for X whenever $\mu(X) \leq 3$. This paper contains a number of estimates of μ; some elementary, some less so. With these estimates in hand, we conclude this paper by checking or rechecking this conjecture in a number of examples: uniruled fourfolds, rationally connected fivefolds, fourfolds fibred by surfaces with $p_g = 0$, Hilbert schemes of a small number points on surfaces with $p_g = 0$, and generic hypersurfaces.

We will work over $\mathbb{C}$. Let $H^*(-)$ denote singular cohomology with rational coefficients. The motivic dimension of a smooth projective variety X can be defined most succinctly in terms of the length of the coniveau filtration on $H^*(X)$. We prefer to spell this out. The *motivic dimension* $\mu(X)$ of X is the smallest integer n, such that any $\alpha \in H^i(X)$ vanishes on the complement of a Zariski closed set all of whose components have codimension at least $(i - n)/2$. The meaning of this number is further clarified by the following:

Lemma 0.1 *Any $\alpha \in H^i(X)$ can be decomposed as a finite sum of elements of the form $f_{j*}(\beta_j)$, where $f_j : Y_j \to X$ are desingularizations of subvarieties and β_j are classes of degree at most $\mu(X)$ on Y_j. In fact, $\mu(X)$ is the smallest integer such that the previous statement holds for all $i \leq \dim X$.*

Proof The first statement follows from [D3, Cor. 8.2.8], and the second from this and the Hard Lefschetz theorem. $\square$

2000 *Mathematics Subject Classification*. Primary 14C30; Secondary 14F42.
Author partially supported by the NSF.

Corollary 0.2 $\mu(X) = 0$ *if and only if all the cohomology of X is generated by algebraic cycles. A surface satisfies $\mu(X) \leq 1$ if and only if $p_g(X) = 0$. We have*

$$dim X \geq \mu(X) \geq level(H^*(X)) = \max\{|p - q| \mid h^{pq}(X) \neq 0\},$$

and the last inequality is equality if the generalized Hodge conjecture holds for X [G].

Proof The first statement is immediate. As for the second, the Lefschetz $(1,1)$ theorem implies that $H^2(X)$ is spanned by divisor classes if and only if $p_g = 0$. The inequality $\mu(X) \geq level(H^*(X))$ follows from the fact that Gysin maps are morphisms of Hodge structures up to Tate twists. $\square$

There is an alternative definition of motivic dimension which may better explain the name, although it is certainly less elementary and for the purposes of this paper less useful. Suppose that Mot is the category of homological motives, and $Mot_{\leq d}$ is the full subcategory generated by motives of varieties of dimension at most d and their Tate twists. Let us say that the true motivic dimension $\mu^{true}(X) \leq d$ if the motive of X lies in $Mot_{\leq d}$. Assuming Grothendieck's standard conjectures, it is possible to show that $\mu^{true} = \mu$.

My thanks to Su-Jeong Kang for her detailed comments.

1 Elementary estimates

It will be convenient to extend the notion of motivic dimension to arbitrary complex algebraic varieties, and for this we switch to homology. Let $H_i(-)$ denote Borel-Moore homology with $\mathbb{Q}$ coefficients, which is dual to compactly supported cohomology. Translating the above definition into homology leads to an integer $\mu^{big}(X)$, which we call the big motivic dimension that makes sense for any variety X. So $\mu^{big}(X)$ is the smallest integer such that every $\alpha \in H_i(X)$ lies in the image of some $f_* H_i(Y) \rightarrow H_i(X)$, where Y is a Zariski closed set whose components have dimension at most $(i + \mu^{big}(X))/2$. Of course, $\mu^{big}(X)$ coincides with $\mu(X)$ when X is smooth and proper, but it seems somewhat difficult to study in general. It turns out to be more useful to restrict attention to certain cycles. The identification $H_i(X) \cong H_c^i(X)^*$ gives homology a mixed Hodge structure with weights $\geq -i$, i.e. $W_{-i-1}H_i(X) = 0$ [D3]. We define the motivic dimension $\mu(X)$ by the replacing $H_i(X)$ by $W_{-i}H_i(X) = Gr_{-i}^W H_i(X)$ in the above definition of μ^{big}. We have $\mu(X) \leq \mu^{big}(X)$ with equality when X is smooth and proper.

Proposition 1.1

(a) *If $f : X' \rightarrow X$ is proper and surjective $\mu(X) \leq \mu(X')$.*
(b) *If $Z \subset X$ is Zariski closed then $\mu(X) \leq \max(\mu(Z), \mu(X - Z))$*
(c) *If $\tilde{X}$ is a desingularization of a partial compactification $\bar{X}$ of X, then*
$$\mu(X) \leq \mu(\bar{X}) \leq \mu(\tilde{X}).$$
(d) *$\mu(X_1 \times X_2) \leq \mu(X_1) + \mu(X_2)$.*
(e) *If $V \rightarrow X$ is a vector bundle then $\mu(V) \leq \mu(X)$ and $\mu(\mathbb{P}(V)) \leq \mu(X)$.*

Proof By [J, lemma 7.6, p. 110], any element $\alpha \in W_{-i}H_i(X)$ lifts to an element of $\alpha' \in W_{-i}H_i(X')$. This in turn lies in $f_* W_{-i}H_i(Y')$ for some $f : Y' \rightarrow X'$ satisfying $\dim Y' \leq (i + \mu(X'))/2$. Therefore α lies in the image of $W_{-i}H_i(f(Y'))$. Since $\dim f(Y') \leq (i + \mu(X'))/2$, therefore (a) holds.

Suppose $i \leq m = \max(\mu(Z), \mu(X - Z))$. We have an exact sequence of mixed Hodge structures

$$H_i(Z) \to H_i(X) \to H_i(X - Z) \to H_{i-1}(Z)$$

which can be deduced from [D3, prop. 8.3.9]. This implies by [D3, thm 2.3.5] that

$$W_{-i}H_i(Z) \to W_{-i}H_i(X) \to W_{-i}H_i(X - Z) \to 0 \qquad (1.1)$$

is exact. Given $\alpha \in W_{-i}H_i(X)$, let β denote its image in $W_{-i}H_i(X - Z)$. Then $\beta = f_*(\gamma)$ for some $f : Y \hookrightarrow X - Z$ with $\dim Y \leq (i + m)/2$ and $\gamma \in W_{-i}H_i(Y)$. Let $\bar{f} : \bar{Y} \to X$ denote the closure of Y. A sequence analogous to (1.1) shows that $W_{-i}H_i(\bar{Y})$ surjects onto $W_{-i}H_i(Y)$, therefore γ extends to a class $\bar{\gamma} \in W_{-i}H_i(\bar{Y})$. The difference $\alpha - \bar{f}_*(\bar{\gamma})$ lies in the image of $W_{-i}H_i(Z)$, and therefore in $g_*H_i(T)$ for some $g : T \to Z$ with $\dim T \leq (i + m)/2$. This proves (b).

Let $\bar{X}$ be a partial compactification of X. Then as above, we see that any class in $W_{-i}H_i(X)$ extends to $\bar{X}$. Therefore $\mu(X) \leq \mu(\bar{X})$. The remaining inequality of (c) follows from (a).

Statement (d) follows from the Künneth formula

$$W_{-i}H_i(X_1 \times X_2) = \bigoplus_{j+k=i} W_{-j}H_j(X_1) \otimes W_{-k}H_k(X_2)$$

Finally for (e), let $r = rk(V)$. Suppose that the Gysin images of $W_{i-2r}(Y_j)$ span $W_{-i+2r}H_{i-2r}(X)$ and satisfy $\dim Y_j \leq (\mu(X) + i - 2r)/2$. Then $W_{-i}H_i(V|_{Y_j})$ will span $W_{-i}H_i(V)$ by the Thom isomorphism theorem. The inequality $\mu(\mathbb{P}(V)) \leq \mu(X)$ can be proved by a similar argument. We omit the details since we will show something more general in corollary 2.8. □

Corollary 1.2 *If $X = \cup X_i$ is given as a finite disjoint union of locally closed subsets, $\mu(X) \leq \max\{\mu(X_i)\}$. In particular, $\mu(X) = 0$ if X is a disjoint union of open subvarieties of affine spaces.*

The last statement, which is of course well known, implies that flag varieties and toric varieties have $\mu = 0$.

Corollary 1.3 *If X_1 and X_2 are birationally equivalent smooth projective varieties, then $\mu(X_2) \leq \max(\mu(X_1), \dim X_1 - 2)$*

Proof Since an iterated blow up of X_1 dominates X_2, it is enough to prove this when X_2 is the blow up of X_1 along a smooth centre Z. Then

$$\mu(X_2) \leq \max(\mu(X_1 - Z), \mu(\mathbb{P}(N))) \leq \max(\mu(X_1), \mu(Z))$$

where N is the normal bundle of Z. □

Recall that a variety is uniruled if it has a rational curve passing through the general point.

Corollary 1.4 *If X is uniruled, then $\mu(X) \leq \dim X - 1$.*

Proof By standard arguments [Ko], X is dominated by a blow up of $Y \times \mathbb{P}^1$, where $\dim Y = \dim X - 1$. □

Corollary 1.5 *If X is a smooth projective variety with a $\mathbb{C}^*$-action, $\mu(X)$ is less than or equal to the dimension of the fixed point set.*

Proof Bialynicki-Birula [BB] has shown that X can be decomposed into a disjoint union of vector bundles over components of the fixed point set. The corollary now follows from the previous results. $\qquad\square$

From this, one recovers the well known fact that the Hodge numbers $h^{pq}(X)$ vanish when $|p - q|$ exceeds the dimension of the fixed set.

Proposition 1.6 *Suppose that X is a smooth projective variety such that the Chow group of zero cycles $CH_0(X) \cong \mathbb{Z}$. Then $\mu(X) \leq \dim X - 2$.*

Proof This follows from the theorem of Bloch-Srinivas [BS] that a positive multiple of the diagonal $\Delta \subset X \times X$ is rationally equivalent to a sum $\xi \times X + \Gamma$, where $\xi \in Z_0(X)$ is a zero cycle, and Γ is supported on $X \times D$ for some divisor $D \subset X$. This implies that the identity map on cohomology $H^i(X)$ factors through the Gysin map $H^{i-2}(\tilde{D}) \to H^i(X)$ for $i > 0$ and a resolution of singularities $\tilde{D} \to D$. $\qquad\square$

Recall that a projective variety is rationally connected if any two general points can be connected by a rational curve. Examples include hypersurfaces in $\mathbb{P}^n$ with degree less than $n + 1$, and more generally Fano varieties [Ko].

Corollary 1.7 *If X is rationally connected then $\mu(X) \leq \dim X - 2$.*

Proof Rational connectedness forces $CH_0(X) \cong \mathbb{Z}$. $\qquad\square$

2 Estimates for fibrations

Theorem 2.1 *Suppose that $f : X \to S$ is a smooth projective morphism. Then*

$$\mu(X) \leq \max_{s \in S(\mathbb{C})} \mu(X_s) + \dim S.$$

Proof We prove this by induction on $d = \dim S$. Let $m = \max_{s \in S(\mathbb{C})} \mu(X_s)$. Let $H_1, H_2, \ldots$ denote irreducible components of the relative Hilbert scheme $Hilb_{X/S}$ which surject onto S. Choose a desingularization $\tilde{\mathcal{F}}_k \to \mathcal{F}_{k,red}$ of the reduced universal family over each H_k.

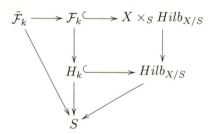

Let $disc(\tilde{\mathcal{F}}_k \to S) \subseteq S$ denote discriminant i.e. the complement of the maximal open set over which this map is smooth. Then $T = \bigcup_k disc(\tilde{\mathcal{F}}_k \to S) \subset S$ is a countable union of proper subvarieties, therefore its complement is nonempty by Baire's theorem. Choose $s \in S - T$. By assumption, there exists a finite collection of subvarieties $Y_{ij} \subset X_s$ such that $\dim Y_{ij} \leq (m + i)/2$ and their images generate $H_i(X_s)$. For each Y_{ij}, we choose one of the families $\mathcal{F}_k$ containing it and rename it $\mathcal{Y}_{ij}$; likewise set $H_{ij} = H_k$ and $\tilde{\mathcal{Y}}_{ij} = \tilde{\mathcal{F}}_k$.

Choose an open set $U \subset S - \bigcup disc(\mathcal{Y}_{ij} \to S)$ containing s. Therefore $X|_U \to U$ and each $\tilde{\mathcal{Y}}_{ij}|_U \to U$ are smooth and thus topological fibre bundles. Consequently the images of $\tilde{\mathcal{Y}}_{ij,t}$ will generate $H_i(X_t)$ for any $t \in U$. After replacing H_{ij} by the normalization of S in H_{ij}, and $\mathcal{Y}_{ij}$ by the fibre product, we can assume that $H_{ij} \to S$ is finite over its image. After shrinking U if necessary and replacing all maps by their restrictions to U, we can assume that the maps $g_{ij} : \tilde{\mathcal{Y}}_{ij} \to U$ are still smooth, and that U is nonsingular and affine. The last assumption implies that for any local system L of $\mathbb{Q}$-vector spaces, we have

$$H_c^i(U, L) = H^{2d-i}(U, L^*)^* = 0 \tag{2.1}$$

for $i < d$ since U is homotopic to a CW complex of dimension at most d [V, thm 1.22]. The Gysin images of $\tilde{\mathcal{Y}}_{ij,t}$ generate the homology of X_t for each $t \in U$, or dually the cohomology of $H^i(X_t)$ injects into $\oplus_j H^i(\tilde{\mathcal{Y}}_{ij,t})$. Since the monodromy actions are semisimple [D3, thm 4.2.6], the map of local systems $R^i f_* \mathbb{Q}|_U \to \oplus_j R^i g_{ij,*} \mathbb{Q}$ is split injective. Thus

$$H_c^k(U, R^i f_* \mathbb{Q}) \to \bigoplus_j H_c^k(U, R^i g_{i,j,*} \mathbb{Q}) \tag{2.2}$$

is injective. Since the Leray spectral sequence degenerates [D1], we get an injection

$$H_c^p(f^{-1}U) \to \bigoplus_{i \leq p-d,j} H_c^p(\tilde{\mathcal{Y}}_{ij}).$$

Note that the bound $i \leq p - d$ follows from (2.1). We therefore have a surjection

$$\bigoplus_{i \leq p-d,j} H_p(\tilde{\mathcal{Y}}_{ij}) \to H_p(f^{-1}U)$$

Since

$$\dim \tilde{\mathcal{Y}}_{ij} \leq \frac{m+i}{2} + d \leq \frac{m+d+p}{2} \tag{2.3}$$

for $i \leq p - d$, we have $\mu(f^{-1}(U)) \leq \mu^{big}(f^{-1}U) \leq m + d$. By induction $\mu(f^{-1}(S - U)) \leq m + d$. Thus $\mu(X) \leq \max(\mu(f^{-1}U), \mu(f^{-1}(S-U))) \leq m+d$ as required. $\square$

Corollary 2.2 *If a smooth projective variety X can be covered by a family of surfaces whose general member is smooth with $p_g = 0$. Then $\mu(X) \leq \dim X - 1$.*

Proof By assumption, there is a family $Y \to T$ of surfaces with $p_g = 0$ and a dominant map $\pi : Y \to X$. After restricting to a subfamily, we can assume that π is generically finite. So that $\dim T = \dim X - 2$, and therefore $\mu(X) \leq \dim X - 1$ by corollary 0.2 and theorem 2.1. $\square$

It is worth noting that many standard examples of surfaces with $p_g = 0$ lie in families. So there are nontrivial examples of varieties admitting fibrations of the above type. We give an explicit class of examples generalizing Enriques surfaces (cf. [BPV, p. 184]).

Example 2.3 Fix $n \geq 2$. Let $i : (\mathbb{P}^1)^n \to (\mathbb{P}^1)^n$ be the involution which acts by $[x_0, x_1] \mapsto [-x_0, x_1]$ on each factor. Choose a divisor $B \subset (\mathbb{P}^1)^n$ defined by a general i-invariant polynomial of multidegree $(4, 4, \ldots 4)$, and let $\pi : X \to (\mathbb{P}^1)^n$ be the double cover branched along B. The involution i can be seen to lift to a fixed point free involution of X. Let S be the quotient. X is covered by a family of an $K3$ surfaces $\pi^{-1}((\mathbb{P}^1)^2 \times t)$ which induces a family of Enriques surfaces on S. Thus $\mu(S) \leq n - 1$. On the other hand X can be checked to be Calabi-Yau, and thus the

Kodaira dimension of S equals 0. Therefore it cannot be uniruled. So this estimate on $\mu(S)$ does not appear to follow from the previous bounds.

In view of proposition 1.1 (d), we may hope for a stronger estimate

$$\mu(X) \leq \max_{s \in S(\mathbb{C})} \mu(X_s) + \mu(S).$$

Unfortunately it may fail without extra assumptions:

Example 2.4 Let S be an Enriques surface which can be realized as the quotient of a K3 surface $\tilde{S}$ by a fixed point free involution σ. Let σ act on $\mathbb{P}^1 \times \mathbb{P}^1$ by interchanging factors. Define $X = (\mathbb{P}^1 \times \mathbb{P}^1 \times \tilde{S})/\sigma$. The natural map $X \to S$ is an etale locally trivial $\mathbb{P}^1 \times \mathbb{P}^1$-bundle. An easy calculation shows that $level(H^2(X)) = 2$, while $p_g(S) = q(S) = 0$. Thus $\mu(X) = 2 > \mu(\mathbb{P}^1 \times \mathbb{P}^1) + \mu(S) = 0$.

Theorem 2.5 *Suppose that $f : X \to S$ is a smooth projective morphism over a quasiprojective base. If the monodromy action of $\pi_1(S)$ on $H^*(X_s)$ is trivial, then*

$$\mu(X) \leq \max_{s \in S(\mathbb{C})} \mu(X_s) + \mu(S).$$

Before proving this, we make some general remarks. If $f : X \to S$ is a not necessarily smooth projective morphism, then there is a filtration $L^\bullet H^i(X) \subset H^i(X)$ called the Leray filtration associated to the Leray spectral sequence. This is a filtration by sub mixed Hodge structures [A1, cor. 4.4]. When f is also smooth the spectral sequence degenerates [D1], so that $Gr_L^p H^{p+q}(X) \cong H^p(S, R^q f_* \mathbb{Q})$ carries a mixed Hodge structure.

Lemma 2.6 *Suppose that $f : X \to S$ is a smooth projective map, and let $m = \max_{s \in S(\mathbb{C})} \mu(X_s)$. Then $Gr_{k+i}^W H_c^k(S, R^i f_* \mathbb{Q})$ injects into $Gr_{k+i}^W Gr_L^k H_c^{k+i}(Y)$ for some Zariski closed $Y \subset X$ satisfying $\dim Y \leq \frac{m+i}{2} + \dim S$.*

Proof As in the proof of theorem 2.1, we can find a nonempty affine open set $U \subset S$ and a morphism of smooth U-schemes $\tilde{\mathcal{Y}} = \cup_j \tilde{\mathcal{Y}}_{ij} \to f^{-1}U$ whose fibres generate the homology of $H_i(X_s)$. The lemma holds when S is replaced by U with $Y = im\tilde{\mathcal{Y}}$ thanks to (2.2) and (2.3). Let $\bar{\mathcal{Y}} \subset X$ denote the closure of $im\mathcal{Y}$. By induction, we have a subset $\mathcal{Y}' \subset f^{-1}(S-U)$ which satisfies the lemma over $S-U$. We have a commutative diagram with exact rows

$$
\begin{array}{ccccc}
H_c^{k+i}(f^{-1}U) & \longrightarrow & H_c^{k+i}(X) & \longrightarrow & H_c^{k+i}(f^{-1}(S-U)) \\
\downarrow & & \downarrow & & \downarrow \\
0 \longrightarrow H_c^{k+i}(\mathcal{Y}) & \overset{j}{\longrightarrow} & H_c^{k+i}(\mathcal{Y}' \coprod \bar{\mathcal{Y}}) & \longrightarrow & H_c^{k+i}(\mathcal{Y}' \coprod (\bar{\mathcal{Y}} - \mathcal{Y}))
\end{array}
$$

Applying $Gr_{k+i}^W Gr_L^k$ and making appropriate identifications results in a commutative diagram

$$
\begin{array}{ccccc}
Gr_{k+i}^W H_c^k(U, R^i f_* \mathbb{Q}) & \longrightarrow & Gr_{k+i}^W H_c^k(S, R^i f_* \mathbb{Q}) & \longrightarrow & Gr_{k+i}^W H_c^k(S-U, R^i f_* \mathbb{Q}) \\
\downarrow{\scriptstyle \iota} & & \downarrow{\scriptstyle \iota''} & & \downarrow{\scriptstyle \iota'} \\
Gr_{k+i}^W H_c^{k+i}(\mathcal{Y}) & \overset{j}{\longrightarrow} & Gr_{k+i}^W H_c^{k+i}(\mathcal{Y}' \coprod \bar{\mathcal{Y}}) & \longrightarrow & Gr_{k+i}^W H_c^{k+i}(\mathcal{Y}' \coprod (\bar{\mathcal{Y}} - \mathcal{Y}))
\end{array}
$$

The top row is exact, but the bottom row need not be. Nevertheless j is injective since $Gr_{k+i}^W Gr_L^k$ preserves injections. The maps ι and ι' are also injective by the above discussion. The injectivity of ι'' follows a diagram chase. Thus $Y = \mathcal{Y}' \cup \bar{\mathcal{Y}}$ does the job. $\qquad\square$

Proof of theorem 2.5 Let $m = \max_{s \in S(\mathbb{C})} \mu(X_s)$. The sheaves $R^i f_* \mathbb{Q}$ are constant, so we have an isomorphism

$$H_c^k(S, R^i f_* \mathbb{Q}) \cong H_c^k(S) \otimes H^i(X_s) \qquad (2.4)$$

as vector spaces. As already noted, the Leray spectral sequence for f degenerates yielding a mixed Hodge structure on the left side. We claim that (2.4) can be made compatible with mixed Hodge structures, at least after taking the associated graded with respect to the weight filtration. We have a surjective morphism of pure polarizable Hodge structures $Gr_*^W H_c^i(X) \to H^i(X_s)$, which admits a right inverse σ since this category is semisimple [D3, lemma 4.2.3]. The image of $H_c^k(S)$ lies in $L^k H_c^k(X)$. Thus $Gr^W Gr_L(f^* \otimes \sigma)$ induces the desired identification

$$Gr_*^W H_c^k(S) \otimes H^i(X_s) \cong Gr_*^W H_c^k(S, R^i f_* \mathbb{Q}). \qquad (2.5)$$

Moreover, this is canonical in the sense that it is compatible with base change with respect to any morphism $T \to S$.

Let $S_k \to S$ be a map such that $\dim S_k \leq (\mu(S) + k)/2$ and such that $W_{-k} H_k(S_k)$ generates $W_{-k} H_k(S)$ (note that S_k may have several components). Dually $Gr_k^W H_c^k(S)$ injects into $Gr_k^W H_c^k(S_k)$. Thus there are injections

$$
\begin{array}{ccc}
Gr_k^W H_c^k(S) \otimes H^i(X_s) & \hookrightarrow & Gr_k^W H_c^k(S_k) \otimes H^i(X_s) \\
\downarrow \cong & & \downarrow \cong \\
Gr_{k+i}^W H_c^k(S, R^i f_* \mathbb{Q}) & \hookrightarrow & Gr_{k+i}^W H_c^k(S_k, R^i f_* \mathbb{Q}) \;.
\end{array}
$$

By lemma 2.6 $Gr_{k+i}^W H_c^k(S_k, R^i f_* \mathbb{Q})$ injects into some $Gr_{i+k}^W Gr_L^k H_c^{k+i}(\mathcal{Y}_{ki})$ with

$$\dim \mathcal{Y}_{ki} \leq \dim S_k + \frac{m+i}{2} \leq \frac{m + \mu(S) + k + i}{2}.$$

Combining this with the degeneration of Leray, shows that the map

$$Gr_p^W H_c^p(X) \to \bigoplus_{k+i=p} Gr_p^W H_c^p(\mathcal{Y}_{ki})$$

is injective, and hence we have a surjection

$$\bigoplus_{k+i=p} Gr_{-p}^W H_p(\mathcal{Y}_{ki}) \to Gr_{-p}^W H_p(X).$$

$\qquad\square$

Corollary 2.7 *With the above notation, suppose that there exists a nonempty Zariski open set $U \subseteq S$ such that $X|_U \to U$ is topologically a product. Then $\mu(X) \leq \mu(X_s) + \mu(S)$.*

Proof $\pi_1(U) \to \pi_1(S)$ is surjective. $\qquad\square$

Corollary 2.8 *If $X \to S$ is a Brauer-Severi morphism (i.e. a smooth map whose fibres are projective spaces) then $\mu(X) \leq \mu(S)$.*

Proof Since $\dim H^i(\mathbb{P}^N) \leq 1$, the monodromy representation is trivial. So $\mu(X) \leq \mu(\mathbb{P}^N) + \mu(S)$. $\qquad\square$

3 Symmetric powers

In this section, we give our take on Abel-Jacobi theory. When X is a smooth projective curve, the symmetric powers $S^n X$ are projective bundles over the Jacobian $J(X)$ for $n \gg 0$. This implies that the motivic dimension of $S^n X$ stays bounded. We consider what happens for more general smooth projective varieties. We note that $S^n X$ are singular in general, but only mildly so. These are in the class of V-manifolds or orbifolds, which satisfy Poincaré duality with rational coefficients, hard Lefschetz and purity of mixed Hodge structures. So for our purposes, we can treat them as smooth. In particular, we work with the original cohomological definition of μ. When X is a surface, the Hilbert schemes provides natural desingularization of the symmetric powers, and we give estimates for these as well.

We can identify $H^*(S^n X)$ with the S_n-invariants of $H^*(X^n)$. Let

$$ sym : H^*(X^n) \to H^*(X^n)^{S_n} \cong H^*(S^n X) $$

denote the symmetrizing operator $\frac{1}{n!} \sum \sigma$.

Lemma 3.1 *Let $f : X \to Y$ be a morphism of smooth projective varieties. If every class in $H^*(X)$ is of the form $(f^*\alpha) \cup \beta$ where β is an algebraic cycle, then $\mu(X) \le \mu(Y)$.*

Proof Let $N^p H^i(X)$ denote the span of the Gysin images of desingularizations of subvarieties of codimension $\ge p$. Then it is enough to show that $N^p H^i(X) = H^i(X)$ for some $p \ge (i - \mu(Y))/2$. Any element of $H^i(X)$ can be written as $f^*\alpha \cup \beta$ where $\beta \in H^{2k}(X)$ is an algebraic cycle and $\alpha \in H^{i-2k}(Y)$. This implies that $\alpha \in N^q H^{i-2k}(Y)$ and $\beta \in N^k H^{2k}(X)$ for some q satisfying $q \ge (i - 2k - \mu(Y))/2$. By [AK2], $f^*\alpha \cup \beta \in N^{q+k} H^i(X)$. Therefore $p = q + k$ gives the desired value. $\square$

Corollary 3.2 *Suppose that G is a finite group. Let $f : X \to Y$ be an equivariant morphism of smooth projective varieties with G-actions satisfying the above assumption, then $\mu(X/G) \le \mu(Y/G)$.*

Proof This is really a corollary of the proof which proceeds as above with the identifications $N^p H^i(X/G) = N^p H^i(X)^G$ etcetera. $\square$

Corollary 3.3 *Let $f : X \to Y$ be a morphism of smooth projective varieties, which is a Zariski locally trivial fibre bundle with fibre F satisfying $\mu(F) = 0$. Then $\mu(S^n X) \le \mu(S^n Y)$ for all n.*

Proof The conditions imply that the hypotheses of the previous corollary holds for $X \to Y$ as well as its symmetric powers. $\square$

Theorem 3.4 *Let X be a smooth projective variety.*

(a) $\mu(S^n X) \le n\mu(X)$.

(b) *If $\mu(X) \le 1$, then the sequence $\mu(S^n X)$ is bounded. If $\dim X \le 2$ then this is bounded above by $h^{10}(X)$.*

(c) *If $level(H^{2*}(X)) > 0$, then $level(H^*(S^n X))$ and $\mu(S^n X)$ are unbounded.*

Proof Inequality (a) follows from proposition 1.1.

Suppose that $\mu(X) \le 1$. Then $H^*(X)$ is spanned by algebraic cycles and classes $f_{j*}\beta$ with $f_j : Y_j \to X$ and $\beta \in H^1(Y_j)$. Let Y be the disjoint union of Y_j, and let

$f^n : S^n Y \to S^n X$ denote the natural map. Fix a basis $\beta_1 \ldots \beta_N$ of $H^1(Y)$. Then $H^*(S^n X)$ is spanned by classes of the form

$$[f^n_* sym(p^*_m(\beta_{i_1} \times \ldots \times \beta_{i_m}))] \cup \gamma \qquad (3.1)$$

where $p_m : Y^n \to Y^m$ is a projection onto the first m factors, and γ is an algebraic cycle. Notice that the expression in (3.1) vanishes if any of the β_{i_j}'s are repeated. Therefore we can assume that $m \leq N$. Let g_j denote the inclusions of components of the algebraic cycle $f^{n*}\gamma$. We can rewrite the expressions in (3.1) as

$$f^n_*(sym(\ldots) \cup f^{n*}\gamma) \in \text{span}\{f^n_* g_{j_*}(g^*_j sym(\ldots))\}.$$

This shows that $H^*(S^n X)$ is spanned by Gysin images of classes of degree at most N. Thus $\mu(S^n X) \leq N$. This proves the first part of (b).

When X is a curve of genus $g = h^{10}(X)$, the Abel-Jacobi map $S^n X \to J(X)$ can be decomposed into a union of projective space bundles. Proposition 1.1 implies that $\mu(S^n X) \leq g = \dim J(X)$.

Suppose X is a surface, then $\mu(X) \leq 1$ forces $p_g(X) = 0$. If $q = h^{10}(X) = 0$ then $\mu(X) = 0$, so $\mu(S^n X) = 0$ by (a). So we may assume that $q > 0$, which implies that the Albanese map $\alpha : X \to Alb(X)$ is nontrivial. If $\dim \alpha(X) = 2$, it is easy to see that the pullback of a generic two form from $Alb(X)$ would be nontrivial. This would imply that the image is a curve. Let C be the normalization of $\alpha(X)$, then we get a map $\phi : X \to C$. Note the the genus g of C is necessarily equal to q since α^* induces an isomorphism on the space 1-forms and it factors as $H^0(\Omega^1_{Alb(X)}) \to H^0(\Omega^1_C) \hookrightarrow H^0(\Omega^1_X)$. With the help of the hard Lefschetz theorem, we can see that every cohomology class on X is of the form $(\phi^*\beta) \cup \gamma$ where γ is an algebraic cycle. Likewise for $S^n X$. Therefore by corollary 3.2 and previous paragraph $\mu(S^n X) \leq \mu(S^n C) \leq g$.

Suppose that $\alpha \in H^{ij}(X)$ is a nonzero class with $i \neq j$ and $i + j$ even. Then $sym(\alpha^{\otimes n})$ provides a nonzero class in $H^{ni,nj}(S^n X)$, which shows that the level and hence the motivic dimension go to infinity. $\square$

We give an example where $\mu(X) > 1$ but $\mu(S^n X)$ stays bounded.

Example 3.5 Let X be a rigid Calabi-Yau threefold. (A number of such examples are known, cf. [S].) Then the Hodge numbers satisfy $h^{10} = h^{20} = h^{21} = 0$ and $h^{30} = 1$. It follows that $H^2(X)$ and $H^4(X)$ are generated by algebraic cycles. Choose a generator $\alpha \in H^{30}(X)$ normalized so that the class $\delta = sym(\alpha \times \bar\alpha) \in H^6(S^2 X)$ is rational. The space of S_2-invariant classes of type $(3,3)$ in $H^3(X) \otimes H^3(X)$ is one dimensional. Therefore δ must coincide with the Künneth component of the diagonal Δ in $H^3(X) \otimes H^3(X)$. The remaining Künneth components of Δ are necessarily algebraic. Therefore δ is algebraic. Thus all factors

$$sym(H^3(X)^{\otimes k} \otimes H^{2i_1}(X) \otimes \ldots \otimes H^{2i_{n-k}}(X))$$

in $H^i(S^n X)$ are spanned by algebraic cycles if k is even, or $\{sym(\xi \times (\text{alg. cycle}) \mid \xi \in H^3(X)\}$ if k is odd. It follows, by a modification of lemma 3.1, that $\mu(S^n X) \leq 3$ for all n.

We review the basic facts about Hilbert schemes of surfaces. Proofs and references can be found in the first 30 or so pages of [Go]. When X is a smooth projective surface, there is a natural desingularization $\pi_n : Hilb^n X \to S^n X$ given by the (reduced) Hilbert scheme of 0-dimensional subschemes of length n. The symmetric product $S^n X$ can be decomposed into a disjoint union of locally closed sets

$$\Delta_{(\lambda_1,\ldots\lambda_k)} = \{\lambda_1 x_1 + \ldots \lambda_k x_k \mid x_i \neq x_j\}$$

indexed by partitions $\lambda_1 \geq \lambda_2 \ldots$ of n, where the elements of $S^n X$ are written additively. Let $\tilde{\Delta}_\lambda \subset X^k$ denote the preimage under $(x_1, \ldots, x_k) \mapsto \lambda_1 x_1 + \ldots \lambda_k x_k$. The map $\tilde{\Delta}_\lambda \to \Delta_\lambda$ is etale with Galois group G. The group can be described explicitly by grouping the terms in the partition as follows:

$$\underbrace{\lambda_1 = \ldots = \lambda_{d_1}}_{d_1} > \underbrace{\lambda_{d_1+1} = \ldots = \lambda_{d_1+d_2}}_{d_2} > \ldots \underbrace{\ldots \lambda_{d_1+d_2+\ldots d_\ell}}_{d_\ell} = \lambda_k > 0.$$

Then $G = S_{d_1} \times \ldots S_{d_\ell}$ and Δ_λ is isomorphic to an open subset $S^{d_1} X \times \ldots S^{d_\ell} X$. We record the following key fact [Go, lemma 2.1.4]

Lemma 3.6 $\pi_n^{-1}\Delta_\lambda \times_{\Delta_\lambda} \tilde{\Delta}_\lambda \to \tilde{\Delta}_\lambda$ *is isomorphic to the restriction of* $\prod H_{\lambda_i} \to X^k$ *to* $\tilde{\Delta}_\lambda \subset X^k$. *Each* $H_{\lambda_i} \to X$ *is a Zariski locally trivial fibre bundle whose fibre can be identified with the subscheme* $Hilb_0^{\lambda_i}\mathbb{C}^2 \subset Hilb^{\lambda_i}\mathbb{C}^2$ *parameterizing schemes supported at the origin of* $\mathbb{C}^2$.

We note that $Hilb_0^{\lambda_i}\mathbb{C}^2$ is smooth and $\mathbb{C}^*$ acts on it with isolated fixed points. Therefore the fibres have $\mu = 0$. We can see that G acts on $\pi_n^{-1}\Delta_\lambda \times_{\Delta_\lambda} \tilde{\Delta}_\lambda$ by permuting the H_{λ_i}'s. Consequently

$$\pi_n^{-1}\overline{\Delta}_\lambda \cong S^{d_1} H_{\lambda_{d_1}} \times S^{d_2} H_{\lambda_{d_1+d_2}} \times \ldots S^{d_\ell} H_{\lambda_{d_1+\ldots d_\ell}}. \tag{3.2}$$

Proposition 3.7 *If X is a surface with $p_g = 0$ then* $\mu(Hilb^n X) \leq \min(n, \sqrt{2nq})$ *for all n, where $q = h^{10}$. In particular, $\mu(Hilb^n X) = 0$ when $q = 0$.*

Proof The closures of each of the strata Δ_λ are dominated by X^k, so that $\mu(\Delta_\lambda) \leq k \leq n$. The maps $\pi^{-1}\Delta_\lambda \to \Delta_\lambda$ are Zariski locally trivial fibrations with fibres having $\mu = 0$. Thus $\mu(Hilb^n X) \leq n$ follows from this together with proposition 1.1 and corollary 2.7.

Each $H_{\lambda_i} \to X$ is a bundle with $\mu = 0$ fibres. Therefore $\mu(S^{d_i} H_{\lambda_{d_1+\ldots d_i}}) \leq q$ by corollary 3.3 and the previous theorem. Combing this with (3.2), we see that Δ_λ has motivic dimension at most ℓq. To estimate ℓ, we use

$$\begin{aligned}
n &= d_1 \lambda_{d_1} + d_2 \lambda_{d_1+d_2} + \ldots d_\ell \lambda_{d_1+\ldots d_\ell} \\
&\geq \lambda_{d_1} + \lambda_{d_1+d_2} + \ldots \lambda_{d_1+\ldots d_\ell} \\
&\geq \ell + (\ell - 1) + \ldots 1
\end{aligned}$$

to obtain $\ell \leq \sqrt{2n}$. This gives the remaining inequality $\mu(Hilb^n X) \leq \sqrt{2nq}$. $\square$

4 Applications to the Hodge conjecture

Jannsen [J] has extended the Hodge conjecture to an arbitrary variety X. This states that any class in $Hom(\mathbb{Q}(i), W_{-2i} H_{2i}(X))$, which should be thought of as a Hodge cycle, is a linear combination of fundamental classes of i-dimensional subvarieties. Lewis [L] has given a similar extension for the generalized Hodge conjecture which would say that an irreducible sub Hodge structure of $W_{-i} H_i(X)$ with level at most ℓ should lie in the Gysin image of a subvariety of dimension bounded by $(\ell + i)/2$. Both statements are equivalent to the usual forms for smooth projective varieties.

Proposition 4.1

(a) If $\mu(X) \leq 3$, then the Hodge conjecture holds for X.

(b) If $\mu(X) \leq 2$, then the generalized Hodge conjecture holds for X.

Proof Suppose that $\mu(X) \leq 3$. Then any Hodge class $\alpha \in H_{2i}(X)$ lies in $f_* W_{-2i} H_{2i}(Y)$ for some subvariety with components satisfying $\dim Y_j \leq (3+2i)/2$. Let $\tilde{Y} = \cup \tilde{Y}_j$ be a desingularization of a compactification of Y. An argument similar to the proof of proposition 1.1 shows that the natural map

$$W_{-2i} H_{2i}(\tilde{Y}) \to f_* W_{-2i} H_{2i}(Y)$$

is a surjective morphism of polarizable Hodge structures. This map admits a section, since the category of such structures is semisimple [D3]. It follows that α can be lifted to a Hodge cycle β on $\tilde{Y}$. This can be viewed as a Hodge cycle in cohomology under the Poincaré duality isomorphism $H_{2i}(\tilde{Y}) = \oplus_j H^{2 \dim Y_j - 2i}(\tilde{Y}_j)(\dim Y_j)$. Since $2 \dim Y_j - 2i \leq 3$, this forces the degree of β to be 0 or 2. Consequently β must be an algebraic cycle. Hence the same is true for its image α.

The second statement is similar. With notation as above, a sub Hodge structure of $H_i(X)$ is the image of a sub Hodge structure of $\oplus_j H^{2 \dim Y_j - i}(\tilde{Y}_j)(\dim Y_j)$ with $2 \dim Y_j - i \leq 2$. Since the generalized Hodge conjecture is trivially true in this range, this structure is contained in the Gysin image of map from a subvariety of expected dimension. $\square$

Corollary 4.2 (Conte-Murre) *The Hodge (respectively generalized Hodge) conjecture holds for uniruled fourfolds (respectively threefolds).*

Corollary 4.3 (Laterveer) *The Hodge (respectively generalized Hodge) conjecture holds for rationally connected smooth projective fivefolds (respectively fourfolds).*

Corollary 4.4 *If X is a smooth projective variety with a $\mathbb{C}^*$-action, then the Hodge (respectively generalized Hodge) conjecture holds if the fixed point set has dimension at most 3 (respectively 2).*

The last result can also be deduced from the main theorem of [AK1], which says in effect that these conjectures factor through the Grothendieck group of varieties. The class of X in the Grothendieck group can be expressed as a linear combination of classes of components of the fixed point set times Lefschetz classes.

Corollary 4.5 *The Hodge (respectively generalized Hodge) conjecture holds for a smooth projective fourfold (respectively threefold) which can be covered by a family of surfaces whose general member is smooth with $p_g = 0$.*

Corollary 4.6 *Let X be a smooth projective surface with $p_g = 0$. Then the Hodge (respectively generalized Hodge) conjecture holds for $S^n X$ for any n if $q \leq 3$ (respectively $q \leq 2$). The Hodge conjecture holds for $Hilb^n X$ if $n \leq 3$, or if $n \leq 7$ and $q = 1$, or if $q = 0$. The generalized Hodge conjecture holds for $Hilb^n X$ if $n = 2$, or if $n \leq 3$ and $q = 1$, or if $q = 0$.*

As a final example, suppose that X is a rigid Calabi-Yau variety. Then we saw that $\mu(S^n X) \leq 3$ in example 3.5. So the Hodge conjecture holds for $S^n X$. (We are being a little circular in our logic, since we essentially verified the conjecture in the course of estimating $\mu(S^n X)$.)

We have a Noether-Lefschetz result in this setting. For the statement, we take "sufficiently general" to mean that the set of exceptions forms a countable union of proper Zariski closed subsets of the parameter space.

Theorem 4.7 *Let $X \subset \mathbb{P}^N$ be a smooth projective variety such that the Hodge (respectively generalized Hodge) conjecture holds. Then there exists an effective constant $d_0 > 0$ such that the Hodge (respectively generalized Hodge) conjecture holds for $H \subset X$ when $H \in \mathbb{P}(\mathcal{O}_X(d))$ is a sufficiently general hypersurface and $d \geq d_0$.*

Remark 4.8 The effectivity of d_0 depends on having enough information about $X \subset \mathbb{P}^N$. As will be clear from the proof, it would be sufficient to know the Chern classes of $X, \mathcal{O}_X(1)$, the Castelnuovo-Mumford regularity of $\mathcal{O}_X$ and T_X, and $h^0(T_X)$. For $X = \mathbb{P}^N$, we have $d_0 = N + 1$.

Proof Let $n = \dim X - 1$. We take d_0 to be the smallest integer for which

(a) $d_0 \geq 2$.
(b) $h^{n0}(H) > h^{n0}(X)$ for nonsingular $H \in \mathbb{P}(\mathcal{O}_X(d))$ with $d \geq d_0$. (This ensures that $H^n(H)/im H^n(X)$ has length n; a fact needed at the end.)
(c) $H \in \mathbb{P}(\mathcal{O}_X(d))$, $d \geq d_0$, has nontrivial moduli, at least infinitesimally.

To clarify these conditions, note that from the exact sequence

$$0 \to \omega_X \to \omega_X(d) \to \omega_H \to 0$$

and Kodaira's vanishing theorem we obtain

$$
\begin{aligned}
h^0(H, \omega_H) &= h^0(X, \omega_X(d)) - h^0(\omega_X) + h^1(\omega_X) \\
&= \chi(\omega_X(d)) - h^0(\omega_X) + h^1(\omega_X).
\end{aligned}
$$

The right side is explicitly computable by Riemann-Roch, so we get an effective lower bound for (b). For (c), we require $H^1(T_H) \neq 0$. Once we choose d_0 so that $H^1(T_X(-d)) = 0$ for $d \geq d_0$, standard exact sequences yield

$$
\begin{aligned}
h^1(T_H) &\geq h^0(\mathcal{O}_H(d)) - h^0(T_X|_H) \\
&\geq h^0(\mathcal{O}_X(d)) - h^0(T_X) - 1.
\end{aligned}
$$

We can compute the threshold which makes the right side positive. When $X = \mathbb{P}^N$, we see by a direct computation that assumptions (a),(b), (c) are satisfied as soon as $d \geq N + 1$.

Let $U_d \subset \mathbb{P}(\mathcal{O}_X(d))$ denote the set of nonsingular hypersurfaces in X. For $H \in U_d$, the weak Lefschetz theorem guarantees an isomorphism $H^i(X) \cong H^i(H)$ for $i < n$, so the (generalized) Hodge conjecture holds for these groups by assumption. This together with the hard Lefschetz theorem takes care of the range $i > n$. So it remains to treat $i = n$. We have an orthogonal decomposition

$$H^n(H) = im H^n(X) \oplus V_H \tag{4.1}$$

as Hodge structures, where V_H is the kernel of the Gysin map $H^n(H) \to H^{n+2}(X)$ [V, sect. 2.3.3]. Equivalently, after restricting to a Lefschetz pencil in $\mathbb{P}(\mathcal{O}_X(d))$ containing H, V_H is just the space of vanishing cycles [loc. cit.]. As H varies, $H^n(H)$ determines a local system over U_d, and (4.1) is compatible with the monodromy action. The action of $\pi_1(U_d)$ respects the intersection pairing $\langle , \rangle$. So the image of a finite index subgroup $\Gamma \subset \pi_1(U_d)$ lies in the identity component $Aut(V_H, \langle , \rangle)^\circ$, which is a symplectic or special orthogonal group according to the parity of n. By [D4, thm 4.4.1], the image of Γ is either finite or Zariski dense in $Aut(V_H, \langle , \rangle)^\circ$.

The first possibility can be ruled out by checking that the Griffiths period map on U_d is nontrivial [CT, lemma 3.3]. The nontriviality of the period map is guaranteed by Green's Torelli theorem [Gr, thm 0.1] and our assumption (c). Thus the Zariski closure $\overline{im(\pi_1(U_d))}^{Zar} \supseteq Aut(V_H, \langle, \rangle)^o$. To finish the argument, we recall that the Mumford-Tate group $MT(V_H) \subset GL(V_H)$ is an algebraic subgroup which leaves all sub Hodge structures invariant. By [D2, prop 7.5], when H is sufficiently general $MT(V_H)$ must contain a finite index subgroup of the image of $\pi_1(U_d)$, and therefore $Aut(V_H, \langle, \rangle)^o$. Thus $MT(V_H)$ contains the symplectic or special orthogonal group. In either case V_H is an irreducible Hodge structure of length n. The (generalized) Hodge conjecture concerns sub Hodge structures of $H^n(H)$ of length less than n. So these must come from X, and therefore be contained in images of Gysin maps of subvarieties of expected codimension by our initial assumptions. $\square$

Remark 4.9 For the Hodge conjecture alone, the bound on d_0 can be improved. When n is odd, we may take $d_0 = 1$ since the Hodge conjecture is vacuous for H^n. When n is even, we can choose the smallest d_0 so that assumptions (a) and (c) hold, since the mondromy action on V_H is irreducible and nontrivial (by the Picard-Lefschetz formula) and therefore it cannot contain a Hodge cycle.

Corollary 4.10 *Let $X \subset \mathbb{P}^N$ be a smooth projective variety.*

(a) *If $\mu(X) \leq 3$, then the Hodge conjecture holds for every sufficiently general hypersurface section of degree $d \gg 0$. This is in particular the case, when X is a rationally connected fivefold.*

(b) *If $\mu(X) \leq 2$, then the generalized Hodge conjecture holds for every sufficiently general hypersurface section of degree $d \gg 0$.*

References

[A1]   D. Arapura, *The Leray spectral sequence is motivic*, Invent. Math 160 (2005), 567–589.

[AK1]   D. Arapura, S-J. Kang, *Coniveau and the Grothendieck group of varieties.* Michigan Math. J. 54 (2006), no. 3, 611–622.

[AK2]   D. Arapura, S-J. Kang, *Functoriality of the coniveau filtration* Canad. Math. Bull (2007)

[BPV]   W. Barth, C. Peters, A. Van de Ven, *Compact complex surfaces*, Springer-Verlag (1984)

[BB]   A. Bialynicki-Birula, *Some theorems on actions of algebraic groups*, Ann. of Math. (1972)

[B]   S. Bloch, *Lectures on algebraic cycles*, Duke U. Press (1979)

[BS]   S. Bloch, V. Srinivas, *Remarks on correspondences and algebraic cycles.* Amer. J. Math. 105 (1983), no. 5, 1235–1253

[CT]   J. Carlson, D. Toledo, *Discriminant complements and kernels of monodromy representations*, Duke Math. J (1999)

[CN]   A. Conte, J. Murre, *The Hodge conjecture for fourfolds admitting a covering by rational curves*, Math. Ann. 238, 79–88.(1978)

[D1]   P. Deligne, *Théorème de Lefschetz et critéres de dégénérescence de suites spectrales* Publ. IHES 35 (1968)

[D2]   P. Deligne, *La conjecture de Weil pour les surfaces K3*, Invent Math (1972)

[D3]   P. Deligne, *Théorie de Hodge II, III*, Publ. IHES 40, 44, (1971, 1974)

[D4]   P. Deligne, *La conjecture de Weil II*, Publ. IHES 52 (1980)

[Go]   L. Göttsche, *Hilbert schemes of Zero dimensional subschemes of smooth varities*, Lect. Notes Math 1572, Springer-Verlag (1994)

[Gr]   M. Green, *The period map for hypersurface sections of high degree of an arbitrary variety.* Compositio (1984)

[G]   A. Grothendieck, *Hodge's general conjecture is false for trivial reasons.* Topology 8 1969 299–303.

[J]   U. Jannsen, *Mixed motives and algebraic K-theory*, Lect notes in math 1400, Springer-Verlag (1990)

[La]   R. Laterveer, *Algebraic varieties with small Chow groups.* J. Math. Kyoto Univ. 38 (1998), no. 4, 673–694.

[L]   J. Lewis, *A survey of the Hodge conjecture. Second edition.* CRM Monograph Series, 10. AMS (1999)

[Ko]   J. Kollár, *Rational curves on algebraic varieties,* Springer-Verlag (1996)

[S]   C. Schoen, *On fiber products of elliptic surfaces with section,* Math. Zeit 197 (1988)

[V]   C. Voisin, *Hodge theory and complex algebraic geometry II,* Cambridge (2003)

Fields Institute Communications
Volume **56**, 2009

$\mathcal{E}$-Factors for the Period Determinants of Curves

Alexander Beilinson

Department of Mathematics
University of Chicago
Chicago, IL 60637, USA
`sasha@math.uchicago.edu`

To Spencer Bloch

The myriad beings of the six worlds –
gods, humans, beasts, ghosts, demons, and devils –
are our relatives and friends.

Tesshu "Bushido", translated by J. Stevens.

Introduction

0.1 Let X be a smooth compact complex curve, M be a holonomic $\mathcal{D}$-module on X (so outside a finite subset $T \subset X$, our M is a vector bundle with a connection ∇). Denote by $dR(M)$ the algebraic de Rham complex of M placed in degrees $[-1, 0]$; this is a complex of sheaves on the Zariski topology X_{Zar}. Its analytic counterpart $dR^{an}(M)$ is a complex of sheaves on the classical topology X_{cl}. Viewed as an object of the derived category of $\mathbb{C}$-sheaves, this is a perverse sheaf, which we denote by $B(M)$; outside T, it is the local system $M^{\nabla}_{X \setminus T}$ of ∇-horizontal sections (placed in degree -1). Set $H^{\cdot}_{\mathrm{dR}}(X, M) := H^{\cdot}(X_{\mathrm{Zar}}, dR(M))$, $H^{\cdot}_{\mathrm{B}}(X, M) := H^{\cdot}(X_{cl}, B(M))$; these are the *de Rham* and *Betti* cohomology. We have the *period isomorphism* $\rho : H^{\cdot}_{\mathrm{dR}}(X, M) \xrightarrow{\sim} H^{\cdot}_{\mathrm{B}}(X, M)$.

The cohomology $H^{\cdot}_{\mathrm{dR}}$ and $H^{\cdot}_{\mathrm{B}}$ have, respectively, algebraic and topological nature that can be tasted as follows. Let $k, k' \subset \mathbb{C}$ be subfields. Then:

- For (X, M) defined over k, we have *de Rham k-structure* $H^{\cdot}_{\mathrm{dR}}(X_k, M_k)$ on $H^{\cdot}_{\mathrm{dR}}(X, M)$;
- A k'-structure on $B(M)$, i.e., a perverse k'-sheaf $B_{k'}$ on X_{cl} together with an isomorphism $B_{k'} \underset{k'}{\otimes} \mathbb{C} \xrightarrow{\sim} B(M)$, yields *Betti k'-structure $H^{\cdot}(X_{cl}, B_{k'})$* on $H^{\cdot}_{\mathrm{B}}(X, M)$.

If both (X_k, M_k) and $B_{k'}$ are at hand, then, computing $\det \rho$ with respect to rational bases, one gets a number whose class $[\det \rho]$ in $\mathbb{C}^{\times}/k^{\times}k'^{\times}$ does not depend on the choice of the bases. In his farewell seminar at Bures [Del], Deligne, guided

---

2000 *Mathematics Subject Classification.* Primary 14F40; Secondary 11G45.
Key words and phrases. ε-factors, period determinants, $\mathcal{D}$-modules.

by an analogy between $[\det \rho]$ and the constant in the functional equation of an L-function, asked if $[\det \rho]$ can be expressed, in presence of an extra datum of a rational 1-form ν, as the product of certain factors of local origin at points of T and $\mathrm{div}(\nu)$. He also suggested the existence of a general geometric format which would yield the product formula (see 0.3 below). Our aim is to establish such a format.

0.2 *Remarks.* (i) A natural class of k'-structures on $B(M)$ comes as follows. Suppose for simplicity that M equals the (algebraic) direct image of $M_{X \setminus T}$ by $X \setminus T \hookrightarrow X$. Let $\pi : \tilde{X} \to X$ be the real blow-up of X at T (so $\tilde{X}$ is a real-analytic surface with boundary $\partial \tilde{X} = \pi^{-1}(T)$, and π is an isomorphism over $X \setminus T$). Then $M^{\nabla}_{X \setminus T}$ extends uniquely to a local system $M^{\nabla}_{\tilde{X}}$ on $\tilde{X}$. Following Malgrange [M], consider the constructible subsheaf $M^{\tau}_{\tilde{X}}$ of $M^{\nabla}_{\tilde{X}}$ of sections of moderate growth (so $M^{\tau}_{\tilde{X}}$ coincides with $M^{\nabla}_{\tilde{X}}$ off $\partial \tilde{X}$, and $M^{\tau}_{\tilde{X}}$ equals $M^{\nabla}_{\tilde{X}}$ if and only if M has regular singularities). By [M] 3.2, one has a canonical isomorphism

$$R\pi_* M^{\tau}_{\tilde{X}} \xrightarrow{\sim} B(M). \tag{0.2.1}$$

Therefore a k'-structure on $M^{\tau}_{\tilde{X}}$ yields a k'-structure on $B(M)$. Notice that the former is the same as a k'-structure on the local system $M^{\nabla}_{\tilde{X}}$, i.e., on $M^{\nabla}_{X \setminus T}$, such that the subsheaf $M^{\tau}_{\tilde{X}}$ is defined over k'.

(ii) By (0.2.1), one has $H_{\mathrm{B}}^{\cdot}(X, M) = H^{\cdot}(\tilde{X}, M^{\tau}_{\tilde{X}})$. The dual vector space equals $H^{\cdot}(\tilde{X}, DM^{\tau}_{\tilde{X}})$, where D is the Verdier duality functor, which is the homology group of cycles with coefficients in $M^{\vee}_{X \setminus T}$ on $X \setminus T$, having rapid decay at T. So ρ, viewed as a pairing $H_{\mathrm{dR}}^{\cdot}(X, M) \times H^{\cdot}(\tilde{X}, DM^{\tau}_{\tilde{X}}) \to \mathbb{C}$, is the matrix of periods of M-valued forms along the cycles of rapid decay. See [BE] for many examples.

(iii) The setting of 0.1 makes sense for proper X of any dimension. The passage B to perverse sheaves commutes with direct image functors for proper morphisms $X \to Y$, so the data $(X_k, M_k, B_{k'})$ are functorial with respect to direct image.

0.3 The next format, which yields the product formula, was suggested in the last exposé of [Del]:

(i) There should exist ε-factorization formalisms for $\det H_{\mathrm{dR}}^{\cdot}$ and $\det H_{\mathrm{B}}^{\cdot}$. These are natural rules which assign to every non-zero meromorphic 1-form ν on X two collections of lines $\mathcal{E}_{\mathrm{dR}}(M)_{(x,\nu)}$ and $\mathcal{E}_{\mathrm{B}}(M)_{(x,\nu)}$ labeled by points $x \in X$. The lines $\mathcal{E}_?(M)_{(x,\nu)}$ have x-local nature; if $x \notin T \cup \mathrm{div}(\nu)$, then $\mathcal{E}_?(M)_{(x,\nu)}$ is naturally trivialized. Finally, one has ε-*factorization*, alias *product formula*, isomorphisms

$$\eta_? : \bigotimes_{x \in T \cup \mathrm{div}(\nu)} \mathcal{E}_?(M)_{(x,\nu)} \xrightarrow{\sim} \det H_?^{\cdot}(X, M). \tag{0.3.1}$$

(ii) The de Rham and Betti ε-factorizations should have, respectively, algebraic and topological origin. Thus, if X, M, ν are defined over k, then the datum $\{\mathcal{E}_{\mathrm{dR}}(M)_{(x,\nu)}\}$ is defined over k, and a k'-structure on $B(M)$ yields a k'-structure on every $\mathcal{E}_{\mathrm{B}}(M)_{(x,\nu)}$. One wants these structures to be compatible with the trivializations of $\mathcal{E}_?(M)_{(x,\nu)}$ off $T \cup \mathrm{div}(\nu)$, and η_{dR}, η_{B} to be defined over k, k'.

(iii) There should be natural ε-*period* isomorphisms $\rho^\varepsilon = \rho^\varepsilon_{(x,\nu)} : \mathcal{E}_{\mathrm{dR}}(M)_{(x,\nu)} \xrightarrow{\sim} \mathcal{E}_{\mathrm{B}}(M)_{(x,\nu)}$ of x-local origin such that the next diagram commutes:

$$
\begin{array}{ccc}
\otimes \mathcal{E}_{\mathrm{dR}}(M)_{(x,\nu)} & \xrightarrow{\eta_{\mathrm{dR}}} & \det H^\cdot_{\mathrm{dR}}(X, M) \\
\otimes \rho^\varepsilon_{(x,\nu)} \downarrow & & \rho \downarrow \\
\otimes \mathcal{E}_{\mathrm{B}}(M)_{(x,\nu)} & \xrightarrow{\eta_{\mathrm{B}}} & \det H^\cdot_{\mathrm{B}}(X, M)
\end{array}
\qquad (0.3.2)
$$

Suppose (X, M, ν) is defined over k. The points in $T \cup \mathrm{div}(\nu)$ are algebraic over k; let $\{O_\alpha\}$ be their partition by the Galois orbits. By (ii), the lines $\underset{x \in O_\alpha}{\otimes} \mathcal{E}_{\mathrm{dR}}(M)_{(x,\nu)}$ carry k-structure. If $B(M)$ is defined over k', then, by (ii), $\mathcal{E}_{\mathrm{B}}(M)_{(x,\nu)}$ carry k'-structure. Writing $\underset{x \in O_\alpha}{\otimes} \rho^\varepsilon_{(x,\nu)}$ in k-k'-bases, we get numbers $[\rho^\varepsilon_{(O_\alpha,\nu)}] \in \mathbb{C}^\times / k^\times k'^\times$. Now (0.3.2) yields the promised product formula

$$
[\det \rho] = \prod_\alpha [\rho^\varepsilon_{(O_\alpha,\nu)}]. \qquad (0.3.3)
$$

We will show that the above picture is, indeed, true.

0.4 Parts of this format were established earlier: the de Rham ε-factorization was constructed already in [Del] (and reinvented later in [BBE]); the Betti counterpart was presented (in the general context of "animation" of Kashiwara's index formula) in [B].[1] It remains to construct ρ^ε. The point is that $\mathcal{E}_?$ satisfy several natural constraints, and compatibility with them determines ρ^ε almost uniquely. Notice that we work completely over $\mathbb{C}$: the k- and k'-structures are irrelevant.

The principal constraints are the global product formula (0.3.1) and its next local counterpart. For ν' close to ν, the points of $T \cup \mathrm{div}(\nu')$ cluster around $T \cup \mathrm{div}(\nu)$. Now the isomorphism $\underset{x' \in T \cup \mathrm{div}(\nu')}{\otimes} \mathcal{E}_?(M)_{(x',\nu')} \xrightarrow{\sim} \underset{x \in T \cup \mathrm{div}(\nu)}{\otimes} \mathcal{E}_?(M)_{(x,\nu)}$ that comes from *global* identifications (0.3.1) can be written as the tensor product of natural isomorphisms of *local origin* at points of $T \cup \mathrm{div}(\nu)$. This *local factorization structure* (which is a guise, with an odd twist, of the geometric class field theory) is fairly rigid: $\mathcal{E}_?(M)$ is determined by a rank 1 local system $\det M_{X \setminus T}$ and a collection of lines labeled by elements of T.[2]

The rest of the constraints for $M \mapsto \mathcal{E}_?(M)$ are listed in 5.1. We show that there is an isomorphism $\rho^\varepsilon : \mathcal{E}_{\mathrm{dR}} \xrightarrow{\sim} \mathcal{E}_{\mathrm{B}}$ compatible with them, which is determined uniquely up to a power of a simple canonical automorphism of $\mathcal{E}_?$, i.e., ρ^ε form a $\mathbb{Z}$-torsor $E_{\mathrm{B/dR}}$. First we recover ρ^ε from η-compatibility (0.3.2) for $(\mathbb{P}^1, \{0, \infty\})$, M with regular singularities at ∞, and $(\mathbb{P}^1, \{0, 1, \infty\})$, M of rank 1 with regular singularities. Having ρ^ε at hand, one has to prove that it is compatible with the constraints for all (X, T, M), of which (0.3.2) is central. The core of the argument is global: we use a theorem of Goldman [G] and Pickrell-Xia [PX1], [PX2], which asserts that the action of the Teichmüller group on the moduli space of unitary local systems with fixed local monodromies is ergodic. As in [G], this implies that, when the genus of X and the order of T are fixed, the possible discrepancy of (0.3.2) depends only on the *local* datum of monodromies at singularities of M. An observation that this discrepancy does not change upon quadratic degenerations of X reduces the proof to a few simple computations.

---

[1]In [Del] it was suggested that in case when $\mathrm{Re}(\nu)$ is exact, $\mathrm{Re}(\nu) = df$, the Betti ε-factorization comes from the Morse theory of f; see 4.6 or [B] 3.8 for a proof.

[2]In the same manner as the ν-dependence of the classical ε-factor of a Galois module V is controlled, via the class field theory, by $\det V$.

0.5   One can ask for an explicit formula for $\rho^\varepsilon(M)$. An analytic approach as in [PS] or [SW] shows that the de Rham ε-factors of a $\mathcal{D}$-module M can be recovered from the $\mathcal{D}^\infty$-module $M^\infty := \mathcal{D}^\infty \underset{\mathcal{D}}{\otimes} M$. Thus the ratio between the ε-factors of M and of another $\mathcal{D}$-module M' (say, with regular singularities) with $B(M) = B(M')$, is certain Fredholm determinant (a variant of τ-function). If x is a regular singular point of M, then $[\rho^\varepsilon_{(x,\nu)}]$ can be written explicitly using the Γ-function, see 6.3 (which is similar to the fact that the classical ε-factors of tamely ramified Galois modules are essentially products of Gauß sums). An example of the product formula is the Euler identity $\int\limits_0^1 t^{\alpha-1}(1-t)^{\beta-1}dt = \frac{\Gamma(\alpha)\Gamma(\beta)}{\Gamma(\alpha+\beta)}$.

0.6   Plan of the article: §1 presents a general story of factorization lines (i.e., of the local factorization structure); in §2–4 the algebraic and analytic de Rham ε-factors, and their Betti counterpart are defined; §5 treats the ε-period map; in §6 the ε-periods are written explicitly in terms of the Γ-function.

A different approach to product formula (0.3.3), based on Fourier transform, was developed by Bloch, Deligne, and Esnault [BDE], [E] (some essential ideas go back to [Del] and [L]; the case of regular singularities was considered earlier, and for X of arbitrary dimension, in [A], [LS], [ST], and [T]).[3] The two constructions are fairly complementary; the relation between them remains to be understood.

Questions & hopes. (o) For Verdier dual M, $M^\vee$ the lines $\mathcal{E}_?(M)_{(x,\nu)}$ and $\mathcal{E}_?(M^\vee)_{(x,-\nu)}$ should be naturally dual,[4] and ρ^ε should be compatible with duality.

(i) The period story should exist for X of any dimension, with mere lines replaced by finer objects (the homotopy points of K-theory spectra). For the Betti side, see [B]; for the de Rham one, see [P].

The meaning of local factorization structure for $\dim X > 1$ is not clear (as of the more general notion of factorization sheaves in the setting of algebraic geometry). Is there an algebro-geometric analog of the recent beautiful work of Lurie on the classification of TQFT?

(ii) There should be a geometric theory of ε-factors (cf. 5.1) for étale sheaves; for an étale sheaf of virtual rank 0 on a curve over a finite field, the corresponding trace of Frobenius function should be equal to the classical ε-factors.[5] Notice that Laumon's construction [L] (which is the only currently available method to establish the product formula for classical ε-factors) has different arrangement: its input is more restrictive (the forms ν are exact), while the output is more precise (the ε-lines are realized as determinants of true complexes).

(iii) What would be a motivic version of the story?

(iv) Γ-function appears in Deninger's vision [Den] of classical local Archimedean ε-factors. Are the two stories related on a deeper level?

0.7   I am grateful to S. Bloch, V. Drinfeld, and H. Esnault whose interest was crucial for this work, to P. Deligne for the pleasure to play in a garden he conceived, to B. Farb for the information about the Goldman and Pickrell-Xia theorems, to

---

[3][BDE] considers the case of M of virtual rank 0 and the Betti structure compatible with the Stokes structures (hence of type considered in 0.3(i)).

[4]For $\mathcal{E}_B$ this is evident from the construction; for $\mathcal{E}_{dR}$ one can hopefully deduce it from (2.10.5) applied to $M \oplus M^\vee$.

[5]The condition of virtual rank 0 is essential: the ε-line for the constant sheaf of rank 1 has non-trivial ± 1 monodromy on the components of ν's with odd order of zero, so the trace of Frobenius function is non-constant on every such component (as opposed to the classical ε-factor).

V. Schechtman and D. Zagier who urged me to write formulas, to the referee for the help, and to IHES for a serene sojourn. The research was partially supported by NSF grant DMS-0401164.

The article is a modest tribute to Spencer Bloch, for all his gifts and joy of books, of the woods, and of our relatives and friends – the numbers.

1 Factorization lines

This section is essentially an exposition of geometric class field theory (mostly) in its algebraic de Rham version.

1.1 We live over a fixed ground field k of characteristic 0; "scheme" means "separated k-scheme of finite type". The category $\mathcal{S}ch$ of schemes is viewed as a site for the étale topology (so "neighborhood" means "étale neighborhood", etc.), "space" means a sheaf on $\mathcal{S}ch$; for a space F and a scheme S elements of $F(S)$ are referred to as S-points of F. All Picard groupoids are assumed to be commutative and essentially small. For a Picard groupoid $\mathcal{L}$, we denote by $\pi_0(\mathcal{L})$, $\pi_1(\mathcal{L})$ the group of isomorphism classes of its objects and the automorphism group of any its object; for $L \in \mathcal{L}$ its class is $[L] \in \pi_0(\mathcal{L})$.

Let X be a smooth (not necessary proper or connected) curve, T its finite subscheme, K a line bundle on X.[6] For a test scheme S, we write $X_S := X \times S$, $T_S := T \times S$, $K_S := K \boxtimes \mathcal{O}_S$; $\pi : X_S \to S$ is the projection. For a Cartier divisor D on X_S we denote by $|D|$ the support of D viewed as a *reduced* closed subscheme.

Consider the next spaces:

(a) $\mathrm{Div}(X)$: its S-points are relative Cartier divisors D on X_S/S such that $|D|$ is finite over S;

(b) 2^T is a scheme whose S-points c are idempotents in $\mathcal{O}(T_S)$. Such c amounts to an open and closed subscheme T_S^c of T_S (the support of c);

(c) $\mathfrak{D} = \mathfrak{D}(X, T) \subset \mathrm{Div}(X) \times 2^T$ consists of those pairs (D, c) that $D \cap T_S \subset T_S^c$;

(d) $\mathfrak{D}^\diamond = \mathfrak{D}^\diamond(X, T; K)$ is formed by triples (D, c, ν_P) where $(D, c) \in \mathfrak{D}$ and ν_P is a trivialization of the restriction of the line bundle $K(D) := K_S(D)$ to the subscheme $P = P_{D,c} := T_S^c \cup |D|$.

Denote by $\pi_0(X)$ the scheme of connected components of X.[7] One has projection $\deg : \mathrm{Div}(X) \to \mathbb{Z}^{\pi_0(X)}$, hence the projections $\mathfrak{D}^\diamond \to \mathfrak{D} \to \mathbb{Z}^{\pi_0(X)} \times 2^T$. Notice that the component $\mathfrak{D}_{c=0}$ equals $\mathrm{Div}(X \setminus T)$, and $\mathfrak{D}_{c=1}$ equals $\mathrm{Div}(X)$.

Remarks. (i) Every S-point of $\mathfrak{D}$ can be lifted S-locally to $\mathfrak{D}^\diamond$.

(ii) Every ν_P as in (d) can be extended S-locally to a trivialization ν of $K(D)$ on a neighborhood $V \subset X_S$ of P. One can view ν_P as an equivalence class of ν's, where ν and ν' are equivalent if the function ν/ν' equals 1 on P. We often write (D, c, ν) for (D, c, ν_P).

(iii) Each space F of the list (a)–(d) is smooth in the next sense: for every closed embedding $S \hookrightarrow S'$, a geometric point $s \in S$, and $\phi \in F(S)$ one can find a neighborhood U' of s in S' and $\phi' \in F(U')$ such that $\phi|_{U'_S} = \phi'|_{U'_S}$.

(iv) The geometric fibers of $\mathrm{Div}(X)$ over $\mathbb{Z}^{\pi_0(X)}$ and of $\mathfrak{D}$, $\mathfrak{D}^\diamond$ over $\mathbb{Z}^{\pi_0(X)} \times 2^T$, are connected (i.e., every two geometric points of any fiber are members of one connected family).

---

[6] Starting from §2, our K equals ω_X.

[7] Which is the spectrum of the integral closure of k in the ring of functions on X.

A comment about the fiber $\mathfrak{D}^\circ_{(D,c)}$ of $\mathfrak{D}^\circ/\mathfrak{D}$ over $(D,c) \in \mathfrak{D}(S)$: Suppose S is smooth, so $P_{D,c}$ is a relative Cartier divisor in X_S/S. Denote by $O^\times_{D,c}$ the Weil $P_{D,c}/S$-descent of $\mathbb{G}_{mP}$, and by $K(D)^\times_{D,c}$ the Weil $P_{D,c}/S$-descent of the $\mathbb{G}_{mP}$-torsor of trivializations of the line bundle $K(D)|_{P_{D,c}}$. Then $O^\times_{D,c}$ is a smooth group S-scheme, and $K(D)^\times_{D,c}$ is an $O^\times_{D,c}$-torsor; for any S-scheme S' an S'-point of $K(D)^\times_{D,c}$ is the same as a trivialization of $K(D)$ over $(P_{D,c})_{S'}$. The latter relative divisor contains $P_{D_{S'},c_{S'}}$ (the corresponding reduced schemes coincide), so we have a canonical surjective morphism $K(D)^\times_{D,c} \to \mathfrak{D}^\circ_{(D,c)}$, hence a canonical $(D,c,\nu_P) \in \mathfrak{D}^\circ(K(D)^\times_{D,c})$. The map $K(D)^\times_{D,c}(S') \to \mathfrak{D}^\circ_{(D,c)}(S')$ is bijective if S' is smooth over S, but *not* in general.

Examples. (i) Suppose $S = X \setminus T$, $(D,c) = (\ell\Delta, 0)$ where Δ is the diagonal divisor, ℓ is any integer. Then $O^\times_{D,c} = \mathbb{G}_{mS}$, and $K(D)^\times_{D,c}$ is the $\mathbb{G}_m$-torsor $\mathcal{K}^{(\ell)}$ of trivializations of the line bundle $K(D)|_\Delta = K \otimes \omega_X^{\otimes -\ell}|_S$. For any S'/S an S'-point of $\mathfrak{D}^\circ_{(\ell\Delta,0)}$ is the same as an S'_{red}-point of $\mathcal{K}^{(\ell)}$, i.e., $\mathfrak{D}^\circ_{(\ell\Delta,0)}$ is the quotient of $\mathcal{K}^{(\ell)}$ modulo the action of the formal multiplicative group $\widehat{\mathbb{G}_m}$.

(ii) For a point $b \in T$ let k_b be its residue field, $T_b \subset T$ be the component of b, and m_b its multiplicity. Consider $(D,c) = (nb, 1_b) \in \mathfrak{D}(S)$, where $S = \operatorname{Spec} k_b$, n is any integer, 1_b is the characteristic function of $b \in T(S)$. Then $P_{D,c} = T_b$, so $O^\times_{D,c} = O^\times_{T_b}$ is an extension of $\mathbb{G}_{mS}$ by the unipotent radical. One has $K(nb)^\times_{T_b} := K(D)^\times_{D,c} \xrightarrow{\sim} \mathfrak{D}^\circ_{(D,c)}$. We set $K^\times_{T_b} := K(0b)^\times_{T_b}$.

1.2   Let $\mathcal{V}$ be a stack, alias a sheaf of categories, on $\mathcal{S}ch$. For a space F we denote by $\mathcal{V}(F)$ the category of Cartesian functors $V : F \to \mathcal{V}$. Explicitly, such V is a rule that assigns to every test scheme S and $\phi \in F(S)$ an object $V_\phi \in \mathcal{V}(S)$ together with a base change compatibility constraint. If $\mathcal{V}$ is a Picard stack, alias a sheaf of Picard groupoids, then $\mathcal{V}(F)$ is naturally a Picard groupoid.

Below we denote by $\bar{F}$ the space with $\bar{F}(S) := F(S_{\mathrm{red}})$. The stack of $\mathcal{V}$-*crystals* $\mathcal{V}_{\mathrm{crys}}$ is defined by formula $\mathcal{V}_{\mathrm{crys}}(S) := \mathcal{V}(\bar{S})$. If F is formally smooth (i.e., satisfies the property from Remark (iii) in 1.1 for every nilpotent embedding $S \hookrightarrow S'$), then $\bar{F}$ is the quotient of F modulo the evident equivalence relation; therefore objects of $\mathcal{V}_{\mathrm{crys}}(F) = \mathcal{V}(\bar{F})$ are the same as objects $V \in \mathcal{V}(F)$ equipped with a *de Rham structure*, i.e., a natural identification $\alpha : V_\phi \xrightarrow{\sim} V_{\phi'}$ for every $\phi, \phi' \in F(S)$ such that $\phi|_{S_{\mathrm{red}}} = \phi'|_{S_{\mathrm{red}}}$, which is transitive and compatible with base change. E.g. if F is a smooth scheme, then a vector bundle crystal on F is the same as a vector bundle on F equipped with a flat connection.

Key examples: Let $\mathcal{L}_k$ be the Picard groupoid of $\mathbb{Z}$-graded k-lines (with "super" commutativity constraint for the tensor structure). Below we call them simply "lines" or "k-lines"; the degree of a line $\mathcal{G}$ is denoted by $\deg(\mathcal{G})$. An $\mathcal{O}$-*line on* S (or $\mathcal{O}_S$-*line*) is an invertible $\mathbb{Z}$-graded vector bundle on S. These objects form a Picard groupoid $\mathcal{L}_\mathcal{O}(S)$; the usual pull-back functors make $\mathcal{L}_\mathcal{O}$ a Picard stack. Below $\mathcal{L}_\mathcal{O}$-crystals are referred to as *de Rham lines*; they form a Picard stack $\mathcal{L}_{\mathrm{dR}}$. Instead of $\mathbb{Z}$-graded lines, we can consider $\mathbb{Z}/2$-graded ones; the corresponding Picard stacks are denoted by $\mathcal{L}'_\mathcal{O}$, $\mathcal{L}'_{\mathrm{dR}}$. We mostly consider $\mathbb{Z}$-graded setting; all the results remain valid, with evident modifications, for $\mathbb{Z}/2$-graded one.

Remarks. (i) Let (X', T') be another pair as in 1.1, and $\pi : (X', T') \to (X, T)$ be a finite morphism of pairs, i.e., $\pi : X' \to X$ is a finite morphism of curves such that

$\pi(T') \subset T$. It yields a morphism of spaces $\pi^* : \mathfrak{D}^\diamond(X, T; K) \to \mathfrak{D}^\diamond(X', T'; \pi^* K)$, $(D, c, \nu) \mapsto (\pi^* D, \pi^* c, \pi^* \nu)$, hence the pull-back functor $\pi_* : \mathcal{V}(\mathfrak{D}^\diamond(X', T'; \pi^* K)) \to \mathcal{V}(\mathfrak{D}^\diamond(X, T; K))$ denoted by π_*; if $\mathcal{V}$ is a Picard stack, then π_* is a morphism of Picard groupoids. If $X' = X$, $T' \subset T$, we refer to π_* as "restriction to (X, T)".

Exercise. If $T' \subset T$, $T'_{\mathrm{red}} = T_{\mathrm{red}}$, then the restriction $\mathcal{L}_{\mathrm{dR}}(\mathfrak{D}^\diamond(X, T'; K)) \to \mathcal{L}_{\mathrm{dR}}(\mathfrak{D}^\diamond(X, T; K))$ is a fully faithful embedding.

We denote the union of the Picards groupoids $\mathcal{L}_{\mathrm{dR}}(\mathfrak{D}^\diamond(X, T''; K))$ for all T'' with $T''_{\mathrm{red}} = T_{\mathrm{red}}$ by $\mathcal{L}_{\mathrm{dR}}(\mathfrak{D}^\diamond(X, \hat{T}; K))$ (here $\hat{T}$ is the formal completion of X at T).

(ii) The space $\mathfrak{D}^\diamond(X, T; K)$, hence $\mathcal{L}_?(\mathfrak{D}^\diamond(X, T; K))$, actually depends only on the restriction of K to $X \setminus T$. Indeed, for any divisor $D_{(T)}$ supported on T, there is a canonical identification $\mathfrak{D}^\diamond(X, T; K) \xrightarrow{\sim} \mathfrak{D}^\diamond(X, T; K(D_{(T)}))$, $(D, c, \nu) \mapsto (D - D^c_{(T)}, \nu)$, where $D^c_{(T)}$ equals $D_{(T)}$ on T^c_S and to 0 outside. We keep K to be a line bundle on X for future notational convenience.

(iii) If U is any open subset of X, then $\mathfrak{D}^\diamond(U, T_U; K_U) \subset \mathfrak{D}^\diamond(X, T; K)$, hence we have the restriction functor $\mathcal{V}(\mathfrak{D}^\diamond(X, T; K)) \to \mathcal{V}(\mathfrak{D}^\diamond(U, T_U; K_U))$.

(iv) Remark (iv) in 1.1 implies that $\pi_1(\mathcal{L}_{\mathrm{dR}}(\mathfrak{D}^\diamond)) = \mathcal{O}^\times(\mathbb{Z}^{\pi_0(X)} \times 2^T)$.

1.3 Let $\mathcal{S}m \subset \mathcal{S}ch$ be the full subcategory of smooth schemes. For $\mathcal{V}$, F as in 1.2 we denote by $\mathcal{V}^{\mathrm{sm}}(F)$ the Picard groupoid of Cartesian functors $F|_{\mathcal{S}m} \to \mathcal{V}|_{\mathcal{S}m}$. One has a restriction functor $\mathcal{V}(F) \to \mathcal{V}^{\mathrm{sm}}(F)$. If F is smooth in the sense of Remark (iii) in 1.1, then this is a faithful functor.

Exercise. Suppose we have $\mathcal{E}, \mathcal{E}' \in \mathcal{L}_{\mathrm{dR}}(\mathfrak{D}^\diamond)$ and a morphism $\phi : \mathcal{E} \to \mathcal{E}'$ in $\mathcal{L}_{\mathcal{O}}(\mathfrak{D}^\diamond)$. Then ϕ is a morphism in $\mathcal{L}_{\mathrm{dR}}(\mathfrak{D}^\diamond)$, if (and only if) the corresponding morphism in $\mathcal{L}^{sm}_{\mathcal{O}}(\mathfrak{D}^\diamond)$ lies in $\mathcal{L}^{sm}_{\mathrm{dR}}(\mathfrak{D}^\diamond)$.[8]

Lemma. $\mathcal{L}_{dR}(F) \xrightarrow{\sim} \mathcal{L}^{sm}_{dR}(F)$.

Proof This follows from the fact that $\mathcal{L}_{\mathrm{dR}}$ is a stack with respect to the h-topology, and h-locally every scheme is smooth. $\square$

Remark. By the lemma and 1.1, one can view $\mathcal{E} \in \mathcal{L}_{\mathrm{dR}}(\mathfrak{D}^\diamond)$ as a rule that assigns to every smooth S and $(D, c) \in \mathfrak{D}(S)$ a de Rham line $\mathcal{E}_{(D,c)} := \mathcal{E}_{(D,c,\nu_P)}$ on $K(D)^\times_{D,c}$ in a way compatible with the base change.

1.4 *For this subsection, X is proper.* Let $\mathrm{Rat}(X, K) = \mathrm{Rat}(X)$ be a space whose S-points are rational sections ν of the line bundle K_S such that $|\mathrm{div}(\nu)|$ does not contain a connected component of any geometric fiber of X_S/S. There is a natural morphism $\mathrm{Rat}(X) \to \mathfrak{D}^\diamond_{c=1}$, $\nu \mapsto (-\mathrm{div}(\nu), 1, \nu)$, so every $\mathcal{E} \in \mathcal{L}_?(\mathfrak{D}^\diamond)$ yields naturally an object of $\mathcal{L}_?(\mathrm{Rat}(X))$, which we denote again by $\mathcal{E}$.

The next fact is a particular case of [BD] 4.3.13:

Proposition. *Every function on $\mathrm{Rat}(X)$ is constant. All $\mathcal{O}$- and de Rham lines on $\mathrm{Rat}(X)$ are constant.*

Proof Let L be an auxiliary ample line bundle on X; set $V_1^{(m)} := \Gamma(X, K \otimes L^{\otimes m})$, $V_2^{(m)} := \Gamma(X, L^{\otimes m})$. Let $U^{(m)} \subset \mathbb{P}(V_1^{(m)} \times V_2^{(m)})$ be the open subset of those $\phi = (\phi_1, \phi_2)$ that neither ϕ_1 nor ϕ_2 vanishes on any connected component of X. Consider a map $\theta^{(m)} : U^{(m)} \to \mathrm{Rat}(X)$, $(\phi_1, \phi_2) \mapsto \phi_1/\phi_2$. We will check that for m

---

[8]Hint: Use Remark (iii) in 1.1 for embeddings $S \hookrightarrow S'$ where S' is smooth.

large the $\theta^{(m)}$-pull-back of any $\mathcal{O}$- or de Rham line on $\mathrm{Rat}(X)$ is trivial, and every function on $U^{(m)}$ is constant. This implies the proposition, since geometric fibers of $\theta^{(m)}$ are connected and the images of $\theta^{(m)}$ form a directed system of subspaces whose inductive limit equals $\mathrm{Rat}(X)$ (i.e., every $\nu \in \mathrm{Rat}(X)(S)$ factors S-locally through $\theta^{(m)}$ for sufficiently large m).

Notice that the complement to $U^{(m)}$ in $\mathbb{P}(V_1^{(m)} \times V_2^{(m)})$ has codimension ≥ 2 for m large. Therefore every function and every $\mathcal{O}$- and de Rham line extend to $\mathbb{P}(V_1^{(m)} \times V_2^{(m)})$. Thus for m large every function on $U^{(m)}$ is constant and every de Rham line is trivial.

The case of an $\mathcal{O}$-line requires an extra argument. Any $(\psi_1, \psi_2) \in U^{(n)}$ yields an embedding $\mathbb{P}(\Gamma(X, L^{\otimes k})) \hookrightarrow \mathbb{P}(V_1^{(m)} \times V_2^{(m)})$, $\gamma \mapsto (\gamma\psi_1, \gamma\psi_2)$; here $m = n + k$. For k large, the preimage of $U^{(m)}$ in $\mathbb{P}(\Gamma(X, L^{\otimes k}))$ is an open dense subset of codimension ≥ 2, and $\theta^{(m)}$ is constant on it. Thus for any $\mathcal{O}$-line $\mathcal{L}$ on $\mathrm{Rat}(X)$ the restriction of the corresponding line on $\mathbb{P}(V_1^{(m)} \times V_2^{(m)})$ to $\mathbb{P}(\Gamma(X, L^{\otimes k}))$ is trivial, hence the line itself is trivial, so $\theta^{(m)*}\mathcal{L}$ is trivial, and we are done. □

Therefore for any $\mathcal{E}$ in $\mathcal{L}_{\mathcal{O}}(\mathfrak{D}^\diamond)$ or $\mathcal{L}_{\mathrm{dR}}(\mathfrak{D}^\diamond)$ the lines $\mathcal{E}_\nu$ for all rational non-zero ν are canonically identified. We denote this line simply by $\mathcal{E}(X)$.

1.5 A finite subset of $\{(D_\alpha, c_\alpha, \nu_\alpha)\}$ of $\mathfrak{D}^\diamond(S)$, is said to be *disjoint* if the sub-schemes P_{D_α, c_α} are pairwise disjoint. Then we have $\Sigma(D_\alpha, c_\alpha, \nu_\alpha) := (\Sigma D_\alpha, \Sigma c_\alpha, \Sigma \nu_\alpha) \in \mathfrak{D}^\diamond(S)$, where $\Sigma\nu_\alpha$ equals ν_α on P_{D_α, c_α}.

For $\mathcal{E}$ in $\mathcal{L}_?(\mathfrak{D}^\diamond)$, where $\mathcal{L}_?$ is a Picard stack, a *factorization structure* on $\mathcal{E}$ is a rule which assigns to every disjoint family as above a *factorization isomorphism*

$$\otimes_\alpha \mathcal{E}_{(D_\alpha, c_\alpha, \nu_\alpha)} \xrightarrow{\sim} \mathcal{E}_{\Sigma(D_\alpha, c_\alpha, \nu_\alpha)}. \tag{1.5.1}$$

These isomorphisms should be compatible with base change and satisfy an evident transitivity property. One defines factorization structure on objects of $\mathcal{L}_?(\mathfrak{D}(X, T))$, $\mathcal{L}_?(\mathrm{Div}(X))$, $\mathcal{L}_?(\mathrm{Div}^{\mathrm{eff}}(X))$, or $\mathcal{L}_?(2^T)$ in the similar way.

Objects of $\mathcal{L}_?(\mathfrak{D}^\diamond)$ equipped with a factorization structure are called *(K-twisted) factorization objects* of $\mathcal{L}_?$ on $(X, T; K)$; they form a Picard groupoid $\mathcal{L}_?^\Phi(X, T; K)$. In particular, we have Picard groupoids $\mathcal{L}_{\mathcal{O}}^\Phi(X, T; K)$, $\mathcal{L}_{\mathrm{dR}}^\Phi(X, T; K)$ of $\mathcal{O}$- and de Rham factorization lines.

Proposition. *Factorization objects have local nature:* $U \mapsto \mathcal{L}_?^\Phi(U, T_U; K_U)$ *is a Picard stack on* $X_{\text{ét}}$.

Proof Let $\pi : U \to X$ be an étale map. For $\mathcal{E} \in \mathcal{L}_?^\Phi(X, T; K)$ one defines its pull-back $\pi^*\mathcal{E}$ as follows. Take any $(D, c, \nu) \in \mathfrak{D}^\diamond(U, T_U; K_U)$. It suffices to define $\mathcal{E}_{(D,c,\nu)}$ étale locally on S. Write $(D, c, \nu) = \Sigma(D_\alpha, c_\alpha, \nu_\alpha)$ with connected $P_\alpha = P_{D_\alpha, c_\alpha}$. Then there is a uniquely defined $(D'_\alpha, c'_\alpha, \nu'_\alpha) \in \mathfrak{D}^\diamond(X, T; K)$ such that D_α is a connected component of the pull-back of D'_α to U and π yields an isomorphism $P_\alpha \xrightarrow{\sim} P'_\alpha$ which identifies $\nu_{\alpha P_\alpha}$ with $\nu'_{\alpha P'_\alpha}$.

Set $\pi^*\mathcal{E}_{(D,c,\nu)} := \otimes \mathcal{E}_{(D'_\alpha, c'_\alpha, \nu'_\alpha)}$. Due to factorization structure on $\mathcal{E}$, this definition is compatible with base change, and $\pi^*\mathcal{E} \in \mathcal{L}_?^\Phi(\mathfrak{D}^\diamond(U, T_U; K_U))$ so defined has an evident factorization structure. Thus $\mathcal{L}_?^\Phi$ is a presheaf of Picard groupoids on $X_{\text{ét}}$. We leave it to the reader to check the gluing property. □

NB: The pull-back functor for open embeddings is defined regardless of factorization structure (see Remark (iii) in 1.2).

Remarks. (i) The evident forgetful functor $\mathcal{L}_{\mathrm{dR}}^{\Phi}(X,T;K) \to \mathcal{L}_{\mathcal{O}}^{\Phi}(X,T;K)$ is faithful. By 1.2 (and Remark (iii) in 1.1), for $\mathcal{E} \in \mathcal{L}_{\mathcal{O}}^{\Phi}(X,T;K)$ a de Rham structure on $\mathcal{E}$, i.e., a lifting of $\mathcal{E}$ to $\mathcal{L}_{\mathrm{dR}}^{\Phi}(X,T;K)$, amounts to a rule which assigns to every scheme S and a pair of points $(D,c,\nu_P),(D',c',\nu_P') \in \mathfrak{D}^{\circ}(S)$ which coincide on S_{red}, a natural identification (notice that $c = c'$)

$$\alpha^{\varepsilon} : \mathcal{E}_{(D,c,\nu_P)} \overset{\sim}{\to} \mathcal{E}_{(D',c,\nu_P')}. \tag{1.5.2}$$

The α^{ε} should be transitive and compatible with base change and factorization.

(ii) Remarks in 1.3 and (i)–(iii) in 1.2 remain valid for factorization lines. Thus we have a Picard groupoid $\mathcal{L}_{\mathrm{dR}}^{\Phi}(X,\hat{T};K)$, etc.

(iii) There is a natural Picard functor

$$\underset{b \in T_{\mathrm{red}}}{\Pi} \mathcal{L}_?(b) \to \mathcal{L}_?^{\Phi}(X,T;K), \tag{1.5.3}$$

which assigns to $E = (E_b) \in \Pi\mathcal{L}_?(b)$ a factorization object $\mathcal{E}$ with $\mathcal{E}_{(D,c,\nu_P)} = \mathrm{Nm}_{T_{\mathrm{red}S}^c/S}(E)$; here $T_{\mathrm{red}S}^c$ is the preimage of T_{red} by the projection $p : T_S^c \to T$.

(iv) By Remark (iv) in 1.2 there is a natural isomorphism

$$\mathcal{O}^{\times}(T_{\mathrm{red}}) \times \mathcal{O}^{\times}(\pi_0(X)) \overset{\sim}{\to} \pi_1(\mathcal{L}_{\mathrm{dR}}^{\Phi}(X,T;K)). \tag{1.5.4}$$

Here $(\alpha,\beta) \in \mathcal{O}^{\times}(T_{\mathrm{red}}) \times \mathcal{O}^{\times}(\pi_0(X))$ acts on $\mathcal{E}_{(D,c,\nu)}$ as multiplication by the locally constant function $\mathrm{Nm}_{T_{\mathrm{red}S}^c/S}(\alpha)\mathrm{Nm}_{\pi_0(X)_S/S}(\beta^{deg(D)})$. Notice that the embedding $\mathcal{O}^{\times}(T_{\mathrm{red}}) \hookrightarrow \pi_1(\mathcal{L}_{\mathrm{dR}}^{\Phi}(X,T;K))$ comes from (1.5.3).

1.6 As in 1.1, every $(D,c) \in \mathfrak{D}(S)$, S is smooth, yields a morphism $K(D)_{D,c}^{\times} \to \mathfrak{D}^{\circ}$, hence a Picard functor $\mathcal{L}_?^{\Phi}(\mathfrak{D}^{\circ}) \to \mathcal{L}_?(K(D)_{D,c}^{\times})$, $\mathcal{E} \mapsto \mathcal{E}_{(D,c)}$. In particular, following Examples in 1.1, for $\ell \in \mathbb{Z}$ we have $\mathcal{E}^{(\ell)} := \mathcal{E}_{(\ell\Delta,0)} \in \mathcal{L}_?(\mathcal{K}^{(\ell)})$, and for $b \in T$, $n \in \mathbb{Z}$, we have $\mathcal{E}_{T_b}^{(n)} := \mathcal{E}_{(nb,1_b)} \in \mathcal{L}_?(K(nb)_{T_b}^{\times})$. Set $\mathcal{E}_{T_b} := \mathcal{E}_{T_b}^{(0)} \in \mathcal{L}_?(K_{T_b}^{\times})$. Notice that $\mathcal{E}^{(0)} \in \mathcal{L}_?(\mathcal{K}^{(0)})$ is canonically trivialized.

If $\mathcal{E} \in \mathcal{L}_{\mathcal{O}}^{\Phi}(\mathfrak{D}^{\circ})$, then the $\mathcal{O}$-lines $\mathcal{E}^{(\ell)}$ carry a canonical connection along the fibers of the projection $\mathcal{E}^{(\ell)} \to X \setminus T$ (see Example (i) in 1.1). A de Rham structure on $\mathcal{E}$ provides a flat connection ∇^{ε} on $\mathcal{E}^{(\ell)}$ that extends this relative connection. Since the degrees of lines are locally constant, the factorization implies

$$\deg(\mathcal{E}^{(\ell)}) = \ell \deg(\mathcal{E}^{(1)}), \quad \deg(\mathcal{E}_{T_b}^{(n+1)}) = \deg(\mathcal{E}_{T_b}^{(n)}) + \deg(\mathcal{E}^{(1)}). \tag{1.6.1}$$

Let $\mathcal{L}_{\mathrm{dR}}(X,T) \subset \mathcal{L}_{\mathrm{dR}}(X \setminus T)$ be the Picard subgroupoid of those de Rham lines whose connection at every $b \in T_{\mathrm{red}}$ has pole of order less or equal the multiplicity of T at b.

For a $\mathbb{G}_m$-torsor $\mathcal{K}$ over a scheme Y we denote by $\mathcal{L}_{\mathrm{dR}}(Y;\mathcal{K})$ the Picard groupoid of de Rham lines $\mathcal{G}$ on $\mathcal{K}$ such that $\mathcal{G}^{\otimes 2}$ is constant along the fibers (i.e., comes from a de Rham line on Y) and the fiberwise monodromy of $\mathcal{G}$ equals $(-1)^{\deg\mathcal{G}}$. Thus we have Picard groupoids $\mathcal{L}_{\mathrm{dR}}(X \setminus T)^{(\ell)} := \mathcal{L}_{\mathrm{dR}}(X \setminus T;\mathcal{K}^{(\ell)})$, $\ell \in \mathbb{Z}$; let $\mathcal{L}_{\mathrm{dR}}(X,T)^{(\ell)} \subset \mathcal{L}_{\mathrm{dR}}(X \setminus T)^{(\ell)}$ be the Picard subgroupoids of those $\mathcal{G}$ that $\mathcal{G}^{\otimes 2} \in \mathcal{L}_{\mathrm{dR}}(X,T)$.

Choose a trivialization ν_T of the restriction of K to T, i.e., a collection $\{\nu_{T_b}\}$ of k_b-points in $K_{T_b}^{\times}$. For a factorization line $\mathcal{E}$ set $\mathcal{E}_{\nu_{T_b}} := \mathcal{E}_{(0,1_b,\nu_{T_b})} =$ the fiber of $\mathcal{E}_{T_b}$ at ν_{T_b}. The next theorem is the main result of this section. The proof for $T = \emptyset$ is in 1.7–1.9; the general case is treated in 1.10–1.11. In 1.11 one finds its reformulation free from the auxiliary ν_T.

Theorem. *For $\mathcal{E} \in \mathcal{L}_{dR}^{\Phi}(X, T; K)$ one has $\mathcal{E}^{(1)} \in \mathcal{L}_{dR}(X, T)^{(1)}$, and the functor*

$$\mathcal{L}_{dR}^{\Phi}(X, T; K) \to \mathcal{L}_{dR}(X, T)^{(1)} \times \prod_{b \in T_{red}} \mathcal{L}_{k_b}, \quad \mathcal{E} \mapsto (\mathcal{E}^{(1)}, \{\mathcal{E}_{\nu_{T_b}}\}), \tag{1.6.2}$$

is an equivalence of Picard groupoids.

If $K = \omega_X$, then the $\mathbb{G}_m$-torsor $\mathcal{K}^{(1)}$ is trivialized by a canonical section ν_1 (its value at $x \in X \setminus T$ is the element in $\omega(x)/\omega$ with residue 1). The functor $\nu_1^* : \mathcal{L}_{dR}(X, T)^{(1)} \to \mathcal{L}_{dR}(X, T)$ is evidently an equivalence, so the theorem can be reformulated as follows (here $\mathcal{E}_{X \setminus T}^{(1)} := \nu_1^* \mathcal{E}^{(1)} = \mathcal{E}_{(\Delta, 0, \nu_1)}$):

Theorem$'$. *One has a Picard groupoid equivalence*

$$\mathcal{L}_{dR}^{\Phi}(X, T; \omega_X) \overset{\sim}{\to} \mathcal{L}_{dR}(X, T) \times \prod_{b \in T_{red}} \mathcal{L}_{k_b}, \quad \mathcal{E} \mapsto (\mathcal{E}_{X \setminus T}^{(1)}, \{\mathcal{E}_{\nu_{T_b}}\}). \tag{1.6.3}$$

Variant. More generally, we can fix a divisor $\Sigma d_b b$ supported on T and take for ν_T a trivialization of $K(\Sigma d_b b)$ on T. The corresponding assertion is equivalent to the above theorem by Remark (ii) in 1.2.

Example. If $K = \omega_X$ and $T = T_{red}$, then a convenient choice is $d_b \equiv 1$, for $K(b)_b^{\times}$ is canonically trivialized by ν_1 as above. We denote the fiber of $\mathcal{E}_{T_b}^{(1)}$ at ν_1 by $\mathcal{E}_b^{(1)}$.

1.7 *For the subsections 1.7–1.9 we assume that $T = \emptyset$, so $\mathfrak{D} = Div(X)$.*

For a Picard stack $\mathcal{L}_?$ and a commutative monoid space D denote by $\mathrm{Hom}(\mathrm{D}, \mathcal{L}_?)$ the Picard groupoid of symmetric monoidal morphisms $\mathrm{D} \to \mathcal{L}_?$ (we view D as a "discrete" symmetric monoidal stack). Thus an object of $\mathrm{Hom}(\mathrm{D}, \mathcal{L}_?)$ is $\mathcal{F} \in \mathcal{L}_?(\mathrm{D})$ together with a *multiplication structure* which is a rule that assigns to every finite collection $\{D_\alpha\}$ of S-points of D a *multiplication isomorphism* $\otimes \mathcal{F}_{D_\alpha} \overset{\sim}{\to} \mathcal{F}_{\Sigma D_\alpha}$ (where Σ is the operation in D); the isomorphisms should be compatible with base change and satisfy an evident transitivity property. If D^{gr} is the group completion of D, then $\mathrm{Hom}(\mathrm{D}^{gr}, \mathcal{L}_?) \overset{\sim}{\to} \mathrm{Hom}(\mathrm{D}, \mathcal{L}_?)$.

We are interested in D equal to the monoid space of effective divisors $\mathrm{Div}^{eff}(X) = \sqcup \mathrm{Sym}^n(X) \subset \mathrm{Div}(X)$; one has $\mathrm{Div}^{eff}(X)^{gr} = \mathrm{Div}(X)$. A multiplication structure on $\mathcal{F}$ being restricted to disjoint divisors makes a factorization structure. Pulling $\mathcal{F}$ back to $\mathfrak{D}^{\diamond}$ is a Picard functor

$$\mathrm{Hom}(\mathrm{Div}(X), \mathcal{L}_?) \to \mathcal{L}_?^{\Phi}(X; K). \tag{1.7.1}$$

Let $\mathcal{L}_{dR}^0 \subset \mathcal{L}_{dR}$ be the Picard stack of degree 0 de Rham lines.

Proposition. *One has $\mathrm{Hom}(\mathrm{Div}(X), \mathcal{L}_{dR}^0) \overset{\sim}{\to} \mathcal{L}_{dR}^{0\Phi}(X; K)$.*

Proof (a) Let us show that for any $\mathcal{E} \in \mathcal{L}_{dR}^{0\Phi}(X; K)$ the de Rham lines $\mathcal{E}^{(\ell)}$ on $\mathcal{K}^{(\ell)}$ come from de Rham lines on X.

The claim is X-local, so we trivialize K by section a ν_0 and pick a function t on X with dt invertible. Then $\mathcal{K}^{(\ell)}$ is trivialized by section $\nu_0^{(\ell)} := \nu_0 dt^{\otimes -\ell}$; let z be the corresponding fiberwise coordinate on $\mathcal{K}^{(\ell)}$. Choose X-locally a trivialization $e^{(\ell)}$ of $\mathcal{E}^{(\ell)}$; let $\theta^{(\ell)} \in \omega_{\mathcal{K}^{(\ell)}/X}$ be the restriction of $\nabla(e^{(\ell)})/e^{(\ell)}$ to the fibers. Then

$\theta^{(\ell)} = \Sigma f_k^{(\ell)}(x) z^k d\log z$, where $f_k^{(\ell)}(x)$, $k \in \mathbb{Z}$, are functions on X; we want to show that $f_k^{(\ell)}(x) = 0$ for $k \neq 0$,[9] and $f_0^{(\ell)}(x) \in \mathbb{Z}$.

Let $S \subset X \times X$ be a sufficiently small neighborhood of the diagonal, x_1, x_2 be the coordinate functions on S that correspond to t, $D^{(\ell_1, \ell_2)} \in \operatorname{Div}(X)(S)$ be the divisor $\ell_1 \Delta_1 + \ell_2 \Delta_2$. Then $(t-x_1)^{-\ell_1}(t-x_2)^{-\ell_2}\nu_0$ is a trivialization of $K(D^{(\ell_1,\ell_2)})$ near $|D^{(\ell_1,\ell_2)}|$. Denote by $\nu_0^{(\ell_1,\ell_2)}$ the corresponding section of $K(D^{(\ell_1,\ell_2)})^\times_{D^{(\ell_1,\ell_2)},0}$ (see 1.1); set $\mathcal{K}^{(\ell_1,\ell_2)} := \mathbb{G}_m \nu_0^{(\ell_1,\ell_2)} \subset K(D^{(\ell_1,\ell_2)})^\times_{D^{(\ell_1,\ell_2)},0}$, $\mathcal{E}^{(\ell_1,\ell_2)} := \mathcal{E}_{(D^{(\ell_1,\ell_2)},0)}|_{\mathcal{K}^{(\ell_1,\ell_2)}}$.

Outside the diagonal in S one has an embedding $i^{(\ell_1,\ell_2)} : \mathcal{K}^{(\ell_1,\ell_2)} \hookrightarrow pr_1^* \mathcal{K}^{(\ell_1)} \times pr_2^* \mathcal{K}^{(\ell_2)}$ defined by the factorization; explicitly, it identifies $z\nu_0^{(\ell_1,\ell_2)}(x_1,x_2)$ with $(z(x_1-x_2)^{-\ell_2}\nu_0^{(\ell_1)}(x_1), z(x_2-x_1)^{-\ell_1}\nu_0^{(\ell_2)}(x_2))$. Restricting to $\mathcal{K}^{(\ell_1,\ell_2)}$ the image of $e^{(\ell_1)} \boxtimes e^{(\ell_2)}$ by the factorization isomorphism (1.5.1), we get a trivialization $e^{(\ell_1,\ell_2)}$ of $\mathcal{E}^{(\ell_1,\ell_2)}$ outside the diagonal in S. Let $m(\ell_1, \ell_2)$ be its order of pole at the diagonal, so $(x_1 - x_2)^{m(\ell_1,\ell_2)} e^{(\ell_1,\ell_2)}$ is a trivialization of $\mathcal{E}^{(\ell_1,\ell_2)}$ on S. Therefore the restriction $\theta^{(\ell_1,\ell_2)}$ of $\nabla(e^{(\ell_1,\ell_2)})/e^{(\ell_1,\ell_2)}$ to the fibers is a regular relative form, which equals $i^{(\ell_1,\ell_2)*}(pr_1^*\theta^{(\ell_1)} + pr_2^*\theta^{(\ell_2)}) = \Sigma(f_k^{(\ell_1)}(x_1)(x_1-x_2)^{-k\ell_2} + f_k^{(\ell_2)}(x_1)(x_2 - x_1)^{-k\ell_1})z^k d\log z$.

Since $\theta^{(\ell,-\ell)}$ has no pole at the diagonal, the above formula implies that $f_k^{(\ell)} = 0$ for $k\ell < 0$. Similarly, the formula for $\theta^{(\ell,2\ell)}$ shows that $f_k^{(\ell)} = 0$ for $k\ell > 0$. To see that $f_0^{(\ell)} \in \mathbb{Z}$, notice that the above picture for $\ell_1 = \ell_2 = \ell$ is symmetric with respect to the transposition involution σ of $X \times X$, hence descends to $S/\sigma \subset \operatorname{Sym}^2 X$. Thus $m(\ell, \ell)$ is even. One has $\nabla(e^{(\ell)})/e^{(\ell)} = f_0^{(\ell)} d\log z + g^{(\ell)}(x)dx$ where $g^{(\ell)}(x)$ is a regular function. Then $\nabla(e^{(\ell,\ell)})/e^{(\ell,\ell)} + d\log(x_1 - x_2)^{m(\ell,\ell)}$ is a regular 1-form. It equals $(f_0^{(\ell)}(x_1) + f_0^{(\ell)}(x_2))(d\log z - d\log(x_1 - x_2)) + m(\ell,\ell)d\log(x_1 - x_2) + g^{(\ell)}(x_1)dx_1 + g^{(\ell)}(x_2)dx_2$. Therefore $f_0^{(\ell)}(x) = m(\ell,\ell)/2 \in \mathbb{Z}$, and we are done.

(b) The next properties of de Rham lines will be repeatedly used. Let $\pi : K \to S$ be a smooth morphism of smooth schemes with dense image.

Lemma. *(i) The functor $\pi^* : \mathcal{L}_{dR}(S) \to \mathcal{L}_{dR}(K)$ is faithful. If the geometric fibers of π are connected (say, π is an open embedding), then π^* is fully faithful.*
(ii) If, in addition, π is surjective, then a de Rham line $\mathcal{E}$ on K comes from S if (and only if) this is true over a neighborhood U of the generic point(s) of S. $\square$

(c) As was mentioned, $\operatorname{Div}(X)$ is the group completion of $\operatorname{Div}^{\mathrm{eff}}(X) = \sqcup \operatorname{Sym}^n(X)$. So we have the projection $\operatorname{Div}^{\mathrm{eff}}(X) \times \operatorname{Div}^{\mathrm{eff}}(X) \to \operatorname{Div}(X)$, $(D_1, D_2) \mapsto D_1 - D_2$, which identifies $\operatorname{Div}(X)$ with the quotient of $\operatorname{Div}^{\mathrm{eff}}(X) \times \operatorname{Div}^{\mathrm{eff}}(X)$ with respect to the diagonal action. Therefore a line $\mathcal{E}$ on $\operatorname{Div}(X)$ is the same as a collection of lines $\mathcal{E}^{n_1,n_2}$ on $\operatorname{Sym}^{n_1,n_2}(X) := \operatorname{Sym}^{n_1}(X) \times \operatorname{Sym}^{n_2}(X)$ together with $\operatorname{Div}^{\mathrm{eff}}(X)$-equivariance structure, which is the datum of identifications of their pull-backs by $\operatorname{Sym}^{n_1,n_2}(X) \leftarrow \operatorname{Sym}^{n_1,n_2}(X) \times \operatorname{Sym}^{n_3}(X) \to \operatorname{Sym}^{n_1+n_3,n_2+n_3}(X)$, $(D_1, D_2) \leftarrow (D_1, D_2; D_3) \to (D_1 + D_3, D_2 + D_3)$ that satisfy a transitivity property.

Let us prove the proposition. We need to show that any $\mathcal{E} \in \mathcal{L}_{\mathrm{dR}}^{0\Phi}(X; K)$, viewed as a mere de Rham line on $\mathfrak{D}^\circ$, is the pull-back by $\mathfrak{D}^\circ \to \operatorname{Div}(X)$ of a uniquely defined line in $\mathcal{L}_{\mathrm{dR}}^0(\operatorname{Div}(X))$, which we denote by $\mathcal{E}$ or $\mathcal{E}_{\mathrm{Div}}$, and that the factorization structure on $\mathcal{E}$ comes from a uniquely defined multiplication structure on $\mathcal{E}_{\mathrm{Div}}$.

---

[9]The proof uses only the factorization $\mathcal{O}$-line structure on $\mathcal{E}$.

We use the fact that for any $D \in \mathrm{Div}(X)(S)$, S is smooth, the projection $K(D)_D^\times := K(D)_{D,0}^\times \to S$ satisfies the conditions of (i), (ii) of the lemma.

To define $\mathcal{E}_{\mathrm{Div}}$ on $S = \mathrm{Sym}^{n_1,n_2}(X)$, we apply (ii) of the lemma to $\mathcal{E}$ on $K = K(D_1 - D_2)_{D_1+D_2}^\times$. Let U be the complement to the diagonal divisor in $X^{n_1} \times X^{n_2}$. Over U our $\mathcal{E}$ equals $(\mathcal{E}^{(1)})^{\boxtimes n_1} \boxtimes (\mathcal{E}^{(-1)})^{\boxtimes n_2}$, and we are done by (a). The $\mathrm{Div}^{\mathrm{eff}}(X)$-equivariance structure on the datum of $\mathcal{E}_{\mathrm{Div}}^{n_1,n_2}$ is automatic by (i) of the lemma (applied to $K(D_1 - D_2)_{D_1+D_2+2D_3}$).

The factorization structure on $\mathcal{E}$ yields one on $\mathcal{E}_{\mathrm{Div}}$. Let us show that it extends uniquely to a multiplication structure. It suffices to define the multiplication $\otimes \mathcal{E}_{D_\alpha} \to \mathcal{E}_{\Sigma D_\alpha}$ over each $\Pi\mathrm{Sym}^{n_{1\alpha},n_{2\alpha}}(X)$ in a way compatible with the $\mathrm{Div}^{\mathrm{eff}}(X)$-equivariance structure. Our multiplication equals factorization over the open dense subset where all $D_{i\alpha}$ are disjoint, so we have it everywhere by (i) of the lemma. The compatibility with $\mathrm{Div}^{\mathrm{eff}}(X)$-equivariance holds over the similar open subset of $\Pi(\mathrm{Sym}^{n_{1\alpha},n_{2\alpha}}(X) \times \mathrm{Sym}^{n_{3\alpha}}(X))$, hence everywhere, and we are done. $\qquad\square$

Corollary. *The functor $\mathcal{L}_{dR}^{0\Phi}(X; K) \to \mathcal{L}_{dR}^0(X)$, $\mathcal{E} \mapsto \mathcal{E}^{(1)}$, is an equivalence.*

Proof Its composition with the equivalence of the proposition is a functor $\mathrm{Hom}(\mathrm{Div}^{\mathrm{eff}}(X), \mathcal{L}_{dR}^0) \to \mathcal{L}_{dR}^0(X)$ which assigns to $\mathcal{F}$ its restriction to the component $X = \mathrm{Sym}^1(X)$ of $\mathrm{Div}^{\mathrm{eff}}(X)$. This functor is clearly invertible: its inverse assigns to $\mathcal{P} \in \mathcal{L}_{dR}^0(X)$ the de Rham line $\mathrm{Sym}(\mathcal{P})$ on $\mathrm{Div}^{\mathrm{eff}}(X) = \sqcup\mathrm{Sym}^n(X)$, $\mathrm{Sym}(\mathcal{P})_{\mathrm{Sym}^n(X)} := \mathrm{Sym}^n(\mathcal{P})$ equipped with an evident multiplication structure. $\qquad\square$

1.8 An example of a de Rham factorization line $\mathcal{E}$ with $\deg(\mathcal{E}^{(1)}) = 1$:

Suppose $X = \mathbb{A}^1$, $T = \emptyset$, $K = \mathcal{O}_X$. We construct $\mathcal{E}$ in the setting of Remark in 1.3. For a smooth S and $D \in \mathrm{Div}(S)$ the line $\mathcal{O}_X(D)$ is naturally trivialized by a section ν_D, $\nu_{\Sigma n_i x_i} = \Pi(t - x_i)^{-n_i}$. Then ν_D trivializes the $\mathcal{O}_{D,0}^\times$-torsor $K(D)_{D,0}^\times$, so the canonical character $f \mapsto f(D)$ of $\mathcal{O}_{D,0}^\times$ yields an invertible function $\phi_D \in \mathcal{O}^\times(K(D)_{D,0}^\times)$, $\phi_D(\nu) := (\nu/\nu_D)(D)$. Our $\mathcal{E}_{(D,0)}$ comes from the Kummer torsor for $\phi_D^{-1/2}$ placed in degree $\deg(D)$, i.e., it equals $\mathcal{O}_{K(D)_{D,0}^\times}[\deg(D)]$ as an $\mathcal{O}$-line, the connection is given by the 1-form $\frac{1}{2}d\log\phi_D$.

Notice that if $D' \in \mathrm{Div}(S)$ is another divisor such that $|D| \cap |D'| = \emptyset$, then the invertible function ν_D on $X_S \setminus |D|$ yields $(D, D') := \nu_D(D') \in \mathcal{O}^\times(S)$. One has

$$(D, D') = (-1)^{\deg(D)\deg(D')}(D', D). \tag{1.8.1}$$

Let us define the factorization structure on $\mathcal{E}$. Suppose $D = \sqcup D_\alpha$, so $K(D)_{D,0}^\times = \Pi K(D_\alpha)_{D_\alpha,0}^\times$. Any linear order on the set of indices α yields an evident identification of the "constant" $\mathcal{O}$-lines $\otimes \mathcal{E}_{(D_\alpha,0)} \xrightarrow{\sim} \mathcal{E}_{(D,0)}$. The factorization isomorphism (1.5.1) is its product with $\underset{\alpha<\alpha'}{\Pi}(D_\alpha, D_{\alpha'})$. The choice of order is irrelevant due to the "super" commutativity constraint and (1.8.1). Both transitivity property and horizontality follow since $\Pi\phi_{D_\alpha} = \phi_D \underset{\alpha\neq\alpha'}{\Pi}(D_\alpha, D_{\alpha'})$.

1.9 *Proof of the theorem in 1.6 in case $T = \emptyset$.* Let us check that for $\mathcal{E} \in \mathcal{L}_{dR}^\Phi(X; K)$ one has $\mathcal{E}^{(1)} \in \mathcal{L}_{dR}(X)^{(1)}$. The claim is X-local, so we can assume that K is trivialized and there is a function t on X with dt invertible, i.e., $t : X \to \mathbb{A}^1$ is étale. Let $\mathcal{E}' \in \mathcal{L}_{dR}^\Phi(X; K)$ be the pull-back of the factorization line from 1.8. Then

$\mathcal{L}_{\mathrm{dR}}^{\Phi}(X;K)$ is generated by $\mathcal{L}_{\mathrm{dR}}^{0\Phi}(X;K)$ and $\mathcal{E}'$. Our claim holds for $\mathcal{E} \in \mathcal{L}_{\mathrm{dR}}^{0\Phi}(X;K)$ by 1.7 and it is evident for $\mathcal{E}'$; we are done.

Let us show that $\mathcal{L}_{\mathrm{dR}}^{\Phi}(X;K) \to \mathcal{L}_{\mathrm{dR}}(X)^{(1)}$, $\mathcal{E} \mapsto \mathcal{E}^{(1)}$, is an equivalence. Notice that the preimage of $\mathcal{L}_{\mathrm{dR}}^{0}(X) \subset \mathcal{L}_{\mathrm{dR}}(X)$ equals $\mathcal{L}_{\mathrm{dR}}^{0\Phi}(X;K)$, and, by the corollary in 1.7, $\mathcal{L}_{\mathrm{dR}}^{0\Phi}(X;K) \overset{\sim}{\to} \mathcal{L}_{\mathrm{dR}}^{0}(X)$. Since X-locally there is $\mathcal{E}$ with $\deg(\mathcal{E}^{(1)}) = 1$ by 1.8 and $\deg : \pi_0(\mathcal{L}_{\mathrm{dR}}(X))/\pi_0(\mathcal{L}_{\mathrm{dR}}^{0}(X)) \overset{\sim}{\to} \mathbb{Z}$, we are done. $\square$

1.10   Suppose now $T \neq \emptyset$. Pick $b \in T_{\mathrm{red}}$, and consider the $O_{T_b}^{\times}$-torsor $K_{T_b}^{\times}$ (see Example (ii) in 1.1; we follow the notation of loc.cit.).

Let $(\omega)_b := \omega_X(\infty b)/\omega_X$ be the k_b-vector space of polar parts of rational 1-forms at b, $(\omega)_b^{\leq n}$ be the subspace of polar parts of order $\leq n$. The Lie algebra of $O_{T_b}^{\times}$ equals $\mathcal{O}(T_b)$. The space $\Omega^1(K_{T_b}^{\times})^{\mathrm{inv}}$ of translation invariant 1-forms on $K_{T_b}^{\times}$ is its dual. The residue pairing $(\omega)_b^{\leq m_b} \times \mathcal{O}(T_b) \to k_b$, $(\psi, f) \mapsto \mathrm{Res}_b(f\psi)$, identifies it with $(\omega)_b^{\leq m_b}$. So one has

$$\Omega^1(K_{T_b}^{\times})^{\mathrm{inv}} \overset{\sim}{\to} (\omega)_b^{\leq m_b}. \tag{1.10.1}$$

Let U be a smooth affine curve over k_b, $u \in U$ a closed point; as above, we set $(\omega)_u := \omega_U(\infty u)/\omega_U$. Let $\xi : U^{\circ} := U \setminus \{u\} \to K_{T_b}^{\times}$ be a k_b-morphism, which amounts to a trivialization ν^{ξ} of K on $T_b U^{\circ} \subset X_{U^{\circ}}$. Denote by (ξ) the composition $(\omega)_b^{\leq m_b} \overset{\sim}{\to} \Omega^1(K_{T_b}^{\times})^{\mathrm{inv}} \overset{\xi^*}{\longrightarrow} \omega(U^{\circ}) \to \omega(U^{\circ})/\omega(U) = (\omega)_u$.

Lemma. *(i) After a possible localization of U at u, one can find $D \in Div(X)(U)$ and a trivialization ν of $K(D)$ on a neighborhood $V \subset X_U$ of $|D| \cup T_b U$ such that $|D| \cap X_u$ is supported at b, $|D| \cap T_b U^{\circ} = \emptyset$, and $\nu|_{T_b U^{\circ}} = \nu^{\xi}$.*
(ii) Suppose U is a neighborhood of b, i.e., we have an étale $\pi : U \to X$, $\pi(u) = b$. Then one can find (D, ν) as in (i) with D equal to (the graph of) π if and only if $-(\xi)$ equals π^*, i.e., the composition $(\omega)_b^{\leq m_b} \subset (\omega)_b \overset{\pi^*}{\to} (\omega)_u$.

Proof   (i) Let us extend ν^{ξ} to a rational section ν of K on an open subset V of X_U, $V \supset T_b U$, which is defined at $T_b U^{\circ}$. Shrinking U and V, one can find ν with $D := \mathrm{div}(\nu)$ prime to X_u (if n is the multiplicity of X_u in D, then we replace ν by $f^{-n}\nu$, where f is any rational function which equals 1 on $T_b U^{\circ}$ and whose divisor contains X_u with multiplicity 1).[10] After further localization of U and shrinking of V, we get D in $\mathrm{Div}(X)(U)$ and $|D| \cap X_u$ is supported at b; we are done.

(ii) A map $\phi : U^{\circ} \to K_{T_b}^{\times}$ extends to U if and only if for every $\beta \in \Omega^1(K_{T_b}^{\times})^{\mathrm{inv}}$ the form $\phi^*(\beta) \in \omega(U^{\circ})$ is regular at b. Thus either of the properties of ξ in the assertion of (ii) determines ν^{ξ} uniquely up to multiplication by an invertible function on $T_b U$. It remains to present a trivialization ν of $K(\pi)$ such that the corresponding ξ satisfies $-(\xi) = \pi^*$.

Shrinking X, we trivialize K and pick a function t with dt invertible; set $x := \pi^*(t) \in \mathcal{O}(U)$. Our ν is $(t - x)^{-1}$. The differential of the corresponding ξ is the $\mathrm{Lie}(O_{T_b}^{\times}) = O(T_b)$-valued 1-form $\nu^{-1}d_x\nu = -(1 + t/x + (t/x)^2 + \ldots)dx/x$. So if $\beta \in \Omega^1(K_{T_b}^{\times})^{\mathrm{inv}}$ is identified with $\psi(t) \in (\omega)_b$ by (1.10.1), then $\xi^*(\beta) = -(\mathrm{Res}_b(1 + t/x + (t/x)^2 + \ldots)\psi(t))dx/x = -\psi(x)$, q.e.d. $\square$

---

[10]To find such f (after possible shrinking of U), pick local coordinate t on X at b, and x on U at u (so $t(b) = 0 = x(u)$, $dt(b) \neq 0 \neq dx(u)$); set $f = x(x - t^{m_b})^{-1}$.

1.11 A de Rham line $\mathcal{F}$ on $K_{T_b}^{\times}$ is said to be *translation invariant* if the de Rham line $\cdot^* \mathcal{F} \otimes pr_2^* \mathcal{F}^{\otimes -1}$ lies in $pr_1^* \mathcal{L}_{\mathrm{dR}}(O_{T_b}^{\times}) \subset \mathcal{L}_{\mathrm{dR}}(O_{T_b}^{\times} \times K_{T_b}^{\times})$; here $\cdot$: $O_{T_b}^{\times} \times K_{T_b}^{\times} \to K_{T_b}^{\times}$ is the action map, pr_i are the projections to the factors. Such $\mathcal{F}$'s form a Picard subgroupoid $\mathcal{L}_{\mathrm{dR}}^{\mathrm{inv}}(K_{T_b}^{\times})$ of $\mathcal{L}_{\mathrm{dR}}(K_{T_b}^{\times})$.

Lemma. *(i) We trivialize $K_{T_b}^{\times}$, i.e., identify it with $O_{T_b}^{\times}$. The translation invariance of $\mathcal{F}$ is equivalent to the next properties:*
(a) The de Rham line $pr_1^* \mathcal{F} \otimes pr_2^* \mathcal{F} \otimes \cdot^* \mathcal{F}^{\otimes -1}$ is constant;
(b) For any smooth curve U, a point $u \in U$, and two maps $\xi_1, \xi_2 : U^o := U \setminus \{u\} \to K_T^{\times}$, the de Rham line $\xi_1^* \mathcal{F} \otimes \xi_2^* \mathcal{F} \otimes (\xi_1 \xi_2)^* \mathcal{F}^{\otimes -1}$ on U^o extends to U.
(c) For some (or every) invertible section $e_{\mathcal{F}}$ of $\mathcal{F}$ on $K_{T_b}^{\times}$ one has $\nabla(e_{\mathcal{F}})/e_{\mathcal{F}} \in \Omega^1(K_{T_b}^{\times})^{\mathrm{inv}}$. (ii) There is a natural isomorphism $\pi_0(\mathcal{L}_{\mathrm{dR}}^{\mathrm{inv}}(K_{T_b}^{\times})) \xrightarrow{\sim} \mathbb{Z} \times ((\omega)_b^{\leq m_b}/\mathbb{Z})$ where $\mathbb{Z} \subset (\omega)_b^{\leq m_b}$ are polar parts of 1-forms with simple pole and integral residue.

Proof (i) (a) is evidently equivalent to invariance of $\mathcal{F}$. Since $K_{T_b}^{\times}$ is a rational variety, (a) amounts to the fact that $\cdot^* \mathcal{F} \otimes pr_1^* \mathcal{F}^{\otimes -1} \otimes pr_2^* \mathcal{F}^{\otimes -1}$ extends to a compactification of $K_{T_b}^{\times}$. This can be tested on curves, which is (b). Finally (c) is equivalent to the translation invariance since every invertible function on $O_{T_b}^{\times}$ is the product of a character by a constant, and every line bundle on $K_{T_b}^{\times}$ is trivial.

(ii) One assigns to $\mathcal{F}$ the pair (n, ψ) where $n = \deg(\mathcal{F})$ and ψ is the class of the image of $\nabla(e_{\mathcal{F}})/e_{\mathcal{F}}$ by (1.10.1). $\square$

We say that $\mathcal{G} \in \mathcal{L}_{\mathrm{dR}}(X \setminus T)^{(1)}$ is *compatible* with $\mathcal{F} \in \mathcal{L}_{\mathrm{dR}}^{\mathrm{inv}}(K_{T_b}^{\times})$ if for some neighborhood U of b, a trivialization ν of $\mathcal{K}^{(1)}$ on U, and a map $\xi : U^o := U \setminus \{b\} \to K_{T_b}^{\times}$ as in (ii) of the lemma in 1.10, the de Rham line $\nu^* \mathcal{G} \otimes \xi^* \mathcal{F}$ on U^o extends to U (the validity of this does not depend on the choice of U, ν, and ξ). By loc.cit., compatibility is equivalent to the next condition: Pick U, ν, and $e_{\mathcal{F}}$ as above; let $e_{\mathcal{G}}$ be a non-zero rational section of $\nu^* \mathcal{G}$. Then the image of $\nabla(e_{\mathcal{F}})/e_{\mathcal{F}}$ by (1.10.1) in $(\omega)_b^{\leq m_b}/\mathbb{Z} \subset (\omega)_b/\mathbb{Z}$ equals the class of $\nabla(e_{\mathcal{G}})/e_{\mathcal{G}}$.

Let $\mathcal{L}_{\mathrm{dR}}^{\natural}(X, T; K)$ be the Picard subgroupoid of $\mathcal{L}_{\mathrm{dR}}(X \setminus T)^{(1)} \times \prod_{b \in T_{\mathrm{red}}} \mathcal{L}_{\mathrm{dR}}^{\mathrm{inv}}(K_{T_b}^{\times})$ formed by those collections $(\mathcal{G}, \{\mathcal{F}_{T_b}\})$ that $\mathcal{G}$ is compatible with every $\mathcal{F}_{T_b}$. Then $\mathcal{G}$ lies automatically in $\mathcal{L}_{\mathrm{dR}}(X, T)^{(1)}$. By (ii) of the lemma, the functor $\mathcal{L}_{\mathrm{dR}}^{\natural}(X, T; K) \to \mathcal{L}_{\mathrm{dR}}(X, T)^{(1)} \times \Pi \mathcal{L}_{k_b}$, $(\mathcal{G}, \{\mathcal{F}_{T_b}\}) \mapsto (\mathcal{G}, \{\mathcal{F}_{\nu_{T_b}}\})$, where $\mathcal{F}_{\nu_{T_b}}$ is the fiber of $\mathcal{F}_{T_b}$ at ν_{T_b} from 1.6, is an equivalence of categories. Thus the theorem in 1.6 follows from the next one:

Theorem. *For every $\mathcal{E} \in \mathcal{L}_{\mathrm{dR}}^{\Phi}(X, T; K)$ one has $(\mathcal{E}^{(1)}, \{\mathcal{E}_{T_b}\}) \in \mathcal{L}_{\mathrm{dR}}^{\natural}(X, T; K)$, and the functor*

$$\mathcal{L}_{\mathrm{dR}}^{\Phi}(X, T; K) \to \mathcal{L}_{\mathrm{dR}}^{\natural}(X, T; K), \quad \mathcal{E} \mapsto (\mathcal{E}^{(1)}, \{\mathcal{E}_{T_b}\}), \qquad (1.11.1)$$

is an equivalence of the Picard groupoids.

Proof The assertion is X-local, and we have proved it for $T = \emptyset$. So we can assume that T_{red} is a single k-point b. Thus $T_b = T$ and $\mathfrak{D}$ is the disjoint sum of $\mathfrak{D}_{c=0}$ equal to $\mathrm{Div}(X \setminus T)$ and $\mathfrak{D}_{c=1}$ equal to $\mathrm{Div}(X)$. If needed, we can assume that K is trivialized and there is an étale map $t : X \to \mathbb{A}^1$.

(a) Let us show that $(\mathcal{E}^{(1)}, \mathcal{E}_T) \in \mathcal{L}_{\mathrm{dR}}^{\natural}(X, T; K)$. Notice that $\mathcal{L}_{\mathrm{dR}}^{\Phi}(X, T; K)$ is generated by $\mathcal{L}_{\mathrm{dR}}^{0\Phi}(X, T; K)$, the image of (1.5.3), and the pull-back by t of the factorization line on $\mathbb{A}^1$ from 1.8. Since the assertion is evident for factorization lines of the latter two types, it suffices to consider the case of $\mathcal{E} \in \mathcal{L}^{0\Phi}_{\mathrm{dR}}(X, T; K)$.

We know that $\mathcal{E}^{(1)}$ comes from a de Rham line on $X \setminus T$ (see 1.7). Let us check that $\mathcal{E}_T$ is translation invariant using the criterion of (i)(b) in the lemma. For U, u, ξ_i as in loc.cit., let us choose D_i, ν_i as in (i) of the lemma in 1.10; then $D_3 := D_1 + D_2$, $\nu_3 := \nu_1 \nu_2$ serves $\xi_3 := \xi_1 \xi_2$. The lines $\mathcal{E}_{(D_i, 1, \nu_i)}$ on U are equal to $\mathcal{E}_{D_i} \otimes \xi_i^* \mathcal{E}_T$ on U^o by factorization; here $\mathcal{E}_{D_i} := \mathcal{E}_{(D_i, 0)}$. Since $\mathcal{E}_{D_3} = \mathcal{E}_{D_1} \otimes \mathcal{E}_{D_2}$ by 1.7, the de Rham line $\xi_1^* \mathcal{E}_T \otimes \xi_2^* \mathcal{E}_T \otimes (\xi_1 \xi_2)^* \mathcal{E}_T^{\otimes -1}$ on U^o extends to U, q.e.d.

It remains to check that $\mathcal{E}^{(1)}$ is compatible with $\mathcal{E}_T$. Let ν is a trivialization of $K(\Delta)$ on an open $V \subset X \times X$ that contains (b, b), $U := V \cap (\{b\} \times X)$, $\xi : U^o \to K_T^\times$ the map defined by the restriction of ν to $T \times U^o$. The de Rham line $\mathcal{E}_{(\Delta, 1, \nu)}$ on U equals $\mathcal{E}^{(1)} \otimes \xi^* \mathcal{E}_T$ on U^o by factorization. Since the compatibility means that the latter line extends to U, we are done.

(b) Consider the projection $\pi : \mathfrak{D}^\circ_{c=1} \to K_T^\times \times \mathrm{Div}(X)$, $(D, 1, \nu_P) \mapsto (\nu_P|_{T_S}, D)$. Let us show for any $\mathcal{E} \in \mathcal{L}^{0\Phi}_{\mathrm{dR}}(X, T; K)$ its restriction $\mathcal{E}_1$ to $\mathfrak{D}^\circ_{c=1}$ comes from a uniquely defined de Rham line on $K_T^\times \times \mathrm{Div}(X)$ which we denote by $\mathcal{E}_1$ or $\mathcal{E}_{\mathrm{Div}1}$.

We use the fact that for every $D \in \mathrm{Div}(X)(S)$, S is smooth, the projection $K(D)^\times_{D,1} \to K_T^\times \times S$ satisfies the conditions of (i), (ii) of the lemma in 1.7.

As in part (c) of the proof in 1.7, we need to define $\mathcal{E}_{\mathrm{Div}1}$ on every $K_T^\times \times \mathrm{Sym}^{n_1, n_2}(X)$ and provide the $\mathrm{Div}^{\mathrm{eff}}(X)$-equivariance structure. Consider our $\mathcal{E}$ on $K(D_1 - D_2)^\times_{D_1 + D_2, 1}$. Over $K_T^\times \times \mathrm{Sym}^{n_1, n_2}(X \setminus T)$ it equals $\mathcal{E}_T \boxtimes \mathcal{E}_{(D_1 - D_2, 0)}$ by factorization, hence it descends to $K_T^\times \times \mathrm{Sym}^{n_1, n_2}(X \setminus T)$ by 1.7. By (ii) of the lemma in 1.7, we have $\mathcal{E}_{\mathrm{Div}1}$ over the whole $K_T^\times \times \mathrm{Sym}^{n_1, n_2}(X)$. The $\mathrm{Div}^{\mathrm{eff}}(X)$-equivariance is automatic by (i) of the lemma (applied to $K(D_1 - D_2)^\times_{D_1 + D_2 + 2D_3, 1}$).

(c) Our functor sends $\mathcal{L}^{0\Phi}_{\mathrm{dR}}(X, T; K)$ to the Picard subgroupoid $\mathcal{L}^{0\natural}_{\mathrm{dR}}(X, T; K)$ of $\mathcal{L}^\natural_{\mathrm{dR}}(X, T; K)$ formed by all $(\mathcal{G}, \{\mathcal{F}_b\})$ with $\deg(\mathcal{G}) = \deg(\mathcal{F}_b) = 0$. Let us prove that $\mathcal{L}^{0\Phi}_{\mathrm{dR}}(X, T; K) \to \mathcal{L}^{0\natural}_{\mathrm{dR}}(X, T; K)$ is an equivalence.

We need to show that every $(\mathcal{E}^{(1)}, \mathcal{E}_T) \in \mathcal{L}^{0\natural}_{\mathrm{dR}}(X, T; K)$ comes from a uniquely defined $\mathcal{E} \in \mathcal{L}^{0\Phi}_{\mathrm{dR}}(X, T; K)$. By the corollary in 1.7, $\mathcal{E}^{(1)}$ defines $\mathcal{E}_0 := \mathcal{E}|_{\mathfrak{D}_{c=0}}$, which we can view, by 1.7, as a de Rham line with multiplication structure on $\mathrm{Div}(X \setminus T)$. As in (b), $\mathcal{E}_1 := \mathcal{E}|_{\mathfrak{D}^\circ_{c=1}}$ comes from $K_T^\times \times \mathrm{Div}(X)$. By factorizaion, its restriction to $K_T^\times \times \mathrm{Div}(X \setminus T)$ equals $\mathcal{E}_T \boxtimes \mathcal{E}_0$. It remains to show that $\mathcal{E}_T \boxtimes \mathcal{E}_0$ extends in a unique way to a de Rham line $\mathcal{E}_1$ on $K_T^\times \times \mathrm{Div}(X)$.

As in (c) of the proof in 1.7, we should define $\mathcal{E}_1$ on every $K_T^\times \times \mathrm{Sym}^{n_1, n_2}(X)$ and provide the $\mathrm{Div}(X)^{\mathrm{eff}}(X)$-equivariance structure. Our $\mathcal{E}_1$ is defined on an open dense subset U of triples (ξ, D_1, D_2), $D_i \in \mathrm{Sym}^{n_i}(X \setminus T)$. Let $U' \supset U$ be the open subset of those (ξ, D_1, D_2) that $D_1 + D_2$ contains b with multiplicity at most 1. Then $\mathcal{E}$ extends to U' due to compatibility of $\mathcal{E}^{(1)}$ and $\mathcal{E}_T$. Since the complement to U' has codimension ≥ 2, $\mathcal{E}$ extends to $K_T^\times \times \mathrm{Sym}^{n_1, n_2}(X)$, and we are done.

As in loc.cit., the $\mathrm{Div}(X)^{\mathrm{eff}}(X)$-equivariance is identification of the pull-backs of our line by $K_T^\times \times \mathrm{Sym}^{n_1, n_2}(X) \leftarrow K_T^\times \times \mathrm{Sym}^{n_1, n_2}(X) \times \mathrm{Sym}^{n_3}(X) \to K_T^\times \times \mathrm{Sym}^{n_1 + n_3, n_2 + n_3}(X)$. The two de Rham lines coincide on the dense open subset $K_T^\times \times \mathrm{Sym}^{n_1, n_2}(X \setminus T) \times \mathrm{Sym}^{n_3}(X \setminus T)$, so they are canonically identified everywhere, and we are done. The factorization structure on $\mathcal{E}$ is evident.

(d) By (c), the theorem is reduced to the claim that our functor yields an isomorphism between the quotients

$$\pi_0(\mathcal{L}_{dR}^{\Phi}(X,T;K))/\pi_0(\mathcal{L}_{dR}^{0\Phi}(X,T;K)) \xrightarrow{\sim} \pi_0(\mathcal{L}_{dR}^{\natural}(X,T;K))/\pi_0(\mathcal{L}_{dR}^{0\natural}(X,T;K)).$$

The degree map identifies the right group with $\mathbb{Z} \times \mathbb{Z}$. Our map is evidently injective; looking at the image of (1.5.3) and the pull-back by t of the factorization line on $\mathbb{A}^1$ from 1.8, we see that it is surjective, q.e.d.   □

1.12   *A complement.* A connection on a trivialized line bundle amounts to a 1-form; multiplying the trivialization by f, we add to the form $d\log f$. Here is a similar fact in the factorization story.

Consider the group $\pi_1(\mathcal{L}_{\mathcal{O}}^{\Phi}(X,T;K))$ of invertible functions on $\mathfrak{D}^{\diamond}$ that satisfy factorization property. One has evident embeddings $\mathcal{O}^{\times}(T_{red}) \hookrightarrow \pi_1(\mathcal{L}_{dR}^{\Phi}(X,T;K))$ $\hookrightarrow \pi_1(\mathcal{L}_{\mathcal{O}}^{\Phi}(X,T;K))$ (see Remarks (iii), (iv) in 1.5). Let $\mathcal{L}_{dR}^{\Phi}(X,T;K)^{\mathcal{O}\text{-triv}}$ be the kernel of the Picard functor $\mathcal{L}_{dR}^{\Phi}(X,T;K) \to \mathcal{L}_{\mathcal{O}}^{\Phi}(X,T;K)$. This is a mere abelian group (since the functor is faithful); its elements are pairs $(\mathcal{E}, e)$ where $\mathcal{E}$ is a factorization de Rham line, e is a trivialization of $\mathcal{E}$ as a factorization $\mathcal{O}$-line. Let $\omega(X,T)$ be the space of 1-forms on $X \setminus T$ whose order of pole at any $b \in T$ is less or equal to the multiplicity of T at b.

Proposition. *There is a natural commutative diagram*

$$
\begin{array}{ccc}
\pi_1(\mathcal{L}_{\mathcal{O}}^{\Phi}(X,T;K))/\mathcal{O}^{\times}(T_{red}) & \xrightarrow{\sim} & \mathcal{O}^{\times}(X \setminus T) \\
\downarrow & & \downarrow \\
\mathcal{L}_{dR}^{\Phi}(X,T;K)^{\mathcal{O}\text{-}triv} & \xrightarrow{\sim} & \omega(X,T).
\end{array}
\tag{1.12.1}
$$

Proof (a) The connection on $\mathcal{E}^{(\ell)}$ along the fibers of $\mathcal{K}^{(\ell)}/X \setminus T$ is determined solely by the $\mathcal{O}$-line structure. So the action of any $h \in \pi_1(\mathcal{L}_{\mathcal{O}}^{\Phi}(X,T))$ on $\mathcal{E}^{(\ell)}$ is fiberwise horizontal, i.e., it is multiplication by a function $h^{(\ell)} \in \mathcal{O}^{\times}(X \setminus T)$. The top horizontal arrow is $h \mapsto h^{(1)}$.

For the same reason, for $(\mathcal{E}, e) \in \mathcal{L}_{dR}^{\Phi}(X,T;K)^{\mathcal{O}\text{-triv}}$ the trivializations $e^{(\ell)}$ of $\mathcal{E}^{(\ell)}$ are fiberwise horizontal, i.e., $\nabla(e^{(\ell)})/e^{(\ell)} \in \omega(X \setminus T)$. By the theorem in 1.6, $\nabla(e^{(1)})/e^{(1)} \in \omega(X,T)$. The bottom horizontal arrow is $(\mathcal{E}, e) \mapsto \nabla(e^{(1)})/e^{(1)}$.

The map $\pi_1(\mathcal{L}_{\mathcal{O}}^{\Phi}(X,T;K)) \to \mathcal{L}_{dR}^{\Phi}(X,T;K)^{\mathcal{O}\text{-triv}}$, $f \mapsto (\mathcal{O}_{\mathfrak{D}^{\diamond}}, f1)$, with kernel $\pi_1(\mathcal{L}_{dR}^{\Phi}(X,T;K))$ yields the left vertical arrow. The right one is the $d\log$ map.

The diagram is evidently commutative. It remains to check that its horizontal arrows are isomorphisms.

(b) For every $h \in \pi_1(\mathcal{L}_{\mathcal{O}}^{\Phi}(X,T))$ its restriction to $\mathfrak{D}_{c=0}^{\diamond}$ comes from a multiplicative function h_0 on $\text{Div}(X \setminus T)$. Similarly, if T_{red} is a single k-point b, then the restriction of h to $\mathfrak{D}_{c=1}^{\diamond}$ comes from a function h_1 on $K_T^{\times} \times \text{Div}(X)$ such that for $\xi \in K_T^{\times}$, $D \in \text{Div}(X \setminus T)$ one has $h_1(\xi, D) = h_1(\xi)h_0(D)$. This follows by a simple modification of the argument from, respectively, part (c) of the proof in 1.7 and part (b) of the proof in 1.11. The details are left to the reader.

Let us show that the map $\pi_1(\mathcal{L}_{\mathcal{O}}^{\Phi}(X,T))/\mathcal{O}^{\times}(T_{red}) \to \mathcal{O}^{\times}(X \setminus T)$ is injective. Suppose we have h such that $h^{(1)} = 1$. Since the group space $\text{Div}(X \setminus T)$ is generated by effective divisors of degree 1, one has $h_0 = 1$. It remains to check that h is locally constant on other components of $\mathfrak{D}^{\diamond}$. The assertion is X-local, so we can assume that T_{red} is a single k-point b, and we look at $\mathfrak{D}_{c=1}^{\diamond}$. By above, it suffices to check that the restriction h_T of h_1 to $K_T^{\times}$ is constant. We use (i) of the lemma 1.10;

we follow the notation of loc.cit. For $\xi : U^o \to K_T^\times$, consider $h_{(D,1,\nu)} \in \mathcal{O}^\times(U)$; by factorization, its restriction to U^o equals $\xi^* h_T h_{(D,0,\nu)} = \xi^* h_T$. Since $\xi^* h_T$ is regular at u for every ξ, h_T is constant, q.e.d.

A similar argument shows that the bottom horizontal arrow in (1.12.1) is injective. The details are left to the reader.

(c) Let us construct a section $\mathcal{O}^\times(X \setminus T) \to \pi_1(\mathcal{L}_\mathcal{O}^\Phi(X,T))$, $f \mapsto \tilde{f}$, of the map $h \mapsto h^{(1)}$. Fix a trivialization ν_0 of K on an open subset V_0 of X that contains T. For $(D, c, \nu_P) \in \mathfrak{D}^\circ(S)$ let us define $\tilde{f}_{(D,c,\nu_P)} \in \mathcal{O}^\times(S)$. Pick ν, V corresponding to (D, c, ν_P) as in Remark (ii) in 1.1; we can assume that $V \cap T_S = T_S^c$. Localizing S, we can decompose D in a disjoint sum of D' and D'' such that $D' \subset V \setminus T_S$ and $D'' \subset V_{0S}$. Set

$$\tilde{f}_{(D,c,\nu_P)} := f(D')\{f, \nu_0/\nu\}_{|D''|\cup T_S^c}. \tag{1.12.2}$$

Here $\{f, \nu_0/\nu\}_{|D''|\cup T_S^c} \in \mathcal{O}^\times(S)$ is the Contou-Carrère symbol at $|D''| \cup T_S^c$ (see [CC] or [BBE] 3.3). One readily checks that (1.12.2) does not depend on the auxiliary choices of ν and the decomposition $D = D' + D''$; its compatibility with the factorization is evident. So $\tilde{f} \in \pi_1(\mathcal{L}_\mathcal{O}^\Phi(X,T))$; clearly $\tilde{f}^{(1)} = f$.

Remark. If ν_0' is another trivialization of K near T, $f \mapsto \tilde{f}'$ the corresponding section, then $\tilde{f}/\tilde{f}'$ is an element of $\mathcal{O}^\times(T_{\mathrm{red}}) \subset \pi_1(\mathcal{L}_\mathcal{O}^\Phi(X,T))$ whose value at $b \in T_{\mathrm{red}}$ equals $(\nu_0'/\nu_0)(b)^{n_b}$, where $\mathrm{div}(f) = \Sigma n_b b$.

(d) To finish the proof, let us construct explicitly a section of the bottom horizontal arrow in (1.12.1). For $\phi \in \omega(X,T)$, we construct the corresponding $\mathcal{E}^\phi = (\mathcal{E}^\phi, e) \in \mathcal{L}_{\mathrm{dR}}^\Phi(X,T;K)^{\mathcal{O}\text{-triv}}$ using Remark (i) in 1.5. Since $\mathcal{E}^\phi$ is trivialized as an $\mathcal{O}$-line, α^ε of (1.5.2) is multiplication by a function $\alpha^\phi = \alpha_{(D',c,\nu_P')/(D,c,\nu_P)}^\phi \in \mathcal{O}^\times(S)$. To determine it, we extend ν_P, ν_P' to ν, ν' as in Remark (ii) in 1.1 such that ν equals ν' on V_{red}. Then $\nu/\nu' \in \mathcal{O}^\times(V \setminus P)$ equals 1 on V_{red}, so we have a function $\log(\nu/\nu') \in \mathcal{O}(V \setminus P)$ that vanishes on V_{red}. The residue $\mathrm{Res}_{P/S}(\log(\nu/\nu')\phi) \in \mathcal{O}(S)$ vanishes on S_{red}, and we set

$$\alpha^\phi := \exp(\mathrm{Res}_{P/S}(\log(\nu/\nu')\phi)). \tag{1.12.3}$$

Our α^ϕ does not depend on the auxiliary choice of ν and ν': Indeed, ν, ν' can be changed to $f\nu$, $f'\nu'$ with $f, f' \in \mathcal{O}^\times(V)$ that coincide on V_{red} and equal 1 on T_S^c (see Remark (ii) in 1.1); then $\log(f/f')$ is a regular function on V that vanishes on T_S^c, so $\mathrm{Res}_{P/S}(\log(f/f')\phi) = 0$, and we are done. The transitivity of α^ϕ and compatibility with base change and factorization are evident; we have defined $\mathcal{E}^\phi$.

Remark. Suppose we have $(D, c, \nu_P) \in \mathfrak{D}^\circ(S)$ where S is smooth. The de Rham structure on $\mathcal{E}_{(D,c,\nu_P)}^\phi$ amounts to a flat connection ∇^ϕ on our line bundle, which is the same as a closed 1-form $\theta^\phi = \nabla^\phi(e)/e$ on S. Choose ν as in Remark (ii) in 1.1; then (1.12.3) implies that

$$\theta^\phi = \mathrm{Res}_{P/S}((d_S(\nu)/\nu) \otimes \phi). \tag{1.12.4}$$

Here d_S means derivation along the fibers of the projection $V \subset X_S \to X$, so $d_S(\nu)$ is a section of $\Omega_S^1 \boxtimes K$ over $V \setminus P$, and $d_S(\nu)/\nu$ is a section of the pull-back of Ω_S^1 to $V \setminus P$. Of course, due to the lemma in 1.3, one can use (1.12.4) as an alternative definition of $\mathcal{E}^\phi$.

Example. Consider the $\mathcal{O}$-trivialized de Rham line $(\mathcal{E}_{T_b}^\phi, e)$ on the $\mathcal{O}_{T_b}^\times$-torsor $K_{T_b}^\times$ (see 1.6). Its 1-form $\theta^\phi = \nabla(e)/e$ is translation invariant and corresponds to the functional $f \mapsto \mathrm{Res}_b(f\phi)$ on the Lie algebra $\mathcal{O}(T_b)$ of $\mathcal{O}_{T_b}^\times$ (cf. 1.11).

It remains to check that the bottom horizontal arrow in (1.12.1) sends $(\mathcal{E}^\phi, e)$ to ϕ, i.e., that $\nabla(e^{(1)})/e^{(1)} = \phi$. Pick a local trivialization of K and a local function t on $X \setminus T$ with non-vanishing dt; let x be the corresponding local function on $S = X \setminus T$. Then $\nu = (t - x)^{-1}$ is a trivialization of $K_S(\Delta)$ near the diagonal, so we have the de Rham line $\mathcal{E}_{(\Delta,0,\nu)}^\phi$ on S. Then $d_S(\nu)/\nu = (t - x)^{-1}dx$, so, by (1.12.4), one has $\nabla(e^{(1)})/e^{(1)} = \theta^\phi = (\mathrm{Res}_{t=x}((t - x)^{-1}\phi))dx = \phi$, q.e.d. $\qquad\square$

Corollary. *For $\mathcal{E}, \mathcal{E}' \in \mathcal{L}_{dR}^\Phi(X, T; K)$, a morphism $\mathcal{E} \to \mathcal{E}'$ in $\mathcal{L}_{\mathcal{O}}^\Phi(X, T; K)$ is horizontal, i.e., is a morphism in $\mathcal{L}_{dR}^\Phi(X, T; K)$, if (and only if) the corresponding morphism $\mathcal{E}^{(1)} \to \mathcal{E}'^{(1)}$ of $\mathcal{O}$-lines on $\mathcal{K}^{(1)}$ is horizontal.* $\qquad\square$

Remark. If $K = \omega_X$, then in the corollary one can replace $\phi^{(1)}$ by the morphism $\phi_{X\setminus T}^{(1)} : \mathcal{E}_{X\setminus T}^{(1)} \to \mathcal{E}_{X\setminus T}'^{(1)}$ of $\mathcal{O}$-lines on $X \setminus T$ (see 1.6).

1.13 The next lemma will be used in 2.12. Assume that $X \setminus T$ is affine and $K = \omega_X$. The Lie algebra $\Theta(X \setminus T)$ of vector fields on $X \setminus T$ acts naturally on $\mathfrak{D}^\circ(X, \hat{T}; \omega) := \varprojlim \mathfrak{D}^\circ(X, nT; \omega)$. Therefore we have the notion of $\Theta(X \setminus T)$-action on any $\mathcal{O}$-line $\mathcal{E}$ on $\mathfrak{D}^\circ(X, \hat{T}; \omega)$. If $\mathcal{E}$ carries a factorization structure, then one can ask our action to be compatible with it. Ditto for a de Rham structure.

Suppose that $\mathcal{E}$ is a de Rham line. The flat connection yields then a $\Theta(X \setminus T)$-action on $\mathcal{E}$, which we refer to as the *standard* action. It is evidently compatible with the de Rham structure.

Lemma. *Any $\Theta(X \setminus T)$-action τ on $\mathcal{E}$ compatible with the de Rham structure equals the standard one.*

Proof Let τ_0 be the standard action. Then $\theta \mapsto \tau(\theta) - \tau_0(\theta)$ is a Lie algebra homomorphism from $\Theta(X \setminus T)$ to the Lie algebra of de Rham line endomorphisms of $\mathcal{E}$. The latter Lie algebra is commutative; the former one is perfect. Thus our homomorphism is 0, i.e., $\tau = \tau_0$. $\qquad\square$

1.14 The whole story makes sense in the relative setting. The input is a smooth (not necessary proper) Q-family of curves $q : X \to Q$ (where Q is a scheme), a relative divisor $T \subset X$ such that T_{red} is finite and étale over Q_{red}, and a line bundle K on X. It yields a space $\mathfrak{D}^\circ = \mathfrak{D}^\circ(X/Q, T; K)$ over Q. If $\mathcal{L}_?$ is any sheaf of Picard groupoids on the category $\mathcal{S}ch_{/Q}$ of Q-schemes (equipped with the étale topology), then we have the Picard groupoid $\mathcal{L}_?(\mathfrak{D})$ and $\mathcal{L}_?(\mathfrak{D}^\circ)$ defined as in 1.2, and the Picard groupoid of factorizarion objects $\mathcal{L}_?^\Phi(X/Q, T; K)$ as in 1.5. The $\mathcal{L}_?$'s we need are $\mathcal{L}_\mathcal{O}$, $\mathcal{L}_{dR/Q}$, and $\mathcal{L}_{dR}$, where $\mathcal{L}_\mathcal{O}$, $\mathcal{L}_{dR}$ are as in 1.2, and $\mathcal{L}_{dR/Q}(S)$ is formed by $\mathcal{O}$-lines equipped with an action of the universal relative formal groupoid on S/Q (if S/Q is smooth, then this is the same as a flat relative connection, cf. 1.1). All the results above immediately generalize to the relative setting. Thus, as in 1.4, for proper q every $\mathcal{E}$ in $\mathcal{L}_\mathcal{O}(\mathfrak{D}^\circ)$ or in $\mathcal{L}_{dR/Q}(\mathfrak{D}^\circ)$ yields an $\mathcal{O}$-line $\mathcal{E}(X/Q)$ on Q; if $\mathcal{E}$ lies in $\mathcal{L}_{dR}(\mathfrak{D}^\circ)$, then $\mathcal{E}(X/Q) \in \mathcal{L}_{dR}(Q)$. The theorem in 1.6 remains valid both in the setting of $\mathcal{L}_{dR/Q}$ and $\mathcal{L}_{dR}$, etc.

1.15 Suppose $k = \mathbb{C}$, and X is any complex smooth curve. All the above definitions and results render into the analytic setting without problems. In fact, the story simplifies since $\mathcal{L}_{\mathrm{dR}}^{\Phi}(X, T_{\mathrm{red}}) \xrightarrow{\sim} \mathcal{L}_{\mathrm{dR}}^{\Phi}(X, T)$.[11] It is equivalent to the Betti version of the story with de Rham lines replaced by local systems of $\mathbb{C}$-lines. And the Betti version works if we replace this $\mathbb{C}$ by any commutative ring R of coefficients.

Remark. The fact that $\mathcal{L}_{\mathrm{dR}}^{\Phi}(X, T; K) = \mathcal{L}_{\mathrm{dR}}^{\Phi}(X, T_{\mathrm{red}}; K)$ permits to consider in case $K = \omega_X$ a *canonical* equivalence (1.6.3) as in Example in 1.6.

Every de Rham factorization line $\mathcal{E}$ in the analytic setting carries a canonical automorphism μ which acts on $\mathcal{E}_{(D,c,\nu)}$ as multiplication by the (counterclockwise) monodromy of the de Rham line $\mathcal{E}_{(D,c,z\nu)}$, $z \in \mathbb{C}^{\times}$, around $z = 0$. Notice that μ is multiplicative, i.e., we have a homomorphism $\mu : \pi_0(\mathcal{L}_{\mathrm{dR}}^{\Phi}(X, T)) \to \pi_1(\mathcal{L}_{\mathrm{dR}}^{\Phi}(X, T))$. Same is true for the Betti factorization lines.

2 The de Rham ε-lines: algebraic theory

This section recasts the story of [Del] and [BBE] in the factorization line format.

2.1 We follow the setting and notation of 1.1, so X is a smooth curve over k, $T \subset X$ is a finite subscheme. *From now on we assume that K from 1.1 equals* $\omega = \omega_X$, *so* $\mathfrak{D}^{\diamond} = \mathfrak{D}^{\diamond}(X, T; \omega)$.

Let M be a (left) holonomic $\mathcal{D}$-module on (X, T), i.e., a holonomic module on X which is smooth on $X \setminus T$. We say that T is *compatible* with M if $\det M_{X \setminus T} \in \mathcal{L}_{\mathrm{dR}}(X, T)$ (see 1.6).

Theorem-construction. *M defines naturally a de Rham factorization line $\mathcal{E}_{dR}(M)$ $\in \mathcal{L}_{dR}^{\Phi}(X, \hat{T})$ with $\mathcal{E}_{dR}(M)_{X \setminus T}^{(1)} = (\det M_{X \setminus T})^{\otimes -1}$. It is functorial with respect to isomorphisms of M's, has local origin, and lies in $\mathcal{L}_{dR}^{\Phi}(X, T)$ if T is compatible with M. For proper X there is a canonical isomorphism of k-lines $\eta_{dR} : \mathcal{E}_{dR}(M)(X) \xrightarrow{\sim}$ $\det H_{dR}(X, M)$. The construction is compatible with base change of k, filtrations on M, and direct images for finite morphisms of X's.*

We construct $\mathcal{E}_{\mathrm{dR}}(M)$ as a factorization $\mathcal{O}$-line in 2.5, and define a de Rham structure on it in 2.10. The identification $\mathcal{E}_{\mathrm{dR}}(M)_{X \setminus T}^{(1)} \xrightarrow{\sim} (\det M_{X \setminus T})^{\otimes -1}$ is established in (2.6.1); we check that it is horizontal in 2.11. η_{dR} is defined in 2.7, the compatibilities are discussed in 2.8. The compatibility of T with M becomes relevant only in 2.10. Let us begin with necessary preliminaries.

2.2 *$\mathcal{L}$-groupoids and $\mathcal{L}$-torsors: a dictionary.* Let $\mathcal{L}$ be a Picard groupoid with the product operation $\otimes$. Below "$\mathcal{L}$-groupoid" means "enriched category over $\mathcal{L}$". Thus this is a collection of objects J and a rule which assigns to every $j, j' \in J$ an object $\lambda(j/j') \in \mathcal{L}$, and to every $j, j', j'' \in J$ a *composition* isomorphism $\lambda(j/j') \otimes \lambda(j'/j'') \xrightarrow{\sim} \lambda(j/j'')$ which satisfies associativity property. Then J is automatically a mere groupoid with $\mathrm{Hom}(j', j) := \mathrm{Hom}(1_{\mathcal{L}}, \lambda(j/j'))$.

Let J_1, J_2 be $\mathcal{L}$-groupoids. Their *tensor product* $J_1 \otimes J_2$ is an $\mathcal{L}$-groupoid whose objects $j_1 \otimes j_2$ correspond to pairs $j_1 \in J_1$, $j_2 \in J_2$, $\lambda(j_1 \otimes j_2/j_1' \otimes j_2') :=$ $\lambda(j_1/j_1') \otimes \lambda(j_2/j_2')$, and the composition $\lambda(j_1 \otimes j_2/j_1' \otimes j_2') \otimes \lambda(j_1' \otimes j_2'/j_1'' \otimes j_2'') \to$ $\lambda(j_1 \otimes j_2/j_1'' \otimes j_2'')$ equal to $(\lambda(j_1/j_1') \otimes \lambda(j_2/j_2')) \otimes (\lambda(j_1'/j_1'') \otimes \lambda(j_2'/j_2'')) \to (\lambda(j_1/j_1') \otimes$

---

[11]Since for any $\mathbb{A}^1$-torsor K/S the pull-back functor $\mathcal{L}_{\mathrm{dR}}(S) \to \mathcal{L}_{\mathrm{dR}}(K)$ is an equivalence.

$\lambda(j_1'/j_1'')) \otimes (\lambda(j_2/j_2')) \otimes \lambda(j_2'/j_2'')) \to \lambda(j_1/j_1'') \otimes \lambda(j_2/j_2'')$ where the first arrow is the commutativity constraint, the second one is the tensor product of the composition maps for J_1, J_2.

An $\mathcal{L}$-*morphism* $\phi : J_1 \to J_2$ is an $\mathcal{L}$-enriched functor, i.e., rule which assigns to every $j \in J_1$ an object $\phi(j) \in J_2$, and to every $j, j' \in J_1$ an identification $\phi : \lambda(j/j') \xrightarrow{\sim} \lambda(\phi(j)/\phi(j'))$ compatible with the composition. Such a ϕ yields a morphism of mere groupoids $J_1 \to J_2$. All $\mathcal{L}$-morphisms form naturally an $\mathcal{L}$-groupoid $\mathrm{Hom}_{\mathcal{L}}(J_1, J_2)$. Precisely, there is an $\mathcal{L}$-groupoid structure on $\mathrm{Hom}_{\mathcal{L}}(J_1, J_2)$ together with an $\mathcal{L}$-morphism $\mathrm{Hom}_{\mathcal{L}}(J_1, J_2) \otimes J_1 \to J_2$ that lifts the action map $\mathrm{Hom}_{\mathcal{L}}(J_1, J_2) \times J_1 \to J_2$ of mere groupoids, and such pair is unique (up to a unique 2-isomorphism). The composition $\mathrm{Hom}_{\mathcal{L}}(J_2, J_3) \times \mathrm{Hom}_{\mathcal{L}}(J_1, J_2) \to \mathrm{Hom}_{\mathcal{L}}(J_1, J_3)$ lifts naturally to a morphism of $\mathcal{L}$-groupoids $\mathrm{Hom}_{\mathcal{L}}(J_2, J_3) \otimes \mathrm{Hom}_{\mathcal{L}}(J_1, J_2) \to \mathrm{Hom}_{\mathcal{L}}(J_1, J_3)$, etc.

For an $\mathcal{L}$-groupoid J its *inverse* $J^{\otimes -1}$ is an $\mathcal{L}$-groupoid whose objects are in bijection $j \leftrightarrow j^{\otimes -1}$ with elements of J, and $\lambda(j^{\otimes -1}/j'^{\otimes -1}) = \lambda(j'/j)$.

There are two equivalent ways to define the notion of $\mathcal{L}$-*torsor*: (a) This is a mere groupoid F equipped with a $\mathcal{L}$-action, i.e., a functor $\otimes : \mathcal{L} \times F \to F$ together with an associativity constraint, such that for some (hence every) object $f \in F$ the functor $\mathcal{L} \to F$, $\ell \mapsto \ell \otimes f$, is an equivalence of groupoids; (b) This is an $\mathcal{L}$-groupoid such that the image of λ meets every isomorphism class in $\mathcal{L}$. To pass from (a) to (b), we lift the groupoid structure on F to $\mathcal{L}$-groupoid one with $\lambda(f/f') := f \otimes f'^{\otimes -1}$ (the latter is an object of $\mathcal{L}$ together with an isomorphism $(f \otimes f'^{\otimes -1}) \otimes f' \xrightarrow{\sim} f$; the pair is defined uniquely up to a unique isomorphism).

For a non-empty $\mathcal{L}$-groupoid J and an $\mathcal{L}$-torsor F both $\mathcal{L}$-groupoids $F \otimes J$ and $\mathrm{Hom}_{\mathcal{L}}(J, F)$ are $\mathcal{L}$-torsors. In particular, we have the product and ratio of $\mathcal{L}$-torsors (with $\mathcal{L}$ being a unit). Notice that there is a natural equivalence $F_1 \otimes F_2^{\otimes -1} \xrightarrow{\sim} \mathrm{Hom}_{\mathcal{L}}(F_2, F_1)$ which assigns to $f_1 \otimes f_2^{\otimes -1}$ the $\mathcal{L}$-morphism $f_2' \mapsto \lambda(f_2'/f_2) \otimes f_1$.

Remarks. (i) For any non-empty $\mathcal{L}$-groupoid J the $\mathcal{L}$-morphism $J \to \mathcal{L} \otimes J$, $j \mapsto 1_{\mathcal{L}} \otimes j$, is a universal $\mathcal{L}$-morphism to an $\mathcal{L}$-torsor.

(ii) Every $\mathcal{L}$-morphism between $\mathcal{L}$-torsors is an equivalence. Thus for non-empty J_i's every $\mathcal{L}$-morphism $J_1 \to J_2$ yields an equivalence of $\mathcal{L}$-torsors $\mathrm{Hom}_{\mathcal{L}}(J_2, \mathcal{L}) \xrightarrow{\sim} \mathrm{Hom}_{\mathcal{L}}(J_1, \mathcal{L})$ and $\mathcal{L} \otimes J_1 \xrightarrow{\sim} \mathcal{L} \otimes J_2$; in particular, the $\mathcal{L}$-torsor $\mathrm{Hom}_{\mathcal{L}}(J, \mathcal{L})$ does not change if we replace J by any its non-empty subset. E.g. every $j \in J$ yields an identification of $\mathcal{L}$-torsors $\mathrm{Hom}_{\mathcal{L}}(J, \mathcal{L}) \xrightarrow{\sim} \mathcal{L}$, $\lambda \mapsto \lambda(j)$, and $\mathcal{L} \xrightarrow{\sim} \mathcal{L} \otimes J$, $\ell \mapsto \ell \otimes J$.

(iii) If $F_i = \mathrm{Hom}_{\mathcal{L}}(J_i, \mathcal{L})$ where J_i are any $\mathcal{L}$-groupoids, then $F_1 \otimes F_2^{\otimes -1}$ identifies naturally with the $\mathcal{L}$-torsor whose objects are maps $\mu : J_1 \times J_2 \to \mathcal{L}$, $(j_1, j_2) \mapsto \mu(j_1/j_2)$, together with identifications $\lambda(j_1'/j_1) \otimes \mu(j_1/j_2) \otimes \lambda(j_2/j_2') \xrightarrow{\sim} \mu(j_1'/j_2')$ which are associative with respect to the composition of λ on both J_i. Here $f_1 \otimes f_2^{\otimes -1}$ corresponds to $\mu(j_1/j_2) := f_1(j_1) \otimes f_2(j_2)^{\otimes -1}$.

Suppose a group G acts on a non-empty $\mathcal{L}$-groupoid J. Then it acts on the $\mathcal{L}$-torsor $\mathcal{L} \otimes J$ by transport of structure. We can view this action as a monoidal functor $g \mapsto \lambda_g$ from G (considered as a discrete monoidal category) into the monoidal category $\mathrm{End}_{\mathcal{L}}(\mathcal{L} \otimes J)$, which is naturally equivalent to $\mathcal{L}$. Explicitly, $\lambda_g := \lambda(g/\mathrm{id}_J) \xrightarrow{\sim} \lambda(g(j)/j)$, $j \in J$; the product isomorphism $\lambda_{g_1} \otimes \lambda_{g_2} \xrightarrow{\sim} \lambda_{g_1 g_2}$

is the composition $\lambda(g_1(g_2(j))/g_2(j)) \otimes \lambda(g_2(j)/j)) \overset{\sim}{\to} \lambda((g_1g_2)(j)/j)$. A monoidal functor $G \to \mathcal{L}$ is sometimes called *(central) $\mathcal{L}$-extension $G^\flat$ of G*.[12]

The group G acts on $G^\flat$ in adjoint way. Namely, for $h \in G$ the isomorphism $\mathrm{Ad}_h : \lambda_g \overset{\sim}{\to} \lambda_{hgh^{-1}}$ is the composition $\lambda_g \overset{\sim}{\to} \lambda_g \otimes \lambda_h \otimes \lambda_{h^{-1}} \overset{\sim}{\to} \lambda_h \otimes \lambda_g \otimes \lambda_{h^{-1}} \overset{\sim}{\to} \lambda_{hgh^{-1}}$, the first arrow is the tensoring with the inverse to the composition map $1_\mathcal{L} \overset{\sim}{\leftarrow} \lambda_h \otimes \lambda_{h^{-1}}$, the second one is the commutativity constraint. Equivalently, Ad_h is determined by the condition that the composition $\lambda(g(j)/j) \overset{\sim}{\to} \lambda_g \overset{\mathrm{Ad}_h}{\longrightarrow} \lambda_{ghg^{-1}} \overset{\sim}{\to} \lambda(hgh^{-1}(h(j))/h(j)) = \lambda(h(g(j))/h(j))$ coincides with the action of h.

For commuting $g, h \in G$ we denote by $\{g, h\}^\flat \in \pi_1(\mathcal{L}) := \mathrm{Aut}_\mathcal{L}(1_\mathcal{L})$ their *commutant* in $G^\flat$, i.e., the action of Ad_g on λ_h or the ratio of $\lambda_g \otimes \lambda_h \to \lambda_{gh}$ and $\lambda_g \otimes \lambda_h \to \lambda_h \otimes \lambda_g \to \lambda_{hg} = \lambda_{gh}$ where the first and the last arrows are the product, the middle one is the commutativity constraint (cf. [BBE] A5).

2.3 *A digression on lattices and relative determinants* (see e.g. [Dr] §5).

Let S be a scheme, P be a relative effective divisor in X_S/S finite over S. Let E be any quasi-coherent $\mathcal{O}_{X_S}$-module such that for some neighborhood V of P the restriction of E to $V \setminus P$ is coherent and locally free.

A *P-lattice in E* is an $\mathcal{O}_{X_S}$-submodule L of E which is locally free (hence coherent) on a neighborhood of P, and equals E on $X_S \setminus P$. Denote by $\Lambda_P(E)$ the set of P-lattices in E. We assume that it is non-empty. Then $\Lambda_P(E)$ is directed by the inclusion ordering. Since P is finite over S, for every P-lattices $L \supset L'$ the $\mathcal{O}_S$-module $\pi_*(L/L')$ is locally free of finite rank.[13]

$\Lambda_P(E)$ carries a natural $\mathcal{L}_\mathcal{O}(S)$-groupoid structure. Namely, for P-lattices L, L' one has their *relative determinant* $\lambda_P(L/L') \in \mathcal{L}_\mathcal{O}(S)$; for L, L', L'' there is a canonical composition isomorphism

$$\lambda_P(L/L') \otimes \lambda_P(L'/L'') \overset{\sim}{\to} \lambda_P(L/L'') \tag{2.3.1}$$

which satisfies associativity property. This datum is uniquely determined by a demand that for $L \supset L'$ one has $\lambda_P(L/L') := \det \pi_*(L/L')$, and for $L \supset L' \supset L''$ the composition is the standard isomorphism defined by the short exact sequence $0 \to \pi_*(L'/L'') \to \pi_*(L/L'') \to \pi_*(L/L') \to 0$.

We denote by $\mathcal{D}et_{P/S}(E)$ the $\mathcal{L}_\mathcal{O}(S)$-torsor $\mathrm{Hom}_{\mathcal{L}_\mathcal{O}(S)}(\Lambda_P(E), \mathcal{L}_\mathcal{O}(S))$. Its objects, referred to as *determinant theories on E at P*, are rules λ that assigns to every $L \in \Lambda_P(E)$ a line $\lambda(L) \in \mathcal{L}_\mathcal{O}(S)$ together with identifications $\lambda_P(L/L') \otimes \lambda(L') \overset{\sim}{\to} \lambda(L)$ compatible with (2.3.1). Here one can replace $\Lambda_P(E)$ by any its non-empty subset. $\mathcal{D}et_{P/S}(E)$ is compatible with base change change.

For quasi-coherent $\mathcal{O}_X$-modules E_1, E_2 as above set

$$\mathcal{D}et_{P/S}(E_1/E_2) := \mathcal{D}et_{P/S}(E_1) \otimes \mathcal{D}et_{P/S}(E_2)^{\otimes -1}.$$

By Remark (iii) in 2.2, objects of this $\mathcal{L}_\mathcal{O}(S)$-torsor, referred to as *relative determinant theories on E_1/E_2 at P*, can be viewed as rules μ which assign to every P-lattices L_1 in E_1 and L_2 in E_2 a line $\mu(L_1/L_2) \in \mathcal{L}_\mathcal{O}(S)$ together with natural

---

[12]If $\mathcal{L}$ is the Picard groupoid of A-torsors, A is an abelian group, then $G^\flat$ amounts to a central extension of G by A, which is the reason for the terminology.

[13]Let us show that $\pi_*(L/L')$ is $\mathcal{O}_S$-flat. We can assume that X is affine, L, L' are locally free. Since $\pi_*(L/L') = \pi_*(L)/\pi_*(L')$ and π_*L, π_*L' are $\mathcal{O}_S$-flat, it suffices to check that for any geometric point s of S the map $\pi_*(L')_s \to \pi_*(L)_s$ is injective. This is clear, since $\pi_*(L^{(\prime)})_s = \Gamma(X_s, L_s^{(\prime)})$ and $L'_s \to L_s$ is injective (being an isomorphism at the generic points of X_s).

identifications

$$\lambda_P(L_1'/L_1) \otimes \mu(L_1/L_2) \otimes \lambda_P(L_2/L_2') \xrightarrow{\sim} \mu(L_1'/L_2') \qquad (2.3.2)$$

which are associative with respect to composition (2.3.1) on the two sides. We can also restrict ourselves to L_i in any non-empty subsets of $\Lambda_P(E_i)$.

Remarks. (i) Let $E(\infty P) := \varinjlim E(nP)$ be the localization of E with respect to P. An evident morphism of $\mathcal{L}_{\mathcal{O}}(S)$-groupoids $\Lambda_P(E) \to \Lambda_P(E(\infty P))$ yields an equivalence $\mathcal{D}et_{P/S}(E) \xrightarrow{\sim} \mathcal{D}et_{P/S}(E(\infty P))$; same for relative determinant theories. Thus $\mathcal{D}et_{P/S}(E)$ depends only on the restriction of E to the complement of P. Notice that $\Lambda_P(E(\infty P))$ is directed by both inclusion ordering and the opposite one.

(ii) Let us order the set of pairs of P-lattices (L_1, L_2) by the product of either of the inclusion orderings. Let I be any of its directed subsets. Suppose we have a rule λ that assigns to every $(L_1, L_2) \in I$ a line $\lambda(L_1/L_2) \in \mathcal{L}_{\mathcal{O}}(S)$ together with natural identifications (2.3.2) defined for $(L_1, L_2) \geq (L_1', L_2')$ and associative for $(L_1, L_2) \geq (L_1', L_2') \geq (L_1'', L_2'')$. Then λ extends uniquely to a relative determinant theory.

(iii) The group $\mathrm{Aut}(E_{V \setminus P})$ acts on $\Lambda_P(E(\infty P))$ as on an $\mathcal{L}_{\mathcal{O}}(S)$-groupoid. As in 2.2, this defines an $\mathcal{L}_{\mathcal{O}}(S)$-extension $\mathrm{Aut}(E_{V \setminus P})^\flat$ of $\mathrm{Aut}(E_{V \setminus P})$. Thus for commuting $g', g \in \mathrm{Aut}(E_{V \setminus P})$ we have $\{g', g\}^\flat = \{g', g\}_P^\flat \in \mathcal{O}^\times(S)$. Example: if g' is multiplication by a function $f \in \mathcal{O}^\times(V \setminus P)$, then $\{g', g\}_P^\flat$ equals the Contou-Carrère symbol $\{\det g, f\}_P$ at P (see e.g. [BBE] 3.3).[14] In particular, if $f \in \mathcal{O}^\times(V)$, then $\{g', g\}_P^\flat = f(-\mathrm{div}(\det g))$; here $\mathrm{div}(\det g)$ is the part of the divisor supported on V, i.e., at P.

P-lattices have local nature with respect to P. In particular, if P is the disjoint sum of components P_α, then a P-lattice L amounts to a collection of P_α-lattices L_α, and there is an evident canonical factorization isomorphism

$$\otimes \lambda_{P_\alpha}(L_\alpha/L_\alpha') \xrightarrow{\sim} \lambda_P(L/L') \qquad (2.3.3)$$

compatible with composition (2.3.1). Therefore $\mathcal{D}et_{P/S}(E)$ is the $\mathcal{L}_{\mathcal{O}}(S)$-torsor product of $\mathcal{D}et_{P_\alpha/S}(E)$. So every collection of determinant theories λ_α on E at P_α yields a determinant theory $\otimes \lambda_\alpha$ on M at P, $(\otimes \lambda_\alpha)(L) := \otimes \lambda_\alpha(L_\alpha)$. Same for relative determinant theories.

2.4 We return to the story of 1.1. Suppose we have $(D, c) \in \mathfrak{D}(S)$. Let us apply the format of 2.3 to $E_1 = M_S = M \boxtimes \mathcal{O}_S$, $E_2 = \omega M_S := (\omega \otimes M)_S$, and $P = P_{D,c}$. The connection $\nabla = \nabla_M : M \to \omega M$ yields a relative determinant theory $\mu_P^\nabla = \mu(M)_P^\nabla \in \mathcal{D}et_{P/S}(M/\omega M) := \mathcal{D}et_{P/S}(M_S/\omega M_S)$ defined as follows.

Let L, L_ω be P-lattices in M_S, ωM_S such that $\nabla(L) \subset L_\omega$. Let $\mathcal{C}(L, L_\omega) = \mathcal{C}(L, L_\omega)_{M,P}$ be the complex $M_S/L \xrightarrow{\nabla} \omega M_S/L_\omega$ in degrees -1 and 0, i.e., it is the quotient $dR(M) \boxtimes \mathcal{O}_S/\mathrm{Cone}(L \xrightarrow{\nabla} L_\omega)$. Then $\pi_* \mathcal{C}(L, L_\omega)$ is a complex of quasi-coherent $\mathcal{O}_S$-modules.

---

[14] A short proof: Both expressions are compatible with base change. Since the datum of E, V, P, g, f can be extended S-locally to a smooth base, we can assume that S is smooth. Then it suffices to check the equality at the generic point of S, where the Contou-Carrère symbol is the usual tame symbol. The rest is a standard computation.

Lemma. $\pi_*\mathcal{C}(L, L_\omega)$ *has $\mathcal{O}_S$-coherent cohomology.*

Proof We can assume that $T_S^c = T_S$, since the assertion is S-local. Its validity does not depend on the choice of L, L_ω. Take them to be "constant" T-lattices equal to M, ωM on $X \setminus T$, and we are reduced to the case of $S = \operatorname{Spec} k$, $P = T$.

Let $\bar{X}$ be the smooth projective curve that contains X, $T^\infty := \bar{X} \setminus X$. Let us extend M to a holonomic $\mathcal{D}$-module on $\bar{X}$, which we denote also by M; let L, L_ω be $T \cup T^\infty$-lattices as above. Since $\mathcal{C}(L, L_\omega)_{M, T\cup T^\infty} = \mathcal{C}(L, L_\omega)_{M, T} \oplus \mathcal{C}(L, L_\omega)_{M, T^\infty}$, it suffices to check that $\pi_*\mathcal{C}(L, L_\omega)_{M, T\cup T^\infty}$ has finite-dimensional cohomology. Therefore we are reduced to the case of projective X.

Now the lemma follows since $\mathcal{C}(L, L_\omega) = dR(M)/\mathcal{C}one(L \xrightarrow{\nabla} L_\omega)$, and the cohomology of X with coefficients in $dR(M)$, L, L_ω are finite dimensional. $\qquad\square$

Pairs (L, L_ω) as above form a directed set $I_\mathcal{C}$ as in Remark (ii) in 2.3. Set

$$\mu_P^\nabla(L/L_\omega) := \det \pi_*\mathcal{C}(L, L_\omega). \tag{2.4.1}$$

If $(L', L_\omega') \in I_\mathcal{C}$ is such that $L \supset L'$, $L_\omega \supset L_\omega'$, then the short exact sequence $0 \to \mathcal{C}one(L/L' \xrightarrow{\nabla} L_\omega/L_\omega') \to \mathcal{C}(L', L_\omega') \to \mathcal{C}(L, L_\omega) \to 0$ together with the identification $\det \pi_*\mathcal{C}one(L/L' \xrightarrow{\nabla} L_\omega/L_\omega') = \det \pi_*(L_\omega/L_\omega') \otimes \det \pi_*(L/L')^{\otimes -1}$ yields an isomorphism

$$\lambda_P(L'/L) \otimes \mu_P^\nabla(L/L_\omega) \otimes \lambda_P(L_\omega/L_\omega') \xrightarrow{\sim} \mu_P^\nabla(L'/L_\omega') \tag{2.4.2}$$

which evidently satisfies associativity property. By Remark (ii) in 2.3, we have defined $\mu_P^\nabla = \mu(M/\omega M)_P^\nabla \in \mathcal{D}et_{P/S}(M/\omega M)$.

Remarks. (i) Sometimes it is convenient to consider a smaller directed set formed by pairs (L, L_ω) that equal M, ωM outside T_S^c, or its subset of those (L, L_ω) that are locally constant with respect to S.

(ii) Denote by j_T the embedding $X \setminus T \hookrightarrow X$. Then $j_{T*}M := j_{T*}M_{X\setminus T}$ is holonomic $\mathcal{D}$-module as well; by Remark (i) from 2.3, one has $\mathcal{D}et_{P/S}(M/\omega M) = \mathcal{D}et_{P/S}(j_{T*}M/\omega j_{T*}M)$. Suppose $T_S^c = T_S$. Then L, L_ω are P-lattices in $j_{T*}M$, $\omega j_{T*}M$, and $\mathcal{C}one(\mathcal{C}(L, L_\omega)_{M,P}, \mathcal{C}(L, L_\omega)_{j_{T*}M,P}) \xrightarrow{\sim} dR(\mathcal{C}one(M \to j_{T*}M)) \otimes \mathcal{O}_S$. Thus $\mu(M/\omega M)_P^\nabla = \mu(j_{T*}M/\omega j_{T*}M)_P^\nabla \otimes \det R\Gamma_{\mathrm{dR}\,T}(X, M)$.

The above construction has local nature with respect to P. If P is disjoint sum of components P_α, and L, L_ω correspond to collections of P_α-lattices L_α, $L_{\omega\alpha}$, then $\mathcal{C}(L, L_\omega) = \oplus \mathcal{C}(L_\alpha, L_{\omega\alpha})$. Passing to the determinants, we see that

$$\mu(M/\omega M)_P^\nabla = \otimes \mu(M/\omega M)_{P_\alpha}^\nabla. \tag{2.4.3}$$

2.5 Now we can construct the promised $\mathcal{E}_{\mathrm{dR}}(M)$ as a factorization $\mathcal{O}$-line.

For $(D, c) \in \mathfrak{D}(S)$, any trivialization ν of $\omega(D)$ on a neighborhood V of $P = P_{D,c}$ yields naturally $\mu_P^\nu = \mu(M/\omega M)_P^\nu \in \mathcal{D}et_{P/S}(M/\omega M)$. Namely, the multiplication by ν isomorphism $M_{V\setminus P} \xrightarrow{\sim} \omega M_{V\setminus P}$ identifies the $\mathcal{L}_{\mathcal{O}}(S)$-groupoids $\Lambda_P(M(\infty P)) \xrightarrow{\sim} \Lambda_P(\omega M(\infty P))$; passing to $\mathcal{D}et_{P/S}$, we get μ_P^ν. Explicitly, for every P-lattices L, L_ω one has a canonical identification

$$\mu_P^\nu(L/L_\omega) \xrightarrow{\sim} \lambda_P(\nu L/L_\omega) = \lambda_P(\omega L(D)/L_\omega), \tag{2.5.1}$$

and identifications (2.3.2) are (2.3.1) combined with the isomorphism

$$\nu_{L/L'} : \lambda_P(L/L') \xrightarrow{\sim} \lambda_P(\nu L/\nu L')$$

that comes from multiplication by ν.

The r.h.s. of (2.5.1) does not depend on ν. Thus every L provides an identification e_L between μ_P^ν for all trivializations ν of $\omega(D)$ near P; it is characterized by property that (2.5.1) transforms $e_L(L/L_\omega)$ into the identity map for $\lambda_P(\omega L(D)/L_\omega)$.

Exercise. Let ν_1, ν_2 be any trivializations of $\omega(D)$ on V; set $f := \nu_2/\nu_1 \in \mathcal{O}^\times(V)$. Consider the identifications $e_L, e_{L'} : \mu_P^{\nu_1} \xrightarrow{\sim} \mu_P^{\nu_2}$. Show that

$$e_{L'} = f(\operatorname{div}(L/L'))e_L. \tag{2.5.2}$$

Here $\operatorname{div}(L/L') := \operatorname{div}(\phi_{L'}/\phi_L)$, where ϕ_L, $\phi_{L'}$ are local trivializations of the line bundles $\det(L)$, $\det(L')$ at P; this is a relative Cartier divisor supported at P.

If $\nu_1|_P = \nu_2|_P$, i.e., $f|_P = 1$, then we define a *canonical* identification

$$e : \mu_P^{\nu_1} \xrightarrow{\sim} \mu_P^{\nu_2} \tag{2.5.3}$$

as follows. The function $f(D) \in \mathcal{O}^\times(S)$ equals 1 on S_{red} since $f|_P = 1$. Let $f(D)^{\frac{1}{2}}$ be the branch of the root that equals 1 on S_{red}. Pick a lattice L_0 which equals M off T_S^c and is S-locally constant. Then[15] $e := f(D)^{\frac{\operatorname{rk}(M)}{2}} e_{L_0}$. By (2.5.2), the auxiliary choice of L_0 is irrelevant: indeed, if L_0, L_0' that satisfy our condition, then $f(\operatorname{div}(L_0/L_0')) = 1$, for the divisor $\operatorname{div}(L_0/L_0')$ is S-locally constant and supported on T_S^c, and $f|_{T_S^c} = 1$. The identifications e are evidently transitive.

Remark. Suppose L is an arbitrary lattice. Then

$$e = f(\operatorname{div}(L/L_0))f(D)^{\operatorname{rk}(M)/2} e_L$$

where L_0 is any lattice as above (see (2.5.2) and (2.5.3)).

Our e provides a canonical identification of $\mathcal{O}_S$-lines

$$r := \operatorname{id}_{\mu_P^\nabla} \otimes e^{\otimes -1} : \mu_P^\nabla \otimes (\mu_P^{\nu_1})^{\otimes -1} \xrightarrow{\sim} \mu_P^\nabla \otimes (\mu_P^{\nu_2})^{\otimes -1}. \tag{2.5.4}$$

Therefore for $(D, c, \nu_P) \in \mathfrak{D}^\circ(S)$ the $\mathcal{O}_S$-line $\mu_P^\nabla \otimes (\mu_P^\nu)^{\otimes -1}$ does not depend on the choice of ν such that $\nu|_P = \nu_P$. Set

$$\mathcal{E}_{\mathrm{dR}}(M)_{(D,c,\nu_P)} := \mu_P^\nabla \otimes (\mu_P^\nu)^{\otimes -1}. \tag{2.5.5}$$

The construction is compatible with base change, so $\mathcal{E}_{\mathrm{dR}}(M)$ is an $\mathcal{O}$-line on $\mathfrak{D}^\circ$. By (2.4.3) and similar property of μ_P^ν, it carries a factorization structure. So we have defined $\mathcal{E}_{\mathrm{dR}}(M) \in \mathcal{L}_{\mathcal{O}}^\Phi(X, T)$.

Example. If M is supported at T, then $\mathcal{E}_{\mathrm{dR}}(M)_{(D,c,\nu)} = \det R\Gamma_{\mathrm{dR}\, T^c}(X, M)$.

Summary. Suppose we have $(D, c, \nu_P) \in \mathfrak{D}^\circ(S)$. By (2.5.1), every L and ν such that $\nu|_P = \nu_P$ yields an isomorphism

$$nr_{L,\nu} : \mathcal{E}_{\mathrm{dR}}(M)_{(D,c,\nu_P)} \xrightarrow{\sim} \mu_P^\nabla(L/\omega L(D)). \tag{2.5.6}$$

If $f|_P = 1$, then, by Remark,

$$r_{L,f\nu} = f(\operatorname{div}(L_0/L))f(D)^{-\frac{\operatorname{rk}(M)}{2}} a_{L,\nu}. \tag{2.5.7}$$

In particular, $r_{L,f\nu} = r_{L,\nu}$ for reduced S.

---

[15]The reason for the normalization will become clear in 2.10.

2.6 Lemma. *(i) On $X \setminus T$ there is a canonical isomorphism*

$$\mathcal{E}_{dR}(M)_{X \setminus T}^{(1)} \xrightarrow{\sim} (\det M_{X \setminus T})^{\otimes -1}. \tag{2.6.1}$$

(ii) Suppose $T' \subset T$ is such that M is smooth at $T \setminus T'$. Let $\mathcal{E}_{dR}(M)'$ be the restriction to (X,T) of the ε-line of M on (X,T'). Then there is a canonical identification $\mathcal{E}_{dR}(M) \xrightarrow{\sim} \mathcal{E}_{dR}(M)'$.

Proof (i) Recall that $\mathcal{E}_{X \setminus T}^{(1)} = \mathcal{E}_{(D,0,\nu)}$ where $S = X \setminus T$, $D = \Delta = P$ is the diagonal divisor, and ν is (the principal part of) a form with logarithmic singularity at Δ with residue 1. Take $L = M_{X \setminus T}$. Then $\mathcal{C}(L, \omega L) = 0$, hence $\mathcal{E}_{dR}(M)_{X \setminus T}^{(1)} \xrightarrow{r_{L,\nu}} \mu_P^\nabla(L/\omega L(D)) \xrightarrow{\sim} \lambda_P(\omega L/\omega L(D)) = \det \pi_*(\omega L(D)/\omega L)^{\otimes -1}$. Now (2.6.1) is this isomorphism followed by the residue map $\pi_*(\omega L(D)/\omega L) \xrightarrow{\sim} L = M_{X \setminus T}$.

(ii) Evident. $\square$

2.7 Proposition. *For proper X there is a canonical isomorphism*

$$\eta_{dR} : \mathcal{E}_{dR}(M)(X) \xrightarrow{\sim} \det H_{dR}^{\cdot}(X, M). \tag{2.7.1}$$

Proof By 1.4, (2.7.1) amounts to a natural identification $\eta_{\mathrm{dR}} : \mathcal{E}_{\mathrm{dR}}(M)_\nu \xrightarrow{\sim} \det H_{\mathrm{dR}}^{\cdot}(X, M) \otimes \mathcal{O}_S$ which is defined for every S-family of rational 1-forms ν on X and is compatible with base change.

Set $P = P_{\mathrm{div}(\nu),1} = T \cup |\mathrm{div}(\nu)|$. Since X is proper, for a P-lattice L in $M(\infty P)$ the complex of $\mathcal{O}_S$-modules $R\pi_*(L)$ is perfect; set $\lambda(L) := \det R\pi_*(L)$. Then λ is a determinant theory on M at P in an evident manner. Replacing M by ωM, we get $\lambda_\omega \in \mathcal{D}et_{P/S}(\omega M)$, hence $\lambda \otimes \lambda_\omega^{\otimes -1} \in \mathcal{D}et_{P/S}(M/\omega M)$. One has an isomorphism

$$\mu_P^\nu \xrightarrow{\sim} \lambda \otimes \lambda_\omega^{\otimes -1}, \tag{2.7.2}$$

namely, $\mu_P^\nu(L/L_\omega) := \lambda_P(\nu L/L_\omega) = \lambda_\omega(\nu L) \otimes \lambda_\omega(L_\omega)^{\otimes -1} \xrightarrow{\sim} \lambda(L) \otimes \lambda_\omega(L_\omega)^{\otimes -1} = (\lambda \otimes \lambda_\omega^{\otimes -1})(L/L_\omega)$ where $\xrightarrow{\sim}$ comes from isomorphism $\nu^{-1} : \nu L \xrightarrow{\sim} L$.

For P-lattices L in M, L_ω in ωM with $\nabla(L) \subset L_\omega$ (see 2.4), set $dR(L, L_\omega) := \mathcal{C}one(L \xrightarrow{\nabla} L_\omega) \subset dR(M)$. Since $dR(M)/dR(L, L_\omega) = \mathcal{C}(L, L_\omega)$, we see that $dR(M)$ carries a 3-step filtration with successive quotients L_ω, $L[1]$, $\mathcal{C}(L, L_\omega)$. Applying $\det R\pi_*$, we get an isomorphism

$$\det R\pi_* \mathcal{C}(L, L_\omega) \otimes \lambda(L)^{\otimes -1} \otimes \lambda(L_\omega) \xrightarrow{\sim} \det H_{\mathrm{dR}}^{\cdot}(X, M) \otimes \mathcal{O}_S. \tag{2.7.3}$$

To get η_{dR}, we combine it with (2.7.2) (and (2.5.5)). The construction does not depend on the auxiliary choice of L, L_ω. $\square$

Example. Suppose M is a $\mathcal{D}$-module on $\mathbb{P}^1$ with regular singularities at 0 and ∞, where it is, respectively, the $*$- and the !-extension. Since $R\Gamma_{\mathrm{dR}}(\mathbb{P}^1, M) = 0$, our η_{dR} is an isomorphism

$$\eta_{\mathrm{dR}} : \mathcal{E}_{\mathrm{dR}}(M)_{(0,t^{-1}dt)} \otimes \mathcal{E}_{\mathrm{dR}}(M)_{(\infty,t^{-1}dt)} \xrightarrow{\sim} k. \tag{2.7.4}$$

To compute it explicitly, pick a $t\partial_t$-invariant vector subspace V of $\Gamma(\mathbb{P}^1 \setminus \{0, \infty\}, M)$, which freely generates $M_{\mathbb{P}^1 \setminus \{0, \infty\}}$ as an $\mathcal{O}$-module and such that the only possible integral eigenvalue of $t\partial_t$ on V is 0. Set $L := \mathcal{O}_{\mathbb{P}^1}(-(\infty)) \otimes V$, $L_\omega := t^{-1}dt \otimes L = \omega_{\mathbb{P}^1}((0) + (\infty)) \otimes L$. The condition on V implies that there are $\mathcal{O}$-linear embeddings

$i : L \hookrightarrow M$, $i_\omega : L_\omega \hookrightarrow \omega M$, which extend the evident isomorphisms on $\mathbb{P}^1 \setminus \{0, \infty\}$, such that $\nabla(L) \subset L_\omega$ and $i_\omega(\phi \ell) = \phi i(\ell)$ for any $\phi \in \omega_{\mathbb{P}^1}$, $\ell \in L$. Such i, i_ω are unique. The complex $\mathcal{C}(L, L_\omega) = \mathcal{C}(L, L_\omega)_0 \oplus \mathcal{C}(L, L_\omega)_\infty$ is acyclic, so (2.5.6) provides trivializations ι_0 of $\mathcal{E}_{\mathrm{dR}}(M)_{(0, t^{-1}dt)}$ and ι_∞ of $\mathcal{E}_{\mathrm{dR}}(M)_{(\infty, t^{-1}dt)}$.

Lemma. *One has* $\eta_{dR}(\iota_0 \otimes \iota_\infty) = 1$.

Proof The determinant of the complex $R\Gamma(\mathbb{P}^1, dR(L, L_\omega))$ has two natural trivializations α_1, α_2: the first one comes since the complex is acyclic, the second one from the identification

$$\det R\Gamma(\mathbb{P}^1, dR(L, L_\omega)) = \det R\Gamma(\mathbb{P}^1, L_\omega) \otimes \det R\Gamma(\mathbb{P}^1, L)^{\otimes -1}$$

and the multiplication by $t^{-1}dt$ isomorphism $L \overset{\sim}{\to} L_\omega$. Now (2.7.3) identifies $\iota_0 \otimes \iota_\infty \otimes \alpha_1$ with 1, and $\iota_0 \otimes \iota_\infty \otimes \alpha_2$ with $\eta_{\mathrm{dR}}(\iota_0 \otimes \iota_\infty)$. Since $R\Gamma(\mathbb{P}^1, L) = R\Gamma(\mathbb{P}^1, L_\omega) = 0$, one has $\alpha_1 = \alpha_2$; we are done. $\square$

2.8 The next constraints follow directly from the construction:
(i) For a finite filtration $M.$ on M, there is a canonical isomorphism

$$\mathcal{E}_{\mathrm{dR}}(M) \overset{\sim}{\to} \otimes \mathcal{E}_{\mathrm{dR}}(\mathrm{gr}_i M) \tag{2.8.1}$$

which satisfies transitivity property with respect to refinement of the filtration.

Remark. If $M = \oplus M_\alpha$, then every linear ordering of the indices yields a filtration on M, hence an isomorphism $\mathcal{E}_{\mathrm{dR}}(M) \overset{\sim}{\to} \otimes \mathcal{E}_{\mathrm{dR}}(M_\alpha)$. This isomorphism does not depend on the ordering. Thus $\mathcal{E}_{\mathrm{dR}}$ is a symmetric monoidal functor.

(ii) Let $\pi : (X', T') \to (X, T)$ be a finite morphism of pairs (see Remark (i) in 1.2) which is étale over $X \setminus T$. As in loc. cit., we have a morphism of Picard groupoids $\pi_* : \mathcal{L}_\mathcal{O}^\Phi(X', T') \to \mathcal{L}_\mathcal{O}^\Phi(X, T)$. We also have the $\mathcal{D}$-module direct image functor π_* which is exact. If M' is a holonomic $\mathcal{D}$-module on (X', T'), then $\pi_* M'$ is holonomic $\mathcal{D}$-module on (X, T). Notice that $\pi_* M'$ coincides with the "naive" direct image $\pi. M'$ outside T, and $dR(\pi_* M')$ is canonically quasi-isomorphic to $\pi. dR(M')$ as a dg module over the de Rham dg algebra of X. Therefore one has a canonical isomorphism

$$\mathcal{E}_{\mathrm{dR}}(\pi_* M') \overset{\sim}{\to} \pi_* \mathcal{E}_{\mathrm{dR}}(M') \tag{2.8.2}$$

compatible with the composition of π's and with (2.8.1).

Exercise. Consider the standard isomorphism $H_{\mathrm{dR}}(X, \pi_* M') \overset{\sim}{\to} H_{\mathrm{dR}}(X', M')$. If X is proper, then (2.8.2) yields an isomorphism $\mathcal{E}_{\mathrm{dR}}(\pi_* M')(X) \overset{\sim}{\to} \mathcal{E}_{\mathrm{dR}}(M')(X')$. Show that η_{dR}'s identify the second isomorphism with the determinant of the first one.

2.9 *Another digression on lattices and relative determinants.* For a Clifford algebra explanation of the next constructions, see [BBE] 2.14–2.17. In this subsection and the next one we use $\mathbb{Z}/2$-graded lines instead of $\mathbb{Z}$-graded ones; as in 1.2, the corresponding Picard groupoids are marked by $'$.

Let S, P be as in 2.3. Let E, E° be $\mathcal{O}_X$-modules as in 2.3, V a neighborhood of P such that both E, E° are locally free over $V \setminus P$, $\psi : E_{V \setminus P} \times E^\circ_{V \setminus P} \to \omega_{V \setminus P/S}$ be a non-degenerate pairing. For a P-lattice L in $E(\infty P)$ its ψ-dual L^ψ is the P-lattice in $E^\circ(\infty P)$ such that ψ is a non-degenerate $\omega_{V/S}$-valued pairing between L_V and L^ψ_V. The map $\tau_\psi : \Lambda_P(E(\infty P)) \to \Lambda_P(E^\circ(\infty P))$, $L \mapsto \tau_\psi(L) := L^\psi$, is an order-reversing bijection. It lifts to an isomorphism of $\mathcal{L}'_\mathcal{O}(S)$-groupoids:

Lemma. *For every* $L, L' \in \Lambda_P(E(\infty P))$ *there is a canonical isomorphism*

$$\tau_\psi : \lambda_P(L/L') \xrightarrow{\sim} \lambda_P(L^\psi/L'^\psi) \tag{2.9.1}$$

of $\mathbb{Z}/2$-*graded lines compatible with the composition.*

Proof It suffices to define (2.9.1) for $L' \supset L$ and check the compatibility with composition for $L'' \supset L' \supset L$.

The pairing $\ell, \ell^\circ \mapsto (\ell, \ell^\circ)_\psi := \mathrm{Res}_{P/S}\psi(\ell, \ell^\circ)$ yields a duality between the vector bundles $\pi_*(L'/L)$ and $\pi_*(L^\psi/L'^\psi)$, hence a duality between the determinant lines $\lambda_P(L'/L) \otimes \lambda_P(L^\psi/L'^\psi) \xrightarrow{\sim} \mathcal{O}_S$, $(\wedge \ell_i) \otimes (\wedge \ell_j^\circ) \mapsto (-1)^{\frac{n(n-1)}{2}} \det(\ell_i, \ell_j^\circ)_\psi$ where $n = \mathrm{rk}\, \pi_*(L'/L)$. Then (2.9.1) is characterized by the property that $\mathrm{id}_{\lambda_P(L'/L)} \otimes \tau_\psi$ identifies the latter pairing with the composition $\lambda_P(L'/L) \otimes \lambda_P(L/L') \xrightarrow{\sim} \mathcal{O}_S$. The compatibility of τ_ψ with composition is left to the reader. $\qquad\square$

Remark. Here is a sketch of a Clifford algebra interpretation of τ_ψ. Consider the Clifford $\mathcal{O}_S$-algebra generated by $\pi_* E_{V \setminus P} \oplus \pi_* E_{V \setminus P}^\circ$ equipped with the hyperbolic form $\mathrm{Res}_{P/S}\psi$. Let N be any "invertible" continuous Clifford module. For any P-lattice L the $\mathcal{O}_S$-submodule $N^{L \oplus L^\psi}$ of vectors killed by $L \oplus L^\psi$ lies in $\mathcal{L}'_{\mathcal{O}}(S)$, and $\lambda_P(L/L') = N^{L \oplus L^\psi} \otimes (N^{L' \oplus L'^\psi})^{\otimes -1} = \lambda_P(L^\psi/L'^\psi)$; the composition is τ_ψ.

Passing to $\mathcal{D}et'_{P/S}$, our τ_ψ yields

$$\mu^\psi \in \mathcal{D}et'_{P/S}(E/E^\circ).$$

Thus for $L \in \Lambda_P(E(\infty P))$, $L_\omega \in \Lambda_P(E^\circ(\infty P))$ one has $\mu^\psi(L/L_\omega) = \lambda_P(L^\psi/L_\omega)$, and identifications (2.3.2) are (2.3.1) combined with τ_ψ.

Exercise. Suppose we have another non-degenerate pairing $E_{V \setminus P} \times E_{V \setminus P}^\circ \to \omega_{V \setminus P/S}$; one can write it as $\psi_g(\cdot, \cdot) = \psi(g\cdot, \cdot) = \psi(\cdot, {}^\psi g\cdot)$ where $g \in \mathrm{Aut}(E_{V \setminus P})$ and ${}^\psi g \in \mathrm{Aut}(E_{V \setminus P})$ is the ψ-adjoint to g. Then $L^{\psi_g} = {}^\psi g^{-1}(L^\psi) = (g(L))^\psi$ and

$$\tau_{\psi_g} = {}^\psi g^{-1} \tau_\psi = \tau_\psi g. \tag{2.9.2}$$

E.g. for $f \in \mathcal{O}^\times(V)$ one has $L^{f\psi} = L^\psi$ and $\tau_{f\psi} = f(\mathrm{div}(L/L'))\tau_\psi : \lambda_P(L/L') \xrightarrow{\sim} \lambda_P(L^\psi/L'^\psi)$; here $\mathrm{div}(L/L')$ was defined in 2.5.

As in 2.2 and Remark (iii) in 2.3, we denote by $g \mapsto \lambda_g$ the $\mathcal{L}'_{\mathcal{O}}(S)$-extension $\mathrm{Aut}_{\mathcal{L}'_{\mathcal{O}}(S)}(\Lambda_P(E(\infty P)))^{\flat\prime}$ of the group $\mathrm{Aut}_{\mathcal{L}'_{\mathcal{O}}(S)}(\Lambda_P(E(\infty P)))$; same for E replaced by E°. Passing to $\mathcal{D}et'_{P/S}$, (2.9.2) yields then an isomorphism

$$\mu^{\psi_g} \xrightarrow{\sim} \lambda_{\psi g^{-1}} \otimes \mu^\psi \xrightarrow{\sim} \mu^\psi \otimes \lambda_g. \tag{2.9.3}$$

Suppose now that $E^\circ = E$ and ψ is symmetric. Then τ_ψ is an involution of the $\mathcal{L}'_{\mathcal{O}}(S)$-groupoid $\Lambda_P(E(\infty P))$. Therefore, since $\mu^\psi = \lambda_{\tau_\psi} \in \mathcal{D}et'_{P/S}(E/E) = \mathcal{L}'_{\mathcal{O}}(S)$, the composition yields a canonical identification

$$a_\psi : \mu^\psi \otimes \mu^\psi \xrightarrow{\sim} \mathcal{O}_S. \tag{2.9.4}$$

Explicitly, the isomorphism $\mu^\psi \xrightarrow{\sim} \lambda_P(\tau_\psi(L)/L)$ identifies a_ψ with the pairing $\lambda_P(\tau_\psi(L)/L) \otimes \lambda_P(\tau_\psi(L)/L) \to \mathcal{O}_S$, $l_1 \otimes l_2 \mapsto \tau_\psi(l_1)l_2 = l_1\tau_\psi(l_2)$.

2.10   Let us construct on $\mathcal{E} = \mathcal{E}_{\mathrm{dR}}(M) \in \mathcal{L}^{\Phi}_{\mathcal{O}}(X, T)$ a de Rham structure such that the identification of (2.6.1) is horizontal. By the corollary in 1.12, it is uniquely defined by this property, and the constraints from 2.8 are automatically compatible with the de Rham structure. As in 2.1, we assume that T is compatible with M.

As in Remark (i) in 1.2, we need to present for every scheme S and a pair of points $(D, c, \nu_P), (D', c', \nu'_P) \in \mathfrak{D}^{\circ}(S)$ which coincide on S_{red}, a natural identification (notice that $c = c'$)

$$\alpha^{\varepsilon} : \mathcal{E}_{(D, c, \nu_P)} \xrightarrow{\sim} \mathcal{E}_{(D', c, \nu'_P)}. \tag{2.10.1}$$

The α^{ε} should be transitive and compatible with base change and factorization.

Set $P := T^c_S \cup |D| \cup |D'| = P_{D,c} \cup P_{D',c}$. Localizing S, we find an open neighborhood V of P, $V \cap T_S = T^c_S$, together with a datum ν, ν', κ, where:
(a) ν, ν' are trivializations of $\omega_{V/S}(D)$, $\omega_{V/S}(D')$ which coincide on V_{red} and such that $\nu|_{P_{D,c}} = \nu_P$, $\nu|_{P_{D',c}} = \nu'_P$ (cf. Remark (ii) in 1.1);
(b) $\kappa : M_{V \setminus P} \times M_{V \setminus P} \to \mathcal{O}_{V \setminus P}$ is a non-degenerate symmetric bilinear form.
We construct α^{ε} explicitly using this datum.

The notation from 2.9 are in use. One can view κ as a non-degenerarate pairing $M_{V \setminus P} \times \omega M_{V \setminus P} \to \omega_{V \setminus P / S}$, which yields $\mu^{\kappa}_P = \mu^{\kappa} \in \mathcal{D}et'_{P/S}(M/\omega M)$. We get

$$\mu^{\nabla/\kappa}_P := \mu^{\nabla}_P \otimes (\mu^{\kappa}_P)^{\otimes -1}, \quad \mu^{\kappa/\nu}_P := \mu^{\kappa}_P \otimes (\mu^{\nu}_P)^{\otimes -1} \in \mathcal{L}'_{\mathcal{O}}(S); \tag{2.10.2}$$

recall that $\mu^{\nu}_P \in \mathcal{D}et'_{P/S}(M/\omega M)$ corresponds to the multiplication by ν identification of $M(\infty P)$ and $\omega M(\infty P)$. Let us rewrite (2.5.5) as an identification

$$\mathcal{E}_{(D, c, \nu)} \xrightarrow{\sim} \mu^{\nabla}_P \otimes (\mu^{\kappa}_P)^{\otimes -1} \otimes \mu^{\kappa}_P \otimes (\mu^{\nu}_P)^{\otimes -1} = \mu^{\nabla/\kappa}_P \otimes \mu^{\kappa/\nu}_P. \tag{2.10.3}$$

There is a similar identification for $\mathcal{E}_{(D', c, \nu')}$.

Notice that $\mu^{\kappa/\nu}_P$ is the line that corresponds to the symmetric pairing $\kappa/\nu = \nu^{-1}\kappa : \omega M_{V \setminus P} \times \omega M_{V \setminus P} \to \omega_{V \setminus P / S}$. So, by (2.9.4), one has a canonical trivialization $a_{\kappa/\nu} : \mu^{\kappa/\nu} \otimes \mu^{\kappa/\nu} \xrightarrow{\sim} \mathcal{O}_S$.

Let $\beta : \mathcal{E}_{(D, c, \nu)} \xrightarrow{\sim} \mathcal{E}_{(D', c, \nu')}$ be an isomorphism obtained by means of (2.10.3) from the tensor product of $\beta_1 := \mathrm{id}_{\mu^{\nabla/\kappa}_P}$ and an identification $\beta_2 : \mu^{\kappa/\nu}_P \xrightarrow{\sim} \mu^{\kappa/\nu'}_P$ such that $\beta_2^{\otimes 2} = a^{-1}_{\kappa/\nu'} a_{\kappa/\nu}$ and β_2 equals identity on S_{red}. We set $\alpha^{\varepsilon} := \gamma \beta$, where

$$\gamma := \exp \mathrm{Res}_{P/S}(\log(\nu'/\nu)\phi_{\kappa}) \in \mathcal{O}^{\times}(S). \tag{2.10.4}$$

Here ν'/ν is an invertible function on $V \setminus P$ that equals 1 on V_{red}, so $\log(\nu'/\nu)$ is a nilpotent function on $V \setminus P$, and $\phi_{\kappa} := \frac{1}{2}\nabla_M(\det \kappa^{-1})/\det \kappa^{-1} \in \Gamma(V \setminus P, \omega_{V/S})$ where $\det \kappa^{-1}$ is the trivialization of $\det M^{\otimes 2}_{V \setminus P}$ defined by κ.

Proposition. *α^{ε} does not depend on the auxiliary choice of ν, ν', and κ.*

Proof (a) Let us show that α^{ε} does not depend on κ for fixed ν, ν'. Suppose we have two forms κ and κ_g, so $\kappa_g(\cdot, \cdot) = \kappa(g\cdot, \cdot)$ for a κ-self-adjoint $g \in \mathrm{Aut}(M_{V \setminus P})$. Consider the corresponding β, γ and β_g, γ_g. Then β_g/β and γ/γ_g are functions on S that equal 1 on S_{red}; we want to check that they are equal.

We have an ω-valued symmetric bilinear form $\psi := \kappa/\nu$ on $\omega M_{V \setminus P}$. Our g, viewed as an automorphism of $\omega M_{V \setminus P}$, is ψ-self-adjoint and $\kappa_g/\nu = \psi_g$. Since $\mathrm{Ad}_{\tau_{\psi}}(g) = g^{-1}$, we have the isomorphism $\mathrm{Ad}_{\tau_{\psi}} : \lambda_{g^{-1}} \xrightarrow{\sim} \lambda_g$ (see 2.2). Consider the identification $\mu^{\psi_g} \xrightarrow{\sim} \lambda_{g^{-1}} \otimes \mu^{\psi}$ of (2.9.3). The next lemma follows directly from the definition of a_{ψ} and a_{ψ_g} as the composition in the $b\ell$-extension (see (2.9.4)):

Lemma. a_{ψ_g} *equals the composition* $\mu^{\psi_g} \otimes \mu^{\psi_g} \xrightarrow{\sim} \lambda_{g-1} \otimes \mu^{\psi} \otimes \lambda_{g-1} \otimes \mu^{\psi} \xrightarrow{\sim} \lambda_{g-1} \otimes \lambda_{g-1} \otimes \mu^{\psi} \otimes \mu^{\psi} \xrightarrow{\sim} \mathcal{O}_S$. *Here the second arrow is the commutativity constraint, the third one is tensor product of a map* $b_\psi : \lambda_{g-1} \otimes \lambda_{g-1} \to \mathcal{O}_S$, $\ell_1 \otimes \ell_2 \mapsto \ell_1 \mathrm{Ad}_{\tau_\psi}(\ell_2)$, *and* a_ψ. $\square$

There is a similar assertion for ψ replaced by $\psi' := \kappa/\nu'$. Combining them, we see that $(\beta_g/\beta)^2 = b_\psi/b_{\psi'}$. Since $\tau_{\psi'} = h\tau_\psi$ where h is the multiplication by ν'/ν automorphism (see (2.9.2)), one has $\mathrm{Ad}_{\tau_{\psi'}} = \{h, g\}_P^\flat \mathrm{Ad}_{\tau_\psi} : \lambda_{g-1} \xrightarrow{\sim} \lambda_g$ (see 2.2). Therefore $(\beta_g/\beta)^2 = (\{h, g\}^\flat)^{-1}$; by Remark (iii) in 2.3, this equals the Contou-Carrère symbol $\{\nu'/\nu, \det g\}_P$.

Now $\phi_\kappa - \phi_{\kappa_g} = \frac{1}{2} d\log(\det g)$, hence $\gamma/\gamma_g = \exp \mathrm{Res}_{P/S}(\frac{1}{2}\log(\nu'/\nu)d\log(\det g)) = \{(\nu'/\nu)^{\frac{1}{2}}, \det g\}_P$, and we are done.

(b) It remains to show that α^ε does not depend on the choice of the lifts ν, ν' of ν_P, ν'_P. One can change ν, ν' to $f\nu$, $f'\nu'$ where $f, f' \in \mathcal{O}^\times(V)$ are such that f equals f' on V_{red}, f equals 1 on $P_{D,c}$, f' equals 1 on $P_{D',c}$.

By (a), in the computation we are free to use κ of our choice. We work X-locally, so we can assume that $c = 1$ and pick κ to be a non-degenerate symmetric form on $M_{X \setminus T}$. Then $\phi_\kappa \in \omega(X, T)$ due to the compatibility of T with M, see 2.1. The function $\log(f'/f)$ is regular on V and vanishes on T_S^c. Therefore $\log(f'/f)\phi_\kappa$ is regular at P, hence its residue vanishes. Thus $\gamma(f'\nu', f\nu) = \gamma(\nu', \nu)$. It remains to show that $\beta(f'\nu', f\nu) = \beta(\nu', \nu)$.

Let L_0 be any T-lattice in M, L_0^κ be the κ-orthogonal T-lattice. Since $\omega L_0^\kappa = \tau_{\kappa/\nu}(\omega L_0(D)) = \tau_{\kappa/f\nu}(\omega L_0(D))$, one has $\mu^{\kappa/\nu} \xrightarrow{\sim} \lambda_P(\omega L_0^\kappa/\omega L_0(D)) \xleftarrow{\sim} \mu^{\kappa/f\nu}$. Let $e_{L_0}^\kappa : \mu^{\kappa/\nu} \xrightarrow{\sim} \mu^{\kappa/f\nu}$ be the composition, and $r_{L_0} : \mathcal{E}_{(D,c,\nu)} \xrightarrow{\sim} \mathcal{E}_{(D,c,f\nu)}$ be the tensor product of $\mathrm{id}_{\mu_P^{\nabla/\kappa}}$ and $e_{L_0}^\kappa$ (we use identifications (2.10.3) for ν and $f\nu$). We see that r_{L_0} coincides with the isomorphism $\mathrm{id}_{\mu_P^\nabla} \otimes e_{L_0}^{\otimes -1}$ where $e_{L_0} : \mu_P^\nu \xrightarrow{\sim} \mu_P^{f\nu}$ was defined in 2.5. Thus $f(D)^{-\frac{\mathrm{rk}(M)}{2}} r_{L_0} : \mathcal{E}_{(D,c,\nu)} \xrightarrow{\sim} \mathcal{E}_{(D,c,f\nu)}$ is the canonical isomorphism r from (2.5.4).

We want to check that r and the similar isomorphism for f', ν' identify $\beta(f'\nu', f\nu)$ with $\beta(\nu', \nu)$. Indeed, $e_{L_0}^\kappa$ identifies $a_{\kappa/f\nu}$ with $f(D)^{\mathrm{rk}(M)}a_{\kappa/\nu}$ (see Exercise in 2.9), so r_{L_0} identify $\beta(f'\nu', f\nu)$ with $f(D)^{\frac{\mathrm{rk}(M)}{2}} f'(D')^{-\frac{\mathrm{rk}(M)}{2}}\beta(\nu', \nu)$. By above, this implies the assertion for r. $\square$

The isomorphisms α^ε are transitive (since such are α^ε with fixed κ) and evidently compatible with base change and factorization, so we have defined a de Rham structure on $\mathcal{E}$. The horizontality of (2.6.1) will be checked in Example (i) of 2.11.

Remark. Let us fix a non-degenerate symmetric bilinear form κ on $M_{X \setminus T}$ (like in part (b) of the proof of the lemma). Then the above construction can be reformulated as follows. There is a canonical isomorphism

$$\mathcal{E}_{\mathrm{dR}}(M) \xrightarrow{\sim} \mathcal{E}_1 \otimes \mathcal{E}_2 \otimes \mathcal{E}_3, \tag{2.10.5}$$

where $\mathcal{E}_i$ are the next de Rham factorization lines:
- $\mathcal{E}_{1(D,c,\nu)} := \mu_P^{\nabla/\kappa}$ in (2.10.2); it depends only on c, i.e., $\mathcal{E}_1$ comes from 2^T.
- $\mathcal{E}_{2(D,c,\nu)} := \mu_P^{\kappa/\nu}$ in (2.10.2), so $\mathcal{E}_2 \otimes \mathcal{E}_2$ is canonically trivialized (as a de Rham factorization line) by $a_{\kappa/\nu}$. Notice that $\mathcal{E}_2$ does not depend on the connection ∇_M.
- $\mathcal{E}_3 := \mathcal{E}^{-\phi_\kappa} \in \mathcal{L}_{\mathrm{dR}}^\Phi(X, T)^{\mathcal{O}\text{-triv}}$ (see 1.12), i.e., it is a de Rham factorization

line equipped with an $\mathcal{O}$-trivialization e with $\nabla(e^{(1)})/e^{(1)} = -\phi_\kappa$, where $\phi_\kappa :=$ $\frac{1}{2}\nabla(\det\kappa^{-1})/\det\kappa^{-1} \in \omega(X,T)$ (see (1.12.3)). Isomorphism (2.10.5) is (2.10.3)$\otimes e$.

2.11 As in 1.3, the de Rham structure on $\mathcal{E}$ can be viewed as a datum of integrable connections ∇^ε on $\mathcal{E}_{(D,c,\nu_P)}$ for S smooth. The next explicit construction of ∇^ε is a paraphrase of the above:

Our problem is X-local, so we can fix κ as in the above remark and a T-lattice L in M; we can assume that $c = 1$. Choose any ν as in Remark (ii) in 1.1. We have identification $r_{L,\nu} : \mathcal{E}_{(D,c,\nu_P)} \xrightarrow{\sim} \mu_P^\nabla(L/L(D)) = \mu_P^\nabla(L/\omega L^\kappa) \otimes \lambda_P(\omega L^\kappa/L(D))$ of (2.5.6). Let $\nabla^1 = \nabla^1_{L,\kappa}$ be the "constant" connection on $\mu_P^\nabla(L/\omega L^\kappa)$, and $\nabla^2 = \nabla^2_{L,\nu,\kappa}$ be the connection on $\lambda_P(\omega L^\kappa/L(D))$ for which the pairing $a_{\kappa/\nu} :$ $\lambda_P(\omega L^\kappa/L(D)) \otimes \lambda_P(\omega L^\kappa/L(D)) \to \mathcal{O}_S$, $\ell_1 \otimes \ell_2 \mapsto \tau_{\kappa/\nu}(\ell_1)\ell_2$, of (2.9.4) is horizontal. We get a connection $\nabla_{\nu,\kappa} := \nabla^1 \otimes \nabla^2$ on $\mathcal{E}_{(D,c,\nu)}$. Then

$$\nabla^\varepsilon = \nabla_{\nu,\kappa} - \theta_{\nu,\kappa} \qquad (2.11.1)$$

where $\theta_{\nu,\kappa} := \operatorname{Res}_{P/S}(d_S\nu/\nu \otimes \phi_\kappa) \in \Omega^1_S$. Here d_S is the derivation along the fibers of X_S/X, so $d_S\nu/\nu$ is a section of $\pi_V^*\Omega^1_S$ on $V \setminus P$.

Examples. (i) Let us compute the connection on $\mathcal{E}^{(\ell)}$ (see 1.6). So let S be a copy of $X \setminus T$, $P = \Delta$, $D = \ell\Delta$. Let t be a local coordinate on $X \setminus T$, x be the corresponding coordinate on S, z be the coordinate on $\mathbb{G}_m$, $\nu := z(t - x)^{-\ell}dt$. Our $\mathcal{E}^{(\ell)}$ is the de Rham line $\mathcal{E}_{(D,0,\nu)}$ on $S \times \mathbb{G}_m$.

We take $L = M_{X \setminus T}$, so $L^\kappa = L$ and $\mu_P^\nabla(L/\omega L^\kappa)$ is trivialized. Therefore $\mathcal{E}^{(\ell)} = \lambda_P(\omega M/\omega M(D))$. The choice of t trivializes ω and all the vector bundles $\pi_*\mathcal{O}_{X \times S}(m\Delta)/\mathcal{O}_{X \times S}(n\Delta)$, which provides an identification $\mathcal{E}^{(\ell)} \xrightarrow{\sim} (\det M)^{\otimes -\ell}$. The pairing $a_{\kappa/\nu}$ is equal (up to sign) to $(z^{-n}\det\kappa)^{-\ell}$, $n := \operatorname{rk}(M)$, so $\nabla_{\nu,\kappa} = \nabla_{(\det M)^{\otimes -\ell}} + \ell\phi_\kappa + \frac{\ell n}{2}z^{-1}dz$. One has $d_{S \times \mathbb{G}_m}\nu/\nu = z^{-1}dz + \ell(t - x)^{-1}dx$. Hence $\theta_{\nu,\kappa} = \ell\phi_\kappa$, and

$$\nabla^\varepsilon = \nabla_{(\det M)^{\otimes -\ell}} + \frac{\ell n}{2}z^{-1}dz. \qquad (2.11.2)$$

In case $\ell = 1$ and $z \equiv 1$, (2.11.2) says that (2.6.1) is horizontal with respect to ∇^ε and the connection on $(\det M_{X \setminus T})^{\otimes -1}$.

(ii) Suppose M has regular singularities at $b \in T$. Let t be a parameter at b. Consider $P = b$, $D = \ell b$, and a family of 1-forms $\nu_z := zt^{-\ell}dt$, $z \in \mathbb{G}_m$. Let us compute the connection ∇^ε on the line bundle $\mathcal{E}_{(b,\nu_z)} := \mathcal{E}_{(D,1_b,\nu_z)}$ on $\mathbb{G}_m$.

Let L be any $t\partial_t$-invariant P-lattice in $M(\infty b)$; denote by r the trace of $t\partial_t$ acting on L/tL, $n := \operatorname{rk}(M)$. Let ∇_0 be a connection on $\mathcal{E}_{(b,\nu_z)}$ such that r_{L,ν_z} of (2.5.6) identifies it with the "constant" connection on z-independent line $\mu_b^\nabla(L/\omega L(D))$. Then

$$\nabla^\varepsilon = \nabla_0 + (\frac{\ell n}{2} - r)z^{-1}dz. \qquad (2.11.3)$$

Indeed, consider the above construction with $\nu = \nu_z$ and $\kappa = t^\ell\kappa_0$ where κ_0 is a non-degenerate symmetric bilinear form on L. Then $L^\kappa = L(D)$, so $\nabla_{\nu,\kappa} = \nabla_0$. The trivialization $\det\kappa^{-1}$ of $\det M_{X \setminus \{b\}}^{\otimes 2}$ has pole of order ℓn at b, thus the form ϕ_κ has logarithmic singularity at b with residue $r - \ell n/2$. Since $d_z(\nu_z)/\nu_z = z^{-1}dz$, one has $\theta_{\nu,\kappa} = (r - \ell n/2)z^{-1}dz$, and we are done by (2.11.1).

2.12   Let $q : X \to Q$, T be as in 1.14. Let M be a coherent $\mathcal{D}_{X/Q}$-module which is $\mathcal{O}_Q$-flat and is a vector bundle on $X \setminus T$. We call such M a *flat Q-family of holonomic $\mathcal{D}$-modules* on $(X/Q, T)$. The notion of compatibility of T and M is defined as in 5.1.

Let $dR_{X/Q}(M) = \mathcal{C}one(\nabla)$ be the relative de Rham complex. If for some (hence every) T-lattices L in M, L_ω in $\omega M := \omega_{X/Q} \otimes M$ with $\nabla(L) \subset L_\omega$ the complex $q_*(dR_{X/Q}(M)/\mathcal{C}one(L \xrightarrow{\nabla} L_\omega))$ has $\mathcal{O}_Q$-coherent cohomology, then we call M a *nice Q-family of $\mathcal{D}$-modules*.

Exercises. Suppose $Q = \operatorname{Spec} \mathbb{C}[s]$, $X = \operatorname{Spec} \mathbb{C}[t, s]$, T is the divisor $t = 0$.
(i) Show that M generated by a section m subject to the relation $t\partial_t m = sm$ is nice, and that the $\mathcal{D}_{X/Q}$-module $j_{T*}M = M[t^{-1}]$ is *not* coherent (cf. [BG]).
(ii) Show that M generated by section m subject to the relation $t^n \partial_t m = sm$, $n > 1$, is nice over the subset $s \neq 0$.

By a straightforward relative version of the constructions of this section, every nice family compatible with T gives rise to a relative factorization line $\mathcal{E}_{dR}(M) \in \mathcal{L}^\Phi_{dR/Q}(X/Q, T)$ (see 1.14). The construction is compatible with base change. For proper X/Q, the $\mathcal{O}_Q$-complex $Rq_{dR*}(M) := Rq_* dR_{X/Q}(M)$ is perfect, and we have an isomorphism of $\mathcal{O}$-lines $\eta_{dR} : \mathcal{E}_{dR}(M)(X/Q) \xrightarrow{\sim} \det Rq_{dR*}(M)$.

Suppose that Q is smooth and the relative connection on M is extended to a flat connection (so our nice family is isomonodromic).

Proposition. *The relative connection on $\mathcal{E}_{dR}(M)$ extends naturally to a flat absolute connection which has local origin and is compatible with base change and constraints from 2.8. Thus $\mathcal{E}_{dR}(M) \in \mathcal{L}^\Phi_{dR}(X/Q, T)$. For proper X/Q, η_{dR} is horizontal (for the Gauß-Manin connection on the target).*

Proof Let $L = L((X, T)/Q)$ be the Lie algebra of infinitesimal symmetries of $(X, T)/Q$; its elements are pairs (θ_X, θ_Q) where θ_X, θ_Q are vector fields on X, Q such that θ_X preserves T and $dq(\theta_X) = \theta_Q$. Our L acts on $\mathfrak{D}^\diamond(X/Q, T; \omega)$ and on $\mathcal{E} := \mathcal{E}_{dR}(M)$ by transport of structure. This action is compatible with constraints from 2.8 and the relative de Rham structure.

Variant. Let $T' \subset T$ be a component of T; set $L' := L((X \setminus T', T \setminus T')/Q) \supset L$. Then L' acts naturally on $\mathfrak{D}^\diamond(X/Q, \hat{T}; \omega)$ (see 1.13 where we considered the "vertical" part of this action). If $M = j_{T'*}M$, then this action lifts naturally to $\mathcal{E}_{dR}(M)$ (as follows directly from the construction of $\mathcal{E}_{dR}(M)$). The L'-action extends the L-action (pulled back to $\mathfrak{D}^\diamond(X/Q, \hat{T}; \omega)$) and satisfies similar compatibilities.
Lemma. *The Lie ideals $L_0 \subset L$, $L'_0 \subset L'$ act on $\mathcal{E}$ via ∇^ε.*

Proof of Lemma It suffices to check this Q-pointwise, so, due to compatibility with the base change, we can assume that Q is a point. By the compatibility with the first constraint in 2.8, it suffices to consider the cases when M is supported at T and $M = j_{T*}M$. In the first situation the lemma is evident. If $M = j_{T*}M$, then it suffices to consider the case of L'_0 for $T' = T$, i.e., $L'_0 = \Theta(X \setminus T)$. The L'_0-action is compatible with the de Rham structure, so we are done by the lemma in 1.13.   □

We want to define the connection in a manner compatible with the localization of X, so it suffices to do it in case when X and Q are affine. Then L/L_0 is the

Lie algebra of vector fields on Q. Therefore, by the lemma, ∇^ε extends in a unique manner to an absolute flat connection such that L acts via this connection.

Remark. In the situation with T' the Lie algebra L' acts on $\mathcal{E}$ via the connection as well (by the same lemma).

All the properties stated in the proposition, except the last one, are evident from the construction. Let us show that η_{dR} is horizontal. We work Q-locally, so we can assume that Q is affine and $X \setminus T$ admits a section s. We can enlarge T to $T^+ := T \sqcup T'$, $T' := s(Q)$. By the first constraint from 2.8, the assertion for M reduces to that for $s_* s^* M$ and $j_{T'*} M = M(\infty T')$. The first case is evident. In the second case the Gauß-Manin connection comes from the action of the Lie algebra L' (for T^+ and T'), and we are done by the remark. $\square$

2.13 *Compatibility of η_{dR} with quadratic degenerations of X.* We will show that η_{dR} remains constant (in some sense) when X degenerates quadratically and M stays constant outside the node. Notice that the family is *not* isomonodromic (so 2.12 is not applicable). We will need the result in §5; the reader can presently skip the subsection. Consider the next data (a), (b):

(a) A smooth proper curve Y, a finite subscheme $T \subset Y$, two points $b_+, b_- \in (Y \setminus T)(k)$, a rational 1-form ν on Y invertible off $T \cup \{b_\pm\}$ and having poles of order 1 at $b_\pm$ with $\mathrm{Res}_{b_\pm} \nu = \pm 1$. Let $t_\pm$ be formal coordinates at $b_\pm$ such that $d \log t_\pm = \pm \nu$.

(b) A $\mathcal{D}$-module N on $(Y, T \cup b_\pm)$ which has regular singularities at $b_\pm$, is the $*$-extension at b_+ and the !-extension at b_-. We also have a $t_\pm \partial_{t_\pm}$-invariant $b_\pm$-lattice L in M, and an identification of the $b_\pm$-fibers $\alpha : L_{b_+} \xrightarrow{\sim} L_{b_-}$. Let $A_\pm \in \mathrm{End}(L_{b_\pm})$ be the action of $\pm t_\pm \partial_{t_\pm}$ on the fibers; we ask that $\alpha A_+ = A_- \alpha$, and that the eigenvalues of A_+ (or A_-) and their pairwise differences cannot be non-zero integers. Then the restriction of L to the formal neighborhoods $Y_{\hat{\pm}} = \mathrm{Spf}\, k[[t_\pm]]$ of $b_\pm$ can be identified in a unique way with $L_{b_\pm}[[t_\pm]]$ so that $L_{b_\pm} \subset \Gamma(Y_{\hat{\pm}}, L)$ is $t_\pm \partial_{t_\pm}$-invariant.

Datum (a) yields a proper family of curves X over $Q = \mathrm{Spf}\, k[[q]] = \varinjlim Q_n$, $Q_n = \mathrm{Spec}\, k[q]/q^n$, which has quadratic degeneration at $q = 0$. The 0-fiber X_0 is Y with $b_\pm$ glued to a single point $b_0 \in X_0(k)$; let j_{b_0} be the embedding $Y \setminus \{b_+, b_-\} = X_0 \setminus \{b_0\} \hookrightarrow X_0$. Outside b_0 our X is trivialized, i.e., $\mathcal{O}_{X \setminus \{b_0\}} = \mathcal{O}_{X \setminus \{b_0\}}[[q]]$. The formal completion of the local ring at b_0 equals $k[[t_+, t_-]]$ with $q = t_+ t_-$, and the glueing comes from the embedding $k[[t_+, t_-]] \hookrightarrow k((t_+))[[q]] \times k((t_-))[[q]]$, $t_+ \mapsto (t_+, q/t_-)$, $t_- \mapsto (q/t_+, t_-)$. Set $\mathcal{R} := k((t_+))[[q]] \times k((t_-))[[q]]/k[[t_+, t_-]]$. We have a short exact sequence ($\mathcal{R}$ is viewed as a skyscraper at b_0)

$$0 \to \mathcal{O}_X \to j_{b_0 *} \mathcal{O}_{X \setminus \{b_0\}} = j_{b_0 *} \mathcal{O}_{X_0 \setminus \{b_0\}}[[q]] \to \mathcal{R} \to 0, \qquad (2.13.1)$$

where the right projection assigns to $f = \Sigma f_n q^n$ the image of (f_+, f_-) in $\mathcal{R}$, $f_\pm = \Sigma f_n(t_\pm) q^n \in k((t_\pm))[[q]]$ are the expansions of f at $b_\pm$.

Our family of curves has standard nodal degeneration, so we have the dualizing line bundle $\omega_{X/Q}$. Our ν defines a rational section ν_Q of $\omega_{X/Q}$, which is "constant" on $X \setminus \{b_0\}$ with respect to the above trivialization, and is invertible near b_0.

Below for an $\mathcal{O}_X$-module F a relative connection on F means a morphism $\nabla : F \to \omega_{X/Q} \otimes F$ such that $\nabla(f\phi) = d(f) \otimes \phi + f \nabla(\phi)$, where d is the canonical differentiation $d : \mathcal{O}_X \to \omega_{X/Q}$. Set $dR_{X/Q}(F) := \mathcal{C}one(\nabla)$, $Rq_{\mathrm{dR}*} F := Rq_* dR_{X/Q}(F)$.

Datum (b) yields an $\mathcal{O}$-module M on X equipped with a relative connection ∇. Our M is locally free over $X \setminus T_Q$. Outside b_0 it is constant with respect to the above trivialization: one has $M|_{X \setminus \{b_0\}} = N|_{Y \setminus \{b_+, b_-\}}[[q]] = L|_{Y \setminus \{b_+, b_-\}}[[q]]$.

The restriction M_0 of M to X_0 equals L with fibers $L_{b_\pm}$ identified by α. On the formal neighborhood of b_0 our M equals $M_{b_0}[[t_+, t_-]]$, and the glueing comes from the trivializations of L on $Y_{\hat{\pm}}$ (see (b)) and the gluing of functions. Therefore we have a short exact sequence

$$0 \to M \to (j_{b_0*}L|_{Y \setminus \{b_+, b_-\}})[[q]] \to M_{b_0} \otimes \mathcal{R} \to 0, \qquad (2.13.2)$$

where the right projection assigns to $\ell = \Sigma \ell_n q^n \in (j_{b_0*}L|_{Y \setminus \{b_+, b_-\}})$ the image in of (ℓ_+, ℓ_-) $M_{b_0} \otimes \mathcal{R}$, $\ell_\pm = \Sigma \ell_n (t_\pm) q^n \in L_{b_\pm}((t_\pm))[[q]] = M_{b_0}((t_\pm))[[q]]$ are the expansions of ℓ at $b_\pm$ with respect to the formal trivializations of L on $Y_{\hat{\pm}}$. On $X \setminus \{b_0\}$ the relative connection ∇ comes from the $\mathcal{D}$-module structure on N; on the formal neighborhood of b_0 this is the relative connection on $M_{b_0}[[t_+, t_-]]$ with potential $A t_+^{-1} dt_+ = -A t_-^{-1} dt_-$.

Remarks. (2.13.2) is an exact sequence of $\mathcal{O}_X$-modules equipped with relative connections. The projection $(t_+^{-1}k[t_+^{-1}] \oplus k[t_-^{-1}])[[q]] \to \mathcal{R}$ is an isomorphism. The relative connection on $M_{b_0} \otimes \mathcal{R}$ is $\nabla(m \otimes t_\pm^a q^b) = \nu(A(m) \pm am) t_\pm^a q^b$.

Set $D = -\operatorname{div}(\nu) - b_+ - b_-$. Then $\operatorname{div}(\nu_Q) = -D_Q$ does not intersect b_0, so we have an $\mathcal{O}_Q$-line $\mathcal{E}_{\mathrm{dR}}(M)_{\nu_Q} := \mathcal{E}_{\mathrm{dR}}(M)_{(D_Q, 1_T, \nu_Q)}$. An immediate modification of the construction in 2.7 (to be spelled out in the proof of the proposition below) yields an isomorphism

$$\eta_{\mathrm{dR}} : \mathcal{E}_{\mathrm{dR}}(M)_{\nu_Q} \xrightarrow{\sim} \det Rq_{\mathrm{dR}*}M. \qquad (2.13.3)$$

Our aim is to compute it explicitly. Notice that since our family (X, T_Q, M, ν_Q) is trivialized outside b_0, one has a canonical identification

$$\mathcal{E}_{\mathrm{dR}}(M)_{\nu_Q} \xrightarrow{\sim} \mathcal{E}_{\mathrm{dR}}(N)_{(D, 1_T, \nu)}[[q]]. \qquad (2.13.4)$$

(i) Consider an embedding $\mathcal{C}one(A)[[q]] \hookrightarrow dR_{X/Q}(M_{b_0} \otimes \mathcal{R})$ whose components are $M_{b_0}[[q]] \hookrightarrow M_{b_0} \otimes \mathcal{R}$, $m \otimes f(q) \mapsto m \otimes (0, f(q))$, and $M_{b_0}[[q]] \hookrightarrow \omega_{X/Q} \otimes M_{b_0} \otimes \mathcal{R}$, $m \otimes f(q) \mapsto \nu \otimes m \otimes (0, f(q))$. By the condition on A, this is a quasi-isomorphism. Thus (2.13.2) yields an isomorphism

$$\det Rq_{\mathrm{dR}*}M \xrightarrow{\sim} \det R\Gamma_{\mathrm{dR}}(Y, j_{b_\pm *}N) \otimes \det \mathcal{C}one(A)[[q]]. \qquad (2.13.5)$$

Let L^- be a $b_\pm$-lattice in N that equals L outside b_- and t_-L at b_-. Set $C_! := \mathcal{C}one(\nabla : L^- \to \omega_Y(\log b_+)L)$, $C_* := \mathcal{C}one(\nabla : L \to \omega_Y(\log b_\pm)L)$. Recall that N is the !-extension at b_- and the *-extension at b_+, so the condition on A assures that the embeddings $C_! \hookrightarrow dR(N)$ and $C_* \hookrightarrow dR(j_{b_\pm *}N)$ are quasi-isomorphisms. Since $C_*/C_!$ equals $\mathcal{C}one(A)$ (viewed as a skyscraper at b_-), we see that $\mathcal{C}one(dR(N) \to dR(j_{b_\pm *}N)) \xrightarrow{\sim} \mathcal{C}one(A)$, hence

$$\det R\Gamma_{\mathrm{dR}}(Y, N) \xrightarrow{\sim} \det R\Gamma_{\mathrm{dR}}(Y, j_{b_\pm *}N) \otimes \det \mathcal{C}one(A). \qquad (2.13.6)$$

Combining it with (2.13.5), we get an isomorphism

$$\det Rq_{\mathrm{dR}*}M \xrightarrow{\sim} \det R\Gamma_{\mathrm{dR}}(Y, N)[[q]]. \qquad (2.13.7)$$

(ii) Consider a $\mathcal{D}$-module on $\mathbb{P}^1 \setminus \{0, \infty\}$ which equals $\mathcal{O}_{\mathbb{P}^1 \setminus \{0, \infty\}} \otimes M_{b_0}$ as an $\mathcal{O}$-module, $\nabla(f \otimes m) = df \otimes m + f \otimes A(m)$. Let $\bar{N}$ be the !-extension to ∞ and the *-extension to 0 of N. Consider the embeddings $t_+ : Y_{\hat{+}} \hookrightarrow \mathbb{P}^1$, $t_-^{-1} : Y_{\hat{-}} \hookrightarrow \mathbb{P}^1$ which identify $Y_{\hat{\pm}}$ with the formal neighborhoods of 0 and ∞. The trivializations of L on $Y_{\hat{\pm}}$ from (b) identify the pull-back of $\bar{N}$ with $N|_{Y_{\hat{\pm}}}$. Since the pull-back of $t^{-1}dt$ equals $\nu|_{Y_{\hat{\pm}}}$, we get the identifications $\mathcal{E}_{\mathrm{dR}}(N)_{(b_+, \nu)} \xrightarrow{\sim} \mathcal{E}_{\mathrm{dR}}(\bar{N})_{(0, t^{-1}dt)}$,

$\mathcal{E}_{dR}(N)_{(b_-,\nu)} \xrightarrow{\sim} \mathcal{E}_{dR}(\bar{N})_{(\infty,t^{-1}dt)}$. Combined with (2.7.4) (for M in loc. cit. equal to $\bar{N}$), they produce an isomorphism

$$\mathcal{E}_{dR}(N)_{(b_+,\nu)} \otimes \mathcal{E}_{dR}(N)_{(b_-,\nu)} \xrightarrow{\sim} k. \tag{2.13.8}$$

Since $\mathcal{E}_{dR}(N)_\nu = \mathcal{E}_{dR}(N)_{(D,1_T,\nu)} \otimes \mathcal{E}_{dR}(N)_{(b_+,\nu)} \otimes \mathcal{E}_{dR}(N)_{(b_-,\nu)}$, we rewrite it as

$$\mathcal{E}_{dR}(N)_{(D,1_T,\nu)} \xrightarrow{\sim} \mathcal{E}_{dR}(N)_\nu. \tag{2.13.9}$$

Proposition. *The diagram*

$$\begin{array}{ccc}
\mathcal{E}_{dR}(M)_{\nu_Q} & \xrightarrow{\eta_{dR}} & \det Rq_{dR*}M \\
\downarrow & & \downarrow \\
\mathcal{E}_{dR}(N)_\nu[[q]] & \xrightarrow{\eta_{dR}} & \det R\Gamma_{dR}(Y,N)[[q]],
\end{array} \tag{2.13.10}$$

where the vertical arrows are (2.13.9)∘(2.13.4) and (2.13.7), commutes.

Proof We check the assertion modulo q^{n+1}. Thus we restrict our picture to $Q_n := \operatorname{Spec} R_n$, $R_n := k[q]/q^{n+1}$; we get a Q_n-curve X_n, $M_n = M \otimes R_n$, etc.

Let F be a $\{b_+,b_-\}$-lattice in $j_{b_\pm *}N$ such that $\nabla(F) \subset \nu F = \omega_Y(\log b_\pm) \otimes F$; set $dR(F) := \mathcal{C}one(\nabla : F \to \nu F)$, $R\Gamma_{dR}(Y,F) := R\Gamma(Y, dR(F))$. Then there is a canonical isomorphism

$$\eta_{dR}(F) : \mathcal{E}_{dR}(N)_{(D,1_T,\nu)} \xrightarrow{\sim} \det R\Gamma_{dR}(Y,F) \tag{2.13.11}$$

defined as in the proposition in 2.7. Precisely, pick any $T \cup |D|$-lattices E, E_ω in N, ωN such that $\nabla(E) \subset E_\omega$. Denote by FE the $T \cup \{b_+,b_-\} \cup |D|$-lattice in $j_{b_\pm *}N$ that equals F off $T \cup |D|$ and E off $b_\pm$; similarly, FE_ω equals E_ω off $b_\pm$ and νF off $T \cup |D|$. Now follow the construction from the proposition in 2.7, with L, L_ω, $dR(M)$ from loc. cit. replaced by FE, FE_ω and $dR(F)$. Namely, $dR(F)$ carries a 3-step filtration with successive quotients FE_ω, $FE[1]$, $\mathcal{C}(E,E_\omega)$, and $\eta_{dR}(F)$ is the composition $\mathcal{E}_{dR}(N)_{(D,1_T,\nu)} \xrightarrow{\sim} \det \Gamma(Y, \mathcal{C}(E,E_\omega)) \otimes \lambda(FE_\omega/\nu FE) \xrightarrow{\sim} \det \Gamma(Y, \mathcal{C}(E,E_\omega)) \otimes \det R\Gamma(Y, FE_\omega) \otimes (\det R\Gamma(Y,\nu FE))^{\otimes -1} \xrightarrow{\sim} \det \Gamma(Y, \mathcal{C}(E,E_\omega)) \otimes \det R\Gamma(Y, FE_\omega) \otimes \det R\Gamma(Y, FE[1])) \xrightarrow{\sim} \det R\Gamma(X, dR(F))$.

For example, for $F = L^-$ from (i) above one has $dR(L^-) = C_!$, so $R\Gamma_{dR}(Y,L^-)) = R\Gamma_{dR}(Y,N)$, and we get $\eta_{dR}(L^-) : \mathcal{E}_{dR}(N)_{(D,1_T,\nu)} \xrightarrow{\sim} \det R\Gamma_{dR}(Y,N)$. We can also view L^- as a lattice in N, and compute $\eta_{dR} : \mathcal{E}_{dR}(N)_\nu \xrightarrow{\sim} R\Gamma_{dR}(Y,N)$ using it (as in the proposition in 2.7). Now the lemma in 2.7 implies that $\eta_{dR}(L^-)$ equals the composition $\mathcal{E}_{dR}(N)_{(D,1_T,\nu)} \xrightarrow{(2.13.9)} \mathcal{E}_{dR}(N)_\nu \xrightarrow{\eta_{dR}} R\Gamma_{dR}(Y,N)$.

Exercise. If $F' \subset F$ is a sublattice with $\nabla(F') \subset \nu F'$, then $dR(F)/dR(F') = \mathcal{C}one(\nabla : F/F' \to \nu F/\nu F')$. Thus $dR(F)$ carries a 3-step filtration with successive quotients $dR(F')$, $\nu F/\nu F'$, F/F', hence $\det R\Gamma_{dR}(Y,F) \xrightarrow{\sim} \det R\Gamma_{dR}(Y,F') \otimes \det \Gamma(Y, \nu F/\nu F') \otimes \det \Gamma(Y, F/F')^{\otimes -1}$. The multiplication by ν isomorphism $F/F' \xrightarrow{\sim} \nu F/\nu F'$ cancels the last two factors, i.e., we have $\det R\Gamma_{dR}(Y,F) \xrightarrow{\sim} \Gamma_{dR}(Y,F')$. Show that this isomorphism equals $\eta_{dR}(F')\eta_{dR}(F)^{-1}$.

One can repeat the above story with Y replaced by X_n, $j_{b_\pm *}N$ by $j_{b_0 *}N \otimes R_n$, and E, E_ω by $E \otimes R_n$, $E_\omega \otimes R_n$. For a b_0-lattice G in $j_{b_0 *}N \otimes R_n$ (i.e., an $\mathcal{O}_{X_n}$-submodule, which is R_n-flat and equals $j_{b_0 *}N \otimes R_n$ outside b_0) such that $\nabla(G) \subset \nu G$, we get an isomorphism

$$\eta_{dR}(G) : \mathcal{E}_{dR}(N)_{(D,1_T,\nu)} \otimes R_n \xrightarrow{\sim} \det Rq_{dR*}G. \tag{2.13.12}$$

For $G = M_n$, this is (2.13.3) combined with (2.13.4). If G is a "constant" lattice, $G = F \otimes R_n$, then $Rq_{\mathrm{dR}*}G = R\Gamma_{\mathrm{dR}}(Y, F) \otimes R_n$ and $\eta_{\mathrm{dR}}(G) = \eta_{\mathrm{dR}}(F) \otimes \mathrm{id}_{R_n}$.

By above, the proposition means that $\eta_{\mathrm{dR}}(L^-)\eta_{\mathrm{dR}}(M_n)^{-1} : \det Rq_{\mathrm{dR}*}M \overset{\sim}{\to} \det R\Gamma_{\mathrm{dR}}(Y, N) \otimes R_n$ coincides with (2.13.7). Set $L^{(n)} := \mathcal{I}^{n+1}L$, where $\mathcal{I}$ is the ideal of $\{b_+, b_-\}$ in $\mathcal{O}_Y$. Then $L^{(n)}$ lies in both L^- and M_n. Set $P_N := L^-/L^{(n)}$ and $P_M := M_n/L^{(n)} \otimes R_n$; let B_N, B_M be their endomorphisms $\nu^{-1}\nabla$. One has evident isomorphisms $\mathcal{C}one(B_M) \overset{\sim}{\to} \Gamma(X_n, dR(M_n)/dR(L^{(n)}) \otimes R_n)$, $\mathcal{C}one(B_N) \overset{\sim}{\to} \Gamma(Y, dR(L^-)/dR(L^{(n)}))$. Thus $\det Rq_{\mathrm{dR}*}M = \det R\Gamma_{\mathrm{dR}}(Y, L^{(n)}) \otimes \det \mathcal{C}one(B_M)$, $\det R\Gamma_{\mathrm{dR}}(Y, N) \overset{\sim}{\to} \det R\Gamma_{\mathrm{dR}}(Y, L^{(n)}) \otimes \det \mathcal{C}one(B_N)$, so both $\eta_{\mathrm{dR}}(L^-)\eta_{\mathrm{dR}}(M)^{-1}$ and (2.13.7) can be rewritten as isomorphisms $\det \mathcal{C}one(B_M) \overset{\sim}{\to} \det \mathcal{C}one(B_N) \otimes R_n$.

Both $\det \mathcal{C}one(B_M)$ and $\det \mathcal{C}one(B_N)$ are naturally trivialized (since $\det \mathcal{C}one(B_M) = \det(P_M) \otimes \det(P_M[1])$, etc.). By Exercise, $\eta_{\mathrm{dR}}(L^-)\eta_{\mathrm{dR}}(M_n)^{-1}$ identifies these trivializations. Isomorphism (2.13.7) comes due to the fact that $\mathcal{C}one(B_M)$ and $\mathcal{C}one(B_N) \otimes R_n$ are naturally quasi-isomorphic: we have the evident embeddings $i_M : M_a \otimes R_n \hookrightarrow M_n/L^{(n)}$, $i_N : M_a \hookrightarrow L^-/L^{(n)}$ such that $B_M i_M = i_M A$, $B_N i_N = i_N A$, which yield quasi-isomorphisms $\mathcal{C}one(A) \otimes R_n \hookrightarrow \mathcal{C}one(B_M)$, $\mathcal{C}one(A) \hookrightarrow \mathcal{C}one(B_N)$. Therefore the ratio of $\eta_{\mathrm{dR}}(L^-)\eta_{\mathrm{dR}}(M_n)^{-1}$ and (2.13.7) equals the ratio of the determinants of B_M and B_N acting on the quotients $(M_n/L^{(n)})/M_a \otimes R_n$, $(L^-/L^{(n)})/M_a$. The first quotient is the direct sum of components $M_a \otimes (t_+^i, q^i t_-^{-i}) \otimes R_n$ and $M_a \otimes (q^i t_+^{-i}, t_-^i) \otimes R_n$ for $1 \le i \le n$, the second one is the direct sum of $M_a \otimes (t_+^i, 0)$ and $M_a \otimes (0, t_-^i)$ for $1 \le i \le n$. Both B_M and B_N act on them as $A + i\,\mathrm{id}$ and $A - i\,\mathrm{id}$, so the two determinants are equal. We are done. $\qquad\square$

2.14 Suppose $k = \mathbb{C}$. The definitions and constructions of this section render immediately into the complex-analytic setting of 1.15. Thus every triple (X, T, M), where X is a smooth (not necessarily compact) complex curve, T its finite subset, M a holonomic $\mathcal{D}$-module on (X, T), yields a factorization line $\mathcal{E}_{\mathrm{dR}}(M) \in \mathcal{L}_{\mathrm{dR}}^{\Phi}(X, T)$ in the complex-analytic setting of 1.15. If X and M came from an algebraic setting, then $\mathcal{E}_{\mathrm{dR}}(M)$ is an analytic factorization line produced by the algebraic one (defined previously). If an algebraic family of $\mathcal{D}$-modules is nice (see 2.12), then the corresponding analytic family is nice.

We work in the analytic setting. Let $q : X \to Q$, $i : T \hookrightarrow X$ be as in 1.14; we assume that T is étale over Q (see 1.15). Let M be a flat family of $\mathcal{D}$-modules on $(X/Q, T)$ which admits locally a T-lattice, see 2.12. Consider the sheaf-theoretic restriction $F := i^{\cdot}dR_{X/Q}(M)$ of the relative de Rham complex to T. Since $q|_T\mathcal{O}_Q = \mathcal{O}_T$, this is a complex of $\mathcal{O}_T$-modules.

Lemma. *M is nice if and only if F has $\mathcal{O}_T$-coherent cohomology.*

Proof The assertion is Q-local, so we can assume that T is a disjoint sum of several copies of Q. Since $q_*(dR_{X/Q}(M)/\mathcal{C}one(L \overset{\nabla}{\to} L_\omega))$ is the direct sum of pieces corresponding to the components of T, we are reduced to the situation when X equals $U \times Q$, where $U \subset \mathbb{A}^1$ is a coordinate disc, and $T = \{0\} \times Q$.

M extends in a unique manner to a $\mathcal{D}_{\mathbb{A}_Q^1/Q}$-module on $\mathbb{A}_Q^1$ which is smooth outside T; denote it also by M. So we can assume that $X = \mathbb{A}_Q^1$. Set $\bar{X} := \mathbb{P}_Q^1$; let $\bar{q} : \bar{X} \to Q$ be the projection, so $X = \bar{X} \setminus T^\infty$, $T_Q^\infty := \{\infty\} \times Q$.

Let us extend M to an $\mathcal{O}_{\bar{X}}$-module $\bar{M}$ such that the relative connection has logarithmic singularity at T^∞. Such an $\bar{M}$ exists locally on Q. Replacing $\bar{M}$ by some $\bar{M}(nT^\infty)$, we can assume that the eigenvalues of $-t\partial_t$ in the fiber of $\bar{M}$ over T^∞ do not meet $\mathbb{Z}_{\geq 0}$. Let $dR_{\bar{X}/Q}(\bar{M}) := Cone(\bar{M} \xrightarrow{\nabla} \omega\bar{M}(T^\infty))$ be the relative de Rham complex of $\bar{M}$ with logarithmic singularities at T^∞. One has the usual quasi-isomorphisms

$$R\bar{q}_*(dR_{\bar{X}/Q}(\bar{M})) \xrightarrow{\sim} q_*(dR_{X/Q}(M)) \xrightarrow{\sim} i_T^! dR_{X/Q}(M). \qquad (2.14.1)$$

Let $\bar{L} \subset \bar{M}$, $\bar{L}_\omega \subset \omega\bar{M}(T^\infty)$ be $\mathcal{O}_{\bar{X}}$-submodules that equal L, L_ω on X and coincide with $\bar{M}$, $\omega\bar{M}(T^\infty)$ outside T. Now $dR_{X/Q}(M)/Cone(L \to L_\omega)$ equals $dR_{\bar{X}/Q}(\bar{M})/Cone(\bar{L} \to \bar{L}_\omega)$, so (2.14.1) yields an exact triangle

$$R\bar{q}_*Cone(\bar{L} \to \bar{L}_\omega) \to i_T^! dR_{X/Q}(M) \to q_* dR_{X/Q}(M)/Cone(L \to L_\omega). \qquad (2.14.2)$$

Its left term is $\mathcal{O}_Q$-coherent, so the other two are coherent simultaneously, q.e.d. $\qquad\square$

3 The de Rham ε-lines: analytic theory

From now on we work in the analytic setting over $\mathbb{C}$ using the classical topology.

3.1   Let X be a smooth (not necessary compact) complex curve, T its finite subset. For a holonomic $\mathcal{D}$-module M we denote by $B(M)$ the de Rham complex $dR(M)$ viewed as mere perverse $\mathbb{C}$-sheaf on X, and set $H_B^\cdot(X, M) := H^\cdot(X, B(M))$; thus one has an evident *period* isomorphism $\rho : H_B^\cdot(X, M) \xrightarrow{\sim} H_{dR}^\cdot(X, M)$. Here is the principal result of this section:

Theorem-construction. *Let M, M' be holonomic $\mathcal{D}$-modules on (X, T). Then every isomorphism $\phi : B(M) \xrightarrow{\sim} B(M')$ yields naturally an identification of the de Rham factorization lines $\phi^\epsilon : \mathcal{E}_{dR}(M) \xrightarrow{\sim} \mathcal{E}_{dR}(M')$. The construction has local origin, and is compatible with constraints from 2.8. If X is compact, then the next diagram of isomorphisms commutes:*

$$\begin{array}{ccc}
\mathcal{E}_{dR}(M)(X) & \xrightarrow{\phi^\epsilon} & \mathcal{E}_{dR}(M')(X) \\
\eta_{dR} \downarrow & & \eta_{dR} \downarrow \\
\det H_{dR}^\cdot(X, M) & & \det H_{dR}^\cdot(X, M') \\
\rho \downarrow & & \rho \downarrow \\
\det H_B^\cdot(X, M) & \xrightarrow{\phi} & \det H_B^\cdot(X, M').
\end{array} \qquad (3.1.1)$$

The idea of the proof: By a variant of Riemann-Hilbert correspondence, $B(M)$ amounts to the $\mathcal{D}^\infty$-module M^∞. Thus what we need is to render the story of §2 into the analytic setting of $\mathcal{D}^\infty$-modules, which is done using a version of constructions from [PS], [SW].

An alternative proof of the theorem, which uses 2.13 and §4 instead of analytic Fredholm determinants, is presented in 5.8. Thus the reader can skip the rest of the section and pass directly to §4.

3.2   *A digression on $\mathcal{D}^\infty$-modules and Riemann-Hilbert correspondence.* For the proofs of the next results, see [Bj] III 4, V 5.5, or [Me].

For a complex variety X we denote by $\mathcal{D}^\infty$ or $\mathcal{D}_X^\infty$ the sheaf of differential operators of infinite order on X. If X is a curve and U is an open subset with a coordinate function t, then $\mathcal{D}^\infty(U)$ consists of series $\underset{n \geq 0}{\Sigma} a_n \partial_t^n$, where a_n are holomorphic functions on U such that for every $\epsilon > 0$ the series $\Sigma a_n \epsilon^{-n} n!$ converges absolutely on any compact subset. $\mathcal{D}_X^\infty$ is a sheaf of rings that acts on $\mathcal{O}_X$ in an evident manner;[16] it contains $\mathcal{D}_X$, and $\mathcal{D}_X^\infty$ is a faithfully flat $\mathcal{D}_X$-module.

By Grothendieck and Sato, one can realize $\mathcal{D}^\infty(U)$ as $H_{\Delta(U)}^{\dim X}(U \times U, \mathcal{O} \boxtimes \omega)$ where Δ is the diagonal embedding. If X is a curve and U has no compact components, this means that

$$\mathcal{D}^\infty(U) = (\mathcal{O} \boxtimes \omega)(U \times U \setminus \Delta(U))/(\mathcal{O} \boxtimes \omega)(U \times U). \qquad (3.2.1)$$

Here $k(x,y) \in (\mathcal{O} \boxtimes \omega)(U \times U \setminus \Delta(U))$ acts on $\mathcal{O}(U)$ as $f \mapsto k(f)$, $k(f)(x) :=$ $\mathrm{Res}_{y=x} k(x,y) f(y)$.

For a (left) $\mathcal{D}$-module M set $M^\infty := \mathcal{D}^\infty \underset{\mathcal{D}}{\otimes} M$. The embedding $dR(M) \hookrightarrow dR(M^\infty)$ is a quasi-isomorphism, hence $H_{\mathrm{dR}}^\cdot(X, M) \xrightarrow{\sim} H_{\mathrm{dR}}^\cdot(X, M^\infty)$. If M is smooth, then $M \xrightarrow{\sim} M^\infty$.

For a $\mathcal{D}^\infty$-module N a $\mathcal{D}$-*structure* on N is a $\mathcal{D}$-module M together with a $\mathcal{D}^\infty$-isomorphism $M^\infty \xrightarrow{\sim} N$. Our N is said to be holonomic if it admits a $\mathcal{D}$-structure with holonomic M.

For a holonomic $\mathcal{D}$-module M the de Rham complex $dR(M)$ is a perverse $\mathbb{C}$-sheaf, which we denote, as above, by $B(M)$; same for a holonomic $\mathcal{D}^\infty$-module. Therefore $B(M) = B(M^\infty)$. The functor B is an equivalence between the category of holonomic $\mathcal{D}^\infty$-modules and that of perverse $\mathbb{C}$-sheaves. The inverse functor assigns to a perverse sheaf F the $\mathcal{D}^\infty$-module[17] $\mathcal{O}_X \underset{\mathbb{C}}{\otimes}{}^! F[\dim X]$. Thus for a holonomic M the $\mathcal{D}^\infty$-module M^∞ carries the same information as $B(M)$.

The functor $M \mapsto M^\infty$ yields an equivalence between the category of holonomic $\mathcal{D}$-modules with regular singularities and that of holonomic $\mathcal{D}^\infty$-modules. Its inverse assigns to a holonomic $\mathcal{D}^\infty$-module N its maximal $\mathcal{D}$-submodule N^{rs} with regular singularities, so one has

$$(N^{\mathrm{rs}})^\infty = N. \qquad (3.2.2)$$

Therefore every holonomic $\mathcal{D}^\infty$-module admits a unique $\mathcal{D}$-structure with regular singularities.

Exercises. Let U be a coordinate disc, t be the coordinate, j be the embedding $U^o := U \setminus \{0\} \hookrightarrow U$.

(i) Recall a description of indecomposable $\mathcal{D}_U$-modules which are smooth of rank n on U^o and have regular singularity at 0. For $s \in \mathbb{C}$ denote by $M_{s,n}$ a $\mathcal{D}$-module whose sections are collections of functions $(f_i) = (f_1, \ldots, f_n)$ having meromorphic singularity at 0, and $\nabla_{\partial_t}((f_i)) = (\partial_t(f_i) + sf_i + f_{i-1})$.[18] Let $M_{0,n}^1$ be a $\mathcal{D}$-submodule of $M_{0,n}$ formed by (f_i) with f_1 regular at 0. Consider an embedding $\mathcal{O}_U \hookrightarrow M_{0,n+1}$, $f \mapsto (0, \ldots, 0, f)$; set $M_{0,n}^2 := M_{0,n+1}^1/\mathcal{O}$, $M_{0,n}^3 := M_{0,n+1}/\mathcal{O}$. E.g., $M_{0,1}^1 = \mathcal{O}_U$, and $M_{0,0}^2 = \delta$ (the δ-function $\mathcal{D}$-module). Then any indecomposable $\mathcal{D}_U$-module M as above is isomorphic to either some $M_{s,n}$, or one of $M_{0,n}^a$, $a = 1, 2, 3$.

Show that the corresponding $\mathcal{D}^\infty$-module M^∞ has the same explicit description with "meromorphic singularity" replaced by "arbitrary singularity".

(ii) For $n > 0$ let $E_{(n)}$ be a $\mathcal{D}$-module of rank 1 generated by $\exp(t^{-n})$, i.e., $E_{(n)}$ is generated by a section e subject to the (only) relation $t^{n+1}\partial_t(e) = -ne$.

---

[16] By [I], $\mathcal{D}_X^\infty$ coincides with the sheaf of all $\mathbb{C}$-linear continuous endomorphisms of $\mathcal{O}_X$.

[17] Here $\mathcal{D}^\infty$ acts via the $\mathcal{O}_X$-factor, and $F \otimes{}^! G := R\Delta^! F \boxtimes G$.

[18] $M_{s,n}$ depends only on s modulo $\mathbb{Z}$-translation: one has $M_{s,n} \xrightarrow{\sim} M_{s-1,n}$, $(f_i) \mapsto (tf_i)$.

Show that there is an isomorphism of $\mathcal{D}^\infty$-modules

$$E^\infty_{(n)} \xrightarrow{\sim} M^{2\infty}_{0,1} \oplus (\delta^\infty)^{n-1}. \tag{3.2.3}$$

Here is an explicit formula for (3.2.3). Let $g(z)$, $h_1(z), \ldots, h_{n-1}(z)$ be entire functions such that $(\partial_z - nz^{n-1})g(z) = z^{-2}(\exp(z^n) - 1 - z^n)$ and $(\partial_z - nz^{n-1})h_i(z) = z^{i-1}$. Then (3.2.3) assigns e a vector whose $(M^2_{0,1})^\infty$-component is $(\exp(t^{-n}), g(t^{-1}))$ and the δ^∞-components are $h_i(t^{-1}) \in (M^2_{0,0})^\infty = \delta^\infty$.

3.3 *A digression on Fredholm determinants* (cf. [PS] 6.6). Recall that a *Fréchet space* is a complete, metrizable, locally convex topological $\mathbb{C}$-vector space. The category $\mathcal{F}$ of those is a quasi-abelian (hence exact) Karoubian $\mathbb{C}$-category. A morphism $\phi : F \to F'$ is said to be Fredholm if it is Fredholm as a morphism of abstract vector spaces, i.e., if $\operatorname{Ker}\phi$ and $\operatorname{Coker}\phi$ have finite dimension. Then ϕ is a split morphism, i.e., $\operatorname{Ker}\phi$, $\operatorname{Im}\phi$ are direct summands of, respectively, F and F', and $F/\operatorname{Ker}\phi \xrightarrow{\sim} \operatorname{Im}\phi$. Denote by $\mathfrak{F} \subset \mathcal{F}$ the subcategory of Fredholm morphisms $\mathfrak{F}(F, F') \subset \operatorname{Hom}(F, F')$.

A Fredholm ϕ yields the determinant line $\lambda_\phi := \det(\operatorname{Coker}\phi) \otimes \det^{\otimes -1}(\operatorname{Ker}\phi) \in \mathcal{L} := \mathcal{L}_{\mathbb{C}}$ (see 1.2). Sometimes we denote λ_ϕ by $\lambda(F' \xrightarrow{\phi} F)$ or, if $F = F'$, by $\lambda(F)_\phi$. If ϕ is invertible, then λ_ϕ has an evident trivialization; denote it by $\det(\phi) \in \lambda_\phi$.

For any Fredholm ϕ one can find finite-dimensional $F_0 \subset F$, $F'_0 \subset F'$ such that $\phi(F'_0) \subset F_0$ and the induced map $F'/F'_0 \to F/F_0$ is an isomorphism (equivalently, $F_0 + \phi(F') = F$, $F'_0 = \phi^{-1}(F_0)$). Then the exact sequence $0 \to \operatorname{Ker}\phi \to F'_0 \to F_0 \to \operatorname{Coker}\phi \to 0$ yields a natural isomorphism

$$\lambda_\phi \xrightarrow{\sim} \det(F_0) \otimes \det^{\otimes -1}(F'_0). \tag{3.3.1}$$

If ϕ is invertible, then (3.3.1) identifies $\det(\phi) \in \lambda_\phi$ with the usual determinant of $\phi|_{F'_0} : F'_0 \xrightarrow{\sim} F_0$ in $\operatorname{Hom}(\det F'_0, \det(F_0)) = \det(F_0) \otimes \det^{\otimes -1}(F'_0)$.

For Fredholm $F'' \xrightarrow{\phi'} F' \xrightarrow{\phi} F$ there is a canonical "composition" isomorphism

$$\lambda_\phi \otimes \lambda_{\phi'} \xrightarrow{\sim} \lambda_{\phi\phi'}, \quad a \otimes b \mapsto ab, \tag{3.3.2}$$

which satisfies the associativity property. Therefore $\phi \mapsto \lambda_\phi$ is a (central) $\mathcal{L}$-extension $\mathfrak{F}^\flat$ of $\mathfrak{F}$ (see e.g. [BBE], Appendix to §1, for terminology). To construct (3.3.2), choose F_0, F'_0, F''_0 for ϕ, ϕ' as above; then (3.3.1) identifies the composition with an evident map $\det(F_0) \otimes \det^{\otimes -1}(F'_0) \otimes \det(F'_0) \otimes \det^{\otimes -1}(F''_0) \xrightarrow{\sim} \det(F_0) \otimes \det^{\otimes -1}(F''_0)$. For invertible ϕ, ϕ' one has $\det(\phi)\det(\phi') = \det(\phi\phi')$.

Suppose F, F' are equipped with finite split filtrations $F_\bullet$, $F'_\bullet$, $\phi : F' \to F$ preserves the filtrations, and $\operatorname{gr}\phi : \operatorname{gr}F' \to \operatorname{gr}F$ is Fredholm. Then ϕ is Fredholm, and there is a canonical isomorphism

$$\lambda_\phi \xrightarrow{\sim} \otimes\lambda_{\operatorname{gr}_i\phi}. \tag{3.3.3}$$

The identification is transitive with respect to refinement of the filtration. If $\operatorname{gr}\phi$ is invertible, it identifies $\det(\operatorname{gr}\phi) = \otimes \det(\operatorname{gr}_i\phi)$ with $\det(\phi)$. For example, for a finite collection of Fredholm morphisms $\{\phi_\alpha\}$, every linear ordering of indices α produces a filtration, hence an isomorphism $\lambda_{\oplus\phi_\alpha} \xrightarrow{\sim} \otimes\lambda_{\phi_\alpha}$; it does not depend on the ordering.

Let $\mathcal{I}^{\operatorname{fin}} \subset \mathcal{I}^{\operatorname{tr}} \subset \mathcal{I}^{\operatorname{com}}$ be the two-sided ideals of finite rank, nuclear, and compact morphisms in $\mathcal{F}$. We have the quotient categories $\mathcal{F}/\mathcal{I}^?$: their objects are Fréchet spaces, and morphisms $\operatorname{Hom}_{/\mathcal{I}^?}(F, F')$ equal $\operatorname{Hom}(F, F')/\mathcal{I}^?(F, F')$. A morphism ϕ is Fredholm if and only if ϕ is invertible in either $\mathcal{F}/\mathcal{I}^?$. Therefore the groupoids $\mathrm{Isom}(\mathcal{F}/\mathcal{I}^?)$ of isomorphisms in $\mathcal{F}/\mathcal{I}^?$ are quotients of $\mathfrak{F}$ modulo the $\mathcal{I}^?$-*equivalence* relation $\phi - \phi' \in \mathcal{I}^?$.

Exercise. Let $G^?(F) \subset \mathrm{Aut}(F)$ be the (normal) subgroup of automorphisms ψ of F that are $\mathcal{I}^?$-equivalent to id_F. The next sequence is exact:

$$1 \to G^?(F)/G^{\mathrm{fin}}(F) \to \mathrm{Aut}_{/\mathcal{I}^{\mathrm{fin}}}(F) \to \mathrm{Aut}_{/\mathcal{I}^?}(F) \to 1 \qquad (3.3.4)$$

Proposition. $\mathfrak{F}^\flat$ *descends naturally to an $\mathcal{L}$-extension $\mathrm{Isom}^\flat(\mathcal{F}/\mathcal{I}^{tr})$ of the groupoid $\mathrm{Isom}(\mathcal{F}/\mathcal{I}^{tr})$.*

Proof We first descend $\mathfrak{F}^\flat$ to $\mathrm{Isom}(\mathcal{F}/\mathcal{I}^{\mathrm{fin}})$, and then to $\mathrm{Isom}(\mathcal{F}/\mathcal{I}^{\mathrm{tr}})$.

(i) To descend $\mathfrak{F}^\flat$ to $\mathrm{Isom}(\mathcal{F}/\mathcal{I}^{\mathrm{fin}})$, means to define for every $\mathcal{I}^{\mathrm{fin}}$-equivalent $\phi, \psi \in \mathfrak{F}(F, F')$ a natural identification $\tau = \tau_{\phi,\psi} : \lambda_\phi \overset{\sim}{\to} \lambda_\psi$ which satisfies the transitivity property and is compatible with composition.

The condition on ϕ, ψ means that we can find finite-dimensional $F_0 \subset F$, $F_0' \subset F'$ such that $\phi(F_0'), (\phi-\psi)(F') \subset F_0$, and the map $F'/F_0' \to F/F_0$ induced by ϕ (or ψ) is an isomorphism. Then τ is the composition $\lambda_\phi \overset{\sim}{\to} \det(F_0) \otimes \det^{\otimes -1}(F_0') \overset{\sim}{\to} \lambda_\psi$ of isomorphisms (3.3.1) for ϕ, ψ. The construction does not depend on the choice of auxiliary datum, and satisfies the necessary compatibilities.

(ii) Recall that for $\psi \in \mathrm{End} F$ that is $\mathcal{I}^{\mathrm{tr}}$-equivalent to id_F, its *Fredholm determinant* $\det_{\mathfrak{F}}(\psi) \in \mathbb{C}$ is defined (see e.g. [Gr2]) as the sum of a rapidly converging series

$$\det_{\mathfrak{F}}(\psi) := \sum_{k \geq 0} \mathrm{tr} \Lambda^k(\psi - \mathrm{id}_F), \qquad (3.3.5)$$

where $\Lambda^k(\psi - \mathrm{id}_F)$ is the kth exterior power of $\psi - \mathrm{id}_F$. If $\psi - \mathrm{id}_F$ is of finite rank, then the sum is finite, and $\det_{\mathfrak{F}}(\psi)$ is the usual determinant.[19]

The central $\mathbb{C}^\times$-extension $\mathrm{Aut}^\flat(F)$ of $\mathrm{Aut}(F)$ is trivialized by the section $\psi \mapsto \det(\psi)$. The Fredholm determinant is multiplicative and invariant with respect to the adjoint action of $\mathrm{Aut}(F)$. We get a trivialization $\psi \mapsto \tau^{\mathrm{an}}(\psi) := \det_{\mathfrak{F}}^{-1}(\psi) \det(\psi)$ of the $\mathbb{C}^\times$-extension $G^{\mathrm{tr}}(F)^\flat$ which is invariant for the adjoint $\mathrm{Aut}(F)$-action.

Since for $\psi \in G^{\mathrm{fin}}(F)$ one has $\tau^{\mathrm{an}}(\psi) = \tau_{\mathrm{id}_F, \psi} \in \lambda_\psi$, our τ^{an} can be viewed as a trivialization of the extension $\mathrm{Aut}^\flat_{/\mathcal{I}^{\mathrm{fin}}}(F)$ over the normal subgroup $G^{\mathrm{tr}}(F)/G^{\mathrm{fin}}(F)$. It is invariant with respect to the adjoint action of $\mathrm{Aut}_{/\mathcal{I}^{\mathrm{fin}}}(F)$. Thus, by (3.3.4), τ^{an} defines a descent of $\mathrm{Aut}^\flat_{/\mathcal{I}^{\mathrm{fin}}}(F)$ to an $\mathcal{L}$-extension $\mathrm{Aut}^\flat_{/\mathcal{I}^{\mathrm{tr}}}(F)$ of $\mathrm{Aut}_{/\mathcal{I}^{\mathrm{tr}}}(F)$.

More generally, for every $F, F' \in \mathcal{F}$, the set $\mathrm{Isom}_{/\mathcal{I}^{\mathrm{fin}}}(F, F')$ is a $(G^{\mathrm{tr}}(F')/G^{\mathrm{fin}}(F'), G^{\mathrm{tr}}(F)/G^{\mathrm{fin}}(F))$-bitorsor over $\mathrm{Isom}_{/\mathcal{I}^{\mathrm{tr}}}(F, F')$, and we define the $\mathcal{L}$-extension $\mathrm{Isom}_{/\mathcal{I}^{\mathrm{tr}}}(F, F')^\flat$ as the quotient of $\mathrm{Isom}^\flat_{/\mathcal{I}^{\mathrm{fin}}}(F, F')$ by the τ^{an}-lifting of either $G^{\mathrm{tr}}(F)/G^{\mathrm{fin}}(F)$- or $G^{\mathrm{tr}}(F')/G^{\mathrm{fin}}(F')$-action. $\qquad \square$

Remark. The above constructions are compatible with constraint (3.3.3).

3.4 For a topological space X whose topology has countable base, a *Fréchet sheaf* on X means a sheaf of Fréchet vector spaces. A *Fréchet algebra* $\mathcal{A}$ is a sheaf of topological algebras which is a Fréchet sheaf; a *Fréchet $\mathcal{A}$-module* is a Fréchet sheaf equipped with a continuous (left) $\mathcal{A}$-action.

The problem of finding Fréchet structures on a given $\mathcal{A}$-module M is delicate. Here is a simple uniqueness assertion. Suppose that M satisfies the next condition:

---

[19] i.e., $\det(\psi|_{F_0})$ where F_0 is any finite-dimensional subspace containing the image of $\psi - \mathrm{id}_F$.

those open subsets U of X that $M(U)$ is a finitely generated $\mathcal{A}(U)$-module form a base of the topology of X.

Lemma. *Every morphism of $\mathcal{A}$-modules $\phi : M \to N$ is continuous with respect to any Fréchet structures on M, N. Thus M admits at most one Fréchet structure.*

Proof It suffices to check that the maps $\phi_U : M(U) \to N(U)$ are continuous for all U as above. Thus there is a surjective $\mathcal{A}(U)$-linear map $\pi_U : \mathcal{A}(U)^n \twoheadrightarrow M(U)$. The maps π_U and $\phi_U \pi_U$ are evidently continuous. Since $\mathcal{A}(U)^n / \mathrm{Ker}(\pi_U) \to M(U)$ is a continuous algebraic isomorphism of Fréchet spaces, it is a homeomorphism, and we are done. $\square$

Example. Every locally free $\mathcal{A}$-module of finite rank is a Fréchet $\mathcal{A}$-module.

From now on our X is a complex curve. The two basic examples of Fréchet algebras on X are $\mathcal{O}_X$ and $\mathcal{D}_X^\infty$. For an open $U \subset X$ the topology on the space of holomorphic functions $\mathcal{O}(U)$ is that of uniform convergence on compact subsets of U. If t is a coordinate function on U, then the topology on $\mathcal{D}^\infty(U)$ is given by a collection of semi-norms $||\Sigma a_n \partial_t^n||_{K\epsilon} := \max_{x \in K} \Sigma |a_n(x)| \epsilon^{-n} n!$; here K is any compact subset of U and ϵ is any small positive real number. Equivalently, one can use (3.2.1): then $(\mathcal{O} \boxtimes \omega)(U \times U)$ is a closed subspace of $(\mathcal{O} \boxtimes \omega)(U \times U \setminus \Delta(U))$, and the topology on $\mathcal{D}^\infty(U)$ is the quotient one.

Proposition. *Any holonomic $\mathcal{D}^\infty$-module N on X admits a unique structure of a Fréchet $\mathcal{D}_X^\infty$-module.*

Proof Uniqueness: As follows easily from Exercise (i) in 3.2, N satisfies the condition of the previous lemma. Existence: The problem is local, so it suffices to define some Fréchet structure compatible with the $\mathcal{D}^\infty(U)$-action on $N(U)$, where U is a disc and N is smooth outside the center 0 of U. Then $N = M^\infty$ where M is a $\mathcal{D}$-module with regular singularities; we can assume that M is indecomposable. If $M \simeq M_{s,n}$ (see Exercise (i) in 3.2), then, by loc. cit., $M^\infty(U) \xrightarrow{\sim} M^\infty(U^o) \simeq \mathcal{O}(U^o)^n$, and we equip it with the topology of $\mathcal{O}(U^o)^n$. Otherwise $M^\infty(U)$ is a subquotient of some $M_{0,n}^\infty(U)$, and we equip it with the corresponding Fréchet structure. $\square$

Question. Can one find a less ad hoc proof (that would not use (3.2.2))? Is the assertion of the proposition remains true for all perfect $\mathcal{D}^\infty$-modules (or perfect $\mathcal{D}^\infty$-complexes) on X of any dimension?[20]

3.5 Let $K \subset X$ be a compact subset which does not contain a connected component of X; denote by j_K the embedding $X \setminus K \hookrightarrow X$. Let E be a Fréchet $\mathcal{O}_X$-module. Suppose that for some open neighborhood U of K, $E|_{U \setminus K}$ is a locally free $\mathcal{O}_{U \setminus K}$-module of finite rank. A K-*lattice* in E is an $\mathcal{O}_X$-module L, which is locally free on U, together with an $\mathcal{O}_X$-linear morphism $L \to E$ such that $L|_{X \setminus K} \xrightarrow{\sim} E|_{X \setminus K}$. Then L is a Fréchet $\mathcal{O}_X$-module, and $L \to E$ is a continuous morphism. Set $\Gamma(E/L) := H^0 R\Gamma(X, \mathcal{C}one(L \to E)) = H^0 R\Gamma(U, \mathcal{C}one(L \to E))$.

Shrinking U if needed, we can assume that the closure $\bar{U}$ of U is compact with smooth boundary $\partial \bar{U}$. We denote by ∂U a contour in $U \setminus K$ homologous to the boundary of $\bar{U}$ in $\bar{U} \setminus K$.

---

[20] For a perfect $\mathcal{D}^\infty$-complex N, [PSch] define a natural ind-Banach structure on its complex of solutions $R\mathcal{H}om_{\mathcal{D}^\infty}(N, \mathcal{O}_X)$. It is not clear if this result helps to see the topology on N.

Remarks. (i) For all our needs it suffices to consider the situation when U is a disjoint union of several discs.

(ii) If $\text{Int}(K) \neq \emptyset$, then the morphism $L \to E$ need not be injective. The map $L(U) \to E(U)$ is injective though, i.e., $H^{-1}R\Gamma(U, \mathcal{C}one(L \to E)) = 0$.

Example. If L is any locally free $\mathcal{O}_X$-module of finite rank, then L is a K-lattice in $j_K.L := j_K.(L|_{X \setminus K})$.

Proposition. *(i) $L(U)$ is a direct summand of the Fréchet space $E(U)$.*
(ii) If $H^1(U, L) = 0$ (which happens, e.g., if none of the connected components of U is compact), then $E(U)/L(U) \xrightarrow{\sim} \Gamma(E/L)$.
(iii) Let (E', L') be a similar pair, and $\phi : E' \to E$ be any morphism of Fréchet sheaves. Then the map $E'(U) \to E(U)$, viewed as a morphism in $\mathcal{F}/\mathcal{I}^{tr}$, sends the subobject $L'(U)$ to $L(U)$.

Proof (i) We want to construct a left inverse to the morphism of Fréchet spaces $L(U) \to E(U)$. It suffices to define a left inverse to the composition $L(U) \to E(U) \to E(U \setminus K) = L(U \setminus K)$, i.e., to the restriction map $L(U) \to L(U \setminus K)$.

We can assume that U is connected and non compact.[21] Then one can find a Cauchy kernel on $U \times U$, which is a section κ of $L \boxtimes \omega L^*(\Delta)$ with residue at the diagonal equal to $-\text{id}_E$. The promised left inverse is $f \mapsto \kappa(f)$, $\kappa(f)(x) = \int_{\partial U} \kappa(x, y) f(y)$.

(ii) Follows from the exact cohomology sequence.

(iii) We want to check that the composition $L'(U) \to E'(U) \to E(U) \to E(U)/L(U)$ is nuclear. We can assume that U is connected and non compact. Choose an open $V \supset K$ whose closure $\bar{V}$ is compact and lies in U. Our map equals the composition $L'(U) \to L'(V) \to E'(V) \to E(V) \to E(V)/L(V) \xleftarrow{\sim} E(U)/L(U)$ (for $\xleftarrow{\sim}$, see (ii)). We are done, since the first arrow is nuclear (see e.g. [Gr1]). $\square$

Corollary. *(a) The isomorphism from (ii) yields a natural Fréchet space structure on $\Gamma(E/L)$, which does not depend on the auxiliary choice of U.*
(b) Every ϕ as in (iii) yields naturally a morphism $\phi_{E'/L', E/L} : \Gamma(E'/L') \to \Gamma(E/L)$ in $\mathcal{F}/\mathcal{I}^{tr}$. In particular, the spaces $\Gamma(E/L)$ for all K-lattices L in E are canonically identified as objects of $\mathcal{F}/\mathcal{I}^{tr}$. $\square$

Proof (a) follows since, for $U' \subset U$ as in (ii), the restriction map $E(U)/L(U) \to E(U')/L(U')$ is a continuous algebraic isomorphism, hence a homeomorphism, and U's form a directed set. The first assertion in (b) follows from (iii); for the second one, consider $\phi = \text{id}_E$. $\square$

3.6 The set $\Lambda_K(E)$ of K-lattices in E has natural structure of an $\mathcal{L}$-groupoid (see 2.2). Namely, by the corollary in 3.5, for every $L, L' \in \Lambda_K(E)$ one has a canonical identification $\text{id}_{E/L, E/L'} : \Gamma(E/L) \xrightarrow{\sim} \Gamma(E/L')$ in $\mathcal{F}/\mathcal{I}^{tr}$. We set

$$\lambda_K(L/L') := \lambda_{\text{id}_{E/L, E/L'}}. \qquad (3.6.1)$$

The composition

$$\lambda_K(L/L') \otimes \lambda_K(L'/L'') \xrightarrow{\sim} \lambda_K(L/L'') \qquad (3.6.2)$$

comes from (3.3.2). $\square$

---

[21] If U is compact, then $L(U)$ is finite dimensional, and the assertion follows from the Hahn-Banach theorem.

Suppose $\Lambda_K(E)$ is non-empty. We get an $\mathcal{L}$-torsor $\mathcal{D}et_K(E) := \mathrm{Hom}_{\mathcal{L}}(\Lambda_K(E), \mathcal{L})$ of *determinant theories on E at K*. For E_1, E_2 we get an $\mathcal{L}$-torsor $\mathcal{D}et_K(E_1/E_2) := \mathcal{D}et_K(E_1) \otimes \mathcal{D}et_K(E_2)^{\otimes -1}$ of *relative determinant theories on E_1/E_2 at K*.

If $K = \sqcup K_\alpha$, then a K-lattice L amounts to a collection of K_α-lattices L_α, and one has an evident canonical isomorphism

$$\otimes \lambda_{K_\alpha}(L_\alpha/L'_\alpha) \xrightarrow{\sim} \lambda_K(L/L') \tag{3.6.3}$$

compatible with composition isomorphisms (3.6.2). Thus one has a canonical identification $\otimes \mathcal{D}et_{K_\alpha}(E) \xrightarrow{\sim} \mathcal{D}et_K(E)$, $(\otimes \lambda_\alpha)(L) = \otimes \lambda_\alpha(L_\alpha)$.

Below we fix a neighborhood U of K as in 3.5; we assume that it has no compact components, so for every K-lattice L the $\mathcal{O}(U)$-module $L(U)$ is free.

$\Lambda_K(E)$ carries a natural topology of compact convergence on $U \setminus K$. Namely, to define a neighborhood of L we pick an $\mathcal{O}(U)$-base $\{\ell_i\}$ of $L(U)$, a compact $C \subset U \setminus K$ and a number $\epsilon > 0$. The neighborhood is formed by those L' which admit a base $\{\ell'_i\}$ of $L'(U)$, $\ell'_i = \Sigma a_{ij}\ell_i$, with (a_{ij}) ϵ-close to the unit matrix on C.

The $\mathcal{L}$-groupoid structure on $\Lambda_K(E)$ is continuous, i.e., λ_K form naturally a line bundle on $\Lambda_K(E) \times \Lambda_K(E)$, and the composition is continuous (cf. [PS] 6.3, 7.7). Namely, if P is a closed linear subspace of $E(U)$, then the subset $\Lambda_K^P(E)$ of those K-lattices L that $L(U)$ is complementary to P, i.e., $P \xrightarrow{\sim} \Gamma(E/L)$, is open in $\Lambda_K(E)$. For $L, L' \in \Lambda_K^P(E)$ the morphism $\mathrm{id}_{E/L, E/L'}$ in $\mathcal{F}/\mathcal{I}^{\mathrm{tr}}$ is represented by id_P, hence the restriction of λ_K to $\Lambda_K^P(E) \times \Lambda_K^P(E)$ is trivialized by the section $\delta^P := \det(\mathrm{id}_P)$. The topology on λ_K is uniquely determined by the condition that all these local trivializations are continuous, and the composition is continuous.

Remarks. (i) If K' is a compact such that $K \subset K' \subset U$, then every K-lattice L in E is a K'-lattice, and $\lambda_{K'}(L/L') = \lambda_K(L/L')$. Hence $\mathcal{D}et_K(E) = \mathcal{D}et_{K'}(E)$.

(ii) By (iii) of the proposition in 3.5, the subobjects $L(U)$, $L'(U)$ of $E(U)$ coincide if viewed in $\mathcal{F}/\mathcal{I}^{\mathrm{tr}}$. Let $\phi_U : L(U) \xrightarrow{\sim} L'(U)$ be the corresponding isomorphism in $\mathcal{F}/\mathcal{I}^{\mathrm{tr}}$. Then there is a canonical identification

$$\lambda_K(L/L') \xrightarrow{\sim} \lambda_{\phi_U}^{\otimes -1} \tag{3.6.4}$$

compatible with the composition maps. Indeed, by (3.3.3) and Remark at the end of 3.3 applied to the filtrations $L^{(\prime)} \subset E(U)$, one has $\lambda_{\mathrm{id}_{E/L, E/L'}} \otimes \lambda_{\phi_U} \xrightarrow{\sim} \lambda_{\mathrm{id}_{E(U)}} = \mathbb{C}$.

(iii) Every K-lattice L in E can be viewed as a K-lattice in $j_K.E := j_K.j_K^* E$ (via $L \to E \to j_K.E$). Then (3.6.4) shows that $\lambda_K(L/L')$ does not depend on whether we consider L, L' as K-lattices in E or in $j_K.E$. Thus $\mathcal{D}et_K(E) = \mathcal{D}et_K(j_K.E)$.

(iv) Let S be an analytic space, L_S be an S-family of K-lattices in E.[22] Then the pull-back of λ_K to $S \times S$ is naturally a holomorphic line bundle, so that the pull-back to S of any local trivialization δ^P is holomorphic.

(v) If L, L' are meromorphically equivalent, then $\lambda_K(L/L')$ coincides with the relative determinant line from 2.3 (where P is a finite subset in K such that L equals L' off P). Indeed, in view of (3.6.2), it suffices to identify the lines in case $L \supset L'$, where the identification comes from $L(U)/L'(U) \xrightarrow{\sim} \Gamma(U, L/L')$. If L, L' vary holomorphically as in (iv), then this identification is holomorphic.

(vi) For every $f \in \mathcal{O}^\times(U \setminus K)$ the lines $\lambda_K(fL/L)$ for all $L \in \Lambda_K(j_K.E)$ are canonically identified. Namely, one defines the isomorphism $\lambda_K(fL'/L') \xrightarrow{\sim} \lambda_K(fL/L)$ as $\lambda_K(fL'/L') \xrightarrow{\sim} \lambda_K(fL'/fL) \otimes \lambda_K(fL/L) \otimes \lambda_K(L'/L)^{\otimes -1} \xrightarrow{\sim} \lambda_K(fL/L)$

---

[22]I.e., L_S is an $\mathcal{O}_{X_S}$-module together with a morphism $L \to E_S$ such that L_S is locally free on U_S, and $L \to E_S$ is an isomorphism off K_S.

where the first arrow is inverse to the composition, and the second comes from the multiplication by f identification $\lambda_K(L'/L) \overset{\sim}{\to} \lambda_K(fL'/fL)$.

(vii) Let $g \in \mathcal{O}^\times(U)$ be an invertible function. The multiplication by g automorphism of $E|_U$ preserves every K-lattice. Let $g(L/L') \in \mathbb{C}^\times$ be the corresponding automorphism of $\lambda_K(L/L')$.

Example. Suppose that $\lambda_K(L/L')$ has degree 0. Choose a Fréchet isomorphism $\alpha : \Gamma(E/L) \overset{\sim}{\to} \Gamma(E/L')$ which represents $\mathrm{id}_{E/L, E/L'}$. Then $g(L/L')$ is the Fredholm determinant $\det_{\mathfrak{F}}(\alpha^{-1} g_{E/L'} \alpha g_{E/L}^{-1})$, where $g_{E/L}$, $g_{E/L'}$ are multiplication by g automorphisms of $\Gamma(E/L)$, $\Gamma(E/L')$.

Here is a formula for $g(L/L')$ (cf. [PS] 6.7, [SW] 3.6). Consider the line bundle $\det E|_{U \setminus K}$. Then $L \in \Lambda_K(j_K . E)$ yields a K-lattice $\det L|_U \in \Lambda_K(j_K . \det E|_{U \setminus K})$. We can assume that U has no compact components. Then the line bundle $\det L|_U$ is trivial; let θ_L be any its trivialization. For two lattices L, L' we get a function $\theta_L/\theta_{L'} \in \mathcal{O}^\times(U \setminus K)$. Consider the analytic symbol $\{g, \theta_{L'}/\theta_L\} \in H^1(U \setminus K, \mathbb{C}^\times)$. Then (see 3.5 for the notation ∂U)

$$g(L/L') = \{g, \theta_{L'}/\theta_L\}(\partial U). \tag{3.6.5}$$

To check (3.6.5), consider first the case when L, L' are meromorphically equivalent. Then $\theta_{L'}/\theta_L$ is a meromorphic function on U, and both parts of (3.6.5) evidently coincide with $g(\mathrm{div}(\theta_{L'}/\theta_L))$. The general case follows since for any L, L' one can find (possibly enlarging K, as in (ii)) an L'' meromorphically equivalent to L which is arbitrary close to L', and both parts of (3.6.5) depend continuously on L'.

3.7 Let N be a holonomic $\mathcal{D}^\infty$-module on X. By 3.4, it carries a canonical Fréchet structure. For K as in 3.5, 3.6, let us define a relative determinant theory $\mu_K^\nabla = \mu(N/\omega N)_K^\nabla \in \mathcal{D}et_K(N/\omega N)$ (cf. 2.4).

If L, L_ω are K-lattices in N, ωN such that $\nabla(L) \subset L_\omega$, then ∇ yields a morphism of sheaves $N/L \to \omega N/L_\omega$, and we denote by $\mathcal{C}(L, L_\omega)_{N,K}$ its cone.

Example. Let $M^\infty \overset{\sim}{\to} N$ be a $\mathcal{D}$-structure on N (see 3.2), and P be any finite subset of K such that M is smooth on $K \setminus P$. Every P-lattices L, L_ω in M, ωM can be viewed as K-lattices in N, ωN. If $\nabla(L) \subset L_\omega$, then we get an evident morphism of complexes of sheaves (see 2.4)

$$\mathcal{C}(L, L_\omega)_{M,P} \to \mathcal{C}(L, L_\omega)_{N,K}. \tag{3.7.1}$$

Proposition. *(3.7.1) is a quasi-isomorphism.*

Proof Our complexes are supported on a finite set P, so it suffices to check that $R\Gamma(X, (3.7.1))$ is a quasi-isomorphism. Notice that $dR(L, L_\omega) := \mathcal{C}one(L \overset{\nabla}{\to} L_\omega)$ is a subcomplex of both $dR(M)$ and $dR(N)$, and $\mathcal{C}(L, L_\omega)_{M,P} = dR(M)/dR(L, L_\omega)$, $\mathcal{C}(L, L_\omega)_{N,T'} = dR(N)/dR(L, L_\omega)$. Since $dR(M) \to dR(N)$ is a quasi-isomorphism (see 3.2), we are done. $\square$

By the corollary in 3.5, for K-lattices L, L_ω in N, ωN one has a $\mathcal{F}/\mathcal{I}^{\mathrm{tr}}$-morphism

$$\nabla_{N/L, \omega N/L_\omega} : \Gamma(N/L) \to \Gamma(\omega N/L_\omega). \tag{3.7.2}$$

Corollary. *(3.7.2) is a Fredholm map.*

Proof By loc. cit., the validity of the assertion does not depend on the choice of L, L_ω. So we can assume to be in the situation of Example, and we are done by 2.4 and the proposition. $\square$

We define $\mu_K^\nabla \in \mathcal{D}et_K(N/\omega N)$ as a relative determinant theory such that for any $L \in \Lambda_K(N)$, $L_\omega \in \Lambda_K(\omega N)$ one has

$$\mu_K^\nabla(L/L_\omega) := \lambda_{\nabla_{N/L,\omega N/L_\omega}}, \tag{3.7.3}$$

and the structure isomorphisms $\lambda_K(L'/L)\otimes\mu_K^\nabla(L/L_\omega)\otimes\lambda_K(L_\omega/L_\omega') \xrightarrow{\sim} \mu_K^\nabla(L'/L_\omega')$ are compositions (3.3.2) for $\mathrm{id}_{N/L',N/L}\nabla_{N/L,\omega N/L_\omega}\mathrm{id}_{\omega N/L_\omega,\omega N/L_\omega'} = \nabla_{N/L',\omega N/L_\omega'}$.

3.8 Let ν be any invertible holomorphic 1-form defined on $U \setminus K$, where U as in 3.5. The multiplication by ν isomorphism $j_K.N|_U \xrightarrow{\sim} j_K.\omega N|_U$ yields an identification of the $\mathcal{L}$-groupoids $\Lambda_K(j_K.N) \xrightarrow{\sim} \Lambda_K(j_K.\omega N)$, hence a relative determinant theory (see Remark (iii) in 3.6)

$$\mu_K^\nu = \mu(N/\omega N)_K^\nu \in \mathcal{D}et_K(j_K.N/j_K.\omega N) = \mathcal{D}et_K(N/\omega N). \tag{3.8.1}$$

Set

$$\mathcal{E}_{\mathrm{dR}}(N)_{(K,\nu)} := \mu_K^\nabla \otimes (\mu_K^\nu)^{\otimes -1} \in \mathcal{L}. \tag{3.8.2}$$

Thus for every $L \in \Lambda_K(j_K.N)$ one has a canonical isomorphism

$$\mathcal{E}_{\mathrm{dR}}(N)_{(K,\nu)} \xrightarrow{\sim} \mu_K^\nabla(L/\nu L); \tag{3.8.3}$$

for two lattices L, L' the corresponding identification $\mu_K^\nabla(L/\nu L) \xrightarrow{\sim} \mu_K^\nabla(L'/\nu L')$ is

$$\mu_K^\nabla(L/\nu L) \xleftarrow{\sim} \lambda_K(L/L') \otimes \mu_K^\nabla(L'/\nu L') \otimes \lambda_K(\nu L/\nu L')^{\otimes -1} \xrightarrow{\sim} \mu_K^\nabla(L'/\nu L'), \tag{3.8.4}$$

where the first arrow is the composition, the second one comes from the multiplication by ν identification $\lambda_K(L/L') \xrightarrow{\sim} \lambda_K(\nu L/\nu L')$.

The construction does not depend on the auxiliary choice of U. When ν varies holomorphically, $\mathcal{E}_{\mathrm{dR}}(N)_{(K,\nu)}$ form a holomorphic line bundle on the parameter space (by (3.8.2) and Remark (vi) in 3.6). If $K = \sqcup K_\alpha$, then (3.6.3) yields a factorization (here ν_α are the restrictions of ν to neighborhoods of K_α)

$$\otimes \mathcal{E}_{\mathrm{dR}}(N)_{(K_\alpha,\nu_\alpha)} \xrightarrow{\sim} \mathcal{E}_{\mathrm{dR}}(N)_{(K,\nu)}. \tag{3.8.5}$$

3.9 The above constructions are compatible with those from 2.5. Precisely, let M, P be as in Example in 3.7, and suppose that ν is meromorphic on U with $D := \mathrm{div}(\nu)$ supported on P. Then $\Lambda_P(M) \subset \Lambda_K(N)$, $\Lambda_P(M(\infty P)) \subset \Lambda_K(j_K.N)$ (see loc. cit.). These embeddings are naturally compatible with the $\mathcal{L}$-groupoid structures, so

$$\mathcal{D}et_P(M) = \mathcal{D}et_K(M), \quad \mathcal{D}et_P(M/\omega M) = \mathcal{D}et_K(N/\omega N). \tag{3.9.1}$$

By the proposition in 3.7, (3.7.1) provides an identification

$$\mu_P^\nabla(M/\omega M) \xrightarrow{\sim} \mu_K^\nabla(N/\omega N). \tag{3.9.2}$$

Joint with an evident isomorphism $\mu_P^\nu(M/\omega M) \xrightarrow{\sim} \mu_K^\nu(N/\omega N)$, it yields

$$\mathcal{E}_{\mathrm{dR}}(M)_{(D,c,\nu_P)} \xrightarrow{\sim} \mathcal{E}_{\mathrm{dR}}(N)_{(K,\nu)}. \tag{3.9.3}$$

If ν varies holomorphically, then (3.9.3) is holomorphic.

3.10   Proof of the theorem in 3.1. By 3.2, $\phi : B(M) \overset{\sim}{\to} B(M')$ amounts to an isomorphism of $\mathcal{D}^\infty$-modules $M^\infty \overset{\sim}{\to} M'^\infty$. Threfore we can view M and M' as two $\mathcal{D}$-structures on a holonomic $\mathcal{D}^\infty$-module N. Choose the multiplicity of T to be compatible with both M and M' (see 2.1). We want to define an isomorphism $\phi^\epsilon : \mathcal{E}_{\mathrm{dR}}(M) \overset{\sim}{\to} \mathcal{E}_{\mathrm{dR}}(M')$ in $\mathcal{L}_{\mathrm{dR}}^\Phi(X, T)$.

Let S be an analytic space, $(D, c, \nu_P) \in \mathfrak{D}^\diamond(S)$ (see 1.1). We work locally on S, so we have $T^c \subset T$. Choose a compact K, its open neighborhood U, and a meromorphic ν on U_S such that K, U satisfy the conditions from 3.5, $P \subset K_S$, $D = -\operatorname{div}(\nu)$, and $\nu_P = \nu|_P$. We define ϕ^ϵ at (D, c, ν_P) as the composition $\mathcal{E}_{\mathrm{dR}}(M)_{(D,c,\nu_P)} \overset{\sim}{\to} \mathcal{E}_{\mathrm{dR}}(N)_{(K,\nu)} \overset{\sim}{\leftarrow} \mathcal{E}_{\mathrm{dR}}(M')_{(D,c,\nu_P)}$; here $\overset{\sim}{\to}$ are (3.9.3). Equivalently, choose $L \in \Lambda_P(M)$, $L' \in \Lambda_P(M')$; then ϕ^ϵ is the composition $\mathcal{E}_{\mathrm{dR}}(M)_{(D,c,\nu_P)} \overset{\sim}{\to} \mu_K^\nabla(L/\nu L) \overset{\sim}{\to} \mu_K^\nabla(L'/\nu L') \overset{\sim}{\leftarrow} \mathcal{E}_{\mathrm{dR}}(M')_{(D,c,\nu_P)}$, the first and the last arrows are compositions of (3.9.2) and (2.5.6), the middle one is (3.8.4).

The last description shows that ϕ^ϵ does not depend on the auxiliary choice of ν. Indeed, ν is defined up to multiplication by an invertible function g on U which equals 1 on P. By (3.8.4), replacing ν by $g\nu$ multiplies the isomorphism $\mu_K^\nabla(L/\nu L) \overset{\sim}{\to} \mu_K^\nabla(L'/\nu L')$ by $g(L/L')$ (see Remark (vii) in 3.6). Since T is compatible with M and M', the 1-form $d\log(\theta_{L'}/\theta_L)$ (see loc. cit.) has pole at P of order $\leq$ the multiplicity of P, so (3.6.5) implies that $g(L/L') = 1$.

The construction is compatible with factorization, so we have defined ϕ^ϵ as isomorphism in $\mathcal{L}_{\mathcal{O}}^\Phi(X, T)$. One has $M|_{X \setminus T} = M'|_{X \setminus T}$, and $\phi^\epsilon|_{X \setminus T}$ is the corresponding evident identification. By the corollary in 1.12, this implies that ϕ^ϵ is horizontal, i.e., it is an isomorphism in $\mathcal{L}_{\mathrm{dR}}^\Phi(X, T)$.

The compatibility of ϕ^ϵ with constraints from 2.8 is evident. Finally, the commutativity of (3.1.1) follows from the next proposition:

3.11   Proposition. *Suppose that X is compact, N is a holonomic $\mathcal{D}^\infty$-module smooth on $X \setminus K$, and ν is a holomorphic invertible 1-form on $X \setminus K$. Then there is a canonical isomorphism*

$$\eta_{dR} : \mathcal{E}_{dR}(N)_{(K,\nu)} \overset{\sim}{\to} \det R\Gamma_{dR}(X, N). \tag{3.11.1}$$

If ν is meromorphic and M is a $\mathcal{D}$-structure on N, then the next diagram of isomorphisms commutes (see 1.4 and 2.7 for the left column, the top arrow is (3.9.2)):

$$\begin{array}{ccc} \mathcal{E}_{dR}(M)_\nu & \overset{\sim}{\to} & \mathcal{E}_{dR}(N)_{(K,\nu)} \\ \eta_{dR} \downarrow & & \eta_{dR} \downarrow \\ \det H_{dR}^\cdot(X, M) & \overset{\sim}{\to} & \det H_{dR}^\cdot(X, N). \end{array} \tag{3.11.2}$$

Proof (cf. 2.7). For $L \in \Lambda_K(j_K.N)$ set $\lambda(L) := \det R\Gamma(X, L)$. Then λ is a determinant theory on $j_K.N$ at K in an evident way. Replacing N by ωN, we get $\lambda_\omega \in \mathcal{D}et_K(j_K.\omega N)$, hence $\lambda \otimes \lambda_\omega^{\otimes -1} \in \mathcal{D}et_K(j_K.N/j_K.\omega N)$. One has an isomorphism

$$\mu_K^\nu \overset{\sim}{\to} \lambda \otimes \lambda_\omega^{\otimes -1}, \tag{3.11.3}$$

namely, $\mu_K^\nu(L/L_\omega) := \lambda_K(\nu L/L_\omega) = \lambda_\omega(\nu L) \otimes \lambda_\omega(L_\omega)^{\otimes -1} \overset{\sim}{\to} \lambda(L) \otimes \lambda_\omega(L_\omega)^{\otimes -1} = (\lambda \otimes \lambda_\omega^{\otimes -1})(L/L_\omega)$ where $\overset{\sim}{\to}$ comes from the isomorphism $\nu^{-1} : \nu L \overset{\sim}{\to} L$.

For K-lattices L in N, L_ω in N_ω such that $\nabla(L) \subset L_\omega$, set $dR(L, L_\omega) := \mathcal{C}one(L \overset{\nabla}{\to} L_\omega)$. Since $dR(N)/dR(L, L_\omega) = \mathcal{C}(L, L_\omega)$, our $dR(N)$ carries a 3-step filtration with successive quotients L_ω, L, $C(L, L_\omega)$. Applying $\det R\Gamma$, we get an isomorphism

$$\det R\Gamma(X, C(L, L_\omega)) \otimes \lambda(L)^{\otimes -1} \otimes \lambda(L_\omega) \xrightarrow{\sim} \det R\Gamma_{\mathrm{dR}}(X, N). \qquad (3.11.4)$$

To get η_{dR}, we combine (3.11.4) with (3.11.3) (and (3.7.3)). The construction does not depend on the auxiliary choice of L, L_ω. $\qquad\square$

4 The Betti ε-line

We present a construction from [B] in a format adapted for the current subject. *In 4.2–4.5 X is considered as a mere real-analytic surface.*

4.1   Let $\mathcal{L}$ be any Picard groupoid. For a (non-unital) Boolean algebra[23] $\mathcal{C}$, an $\mathcal{L}$-*valued measure* λ on $\mathcal{C}$ is a rule that assigns to every $S \in \mathcal{C}$ an object $\lambda(S) \in \mathcal{L}$, and to every finite collection $\{S_\alpha\}$ of *pairwise disjoint* elements of $\mathcal{C}$ an identification $\otimes \lambda(S_\alpha) \xrightarrow{\sim} \lambda(\cup S_\alpha)$ (referred to as *integration*); the latter should satisfy an evident transitivity property. Such λ form naturally a Picard groupoid $\mathcal{M}(\mathcal{C}, \mathcal{L})$.

Remarks. (i) For an abelian group A denote by $\mathcal{M}(\mathcal{C}, A)$ the group of A-valued measures on $\mathcal{C}$. Then $\pi_1(\mathcal{M}(\mathcal{C}, \mathcal{L})) = \mathcal{M}(\mathcal{C}, \pi_1(\mathcal{L}))$, and one has a map $\pi_0(\mathcal{M}(\mathcal{C}, \mathcal{L})) \to \mathcal{M}(\mathcal{C}, \pi_0(\mathcal{L}))$ which assigns to $[\lambda]$ a $\pi_0(\mathcal{L})$-valued measure $|\lambda|$, $|\lambda|(S) := [\lambda(S)]$ (see 1.1 for the notation).

(ii) Let $\mathcal{I} \subset \mathcal{C}$ be an ideal. Then $\mathcal{M}(\mathcal{C}/\mathcal{I}, \mathcal{L})$ identifies naturally with the Picard groupoid of pairs (λ, τ) where $\lambda \in \mathcal{M}(\mathcal{C}, \mathcal{L})$ and τ is a trivialization of its restriction $\lambda|_\mathcal{I}$ to $\mathcal{I}$, i.e., an isomorphism $1_{\mathcal{M}(\mathcal{I}, \mathcal{L})} \xrightarrow{\sim} \lambda|_\mathcal{I}$ in $\mathcal{M}(\mathcal{I}, \mathcal{L})$.

(iii) Suppose $\mathcal{C}$ is finite, i.e., $\mathcal{C} = (\mathbb{Z}/2)^T :=$ the Boolean algebra of all subsets of a finite set T. Then an $\mathcal{L}$-valued measure λ on $\mathcal{C}$ is the same as a collection of objects $\lambda_t = \lambda(\{t\})$, $t \in T$. Thus $\mathcal{M}(T, \mathcal{L}) := \mathcal{M}(\mathbb{Z}/2^T, \mathcal{L}) \xrightarrow{\sim} \mathcal{L}^T$.

4.2   For an open $U \subset X$ we denote by $\mathcal{C}(U)$ the (non-unital) Boolean algebra of relatively compact subanalytic subsets of U. For $U' \subset U$ one has $\mathcal{C}(U') \subset \mathcal{C}(U)$, and $U \mapsto \mathcal{M}(\mathcal{C}(U), \mathcal{L})$ is a sheaf of Picard groupoids on X.

For a commutative ring R, let $\mathcal{L}_R$ be the Picard groupoid of $\mathbb{Z}$-graded super R-lines. Its objects are pairs $L = (L, \deg(L))$ where L is an invertible R-module, $\deg(L)$ a locally constant $\mathbb{Z}$-valued function on $\mathrm{Spec} R$; the commutativity constraint is "super" one. Every perfect R-complex F yields a graded super line $\det F \in \mathcal{L}_R$. For a finite filtration $F.$ on F by perfect subcomplexes, one has a canonical isomorphism $\det F \xrightarrow{\sim} \otimes \det \mathrm{gr}_i F$; it satisfies transitivity property with respect to refinement of the filtration. For a finite collection $\{F_\alpha\}$, every linear ordering of α's yields a filtration on $\oplus F_\alpha$, and the corresponding isomorphism $\det \oplus F_\alpha \xrightarrow{\sim} \otimes \det F_\alpha$ is independent of the ordering; thus $\det$ is a symmetric monoidal functor.

Let $F = F_U$ be a perfect constructible complex of R-sheaves on U. Then for every locally closed subanalytic subset $i_C : C \hookrightarrow U$ the R-complex $R\Gamma(C, Ri_C^! F)$ is perfect, so we have $\det R\Gamma(C, Ri_C^! F) \in \mathcal{L}_R$.

---

[23]Recall that a Boolean algebra is the same as a commutative $\mathbb{Z}/2$-algebra each of whose elements is idempotent; the basic Boolean operations are $x \cap y = xy$, $x \cup y = x + y + xy$; elements x, y are said to be disjoint if $x \cap y = 0$. The Boolean algebras we meet are already realized as Boolean algebras of subsets of some set.

Suppose we have a finite closed subanalytic filtration[24] $C_{\leq \cdot}$ on C (therefore $C_i := C_{\leq i} \setminus C_{\leq i-1}$ are locally closed and form a partition of C). It yields a finite filtration[25] on $R\Gamma(C, Ri_C^! F)$ with $\mathrm{gr}_i R\Gamma(C, Ri_C^! F) = R\Gamma(C_i, Ri_{C_i}^! F)$, hence an identification

$$\otimes \det R\Gamma(C_i, Ri_{C_i}^! F) \xrightarrow{\sim} \det R\Gamma(C, Ri_C^! F). \tag{4.2.1}$$

It satisfies transitivity property with respect to refinement of the filtration.

Lemma. *There is a unique (up to a unique isomorphism) pair $(\lambda(F), \iota)$, where $\lambda(F) \in \mathcal{M}(\mathcal{C}(U), \mathcal{L}_R)$, ι is a datum of isomorphisms*

$$\iota_C : \lambda(F)(C) \xrightarrow{\sim} \det R\Gamma(C, Ri_C^! F)$$

defined for any locally closed C, such that for every filtration $C_{\leq \cdot}$ on C as above, ι identifies (4.2.1) with the integration $\otimes \lambda(F)(C_i) \xrightarrow{\sim} \lambda(F)(C)$.

Proof Suppose we have a compact subanalytic subset of U equipped with a subanalytic stratification whose strata C_α are smooth and connected. The strata generate a Boolean subalgebra $\mathcal{C}(\{C_\alpha\})$ of $\mathcal{C}(U)$; call a subalgebra of such type *nice*. Every finite subset of $\mathcal{C}(U)$ lies in a nice subalgebra; in particular, the set of nice subalgebras is directed. To prove the lemma, it suffice to define the restriction of $(\lambda(F), \iota)$ to every nice $\mathcal{C}(\{C_\alpha\})$; their compatibility is automatic.

By Remark (iii) in 4.1, $\lambda(F)|_{\mathcal{C}(\{C_\alpha\})}$ is the measure defined by condition $\lambda(F)(C_\alpha) = \det R\Gamma(C_\alpha, Ri_{C_\alpha}^! F)$. For a locally closed C in $\mathcal{C}(\{C_\alpha\})$ one defines ι_C using (4.2.1) for a closed filtration on C whose layers are strata of increasing dimension; its independence of the choice of filtration follows since det is a symmetric monoidal functor in the way described above. The compatibility with (4.2.1) is checked by induction by the number of strata involved. □

Example. Suppose $C', C'' \in \mathcal{C}(U)$ are such that C', C'', and $C := C' \cup C''$ are locally closed, $C' \cap C'' = \emptyset$. By the lemma, there is a canonical isomorphism

$$\det R\Gamma(C', Ri_{C'}^! F) \otimes \det R\Gamma(C'', Ri_{C''}^! F) \xrightarrow{\sim} \det R\Gamma(C, Ri_C^! F). \tag{4.2.2}$$

If, say, C' is closed in C, then this is (4.2.1) for the filtration $C' \subset C$. To construct (4.2.2) when neither C' nor C'' are closed (e.g. $C = X$ is a torus, and C', C'' are non-closed annuli), consider a 3-step closed filtration $C' \subset \bar{C}' \subset C$, where $\bar{C}'$ is the closure of C' in C; set $P := \bar{C}' \setminus C' = \bar{C}' \cap C''$, $Q := C \setminus \bar{C}' = C'' \setminus P$. By (4.2.1), $\det R\Gamma(C', Ri_{C'}^! F) \otimes \det R\Gamma(P, Ri_P^! F) \xrightarrow{\sim} \det R\Gamma(\bar{C}', Ri_{\bar{C}'}^! F)$, $\det R\Gamma(P, Ri_P^! F) \otimes \det R\Gamma(Q, Ri_Q^! F) \xrightarrow{\sim} \det R\Gamma(C'', Ri_{C''}^! F)$, and $\det R\Gamma(\bar{C}', Ri_{\bar{C}'}^! F) \otimes \det R\Gamma(Q, Ri_Q^! F) \xrightarrow{\sim} \det R\Gamma(C, Ri_C^! F)$. Combining them, we get (4.2.2).

4.3 Let $U \subset X$ be an open subset, and $\mathcal{N} = \mathcal{N}_U \subset T_U$ be a continuous family of proper cones in the tangent bundle (so for each $x \in U$ the fiber $\mathcal{N}_x$ is a proper closed sector with non-empty interior in the tangent plane T_x). For an open $V \subset U$ we denote by $\mathcal{N}_V$ the restriction of $\mathcal{N}$ to V.

One calls $C \in \mathcal{C}(U)$ an $\mathcal{N}$-*lens* if it satisfies the next two conditions:

---

[24]I.e., each $C_{\leq i}$ is a subanalytic subset closed in C.

[25]A filtration on an object C of a derived category is an object of the corresponding filtered derived category identified with C after the forgetting of the filtration.

(a) Every point in U has a neighborhood V such that $C \cap V = C_1 \setminus C_2$ where C_1, C_2 are closed subsets of V that are invariant with respect to some family of proper cones $\mathcal{N}_V' \ni \mathcal{N}_V$.[26]

(b) There is a C^1-function f defined on a neighborhood V of the closure $\bar{C}$ such that for every $x \in V$ and a non-zero $\tau \in \mathcal{N}_x$ one has $\tau(f) > 0$.

Let $\mathcal{I}(U, \mathcal{N}) \subset \mathcal{C}(U)$ be the Boolean subalgebra generated by all $\mathcal{N}$-lenses.

Basic properties of lenses (see [B] 2.4, 2.7): (i) Every $\mathcal{N}$-lens C is locally closed, and Int C is dense in C; the intersection of two $\mathcal{N}$-lenses is an $\mathcal{N}$-lens.
(ii) Every point in U admits a base of neighborhoods formed by $\mathcal{N}$-lenses.
(iii) Suppose we have an $\mathcal{N}$-lens C and a (finite) partition $\{C_\alpha\}$ of C with $C_\alpha \in \mathcal{I}(U, \mathcal{N})$. Then there exists a finer partition $\{C_1, \ldots, C_n\}$ of C such that C_i are $\mathcal{N}$-lenses and each subset $C_{\leq i} := C_1 \cup C_2 \cup \ldots \cup C_i$ is closed in C.

Exercise. Every $C \in \mathcal{C}(U)$ that satisfies (a) lies in $\mathcal{I}(U, \mathcal{N})$.

Suppose F from 4.2 is locally constant (say, a local system of finitely generated projective R-modules). Then for any $\mathcal{N}$-lens C one has $R\Gamma(C, Ri_C^! F) = 0$ (see [B] 2.5). Let $\tau_C : 1 \overset{\sim}{\to} \lambda(F)(C)$ be the corresponding trivialization of the determinant.

Proposition. *The restriction of $\lambda(F)$ to $\mathcal{I}(U, \mathcal{N})$ admits a unique trivialization $\tau_\mathcal{N} : 1_{\mathcal{M}(\mathcal{I}(U,\mathcal{N}), \mathcal{L}_R)} \overset{\sim}{\to} \lambda(F)|_{\mathcal{I}(U,\mathcal{N})}$ such that for every $\mathcal{N}$-lens C the trivialization $\tau_{\mathcal{N}C}$ coincides with τ_C.*

Proof By (iii) above, every finite subset of $\mathcal{I}(U, \mathcal{N})$ lies in the Boolean subalgebra generated by a finite subset of pairwise disjoint $\mathcal{N}$-lenses. This implies uniqueness. To show that $\tau_\mathcal{N}$ exists, it suffices to check the next assertion: For any $\mathcal{N}$-lens C and any finite partition $\{C_\alpha\}$ of C by $\mathcal{N}$-lenses the integration $\otimes \lambda(F)(C_\alpha) \overset{\sim}{\to} \lambda(F)(C)$ identifies $\otimes \tau_{C_\alpha}$ with τ_C.

Choose $\{C_1, \ldots, C_n\}$ as in (iii) above. Since $C_{\leq}$ is a closed filtration, the integration $\otimes \lambda(F)(C_i) \overset{\sim}{\to} \lambda(F)(C)$ identifies $\otimes \tau_{C_i}$ with τ_C (see the lemma in 4.2), and for each α the integration $\otimes \lambda(F)(C_i \cap C_\alpha) \overset{\sim}{\to} \lambda(F)(C_\alpha)$ identifies $\otimes \tau_{C_i \cap C_\alpha}$ with τ_{C_α}. The partition $\{C_i\}$ is finer than $\{C_\alpha\}$, so we are done by the transitivity of integration. $\square$

4.4   Let K be a compact subset of X, W be an open subset that contains K, $U := W \setminus K$. For $\mathcal{N} = \mathcal{N}_U$ as above, let $\mathcal{C}(W, \mathcal{N})$ be the set of $C \in \mathcal{C}(W)$ that satisfy the next two conditions:

(a) For every $C' \in \mathcal{I}(U, \mathcal{N})$ one has $C \cap C' \in \mathcal{I}(U, \mathcal{N})$.

(b) One has $\text{Int}(C) \cap K = \bar{C} \cap K$.

Then $\mathcal{C}(W, \mathcal{N})$ is a Boolean subalgebra of $\mathcal{C}(W)$, and $\mathcal{I}(U, \mathcal{N})$ is an ideal in it.

Exercise. Let K' be a subset of K which is open and closed in K. Choose an open relatively compact subset V of W such that $V \cap K = \bar{V} \cap K = K'$, and $C' \in \mathcal{I}(U, \mathcal{N})$ such that $C' \supset \partial V := \bar{V} \setminus V$. Then $C := V \setminus C' \in \mathcal{C}(W, \mathcal{N})$ and $C \cap K = K'$.

Denote by $\mathcal{C}[K]$ the Boolean algebra of subsets of K which are open and closed in K. By (b), we have a morphism of Boolean algebras $\mathcal{C}(W, \mathcal{N}) \to \mathcal{C}[K]$, $C \mapsto C \cap K$. It yields an identification

$$\mathcal{C}(W, \mathcal{N}) / \mathcal{I}(U, \mathcal{N}) \overset{\sim}{\to} \mathcal{C}[K]. \qquad (4.4.1)$$

---

[26]Here $\ni$ means that $\text{Int}\, \mathcal{N}_V' \supset \mathcal{N}_V \setminus \{0\}$, and $\mathcal{N}_V'$-invariance of C_i means that every C^1-arc $\gamma : [0, 1] \to V$ such that $\gamma(0) \in C_i$ and $\frac{d}{dt}\gamma(t) \in \mathcal{N}_V' \setminus \{0\}$ for every t, lies in C_i.

Let $F = F_W$ be a perfect constructible complex of R-sheaves on W whose restriction to U is locally constant. By the proposition in 4.3, we have a trivialization $\tau_{\mathcal{N}}$ of the restriction of $\lambda(F)$ to $\mathcal{I}(U,\mathcal{N})$. By Remark (ii) in 4.1, (4.4.1), the pair $(\lambda(F)|_{\mathcal{C}(W,\mathcal{N})}, \tau_N)$ can be viewed as a measure $\mathcal{E}(F)_{\mathcal{N}} \in \mathcal{M}(\mathcal{C}[K], \mathcal{L}_R)$. If K is finite, then, by Remark (iii) in 4.1, it amounts to a collection of lines $\mathcal{E}(F)_{(b,\mathcal{N})} := \mathcal{E}(F)_{\mathcal{N}b}$, $b \in K$.

If $C \in \mathcal{C}(W,\mathcal{N})$ is locally closed, then we have identifications

$$\mathcal{E}(F)_{\mathcal{N}}(C \cap K) \overset{\sim}{\to} \lambda(F)(C) \overset{\sim}{\to} \det R\Gamma(C, Ri_C^! F). \tag{4.4.2}$$

In particular, if X is compact and $W = X$, then $X \in \mathcal{C}(X,\mathcal{N})$, and

$$\mathcal{E}(F)_{\mathcal{N}}(K) \overset{\sim}{\to} \det R\Gamma(X, F). \tag{4.4.3}$$

If K is finite, this is a product formula

$$\underset{b \in K}{\otimes} \mathcal{E}(F)_{(b,\mathcal{N})} \overset{\sim}{\to} \det R\Gamma(X, F). \tag{4.4.4}$$

4.5 Suppose we have another datum of $W' \supset U'$, $\mathcal{N}' = \mathcal{N}'_{U'}$ as above, such that $W' \subset W$, $U' \subset U$, and $\mathcal{N}' \supset \mathcal{N}_{U'}$. Then $\mathcal{I}(U',\mathcal{N}') \subset \mathcal{I}(U,\mathcal{N})$, $\mathcal{C}(W',\mathcal{N}') \subset \mathcal{C}(W,\mathcal{N})$, and (4.4.1) identifies the morphism of Boolean algebras $\mathcal{C}(W',\mathcal{N}')/\mathcal{I}(U',\mathcal{N}') \to \mathcal{C}(W,\mathcal{N})/\mathcal{I}(U,\mathcal{N})$ with a morphism $r : \mathcal{C}[K'] \to \mathcal{C}[K]$, $Q \mapsto r(Q) := Q \setminus U = Q \cap K$. Since $\tau_{\mathcal{N}'}$ equals the restriction of $\tau_{\mathcal{N}}$ to $\mathcal{I}(U',\mathcal{N}')$, one has

$$\mathcal{E}(F)_{\mathcal{N}'} = r^* \mathcal{E}(F)_{\mathcal{N}}. \tag{4.5.1}$$

Remarks. (i) Taking for W' a small neighborhood of a component K' of K, $U' = W' \cap U$, $\mathcal{N}' = \mathcal{N}|_{U'}$, we see that $\mathcal{E}(F)_{\mathcal{N}}$ has local nature with respect to K.
(ii) By (4.5.1), $\mathcal{E}(F)_{\mathcal{N}}(W' \cap K)$ depends only on the restriction of $\mathcal{N}$ to U'.

4.6 Suppose now X is a complex curve, $T \subset X$ a finite subset, F is a constructible sheaf on X which is smooth on $X \setminus T$. Let us define a constructible factorization R-line $\mathcal{E}_{\mathrm{B}}(F)$ on (X,T) (see 1.15).

Let S be an analytic space, $(D, c, \nu_P) \in \mathfrak{D}^\diamond(S)$. Let us define a local system of R-lines $\mathcal{E}(F)_{(D,c,\nu_P)}$ on S. Consider a datum $(W, K, \mathcal{N}, \nu_S)$, where W is an open subset of X, K a compact subset of W, $\mathcal{N} = \mathcal{N}_U$ is a continuous family of proper cones in the tangent bundle to $U := W \setminus K$ (viewed as a real-analytic surface, see 4.3), ν_S is an S-family of meromorphic 1-forms on W. We say that our datum is *compatible* if $P = P_{D,c} \subset K_S$, $\mathrm{div}(\nu) = -D$, $\nu|_P = \nu_P$, and the 1-forms $\mathrm{Re}(\nu_s)$ are negative on $\mathcal{N}$. As in 4.4, every compatible datum yields the R-line $\mathcal{E}(F)_{\mathcal{N}}(K)$.

Lemma. *Locally on S compatible data exist; the lines $\mathcal{E}(F)_{\mathcal{N}}(K)$ for all compatible data are naturally identified.*

Proof The existence statement is evident. Suppose that we fix an open subset W_0 of X and an S-family of meromorphic forms ν_S on W_0 such that $P \subset W_{0S}$, $\mathrm{div}(\nu_S) = D$, and $\nu_P = \nu|_P$. Let us consider compatible data with $W \subset W_0$ and the above ν_S. The identification of the lines for these data comes from 4.5. Thus our line depends only on ν_S; in fact, by Remark (i) in 4.5, on the germ of ν_S at P. If we move ν_S, it remains locally constant. Since the space of germs of ν_S is contractible, we are done. $\square$

Locally on S, we define $\mathcal{E}(F)_{(D,c,\nu_P)}$ as $\mathcal{E}(F)_{\mathcal{N}}(K)$ for a compatible datum. The factorization structure is evident. For X compact, we have, by (4.4.3), a canonical identification

$$\eta : \mathcal{E}(F)(X) \xrightarrow{\sim} \det R\Gamma(X,F). \tag{4.6.1}$$

Exercise. Check that $\mathcal{E}$ satisfies the constraints from 2.8.

Remark. Suppose X is compact and a rational form ν has property that $\mathrm{Re}(\nu)$ is exact, $\mathrm{Re}(\nu) = df$. Then the isomorphism $\eta : \mathcal{E}(F)_\nu \xrightarrow{\sim} \det R\Gamma(X,F)$ can be computed using Morse theory: indeed, if $a < a'$ are non-critical values of f, then $f^{-1}((a',a]) \in C(\mathcal{N})$ for $\mathcal{N}$ compatible with ν.

In §5 we apply this construction to $F = B(M)$, the de Rham complex of a holonomic $\mathcal{D}$-module M, and write $\mathcal{E}_B(M) := \mathcal{E}(B(M))$. We use the same notation for the corresponding de Rham factorization line (in the analytic setting).

4.7 For $b \in X$ and a meromorphic ν on a neighborhood of b, $v_b(\nu) = -\ell$, set $\mathcal{E}(F)_{(b,\nu)} := \mathcal{E}(F)_{(\ell b,\nu)}$. Let $F_b^{(!)} := Ri_b^! F = R\Gamma_{\{b\}}(X,F) = R\Gamma_c(X_b,F)$, $F_b^{(*)} := i_b^* F = R\Gamma(X_b,F)$ be the fibers of F at b in !- and *-sense (here X_b is a small open disc around b). Let t be a local parameter at b.

Lemma. *There are canonical identifications*

$$\mathcal{E}(F)_{(b,t^{-1}dt)} = \det F_b^{(!)}, \quad \mathcal{E}(F)_{(b,-t^{-1}dt)} = \det F_b^{(*)}. \tag{4.7.1}$$

Proof Let X_b be a small disc $|t| < r$. Let W be the open disc of radius r', K be the closed disc of radius r'', $r'' < r < r'$. If $\mathcal{N}$ is a sufficiently tight cone around the Euler vector field $\mathrm{Re}(t\partial_t)$, then $X_b \in C(W,\mathcal{N})$ and $\bar{X}_b \in C(W,-\mathcal{N})$. The data $(W,K,\mathcal{N},-t^{-1}dt)$ and $(W,K,-\mathcal{N},t^{-1}dt)$ are compatible, and we are done. $\square$

Thus if F is the *-extension at b, i.e., $F_b^{(!)} = 0$, then $\mathcal{E}(F)_{(b,t^{-1}dt)}$ is canonically trivialized; if F is the !-extension at b, i.e., $F_b^{(*)} = 0$, then $\mathcal{E}(F)_{(b,-t^{-1}dt)}$ is canonically trivialized. Denote these trivializations by $1_b^! \in \mathcal{E}(F)_{(b,t^{-1}dt)}$, $1_b^* \in \mathcal{E}(F)_{(b,-t^{-1}dt)}$.

Exercise. Suppose $X = \mathbb{P}^1$ and F is smooth outside $0, \infty$. Then the composition $\det F_0^{(!)} \otimes \det F_\infty^{(*)} \xrightarrow{\sim} \mathcal{E}(F)_{(0,t^{-1}dt)} \otimes \mathcal{E}(F)_{(\infty,t^{-1}dt)} \xrightarrow{\eta} \det R\Gamma(\mathbb{P}^1,F)$ comes from the standard triangle $R\Gamma_{\{0\}}(\mathbb{P}^1,F) \to R\Gamma(\mathbb{P}^1,F) \to R\Gamma(\mathbb{P}^1 \setminus \{0\},F) = F_\infty^{(*)}$. In particular, if F is *-extension at 0 and !-extension at ∞, then $\eta(1_0^! \otimes 1_\infty^*) = 1$:= the trivialization of $\det R\Gamma(\mathbb{P}^1,F)$ that comes since $R\Gamma(\mathbb{P}^1,F) = 0$.

For $x \in X \setminus T$ one has $F_x^{(!)} = F_x^{(*)}(-1)[-2]$, so (4.7.1) yields a natural identification

$$\mathcal{E}(F)_{X \setminus T}^{(1)} \xrightarrow{\sim} \det F(-1)_{X \setminus T}. \tag{4.7.2}$$

If $R = \mathbb{C}$, then the Tate twist acts as identity. If M is a holonomic $\mathcal{D}$-module, then $B(M)_{X \setminus T} = M_{X \setminus T}^\nabla[1]$, and (4.7.2) can be rewritten as

$$\mathcal{E}_B(M)_{X \setminus T}^{(1)} \xrightarrow{\sim} (\det M_{X \setminus T})^{\otimes -1}. \tag{4.7.3}$$

Remark. For any $\ell \in \mathbb{Z}$ we have the local system of lines $\mathcal{E}(F)_{(b,zt^{-\ell}dt)}$, $z \in \mathbb{C}^\times$. A simple computation (or a reference to the compatibility property in 1.11) together with (4.7.2) shows that its monodromy around $z = 0$ equals $(-1)^{\mathrm{rk}(F)\ell} m_b$ where

$\mathrm{rk}(F) = \deg \det F_{X\backslash T}$ is the rank of F and m_b is the monodromy of $\det F$ around b. Thus (4.7.1) provides two descriptions of this local system for $\ell = 1$ (using the fibers at $z = \pm 1$). For a relation between them, see below.

4.8   We are in the situation of 4.7. Suppose R is a field, outside singular points our F is a local system of rank 1 placed in degree -1, and $F_b^{(*)} = 0$. Let m be the monodromy of F around b; suppose $m \neq 1$. Then $F_b^{(!)}$ vanishes as well, so we have $1_b^! \in \mathcal{E}(F)_{(b, t^{-1}dt)}$, $1_b^* \in \mathcal{E}(F)_{(b, -t^{-1}dt)}$.

Proposition. *The (counterclockwise) monodromy from $t^{-1}dt$ to $-t^{-1}dt$ identifies $1_b^!$ with $(1-m)^{-1}1_b^*$.*

Proof Consider an annulus around b (which lies in U), and cut it like this:

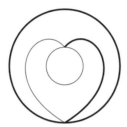

Let D_+ be the larger open disc, D_- the smaller one, $\bar{D}_\pm$ be their closures. Let $\heartsuit$ be a constructible set obtained from the closed heart figure by removing the right part of its boundary together with the lower vertex; the upper vertex stays in $\heartsuit$. Set $A := \bar{D}_+ \backslash \heartsuit$, $B := \heartsuit \backslash D_- \in \mathcal{I}(U, \mathcal{N}_-)$, $C := A \cup B = \bar{D}_+ \backslash D_-$.

Suppose $\epsilon > 0$ is small; let $\theta_\pm$ be the real part of the complex vector field $\exp(i\pi/2 \pm i\epsilon)t\partial_t$, and $\mathcal{N}_\pm$ be a tight cone around $\theta_\pm$. Then $\mathcal{N}_+$ is compatible with $\exp(i\alpha)t^{-1}dt$ for $\alpha \in [0, \pi/2]$, and $\mathcal{N}_-$ is compatible with $\exp(i\beta)t^{-1}dt$ for $\beta \in [\pi/2, \pi]$. Our $\theta_\pm$ are transversal to the lines of the drawing. Then $A \in \mathcal{I}(U, \mathcal{N}_+)$ and $B \in \mathcal{I}(U, \mathcal{N}_-)$ (each of them is the union of two lenses), hence $\heartsuit \in \mathcal{C}(W, \mathcal{N}_\pm)$.

Thus the monodromy in the statement of the proposition is inverse to the composition $\lambda(F)(D_-) \xrightarrow{\sim} \lambda(F)(\heartsuit) \xrightarrow{\sim} \lambda(F)(\bar{D}_+)$ where the first arrow is multiplication by $\tau_{\mathcal{N}_-} \in \lambda(F)(B)$, the second one is multiplication by $\tau_{\mathcal{N}_+} \in \lambda(F)(A)$ (we use tacitly the integration). Notice that $R\Gamma_C(X, F)$ is acyclic; let τ be the corresponding trivialization of $\lambda(F)(C)$. Since the multiplication by τ map $\lambda(F)(D_-) \xrightarrow{\sim} \lambda(F)(\bar{D}_+)$ sends 1_b^* to $1_b^!$, the proposition can be restated as $\tau_{\mathcal{N}_+}\tau_{\mathcal{N}_-} = (1-m)\tau$.

Consider the chain complex (P, d) that computes $R\Gamma_C(X, F)$ by means of the cell decomposition of the drawing. The graded vector space P is a direct sum of rank 1 components P_α labeled by the cells. Set $P_A := \bigoplus_{\alpha \in A} P_\alpha$, $P_B := \bigoplus_{\beta \in B} P_\beta$. Since A is the image of a 2-simplex with one face removed, P_A carries a differential d_A such that (P_A, d_A) is the chain complex of the simplex modulo the face. One defines d_B in a similar way. Both (P_A, d_A) and (P_B, d_B) are acyclic, and the corresponding trivializations of $\det P_A = \lambda(F)(A)$, $\det P_B = \lambda(F)(B)$ equal $\tau_{\mathcal{N}_+}$, $\tau_{\mathcal{N}_-}$.

Let $P' = P'_A \oplus P'_B$ be sum of P_α's for α in the boundary of $\heartsuit$. Then P' is a subcomplex with respect to both d and $d_A \oplus d_B$, on P/P' the differentials d and $d_A \oplus d_B$ coincide, and the complex P/P' is acyclic. We see that P' is acyclic with respect to both $d|_{P'}$ and $(d_A \oplus d_B)|_{P'}$, and τ, $\tau_{\mathcal{N}_+}\tau_{\mathcal{N}_-}$ are the trivializations of $\det P'$ that correspond to these differentials. Our P' sits in degrees 0, 1. Choose base vectors $e_A \in P_A'^0$, $f_A \in P_A'^1$, $e_B \in P_B'^0$, $f_B \in P_B'^1$ such that $d_A(e_A) = f_A$, $d_B(e_B) = f_B$, and $d(e_A) = f_A - f_B$. Then $d(e_B) = -m f_A + f_B$. Therefore $\tau_{\mathcal{N}_+}\tau_{\mathcal{N}_-}/\tau = 1 - m$, q.e.d.   □

4.9   Let b, ν, ℓ be as in the beginning of 4.7; suppose $\ell \neq 1$. Denote by ν_b the principal term of ν at b. Let f be any holomorphic function defined near b and vanishing at b such that $\nu_b = (df)_b$ if $\ell < 1$, and $\nu_b = (d(f^{-1}))_b$ if $\ell > 1$. For a small $z \in \mathbb{C}$, $z \neq 0$, the set $f^{-1}(z)$ is finite of order $|\ell - 1|$. For a finite subset Z of X, denote by $F_Z^{(!)}$, $F_Z^{(*)}$ the direct sum of !-, resp. *-fibers of F at points of Z. Let ϵ be a small positive real number.

Proposition. *For $\ell < 1$, one has canonical identifications*

$$\mathcal{E}(F)_{(b,\nu)} \xrightarrow{\sim} \det F_b^{(!)} \otimes (\det F_{f^{-1}(-\epsilon)}^{(!)})^{\otimes -1} \xrightarrow{\sim} \det F_b^{(*)} \otimes (\det F_{f^{-1}(\epsilon)}^{(*)})^{\otimes -1}. \quad (4.9.1)$$

For $\ell > 1$, one has

$$\mathcal{E}(F)_{(b,\nu)} \xrightarrow{\sim} \det F_b^{(*)} \otimes \det F_{g^{-1}(-\epsilon)}^{(!)} \xrightarrow{\sim} \det F_b^{(!)} \otimes \det F_{g^{-1}(\epsilon)}^{(*)}. \quad (4.9.2)$$

Proof Since $\mathcal{E}(F)_{(b,\nu)}$ depends only on the principal term of ν at b, we can assume that ν equals df or $d(f^{-1})$. Set $\lambda := \lambda(F)$.

Case $\ell = 0$: Then f is a local coordinate at b. Let Q be an open romb around b with vertices at $f = \pm\epsilon, \pm i\epsilon$, I and I' be the parts of its boundary where $\mathrm{Re}(f) \leq 0$, resp. $\mathrm{Re}(f) > 0$; set $C := \bar{Q} \setminus I = Q \cup I'$. Then one has $\mathcal{E}(F)_{(b,\nu)} = \lambda(C) \xrightarrow{\sim} \lambda(\bar{Q}) \otimes \lambda(I)^{\otimes -1} \xrightarrow{\sim} \lambda(Q) \otimes \lambda(I')$. This yields (4.9.1) due to the next identifications:

(a) $\lambda(\bar{Q}) \xrightarrow{\sim} \det R\Gamma_{\bar{Q}}(X, F) \xleftarrow{\sim} \det F_b^{(!)}$ and $\lambda(Q) \xrightarrow{\sim} \det R\Gamma(Q, i_Q^* F) \xrightarrow{\sim} \det F_b^{(*)}$;

(b) $\lambda(I) \xrightarrow{\sim} \det R\Gamma_I(X, F)$, and $R\Gamma_{f^{-1}(-\epsilon)}(X, F) \xrightarrow{\sim} R\Gamma_I(X, F)$;

(c) $\lambda(I') \xrightarrow{\sim} \det R\Gamma(I', Ri_{I'}^! F) \xrightarrow{\sim} \det R\Gamma(I', i_{I'}^* F[-1]) \xrightarrow{\sim} (\det F_{f^{-1}(\epsilon)}^{(*)})^{\otimes -1}$, where the second isomorphism comes from $i_{I'}^* F[-1] \xrightarrow{\sim} Ri_{I'}^! i_{Q!} i_Q^* F \xrightarrow{\sim} Ri_{I'}^! F$.

Case $\ell = 2$: Then f is a local coordinate at b. Define Q, etc., as above. Then $\mathcal{E}(F)_{(b,\nu)} = \lambda(Q \cup I) = \lambda(\bar{Q} \setminus I')$, so (a)–(c) yield (4.9.2).

If $\ell \neq 1$ is arbitrary, then f is a $|\ell - 1|$-sheeted cover of a neighborhood of b over a coordinate disc. The projection formula compatibility $\mathcal{E}(F)_{(b,\nu)} \xrightarrow{\sim} \mathcal{E}(f_* F)_{(0,dt)}$ for $\ell < 1$ and $\mathcal{E}(F)_{(b,\nu)} \xrightarrow{\sim} \mathcal{E}(f_* F)_{(0,d(t^{-1}))}$ for $\ell > 1$ reduces the assertion to the cases of ℓ equal to 0 and 2, and we are done.   □

5 The torsor of ε-periods

5.1   We consider triples (X, T, M) where X is a complex curve, T its finite subset, and M is a holonomic $\mathcal{D}$-module on (X, T) (i.e., a $\mathcal{D}$-module on X smooth off T). For us, a *weak theory of ε-factors* is a rule $\mathcal{E}$ that assigns to every such triple a de Rham factorization line $\mathcal{E}(M)$ on (X, T) (in complex-analytic sense). Our $\mathcal{E}$ should be functorial with respect to isomorphisms of triples, and have local nature, i.e., be compatible with pull-backs by open embeddings. We ask that:

(i) For a nice flat family $(X/Q, T, M)$ (see 2.12) with reduced Q the factorization lines $\mathcal{E}(M_q)$ vary holomorphically, i.e., we have $\mathcal{E}(M) \in \mathcal{L}_{\mathrm{dR}/Q}^{\Phi}(X/Q, T)$. If the family is isomonodromic, then $\mathcal{E}(M) \in \mathcal{L}_{\mathrm{dR}}^{\Phi}(X/Q, T)$.

(ii) $\mathcal{E}(M)$ is multiplicative with respect to finite filtrations of M's: for a finite filtration $M.$ on M there is a natural isomorphism $\mathcal{E}(M) \xrightarrow{\sim} \otimes \mathcal{E}(\mathrm{gr}_i M)$.

(iii) (projection formula) Let $\pi : (X', T') \to (X, T)$ be a finite morphism of pairs étale over $X \setminus T$, so, as in Remarks (i), (ii) in 1.2, (ii) in 1.5, we have a morphism $\pi_* : \mathcal{L}^\Phi_{\mathrm{dR}}(X', T') \to \mathcal{L}^\Phi_{\mathrm{dR}}(X, T)$. Then for any M' on (X', T') one has a natural identification $\mathcal{E}(\pi_* M') \xrightarrow{\sim} \pi_* \mathcal{E}(M')$ compatible with composition of π's.

(iv) (product formula) For compact X there is a natural identification (see 1.4 for the notation) $\eta = \eta(M) : \mathcal{E}(M)(X) \xrightarrow{\sim} \det R\Gamma_{\mathrm{dR}}(X, M)$.

The constraints should be pairwise compatible in the evident sense. (i) should be compatible with the base change. (ii) should be transitive with respect to refinements of the filtration, and the isomorphism $\mathcal{E}(\oplus M_\alpha) \xrightarrow{\sim} \otimes \mathcal{E}(M_\alpha)$ should not depend on the linear ordering of the indices α (which makes $\mathcal{E}$ a symmetric monoidal functor). (iii) should be compatible with the composition of π's.

Weak theories of ε-factors form naturally a groupoid which we denote by $^w\mathbb{E}$. Its key objects are $\mathcal{E}_{\mathrm{dR}}$ and $\mathcal{E}_{\mathrm{B}}$.

Replacing $\det R\Gamma_{\mathrm{dR}}(X, M)$ in (iv) by the trivial line $\mathbb{C}$ and leaving the rest of the story unchanged, we get a groupoid $^w\mathbb{E}^0$. It has an evident Picard groupoid structure, and $^w\mathbb{E}$ is naturally a $^w\mathbb{E}^0$-torsor. Below we denote by $(i)^0$–$(iv)^0$ the structure constraints in $^w\mathbb{E}^0$ that correspond to (i)–(iv) above.

5.2 *Compatibility with quadratic degenerations of X.* Let us formulate an important property of $\mathcal{E} \in {}^w\mathbb{E}$.

Suppose we have data 2.13(a),(b) in the analytic setting; we follow the notation of loc. cit. In 2.13 we worked in the formal scheme setting. Now the whole story of (2.13.1)–(2.13.7) makes sense analytically: we have a proper family of curves X over a small coordinate disc Q and an $\mathcal{O}_X$-module M equipped with a relative connection ∇, etc., so that the picture of 2.13 coincides with the formal completion of the present one at $q = 0$. To construct X, notice that $t_\pm$ from 2.13(a) converge on some true neighborhoods $U_\pm$ of $b_\pm$. Suppose that $t_\pm$ identify $U_\pm$ with coordinate discs of radii $r_\pm$ and $U_+ \cap U_- = U_\pm \cap (T \cup |D|) = \emptyset$. Then Q is the coordinate disc of radius $r_+ r_-$. Let W be an open subset of $Y \times Q$ formed by those pairs (y, q) that if $y \in U_+$, then $r_- |t_+(y)|^2 > |q| r_+$, and if $y \in U_-$, then $r_+ |t_-(y)|^2 > |q| r_-$. Our X is the union of two open subsets $V := U_+ \times U_-$ and W: we glue $(y, q) \in W$ such that $y \in U_\pm$ with $(y_+, y_-) \in V$ such that either y_+ or y_- equals y and $t_+(y_+) t_-(y_-) = q$. The projection $q : X \to Q$ is $(y, q) \mapsto q$ on W and $q(y_+, y_-) = t_+(y_+) t_-(y_-)$ on V. Set $K_{0Q} := X \setminus W = \{(y_+, y_-) \in V : r_+^{-1}|t_+(y_+)| = r_-^{-1}|t_-(y_-)|\}$. The formal trivializations of L at $b_\pm$ from 2.13(b) converge on $U_\pm$, and we define M by gluing $M_{b_0} \otimes \mathcal{O}_V$ and the pull-back N_W of N by the projection $W \to Y$.

Let us define an analytic version of (2.13.7), which is an isomorphism of $\mathcal{O}_Q$-lines

$$\det Rq_{\mathrm{dR}*} M \xrightarrow{\sim} \det R\Gamma_{\mathrm{dR}}(Y, N) \otimes \mathcal{O}_Q. \qquad (5.2.1)$$

Let i, j be the embeddings $K_{0Q} \hookrightarrow X \hookleftarrow W$. Since $V \setminus K_{0Q}$ is disjoint union of two open subsets $V_\pm$, $V_+ := \{(y_+, y_-) : r_+^{-1}|t_+(y_+)| > r_-^{-1}|t_-(y_-)|\}$, the complex $\mathcal{F} := i^* j_* dR_{W/Q}(N_W)$ is the direct sum of the two components $\mathcal{F}_\pm$. Both maps $i^* dR_{X/Q}(M) \to \mathcal{F}_\pm$ are quasi-isomorphisms, so $i_* \mathcal{F}_- \xrightarrow{\sim} Cone(dR_{X/Q}(M) \to j_* dR_{W/Q}(N_W))$, hence $\det Rq_{\mathrm{dR}*} M \xrightarrow{\sim} (\det Rq|_W {}_{\mathrm{dR}*} N_W) \otimes (\det Rq|_{K_{0Q}} \mathcal{F}_-)^{\otimes -1}$. Let $N_X \subset j_* N_W$ be $*$-extension from V_+ side and $!$-extension from V_- side. Then $i^* dR_{X/Q}(N_X) = \mathcal{F}_+ \subset \mathcal{F}_+ \oplus \mathcal{F}_-$, hence $\det Rq_{\mathrm{dR}*} N_X \xrightarrow{\sim} (\det Rq|_W {}_{\mathrm{dR}*} N_W) \otimes$

$(\det Rq|_{K_{0Q}*}\mathcal{F}_-)^{\otimes -1}$. Thus we get a canonical identification $\alpha : \det Rq_{\mathrm{dR}*}M \xrightarrow{\sim}$ $\det Rq_{\mathrm{dR}*}N_X$. Now N_W, hence N_X, are $\mathcal{D}$-modules, i.e., they carry an absolute flat connection, so $Rq_{\mathrm{dR}*}N_X$, $Rq|_{W\mathrm{dR}*}N_W$ carry a natural connection. It is clear from the topology of the construction that the cohomology are smooth, hence constant, $\mathcal{D}_Q$-modules. Since the fiber of $Rq_{\mathrm{dR}*}N_X$ at $q = 0$ equals $R\Gamma_{\mathrm{dR}}(Y, N)$, we get (5.2.1).

For any $\mathcal{E} \in {}^w\mathbb{E}$ we have an $\mathcal{O}_Q$-line $\mathcal{E}(M)_{\nu_Q} := \mathcal{E}(M)_{(D_Q, 1_T, \nu_Q)}$ and a natural isomorphism $\mathcal{E}(M)_{\nu_Q} \xrightarrow{\sim} \mathcal{E}(N)_{(D, 1_T, \nu)} \otimes \mathcal{O}_Q$, cf. (2.13.4). There is a canonical isomorphism $\mathcal{E}(N)_{(D, 1_T, \nu)} \xrightarrow{\sim} \mathcal{E}(N)_\nu$ defined in the same way as (2.13.9) (using η on $\mathbb{P}^1$). Since X is smooth over $Q^o := Q \setminus \{0\}$, we have $\eta : \mathcal{E}(M)_{\nu_Q}|_{Q^o} \xrightarrow{\sim} \det Rq_{\mathrm{dR}*}M|_{Q^o}$. Thus comes a diagram

$$
\begin{array}{ccc}
\mathcal{E}(M)_{\nu_Q}|_{Q^o} & \xrightarrow{\ \eta\ } & \det Rq_{\mathrm{dR}*}M|_{Q^o} \\
\downarrow & & \downarrow \\
\mathcal{E}(N)_\nu \otimes \mathcal{O}_{Q^o} & \xrightarrow{\ \eta\ } & \det R\Gamma_{\mathrm{dR}}(Y, N) \otimes \mathcal{O}_{Q^o}.
\end{array}
\tag{5.2.2}
$$

We say that $\mathcal{E}$ is a *theory of ε-factors* if (5.2.1) commutes for all data 2.13(a),(b). Such $\mathcal{E}$ form a subgroupoid $\mathbb{E}$ of ${}^w\mathbb{E}$ called the *ε-gerbe*; see 5.4 for the reason.

For $\mathcal{E}^0 \in {}^w\mathbb{E}^0$ there is a similar diagram

$$
\begin{array}{ccc}
\mathcal{E}^0(M)_{\nu_Q}|_{Q^o} & & \\
& \searrow & \\
\downarrow & & \mathcal{O}_{Q^o} \\
& \nearrow & \\
\mathcal{E}^0(N)_\nu \otimes \mathcal{O}_{Q^o} & &
\end{array}
\tag{5.2.3}
$$

Those $\mathcal{E}^0$ for which (5.2.3) commutes for every datum 2.13(a),(b) form a Picard subgroupoid $\mathbb{E}^0$ of ${}^w\mathbb{E}^0$. Our $\mathbb{E}$ is an $\mathbb{E}^0$-torsor.

Proposition. *$\mathcal{E}_{dR}$ and $\mathcal{E}_B$ are theories of ε-factors.*

Proof Compatibility of $\mathcal{E}_{\mathrm{dR}}$ with quadratic degenerations follows from 2.13. Namely, the construction from loc. cit., spelled analytically as above, provides $\eta_{\mathrm{dR}} :$ $\mathcal{E}_{\mathrm{dR}}(M)_{\nu_Q} \xrightarrow{\sim} \det Rq_{\mathrm{dR}*}M$ over the whole Q (not only on Q^o), and the proposition in 2.13 says that our diagram commutes on the formal neighborhood of $q = 0$. Hence it commutes everywhere, q.e.d.

Let us treat $\mathcal{E}_B$. Let K_∞ be a compact neighborhood of $T \cup |D|$ in Y that does not intersect $U_\pm$, $K := K_\infty \cup \{b_+, b_-\}$. Let $\mathcal{N}$ be a continuous family of proper cones in the tangent bundle to $Y \setminus K$ such that $\mathrm{Re}(\nu)$ is negative on it. Set $K_{\infty Q} := K_\infty \times Q$, $K_Q := K_{\infty Q} \sqcup K_{0Q} \subset X$; let $\mathcal{N}_W$ be the pull-back of $\mathcal{N}$ by the projection $W \to Y$. Then $(X, K_Q, \mathcal{N}_W, \nu_Q)$ form a Q-family of compatible data as in 4.6. Consider isomorphisms η of (4.7.1) for M and N_X. Our Q-family is constant near $K_{\infty Q}$, so $\mathcal{E}_B(M)_{\mathcal{N}}(K_{\infty Q}) = \mathcal{E}_B(N_X)_{\mathcal{N}}(K_{\infty Q}) = \mathcal{E}_B(N)_{\mathcal{N}}(K_\infty) \otimes \mathcal{O}_Q$. Let C be a locally closed subset of V which consists of those (y_+, y_-) that $2|t_+(y_+)|/r_+ - |t_-(y_-)|/r_- \leq 1$ and $2|t_-(y_-)|/r_- - |t_+(y_+)|/r_+ < 1$; set $Rq_{\mathrm{dR}*}^{(C)}(?) := Rq|_{C*}Ri_C^! dR_{X/Q}(?)$. If $\mathcal{N}$ is sufficiently tight, then $C \in C(X, \mathcal{N}_W)$ (see the proof of the lemma in 4.7), thus $\mathcal{E}_B(?)_{\mathcal{N}}(K_{\infty Q}) = \det Rq_{\mathrm{dR}*}^{(C)}(?)$.

Since the construction of (5.2.1) was local at K_{0Q}, the composition of $\mathcal{E}_B(M) \xrightarrow{\eta_B}$ $\det Rq_{\mathrm{dR}*}M \xrightarrow{\sim} \det R\Gamma_{\mathrm{dR}}(Y, N) \otimes \mathcal{O}_Q = \det Rq_{\mathrm{dR}*}N_X$ in (5.2.2) can be rewritten as

$\mathcal{E}_{\mathrm{B}}(N)_{\mathcal{N}}(K_\infty) \otimes \det Rq_{\mathrm{dR}*}^{(C)}(M) \xrightarrow{\sim} \mathcal{E}_{\mathrm{B}}(N)_{\mathcal{N}}(K_\infty) \otimes \det Rq_{\mathrm{dR}*}^{(C)}(N_X) \xrightarrow{\eta_{\mathrm{B}}} Rq_{\mathrm{dR}*} N_X$. Here $\xrightarrow{\sim}$ comes from the identification

$$\alpha_C : \det Rq_{\mathrm{dR}*}^{(C)}(M) \xrightarrow{\sim} \det Rq_{\mathrm{dR}*}^{(C)}(N_X) \qquad (5.2.4)$$

defined by the same construction as (5.2.1) with $Rq_{\mathrm{dR}*}$ replaced by $Rq_{\mathrm{dR}*}^{(C)}$.

The composition $\mathcal{E}_{\mathrm{B}}(M) \xrightarrow{\sim} \mathcal{E}_{\mathrm{B}}(N) \otimes \mathcal{O}_Q \xrightarrow{\sim} \det R\Gamma_{\mathrm{dR}}(Y, N) \otimes \mathcal{O}_Q = \det Rq_{\mathrm{dR}*} N_X$ in (5.2.2) equals $\mathcal{E}_{\mathrm{B}}(N)_{\mathcal{N}}(K_\infty) \otimes \det Rq_{\mathrm{dR}*}^{(C)}(M) \xrightarrow{\sim} \mathcal{E}_{\mathrm{B}}(N)_{\mathcal{N}}(K_\infty) \xrightarrow{\sim} \mathcal{E}_{\mathrm{B}}(N)_{\mathcal{N}}(K_\infty) \otimes \det Rq_{\mathrm{dR}*}^{(C)}(N_X) \xrightarrow{\eta_{\mathrm{B}}} Rq_{\mathrm{dR}*} N_X$. Here $\xrightarrow{\sim}$ come since $Rq_{\mathrm{dR}*}^{(C)}(M) = Rq_{\mathrm{dR}*}^{(C)}(N_X) = 0$, hence their determinant lines are trivialized (notice that the trivialization of $Rq_{\mathrm{dR}*}^{(C)}(N_X)$ is horizontal, and at $q = 0$ it equals the Betti version of (2.13.8) due to Exercise in 4.7).

We see that commutativity of (5.2.2) means that α_C identifies the above trivializations. To see this, consider the open subspace $V_- \subset V$, and the corresponding 2-step filtration $j_{V_-!} M|_{V_-} \subset M|_V$, and notice that $N_X|_V = \mathrm{gr} M|_V$. The assertion follows now from the construction of α_C. $\qquad\square$

5.3   Let (X, T, M) be as in 5.1. For $\mathcal{E} \in {}^w\mathbb{E}$, $b \in X$, and a meromorphic form ν on a neighborhood of b, $v_b(\nu) = -\ell$, we write $\mathcal{E}(M)_{(b,\nu)} := \mathcal{E}(M)_{(\ell b, \nu)}$.

Remark. If b is a smooth point of M, then $\mathcal{E}(M)_{(b,\nu)}$ does not depend on whether we view b as a point of T or $X \setminus T$ (by 5.1(iii) with $\pi = \mathrm{id}_X$, $T = T' \cup \{b\}$).

Let $\delta(M)_{b,\nu} \in \mathbb{Z}$ be the degree of $\mathcal{E}(M)_{(b,\nu)}$, and $\mu(M)_{b,\nu} \in \mathbb{C}^\times$ be the value of $\mu \in \mathrm{Aut}(\mathcal{E}(M))$ (see 1.15) at (b, ν).

Lemma. *(i) One has $\delta(M)_{b,\nu} = \dim(B(M)_b^{(!)}) + (1 - \ell) rk(M)$.*
(ii) For $c \in \mathbb{C}^\times$ the multiplication by c automorphism of M acts on $\mathcal{E}(M)_{(b,\nu)}$ as multiplication by $c^{\delta(M)_{b,\nu}}$.
(iii) One has $\mu(M)_{b,\nu} = (-1)^{\ell\, rk(M)} m_b(M)^{-1}$ where $m_b(M)$ is the monodromy of $\det M_{X \setminus T}$ around b.
(iv) For smooth M, there is an isomorphism $\mathcal{E}(M)^{(1)} \xrightarrow{\sim} (\det M)^{\otimes -1}$ compatible with constraint 5.1(ii) and pull-backs by open embeddings. In particular, this is an isomorphism of symmetric monoidal functors.
(v) Suppose we have M, M' over discs U, U' which have regular singularity and are either $*$- or !-extension at b, b'. Let $\phi : U \to U'$ be any open embedding, $\phi(b) = b'$, and $\tilde{\phi} : M \to \phi^* M'$ be any its lifting. Then the isomorphism $\mathcal{E}(\tilde{\phi}) : \mathcal{E}(M)_b^{(1)} \xrightarrow{\sim} \mathcal{E}(M')_{b'}^{(1)}$ does not depend on the choice of $(\phi, \tilde{\phi})$.

Proof (i) Let t be a local coordinate at b, $t(b) = 0$. We can view t as an identification of a small disc X_b at b with a neighborhood of $0 \in \mathbb{P}^1$. Let us extend $M|_{X_b}$ to a $\mathcal{D}$-module $M^{(t)}$ on $\mathbb{P}^1$ which is smooth outside $\{0, \infty\}$, and is the $*$-extension with regular singularities at ∞. Such $M^{(t)}$ is unique.

By continuity, $\delta(M)_{b,\nu}$ is the same for all ν with fixed ℓ. We can assume that ν is meromorphic on $\mathbb{P}^1$ with $\mathrm{div}(\nu) \subset \{0, \infty\}$, so $v_0(\nu) = -2 - v_\infty(\nu)$. By 5.1(iv), one has $\delta(M^{(t)})_{0,\nu} + \delta(M^{(t)})_{\infty,\nu} = \chi_{\mathrm{dR}}(\mathbb{P}^1, M^{(t)}) = \dim(B(M^{(t)})_b^{(!)})$. Thus the assertion for (X_b, M, ν) amounts to that for $(\mathbb{P}^1_\infty, M^{(t)}, \nu)$, i.e., we are reduced to the case when M is the $*$-extension with regular singularities. By 5.1(ii), it suffices to treat the case of $rk(M) = 1$; then, by continuity, it suffices to consider $M = \mathcal{O}_X$ (the trivial $\mathcal{D}$-module). By 5.1(iv) applied to $\mathbb{P}^1$ and $t^{-1} dt$, we see that

$\delta(\mathcal{O}_{\mathbb{P}^1})_{0,t^{-1}dt} + \delta(\mathcal{O}_{\mathbb{P}^1})_{\infty,t^{-1}dt} = -2$, hence, since $v_\infty(t^{-1}dt) = v_0(t^{-1}dt)$, one has $\delta(\mathcal{O}_{\mathbb{P}^1})_{b,t^{-1}dt} = -1$. By factorization, $\delta(\mathcal{O}_{\mathbb{P}^1})_{0,t^{-\ell}dt} = \ell\delta(\mathcal{O}_{\mathbb{P}^1})_{0,t^{-1}dt} = -\ell$, q.e.d.

(ii) The $\mathbb{C}^\times$-action on $\mathcal{E}(M)_{(b,\nu)}$, which comes from the action of homotheties on M, is a holomorphic character of $\mathbb{C}^\times$. Thus c acts as multiplication by $c^{\delta'(M)_{b,\nu}}$ for some $\delta'(M)_{b,\nu} \in \mathbb{Z}$. The argument of (i) works for δ replaced by δ', so δ and δ' are given by the same formula, q.e.d.

(iv) By (i), $\mathcal{E}^{(1)}(\mathcal{O}_X)$ is a de Rham line of degree -1, which has local origin. Thus there is a line E of degree -1 and an isomorphism $E \otimes \mathcal{O}_X \xrightarrow{\sim} \mathcal{E}^{(1)}(\mathcal{O}_X)$ compatible with the pull-backs by open embeddings of X's; such a datum is uniquely defined.

The set of isomorphisms $\alpha : E \xrightarrow{\sim} \mathbb{C}[1]$ identifies naturally with the set of isomorphisms of symmetric monoidal functors $\alpha_{\mathcal{E}} : \mathcal{E}(M)^{(1)} \xrightarrow{\sim} (\det M)^{\otimes -1}$ (where M is smooth) that are compatible with the pull-backs by open embeddings. Namely, $\alpha_{\mathcal{E}}$ is a unique isomorphism that equals $\alpha \otimes \mathrm{id}_{\mathcal{O}_X}$ for $M = \mathcal{O}_X$. To see this, notice that a matrix $g \in \mathrm{GL}_n(\mathbb{C}) \xrightarrow{\sim} \mathrm{Aut}(\mathcal{O}_X^n)$ acts on $\mathcal{E}^{(1)}(\mathcal{O}_X^n)$ as multiplication by $\det(g)^{-1}$ (which follows from (ii) and 5.1(ii)).

(iii) Use (iv) and the compatibility property from 1.11.

(v) Let us show that $\mathrm{Aut}(M)$ acts trivially on $\mathcal{E}(M)_b^{(1)}$. Pick any $g \in \mathrm{Aut}(M)$. Since $M_{U \setminus \{b\}}$ admits a g-invariant filtration with successive quotients of rank 1, we are reduced, by 5.1(ii), to the case of M of rank 1. Here g is multiplication by some $c \in \mathbb{C}^\times$, and we are done by (ii) (since, by (i), $\mathcal{E}(M)_b^{(1)}$ has degree 0).

Thus for given ϕ the isomorphism $\mathcal{E}(\tilde{\phi})$ does not depend on the choice of $\tilde{\phi}$. The space of ϕ's is connected, so it suffices to show that the map $\phi \mapsto \mathcal{E}(\tilde{\phi})$ is locally constant. If ϕ varies in a disc Q, then we can find $\tilde{\phi}$ which is an isomorphism of $\mathcal{D}$-modules on $U \times Q$, hence our map is horizontal (see 5.1(i)), q.e.d. $\qquad\square$

5.4 For $\mathcal{E} \in \mathbb{E}$ and (X, T, M) as in 5.1 the canonical automorphism μ of $\mathcal{E}(M)$ (see 1.15) is evidently compatible with constraints 5.1(i)–(iv), i.e., μ is an automorphism of $\mathcal{E}$. Here is the main result of this section:

Theorem. *$\mathrm{Aut}(\mathcal{E})$ is an infinite cyclic group generated by μ. All objects of $\mathbb{E}$ are isomorphic. Thus $\mathbb{E}$ is a $\mathbb{Z}$-gerbe.*

We call $\rho^\varepsilon : \mathcal{E}_{\mathrm{dR}} \xrightarrow{\sim} \mathcal{E}_{\mathrm{B}}$ an ε-*period* isomorphism. By the theorem, ε-period isomorphisms form a $\mathbb{Z}$-torsor $E_{\mathrm{B}/\mathrm{dR}}$ referred to as the ε-*period* torsor.

Since $\mathbb{E}$ is an $\mathbb{E}^0$-torsor, the theorem can be reformulated as follows. By (iii) of the lemma in 5.3, every $\mathcal{E}^0 \in \mathbb{E}^0$ carries a natural automorphism μ^0 that acts on $\mathcal{E}^0(M)_{(b,\nu)}$ as multiplication by $\mu(M)_{b,\nu} \in \mathbb{C}^\times$.[27]

Theorem'. *The map $\mathbb{Z} \to \pi_1(\mathbb{E}^0)$, $1 \mapsto \mu^0$, is an isomorphism, and $\pi_0(\mathbb{E}^0) = 0$.*

The proof occupies the rest of the section.

5.5 Pick any $\mathcal{E}^0 \in {}^w\mathbb{E}^0$. Then for (X, T, M) as in 5.1 one has:

---

[27]μ^0 does *not* equal the canonical automorphism μ of $\mathcal{E}^0(M)$ (which is identity by 5.5(i)).

Lemma. *(i) The factorization line $\mathcal{E}^0(M) \in \mathcal{L}_{dR}^\Phi(X, T)$ is trivial.*
(ii) Every automorphism of M acts trivially on $\mathcal{E}^0(M)$.
(iii) For M of rang 0, the factorization line $\mathcal{E}^0(M)$ is canonically trivialized. The trivialization has local nature and is compatible with constraints $5.1(i)^0$–$(iv)^0$; it is uniquely defined by this property.

Proof (i) One checks that the de Rham line $\mathcal{E}^0(M)^{(1)}$ on $X \setminus T$ is trivial by modifying the argument in the proof of 5.3(iv) in the evident manner (or one can use 5.3(iv) directly, noticing that $\mathcal{E}^0$ is the ratio of two objects of $^w\mathbb{E}$). Similarly, $\mathcal{E}(M)$ has zero degree by 5.3(i). Now use the theorem in 1.6.

(ii) Let us show that $g \in \mathrm{Aut}(M)$ acts trivially on $\mathcal{E}^0(M)_{(b,\nu)}$. Let $M^{(t)}$ be as in the proof of 5.3(i); g acts on it. We can assume that ν is meromorphic on $\mathbb{P}^1$ with $\mathrm{div}(\nu) \subset \{0, \infty\}$. The action of g on $\mathcal{E}^0(M^{(t)})(\mathbb{P}^1) = \mathcal{E}^0(M)_{(0,\nu)} \otimes \mathcal{E}^0(M^{(t)})_{(\infty,\nu)}$ is trivial by $5.1(iv)^0$, it suffices to check that it acts trivially on $\mathcal{E}^0(M^{(t)})_{(\infty,\nu)}$. Thus we are reduced to the situation when our M is the $*$-extension with regular singularities. By constraint $5.1(ii)^0$, it suffices to consider the case when the monodromy of M around b is multiplication by a constant. Then $\mathrm{Aut}(M)$ is generated by the diagonal matrices, and by $5.1(ii)^0$ we are reduced to the case when M has rank 1, where we are done by 5.3(ii).

(iii) is left to the reader. $\qquad\square$

Remarks. (i) By (i) of the lemma, the degree 0 lines $\mathcal{E}^0(M)_{(b,\nu)}$ for all ν with fixed $v_b(\nu) = -\ell$ are canonically identified; we denote this line by $\mathcal{E}^0(M)_{(b,\ell)}$. By (ii) of loc. cit., it depends only on the isomorphism class of M, and by (iii) there is a canonical identification $\mathcal{E}^0(M)_{(b,\ell)} \xrightarrow{\sim} \mathcal{E}^0(j_{b*}M)_{(b,\ell)}$.

(ii) Suppose we have M, M' on discs U, U' which have regular singularity at $b^{(\prime)} \in U^{(\prime)}$. Let $\phi : U \to U'$ be an open embedding, $\phi(b) = b'$, and $\tilde{\phi} : M \to \phi^* M'$ be any its lifting. Then the isomorphisms $\mathcal{E}^0(\tilde{\phi}) : \mathcal{E}^0(M)_{(b,\ell)} \xrightarrow{\sim} \mathcal{E}^0(M')_{(b',\ell)}$ do not depend on the choice of $(\phi, \tilde{\phi})$. To see this, we can assume that M, M' are $*$-extensions at b, b', and then repeat the second part of the proof of 5.3(v).

5.6 For $m \in \mathbb{C}^\times$ let M_m be a $\mathcal{D}$-module of rank 1 on a disc U, which has regular singularity at $b \in U$ with the monodromy m and is $*$-extension at b. By the remark in 5.5, the degree 0 line $\mathcal{G}_{(m,\ell)} := \mathcal{E}^0(M_m)_{(b,\ell)}$ depends only on m and ℓ. By $5.1(i)^0$, $\mathcal{G}_{(m,\ell)}$ form a holomorphic line bundle $\mathcal{G} = \mathcal{G}(\mathcal{E}^0)$ over $\mathbb{G}_m \times \mathbb{Z}$. The factorization structure on $\mathcal{E}^0(M_m)$ provides, by 5.5(i), canonical isomorphisms

$$\otimes \mathcal{G}_{(1,\ell_\alpha)} \xrightarrow{\sim} \mathcal{G}_{(1,\Sigma\ell_\alpha)}, \quad \mathcal{G}_{(1,\ell)} \otimes \mathcal{G}_{(m,\ell')} \xrightarrow{\sim} \mathcal{G}_{(m,\ell+\ell')}. \tag{5.6.1}$$

Suppose we have a finite collection $\{(m_\alpha, \ell_\alpha)\}$ with $\Pi m_\alpha = 1$, $\Sigma \ell_\alpha = 2$. Then for any choice of a subset $\{b_\alpha\} \subset \mathbb{P}^1$ one can find a $\mathcal{D}$-module M on $\mathbb{P}^1$ of rank 1 which is smooth off $\{b_\alpha\}$ and is $*$-extension with regular singularity at b_α with monodromy m_α, and a rational form ν with $\mathrm{div}(\nu) = -\Sigma\ell_\alpha b_\alpha$. Writing $\mathcal{E}(M)(\mathbb{P}^1) = \mathcal{E}(M)_\nu$ in $5.1(iv)^0$, we get

$$\eta : \otimes \mathcal{G}_{(m_\alpha,\ell_\alpha)} \xrightarrow{\sim} \mathbb{C}. \tag{5.6.2}$$

It does not depend on the auxiliary choices (for η is locally constant, and the datum of $\{b_\alpha\}$, ν forms a connected space; M is unique up to an isomorphism).

Now assume that $\mathcal{E}^0 \in \mathbb{E}$. Consider the holomorphic $\mathbb{G}_m$-torsor $\mathcal{G}^\times = \mathcal{G}^\times(\mathcal{E}^0)$ over $\mathbb{G}_m \times \mathbb{Z}$ that corresponds to $\mathcal{G}$.

Lemma. $\mathcal{G}^{\times}$ *has a unique structure of holomorphic commutative group* $\mathbb{G}_m$-*extension of* $\mathbb{G}_m \times \mathbb{Z}$ *such that for any* $g_i \in \mathcal{G}^{\times}_{(m_i,\ell_i)}$, $1 \le i \le n$, *and* $g \in \mathcal{G}^{\times}_{((m_1\ldots m_n)^{-1},2-\ell_1-\ldots-\ell_n)}$ *one has* $\eta(g \otimes (g_1 \cdots g_n)) = \eta(g \otimes g_1 \otimes \ldots \otimes g_n)$.

Proof The above formula defines commutative n-fold product maps $\mathcal{G}^{\times n} \to \mathcal{G}^{\times}$, $(g_1,\ldots,g_n) \mapsto g_1 \cdots g_n$, which lift the n-fold products on $\mathbb{G}_m \times \mathbb{Z}$. We need to check the associativity property, which says that for any $g_1,\ldots,g_n \in \mathcal{G}^{\times}$ and any k, $1 < k < n$, one has $g_1 \cdots g_n = (g_1 \cdots g_k)g_{k+1} \cdots g_n$.

For $m \in \mathbb{C}^{\times}$ set $\mathcal{G}^{\times}_m := \sqcup \mathcal{G}^{\times}_{(m,\ell)} \subset \mathcal{G}^{\times}$. The maps $(\mathcal{G}^{\times}_1)^n \to \mathcal{G}^{\times}_1$, $\mathcal{G}^{\times}_1 \times \mathcal{G}^{\times}_m \to \mathcal{G}^{\times}_m$ coming from the arrows in (5.6.1) are evidently associative and commutative, i.e., they define a commutative group structure on $\mathcal{G}^{\times}_1$ and a $\mathcal{G}^{\times}_1$-torsor structure on $\mathcal{G}^{\times}_m$. Thus $\mathcal{G}^{\times}$ is a $\mathcal{G}^{\times}_1$-torsor over $\mathbb{G}_m$.

Since (5.6.2) comes from a trivialization of $\mathcal{E}(M)$, the above maps are restrictions of the multiple products maps in $\mathcal{G}^{\times}$. Moreover, the n-fold product on $\mathcal{G}^{\times}$ is compatible with the $\mathcal{G}^{\times}_1$-action on $\mathcal{G}^{\times}$: for $h \in \mathcal{G}^{\times}_1$ one has $(hg_1)g_2 \cdots g_n = h(g_1,\ldots,g_n)$. So, while checking the associativity, we have a freedom to change g_i in its $\mathcal{G}^{\times}_1$-orbit. Thus we can assume that $g_i \in \mathcal{G}^{\times}_{(m_i,\ell_i)}$ are such that $\ell_1+\ldots+\ell_k = 1$. Then one can find a quadratic degeneration picture as in 5.2 such that $\tilde{T} = b_{1Q} \sqcup \ldots \sqcup b_{nQ}$, $\tilde{M}$ of rank 1 has regular singularities at b_{iQ} with monodromy m_i, $\mathrm{div}(\tilde{\nu}) = -\Sigma \ell_i b_{iQ}$; the fiber $\tilde{X}_1$ is $\mathbb{P}^1$, and $\tilde{X}_0$ is the union of two copies of $\mathbb{P}^1$ with $\{b_1,\ldots,b_k\}$ in the first copy and $\{b_{k+1},\ldots,b_n\}$ in the second. The compatibility with quadratic degeneration yields the promised associativity, q.e.d. □

5.7 Let $\pi^{(n)} : U' \to U$, $\pi^{(n)}(b') = b$, be a degree n covering of a disc completely ramified at b. Then $\pi^{(n)}_* M_{m'}$ is isomorphic to $\underset{m^n=m'}{\oplus} M_m$, and $v_{b'}(\pi^{(n)*}\nu) + 1 = n(v_b(\nu)+1)$. Therefore 5.1(iii)^0, 5.1(ii)^0 yield a canonical isomorphism

$$\underset{m^n=m'}{\otimes} \mathcal{G}_{(m,\ell)} \xrightarrow{\sim} \mathcal{G}_{(m',n(\ell-1)+1)}. \tag{5.7.1}$$

For example, if $n = 2$, $m' = 1$, $\ell = 1$, then (5.6.2), with $\mathcal{G}_{(1,1)}$ factored off, is a trivialization of $\mathcal{G}_{(-1,1)}$, which we denote by $e_{(-1,1)} \in \mathcal{G}_{(-1,1)}$. Notice that

$$\eta(e^{\otimes 2}_{(-1,1)}) = 1. \tag{5.7.2}$$

This follows from compatibility of η with π_* applied to a covering $\mathbb{P}^1 \to \mathbb{P}^1$, $t \mapsto t^2$, the trivial $\mathcal{D}$-module $\mathcal{O}_{\mathbb{P}^1}$ on the source, and the form $t^{-1}dt$ on the target.

Let $\mathrm{Ext}(\mathbb{G}_m, \mathbb{G}_m)$ be the Picard groupoid of holomorphic commutative group extensions of $\mathbb{G}_m$ by $\mathbb{G}_m$. One has $\pi_0(\mathrm{Ext}(\mathbb{G}_m, \mathbb{G}_m)) = 0$ and $\pi_1(\mathrm{Ext}(\mathbb{G}_m, \mathbb{G}_m)) = \mathrm{Hom}(\mathbb{G}_m, \mathbb{G}_m) = \mathbb{Z}$.

The quotient of $\mathbb{G}_m \times \mathbb{Z}$ modulo the subgroup generated by $(-1,1)$ identifies with $\mathbb{G}_m$ by the projection $(m,\ell) \mapsto (-1)^{\ell}m$. Thus the quotient $\bar{\mathcal{G}}^{\times}(\mathcal{E}^0)$ of $\mathcal{G}^{\times}(\mathcal{E}^0)$ modulo the subgroup generated by $e_{(-1,1)}$ is an object of $\mathrm{Ext}(\mathbb{G}_m, \mathbb{G}_m)$.

$$\bar{\mathcal{G}}^{\times} : \mathbb{E}^0 \to \mathrm{Ext}(\mathbb{G}_m, \mathbb{G}_m) \tag{5.7.3}$$

is a Picard functor. It assigns to $\mu^0 \in \mathrm{Aut}(\mathcal{E}^0)$ (see 5.4) the generator -1 of $\mathbb{Z} = \mathrm{Aut}(\bar{\mathcal{G}}^{\times}(\mathcal{E}^0))$. Therefore we can reformulate the theorem from 5.4 as follows:

Theorem. $\bar{\mathcal{G}}^{\times}$ *is an equivalence of Picard groupoids.*

Let us define a Picard functor

$$\mathrm{Ext}(\mathbb{G}_m, \mathbb{G}_m) \to \mathbb{E}^0 \tag{5.7.4}$$

right inverse to (5.7.3). We need to assign to an extension $\bar{\mathcal{G}}^\times$ an object $\mathcal{E}^0 = \mathcal{E}^0(\bar{\mathcal{G}}^\times)$ of $\mathbb{E}^0$. Suppose we have (X, T, M) as in 5.1. For $b \in T$ let $m(M)_b$ be the monodromy of $\det M|_{X \backslash T}$ around b; for $c \in 2^T$ we denote by $m(M)_c$ the product of $m(M)_b$ for $b \in T^c$ (see 1.1 for the notation). Then

$$\mathcal{E}^0(M)_{(D,c,\nu)} := \bar{\mathcal{G}}_{(-1)^{\deg(D)\mathrm{rk}(M)} m(M)_c}. \tag{5.7.5}$$

Here $\bar{\mathcal{G}}$ is the degree 0 line that corresponds to the $\mathbb{G}_m$-torsor $\bar{\mathcal{G}}^\times$. The factorization structure comes from the product in $\bar{\mathcal{G}}^\times$. Constraints $5.1(\mathrm{i})^0$, $5.1(\mathrm{ii})^0$ are evident. The identification $\mathcal{E}^0(\pi_* M') \xrightarrow{\sim} \pi_* \mathcal{E}(M')$ of $5.1(\mathrm{iii})^0$ comes since both lines are fibers of $\bar{\mathcal{G}}$ over the same point of $\mathbb{C}^\times$. To see this, it suffices to consider the situation of (5.7.1): there the assertion is clear since $\prod_{m^n = m'} (-1)^\ell m = (-1)^{n(\ell-1)+1} m'$. Finally, for compact X one has $m(M)_1 = 1$ and $\deg(D)$ is even, hence $\mathcal{E}^0(M)_{(D,c,\nu)} = \bar{\mathcal{G}}_1 = \mathbb{C}$, which is $5.1(\mathrm{iv})^0$. The constraints are mutually compatible by construction. We leave it to the reader to check that $\mathcal{E}^0$ is compatible with quadratic degenerations of X, so we have defined (5.7.4). Due to an evident identification $\bar{\mathcal{G}}^\times(\mathcal{E}^0) = \bar{\mathcal{G}}^\times$, (5.7.3) is left inverse to (5.7.4), so the theorem amounts to the next statement:

Theorem'. *For any $\mathcal{E}^0 \in \mathbb{E}^0$ there is a natural isomorphism $\iota : \mathcal{E}^0 \xrightarrow{\sim} \mathcal{E}^0(\bar{\mathcal{G}}^\times(\mathcal{E}^0))$.*

5.8 The next step reduces us to the setting of $\mathcal{D}$-modules with regular singularities. For a holonomic $\mathcal{D}$-module M we denote by M^{rs} the holonomic $\mathcal{D}$-module with regular singularities such that $B(M^{\mathrm{rs}}) = B(M)$, or, equivalently, $M^\infty = M^{\mathrm{rs}\infty}$, see 3.2. The functor $M \mapsto M^{\mathrm{rs}}$ sends nice families of $\mathcal{D}$-modules to nice families (as follows from 2.14), it is exact, commutes with π_*, and one has an evident identification $R\Gamma_{\mathrm{dR}}(X, M) \xrightarrow{\sim} R\Gamma_{\mathrm{dR}}(X, M^{\mathrm{rs}})$. Thus for any theory of ε-factors $\mathcal{E}$ the rule $M \mapsto {}^r\mathcal{E}(M) := \mathcal{E}(M^{\mathrm{rs}})$ is again a theory of ε-factors. Clearly r is an endofunctor of $\mathbb{E}$. The same formula defines an endofunctor r^0 of $\mathbb{E}^0$. It is naturally a Picard endofunctor, and r is a companion $\mathbb{E}^0$-torsor endofunctor: one has ${}^{r^0}\mathcal{E}^0 \otimes {}^{r^0}\mathcal{E}'^0 \xrightarrow{\sim} {}^{r^0}(\mathcal{E}^0 \otimes \mathcal{E}'^0)$, ${}^{r^0}\mathcal{E}^0 \otimes {}^r\mathcal{E} \xrightarrow{\sim} {}^r(\mathcal{E}^0 \otimes \mathcal{E})$.

Proposition. *The endofunctor r of $\mathbb{E}$ is naturally isomorphic to $\mathrm{id}_{\mathbb{E}}$. Namely, there is a unique $\kappa : \mathrm{id}_{\mathbb{E}} \xrightarrow{\sim} r$ such that $\kappa_{\mathcal{E},M} : \mathcal{E}(M) \to \mathcal{E}(M^{\mathrm{rs}})$ is the identity map if M has regular singularities. Same is true for $\mathbb{E}$ replaced by $\mathbb{E}^0$. Here $\kappa^0 : \mathrm{id}_{\mathbb{E}^0} \xrightarrow{\sim} r^0$ is an isomorphism of Picard endofunctors, and κ is an isomorphism of the companion $\mathbb{E}^0$-torsor endofunctors.*

Proof (i) Let us define a canonical isomorphism of factorization lines

$$\kappa = \kappa_{\mathcal{E},M} : \mathcal{E}(M) \xrightarrow{\sim} \mathcal{E}(M^{\mathrm{rs}}). \tag{5.8.1}$$

One has $M|_{X \backslash T} = M^{\mathrm{rs}}|_{X \backslash T}$, and over $X \backslash T$ our κ is the identity map. It remains to define $\kappa^{(1)} : \mathcal{E}(M)_b^{(1)} \xrightarrow{\sim} \mathcal{E}(M^{\mathrm{rs}})_b^{(1)}$ for $b \in T$ (see 1.6 and 1.15).

Pick a local parameter t at b. As in the proof of 5.3(i), M yields a $\mathcal{D}$-module $M^{(t)}$ on $\mathbb{P}^1$ with $\mathcal{E}(M)_b^{(1)} = \mathcal{E}(M^{(t)})_{(0,t^{-1}dt)}$. Ditto for M^{rs}. Since $M^{\mathrm{rs}(t)} = M^{(t)\mathrm{rs}}$ equals $M^{(t)}$ outside 0, constraints 5.1(iv) for $M^{(t)}$ and $M^{\mathrm{rs}(t)}$ yield isomorphisms $\mathcal{E}(M^{(t)})_{(0,t^{-1}dt)} \otimes \mathcal{E}(M^{(t)})_{(\infty,t^{-1}dt)} \xrightarrow{\sim} \det R\Gamma_{\mathrm{dR}}(\mathbb{P}^1, M^{(t)}) = \det R\Gamma_{\mathrm{dR}}(\mathbb{P}^1, M^{\mathrm{rs}(t)}) \xleftarrow{\sim} \mathcal{E}(M^{\mathrm{rs}(t)})_{(0,t^{-1}dt)} \otimes \mathcal{E}(M^{(t)})_{(\infty,t^{-1}dt)}$. Factoring out $\mathcal{E}(M^{(t)})_{(\infty,t^{-1}dt)}$, we get $\kappa^{(1)}$.

It remains to show that $\kappa^{(1)}$ does not depend on the auxiliary choice of t. The space of local parameters t is connected, so we need to check that κ is locally constant with respect to it. Let t_s be a family of local parameters at b that are defined on the same disc X_b and depend holomorphically on $s \in S$; then t_s identify X_{bS} with a neighborhood U of $\{0\} \times S$ in $\mathbb{P}^1_S$. Let $M_U^{(t)}$ be the pull-back of M by the projection $U \to X_b$, $(v, s) \mapsto t_s^{-1}(v)$. This is a holonomic $\mathcal{D}_U$-module; let $M^{(t)}$ be a holonomic $\mathcal{D}$-module on $\mathbb{P}^1_S$ which equals $M_U^{(t)}$ on U, is smooth outside $\{0, \infty\} \times S$, and is the $*$-extension with regular singularities at $\{\infty\} \times S$. The restriction of $M^{(t)}$ to any fiber equals $M^{(t_s)}$, i.e., $M^{(t_s)}$ form a nice isomonodromic family. The identifications $\mathcal{E}(M)_b^{(1)} \xrightarrow{\sim} \mathcal{E}(M^{(t_s)})_{(0,t^{-1}dt)}$ are horizontal, ditto for M^{rs}. We are done the compatibility of η with the Gauß-Manin connection.

The same construction (with $R\Gamma_{\mathrm{dR}}(\mathbb{P}^1, \cdot)$ replaced by $\mathbb{C}$) yields for $\mathcal{E}^0 \in \mathbb{E}^0$ a canonical isomorphism

$$\kappa^0 : \mathcal{E}^0(M) \xrightarrow{\sim} \mathcal{E}^0(M^{\mathrm{rs}}). \tag{5.8.2}$$

(ii) κ^0 and κ are evidently compatible with the tensor product of $\mathcal{E}^0$'s and $\mathcal{E}$'s, and with constraints 5.1(i)–(iii) and 5.1(i)^0–(iii)^0. It remains to check compatibility with constraint (iv). We treat the setting of $\mathbb{E}^0$ (which suffices, say, since $r^2 = r$). Suppose we have (X, T, M) with compact X. We need to prove that the composition $\mathcal{E}^0(M)(X) \xrightarrow{\kappa^0} \mathcal{E}^0(M^{\mathrm{rs}})(X) \xrightarrow{\eta(M^{\mathrm{rs}})} \mathbb{C}$ equals $\eta(M)$.

Let $b \in T$ be a point where M has non-regular singularity, and $M^{\mathrm{rs}b}$ be the $\mathcal{D}$-module which equals M outside of b and M^{rs} near b. Let $\kappa_b^0 : \mathcal{E}^0(M) \xrightarrow{\sim} \mathcal{E}^0(M^{\mathrm{rs}b})$ be equal to κ^0 near b and the identity morphism off b.

Lemma. *The composition* $\mathcal{E}^0(M)(X) \xrightarrow{\kappa_b^0} \mathcal{E}^0(M^{\mathrm{rs}b})(X) \xrightarrow{\eta(M^{\mathrm{rs}b})} \mathbb{C}$ *equals* $\eta(M)$.

The lemma implies the proposition: since $(M^{\mathrm{rs}b})^{\mathrm{rs}} = M^{\mathrm{rs}}$ and the composition $\mathcal{E}^0(M) \xrightarrow{\kappa_b^0} \mathcal{E}^0(M^{\mathrm{rs}b}) \xrightarrow{\kappa^0} \mathcal{E}(M^{\mathrm{rs}})$ equals κ^0, we are done by induction by the number of points of T where M has non-regular singularity.

Proof of Lemma. Let ν be a rational form on X such that $\mathrm{Res}_b \nu = 1$. Let t_b be a local parameter at b such that $t_b^{-1}dt_b = \nu$.

Consider a datum 2.13(a) with $Y = \mathbb{P}^1 \sqcup X$, $b_+ = b \in X$, $b_- = \infty \in \mathbb{P}^1$ and ν_Y equal to $t^{-1}dt$ on $\mathbb{P}^1$ and ν on X. Let t_+ be the parameter t_b, t_- be the parameter t^{-1} at ∞. The corresponding family of curves X' as defined in 5.2 (it was denoted by X there) over $Q = \mathbb{A}^1$ is the blow-up of $X \times \mathbb{A}^1$ at $(b, 0)$. We have a datum 2.13(b) with N equal to $M^{(t_b)}$ on $\mathbb{P}^1 \setminus \{\infty\}$ and to M on $X \setminus \{b\}$ (this determines N since it is !-extension with regular singularities at b_- and $*$-extension with regular singularities at b_+). Let L be any $t_\pm \partial_{t_\pm}$-invariant $b_\pm$-lattice in N such that the eigenvalues of $\pm t_\pm \partial_{t_\pm}$ on $L_{b_\pm}$ and their pairwise differences do not contain non-zero integers. Then the spectra of the $\pm t_\pm \partial_{t_\pm}$ actions on $L_{b_\pm}$ coincide, and there is a canonical identification $\alpha : L_{b_+} \xrightarrow{\sim} L_{b_-}$ characterized by the next property: Consider the $t_\pm \partial_{t_\pm}$-invariant embeddings $L_{b_-} \subset \Gamma(\mathbb{P}^1 \setminus \{0\}, L)$ and $L_{b_+} \subset \Gamma(U, L)$ as in 2.13(b). By the definition of $M^{(t_b)}$, its sections over a punctured neighborhood of 0 are identified with sections of M over $U \setminus \{b\}$; by this identification the subspaces $L_{b_\pm}$ correspond to one another, and α is the corresponding isomorphism. Let M' be the corresponding family of $\mathcal{O}_X$-modules with relative connection on X/Q (which was denoted by M in 5.2).

At $q = 1$ our M' equals M, so the top arrow in (5.2.3) at $q = 1$ equals $\eta(M)$. And the composition of its lower arrows equals the composition from the statement of our lemma. Since $\mathcal{E}^0 \in \mathbb{E}^0$, the diagram commutes; we are done. $\square$

5.9 Let us turn to the proof of Theorem$'$ in 5.7. For $\mathcal{E}^0 \in \mathbb{E}^0$ let $\bar{\mathcal{G}}^\times = \bar{\mathcal{G}}^\times$ be the corresponding extension of $\mathbb{G}_m$ (see (5.7.3)) and $\mathcal{E}^{0\prime} \in \mathbb{E}^0$ be the object defined by $\bar{\mathcal{G}}^\times$ (see (5.7.4)). We want to define a natural isomorphism $\iota : \mathcal{E}^0 \overset{\sim}{\to} \mathcal{E}^{0\prime}$.

For (X, T, M) as in 5.1 we define a canonical isomorphism of factorization lines

$$\iota : \mathcal{E}^0(M) \overset{\sim}{\to} \mathcal{E}^{0\prime}(M) \tag{5.9.1}$$

as follows. Due to isomorphism (5.8.2), we can assume that $M = M^{\mathrm{rs}}$. Now ι is a unique isomorphism of local nature which is compatible with constraints $5.1(\mathrm{ii})^0$ and $5.5(\mathrm{iii})$, and is the identity map for $M = M_m$ (see 5.6). Indeed, we can assume that X is a disc, $T = \{b\}$, and, by $5.5(\mathrm{iii})$, that M is $*$-extension at b. Then $M = \oplus M^{(m)}$, where the monodromy around b acts on $M^{(m)}$ with eigenvalues m. By compatibility with $5.1(\mathrm{ii})^0$, we can assume that $M = M^{(m)}$. Pick any filtration on M with successive quotients of rank 1 and define ι as the composition $\mathcal{E}^0(M) \overset{\sim}{\to} \otimes \mathcal{E}^0(\mathrm{gr}_i M) = \otimes \mathcal{E}^{0\prime}(\mathrm{gr}_i M) \overset{\sim}{\to} \mathcal{E}^{0\prime}(M)$ where $\overset{\sim}{\to}$ are constraints $5.1(\mathrm{ii})^0$. The choice of the filtration is irrelevant by 5.1^0 (for the space of filtrations is connected).

Our ι is compatible with constraints $5.1(\mathrm{i})^0$, $5.1(\mathrm{ii})^0$; its compatibility with $5.1(\mathrm{iii})^0$ will be checked in 5.13. We treat $5.1(\mathrm{iv})^0$ first; this takes 5.10–5.12. For (X, T, M) with compact X let $\xi(X, M) \in \mathbb{C}^\times$ be the ratio of $\eta(M)$ for $\mathcal{E}^0$ and the composition of $\eta(M)$ for $\mathcal{E}^{0\prime}$ with ι. We want to show that $\xi(X, M) \equiv 1$.

5.10 By 5.8 and the construction of ι, it suffices to consider M with regular singularities. By $5.5(\mathrm{iii})$, $\xi(X, M)$ depends only on $M|_{X \setminus T}$. Therefore, by compatibility with 5.1^0, $\xi(X, M)$ depends only on the purely topological datum of (the isomorphism class of) a punctured oriented surface $X \setminus T$ (that can be replaced by a compact surface with boundary) and a local system on it.

The compatibility with quadratic degenerations implies that if Y is obtained from X by cutting along a disjoint union of embedded circles and $N = M|_Y$, then $\xi(X, M) = \xi(Y, N)$. Here is an application:

Lemma. *(i) If M admits a filtration such that $\mathrm{gr}_i M$ are $\mathcal{D}$-modules of rank 1, then $\xi(X, M) = 1$.*
(ii) For every (X, T, M) with X connected one can find (X', T, M') such that the restriction of M to a neighborhood of T is isomorphic to that of M', X' is connected of any given genus $g \geq g(X)$, and $\xi(X, M) = \xi(X', M')$.
(iii) For every (X, T, M) one can find (X', T, M') with X' connected such that $\xi(X, M) = \xi(X', M')$ and for every $b \in T$ the restriction of M' to a neighborhood of b is isomorphic to that of M plus a direct sum of copies of a trivial $\mathcal{D}$-module.

Proof (i) By compatibility with $5.1(\mathrm{ii})^0$, we can assume that M is a $\mathcal{D}$-module of rank 1. Our assertion is true if X has genus 0 by the construction. An arbitrary X can be cut into a union of genus 0 surfaces, and we are done.

(ii) Consider $Y = X \sqcup Z$ where Z is a compact smooth connected curve of genus $g - g(X)$; let N be a $\mathcal{D}_Y$-module such that $N|_X = M$ and $N|_Z$ is a trivial $\mathcal{D}$-module of the same rank as M. Pick $x \in X \setminus T$, $z \in Z$, cut off small discs around x, z and connect their boundaries by a tube. This is X'. Take for M' any extension of

M_Y (restricted to the complement of the cut discs) to a local system on X'. Since $\xi(Z, N|_Z) = 1$ by (i), one has $\xi(Y, N) = \xi(X, M)$, hence $\xi(X', M') = \xi(X, M)$.

(iii) Let us construct (X', M'). First, add to M on different components of X appropriate number of copies of the trivial $\mathcal{D}$-module to assure that the rank of M is constant; this does not change $\xi(X, M)$ by (i). Let $X_1, \ldots, X_n$ be the connected components of X. On each $X_i \setminus T$, choose a pair of distinct points x_i, y_i. Cut off small discs around $x_1, \ldots, x_{n-1}$ and $y_2, \ldots, y_n$, and connect the boundary circle at x_i with that at y_{i+1} by a tube. This is our X'. Take for M' any extension of M to a local system on X'. $\qquad\square$

5.11 Proposition. *For X connected, $\xi(X, M)$ depends only on the datum of conjugacy classes of local monodromies of M (the rank of M is fixed).*

Proof According to [PX1], [PX2], the action of the mapping class group on the moduli of unitary local systems of given rank with fixed cojugacy classes of local monodromies, is ergodic (provided that the genus of the Riemann surface is > 1). As in Theorem 1.4.1 in [G], this implies that for connected X with $g(X) > 1$ our $\xi(X, M)$ depends only on $g(X)$, the rank of M, and the datum of conjugacy classes of local monodromies of M (indeed, ξ is invariant with respect to the action of the mapping class group by the compatibility with 5.1^0, and is holomorphic; by the ergodicity, its restriction to the real points of the moduli space of local systems is constant, and we are done). Use 5.10(ii) to eliminate the dependence on $g(X)$ (and the condition on $g(X)$). $\qquad\square$

5.12 For any (X, M), let $Sp(M)$ be the datum of other than 1 eigenvalues (with multiplicity) of the direct sum of local monodromies. We write it as an element $\Sigma n_i z_i$ (z_i are the eigenvalues, n_i are the multiplicities) of the quotient of $Div(\mathbb{C}^\times)$ modulo the subgroup of divisors supported at $1 \in \mathbb{C}^\times$.

Lemma. *$\xi(X, M)$ depends only on $Sp(M)$.*

Proof (i) By 5.10(iii), it suffices to check this assuming that X is connected, and by 5.10(i) we can assume that the rank of M is fixed. By 5.11, it suffices to find for any (X, M) some (X', M') such that $\xi(X, M) = \xi(X', M')$, $Sp(M) = Sp(M')$, and each local monodromy of M' has at most one eigenvalue different from 1. Take any $b \in T$; let m_b be the local monodromy at b. Then one can find a local system $K_{(b)}$ on $\mathbb{P}^1$ with ramification at ∞ and n other points, $n = \mathrm{rk}(M)$, such that its local monodromy at ∞ is conjugate to m_b^{-1}, and $K_{(b)}$ admits a flag of local subsystems such that each $\mathrm{gr}_i K_{(b)}$ has rank 1 and ramifies at ∞ and only one other point. Cut a small disc around b in X and that around ∞ in $\mathbb{P}^1$, and connect the two boundary circles by a tube; we get a surface $X'_{(b)}$. Let $M'_{(b)}$ be a local system on it that extends M and $K_{(b)}$. By 5.10(i), $\xi(\mathbb{P}^1, K_{(b)}) = 1$, so $\xi(X'_{(b)}, M'_{(b)}) = \xi(X, M)$ by the compatibility with quadratic degenerations. Repeating this construction for each point of T, we get (X', M'). $\qquad\square$

The lemma implies that $\xi(X, M) = 1$. Indeed, if $Sp(M) = \Sigma n_i z_i$, then $\Pi z_i^{n_i} = 1$ (for the product does not change if we replace M by $\det M$, where it equals 1 by the Stokes formula). Therefore one can find a $\mathcal{D}$-module M' of rank 1 on $\mathbb{P}^1$ with $Sp(M') = Sp(M)$. Since $\xi(X, M') = 1$, we are done by the lemma.

5.13 It remains to check that ι of (5.9.1) is compatible with 5.1(iii)^0. We want to show that for $\pi : X' \to X$ and a $\mathcal{D}$-module M' the diagram

$$
\begin{array}{ccc}
\mathcal{E}^0(\pi_* M') & \overset{\iota}{\longrightarrow} & \mathcal{E}^{0\prime}(\pi_* M') \\
\downarrow & & \downarrow \\
\pi_* \mathcal{E}^0(M') & \overset{\pi_* \iota}{\longrightarrow} & \pi_* \mathcal{E}^{0\prime}(M'),
\end{array}
\tag{5.13.1}
$$

where the vertical arrows are constraints 5.1(iii)^0 for $\mathcal{E}^0$, $\mathcal{E}^{0\prime}$, commutes. For $b \in X$ let $\psi(M', \pi, b)$ be the ratio of the morphisms $\mathcal{E}^0(\pi_* M')_b^{(1)} \to \pi_* \mathcal{E}^{0\prime}(M')_b^{(1)}$ that come from the two sides of the diagram. We want to show that $\psi(M', \pi, b) \equiv 1$ (see 1.6). It is clear that $\psi(M', \pi, b) = 1$ if π is unramified at b.

Our ψ has X-local nature and it is multiplicative with respect to disjoint unions of X', so it suffices to consider the case when X, X' are discs and $\pi = \pi^{(n)}$ is ramified of index n at b. Choosing a local coordinate t at b, we identify X' and X with neighborhoods of 0 in $\mathbb{P}^1$ so that our covering is the restriction of $\pi : \mathbb{P}^1 \to \mathbb{P}^1$, $t \mapsto t^n$, to X. Let us extend M' to a $\mathcal{D}$-module on $\mathbb{P}^1$, which we again denote by M', such that it is smooth outside 0 and 1. We know that 5.1(iii)^0 is compatible with 5.1(iv)^0 for both $\mathcal{E}^0$ and $\mathcal{E}^{0\prime}$. Since ι is compatible with 5.1(iv)^0 and π is ramified only at 0 and ∞, we know that $\psi(M', \pi, 0)\psi(M', \pi, \infty) = 1$. Since M' is smooth over ∞, this means that $\psi(M', \pi^{(n)}, b)\psi(\mathcal{O}_{X'}, \pi^{(n)}, b)^{\mathrm{rk}(M)} = 1$. We are reduced to the case $M' = \mathcal{O}_{X'}$.

Set $\psi_n := \psi(\mathcal{O}_{X'}, \pi^{(n)}, b)$. By above, $\psi_n^2 = 1$, i.e., $\psi_n = \pm 1$. By the construction of $\mathcal{E}^{0\prime}$ (see 5.7), one has $\psi_2 = 1$. Due to compatibility with the composition, one has $\psi_{mn} = \psi_n \psi_m^n = \psi_m \psi_n^m$, i.e, $\psi_n^{m-1} = \psi_m^{n-1}$. For $m = 2$ we get $\psi_n \equiv 1$, q.e.d. $\square$

6 The Γ-function.

6.1 Let us describe explicitly the ε-period map $\rho^\varepsilon = \rho^\varepsilon(M) : \mathcal{E}_{\mathrm{dR}}(M) \overset{\sim}{\to} \mathcal{E}_{\mathrm{B}}(M)$ for $\rho^\varepsilon \in E_{\mathrm{B}/\mathrm{dR}}$ (see 5.4).

By 5.8, $\rho^\varepsilon(M)$ equals the composition $\mathcal{E}_{\mathrm{dR}}(M) \overset{\kappa}{\to} \mathcal{E}_{\mathrm{dR}}(M^{\mathrm{rs}}) \overset{\sim}{\to} \mathcal{E}_{\mathrm{B}}(M^{\mathrm{rs}}) = \mathcal{E}_{\mathrm{B}}(M)$, where κ is the canonical isomorphism of (5.8.1) and $\overset{\sim}{\to}$ is $\rho^\varepsilon(M^{\mathrm{rs}})$. A different construction of the same κ was given in (3.1.1) in terms of certain analytical Fredholm determinant (a version of τ-function).

From now on we assume that M has regular singularities.

By (1.6.3), Example in 1.6, and Remark in 1.15, $\mathcal{E}_?(M)$ amounts to a datum $(\mathcal{E}_?(M)_{X \backslash T}^{(1)}, \{\mathcal{E}_?(M)_b^{(1)}\})$. Thus ρ^ε is completely determined by the isomorphisms $\mathcal{E}_{\mathrm{dR}}(M)_{X \backslash T}^{(1)} \overset{\sim}{\to} \mathcal{E}_{\mathrm{B}}(M)_{X \backslash T}^{(1)}$ and $\mathcal{E}_{\mathrm{dR}}(M)_b^{(1)} \overset{\sim}{\to} \mathcal{E}_{\mathrm{B}}(M)_b^{(1)}$, $b \in T$.

Let us write a formula for $\rho^\varepsilon = \rho_b^\varepsilon : \mathcal{E}_{\mathrm{dR}}(M)_b^{(1)} \overset{\sim}{\to} \mathcal{E}_{\mathrm{B}}(M)_b^{(1)}$, $b \in T$. Below t is a local parameter at b, and i_b, j_b are the embeddings $\{b\} \hookrightarrow X_b \hookleftarrow X_b^o := X_b \backslash \{b\}$.

If M is supported at b, then $\mathcal{E}_{\mathrm{dR}}(M)_b^{(1)} = \mathcal{E}_{\mathrm{B}}(M)_b^{(1)} = \det R\Gamma_{\mathrm{dR}}(X, M)$ and ρ_b^ε is the identity map. Thus for arbitrary M one has $\mathcal{E}_?(M)_b^{(1)} = \mathcal{E}_?(j_{b*} M)_b^{(1)} \otimes \det R\Gamma_{\mathrm{dR}\,\{b\}}(X, M)$ and

$$
\rho_b^\varepsilon(M) = \rho_b^\varepsilon(j_{b*} M) \otimes \mathrm{id}_{\det R\Gamma_{\mathrm{dR}\,\{b\}}(X, M)}.
\tag{6.1.1}
$$

So it suffices to define ρ_b^ε for $M = j_{b*} M$. Then we have a canonical trivialization $1_b^!$ of $\mathcal{E}_B(M)_b^{(1)} := \mathcal{E}_B(M)_{(b, t^{-1} dt)}$, see 4.7.

Let L be a $t\partial_t$-invariant b-lattice in M. Denote by $\Lambda(L)$ the spectrum of the operator $t\partial_t$ acting on on $L_b = L/tL$. Suppose that it does not contain positive integers. Then the complex $\mathcal{C}(L, \omega L(b))$ (see 2.4) is acyclic; here $L(b) := t^{-1}L$, i.e., $\omega L(b) = t^{-1}dtL$. Denote by $\iota(L)_{zt^{-1}dt}$ the corresponding trivialization of $\mathcal{E}_{\mathrm{dR}}(M)_{(b,zt^{-1}dt)} \xrightarrow{\sim} \det \mathcal{C}(L, \omega L(b))$, $z \neq 0$ (see (2.5.6)); it does not depend on the choice of t. If $L' \supset L$ is another lattice, then $\iota(L)_{zt^{-1}dt}/\iota(L')_{zt^{-1}dt}$ is the determinant of the action of $z^{-1}t\partial_t$ on L'/L (see 2.5). In particular, we have a trivialization $\iota(L)_{t^{-1}dt}$ of $\mathcal{E}_{\mathrm{dR}}(M)_b^{(1)}$.

We write $\rho_b^\varepsilon(\iota(L)_{t^{-1}dt}) = \gamma_{\rho^\varepsilon}^!(L)1_b^!$. For example, if $M = M_t^\lambda$ is the $\mathcal{D}$-module M_t^λ generated by t^λ, $t\partial_t(t^\lambda) = \lambda t^\lambda$, where $\lambda \in \mathbb{C} \setminus \mathbb{Z}_{>0}$, and $L = L_t^\lambda$ is the lattice generated by t^λ, then $\Lambda(L_t^\lambda) = \{\lambda\}$, and we write $\gamma_{\rho^\varepsilon}^!(\lambda) := \gamma_{\rho^\varepsilon}^!(L_t^\lambda)$.

Theorem. *(i) One has* $\gamma_{\rho^\varepsilon}^!(L) = \prod_{\lambda \in \Lambda(L)} \gamma_{\rho^\varepsilon}^!(\lambda)$.

(ii) For $a \in \mathbb{Z}$ one has[28] $\gamma_{\rho^\varepsilon+a}^!(\lambda) = (-1)^a \exp(-2\pi i\lambda a)\gamma_{\rho^\varepsilon}^!(\lambda)$.
(iii) For one ρ^ε in $E_{B/\mathrm{dR}}$ one has

$$\gamma_{\rho^\varepsilon}^!(\lambda) = (2\pi)^{-1/2}(1 - \exp(2\pi i\lambda))\Gamma(\lambda), \qquad (6.1.2)$$

where Γ is the Euler Γ-function and $(2\pi)^{1/2}$ is the positive square root.

Proof (i) By above, for $L' \supset L$ one has $\gamma_{\rho^\varepsilon}^!(L)/\gamma_{\rho^\varepsilon}^!(L') = \det(t\partial_t; L'/L)$. Therefore the validity of (i) does not depend on the choice of L. So we can assume that (M, L) is a successive extension of some $(M_t^\lambda, L_t^\lambda)$, and we are done since all our objects are multiplicative with respect to extensions.

(ii) By 5.3(iii), $\mu(M_t^\lambda)$ acts on $\mathcal{E}(M_t^\lambda)_b^{(1)}$ as multiplication by $-\exp(-2\pi i\lambda)$.
(iii) The claim follows from (ii) and the next lemma:

Lemma. *(i) The function $\gamma_{\rho^\varepsilon}^!(\lambda)$ is holomorphic and invertible for $\lambda \in \mathbb{C} \setminus \mathbb{Z}_{>0}$, and satisfies the next relations: (a) $\gamma_{\rho^\varepsilon}^!(\lambda + 1) = \lambda\gamma_{\rho^\varepsilon}^!(\lambda)$; (b) For every positive integer n one has $\gamma_{\rho^\varepsilon}^!(\frac{\lambda}{n})\gamma_{\rho^\varepsilon}^!(\frac{\lambda+1}{n})\cdots\gamma_{\rho^\varepsilon}^!(\frac{\lambda+n-1}{n}) = n^{\frac{1}{2}-\lambda}\gamma_{\rho^\varepsilon}^!(\lambda)$.*
(ii) Any function γ that satisfies the properties from (i) equals one of the functions $\gamma_a(\lambda) = (2\pi)^{-1/2}(-1)^a \exp(2\pi i\lambda a)(1 - \exp(2\pi i\lambda))\Gamma(\lambda)$ for some integer a.

Proof of Lemma. (i) We check (b); the rest is clear. Let $\pi : X' \to X$ is a covering of a disc completely ramified of index n at b, so for a parameter t' at b' one has $\pi^*(t) = t'^n$, hence $\pi^*(t^{-1}dt) = nt'^{-1}dt'$. Let M' be a $\mathcal{D}$-module on X' which is the $*$-extension with regular singularities at b'. Consider isomorphisms

$$\mathcal{E}_{\mathrm{dR}}(\pi_*M')_{(x,t^{-1}dt)} \xrightarrow{\alpha} \mathcal{E}_{\mathrm{dR}}(M')_{(x',mt'^{-1}dt')} \xrightarrow{\beta} \mathcal{E}_{\mathrm{dR}}(M')_{(x',t'^{-1}dt')} \qquad (6.1.3)$$

where α is the projection formula identification and β is the ∇^ε-parallel transport along the interval $[m, 1]t'^{-1}dt'$. By the construction of $1_b^!$, the Betti version of $\beta\alpha$ transforms $1_b^!$ to $1_{b'}^!$.

Suppose M' equals $M_{t'}^\lambda$ for some $\lambda \in \mathbb{C}$. Then π_*M' equals $M_t^{\lambda/n} \oplus M_t^{(\lambda+1)/n} \oplus \cdots \oplus M_t^{(\lambda+n-1)/n}$. If $L' = L_{t'}^\lambda \subset M_{t'}^\lambda$, then $\pi_*L' = L_t^{\lambda/n} \oplus L_t^{(\lambda+1)/n} \oplus \cdots \oplus L_t^{(\lambda+(n-1))/n}$. It is clear that α sends $\iota(\pi_*L')_{t^{-1}dt}$ to $\iota(L')_{nt'^{-1}dt'}$. Since $\iota(L')$ is horizontal for the connection ∇_0 of (2.11.3) (with ℓ and n in loc. cit. equal to 1), β sends $\iota(L')_{nt'^{-1}dt'}$ to $n^{\frac{1}{2}-\lambda}\iota(L')_{t'^{-1}dt'}$. Now $\rho^\varepsilon(\iota(L')_{t'^{-1}dt'}) = \gamma_{\rho^\varepsilon}^!(\lambda)1_{b'}^!$

---

[28]Recall that $E_{\mathrm{B/dR}}$ is a $\mathbb{Z}$-torsor.

and $\rho^{\varepsilon}(\iota(\pi_{*}L')_{t^{-1}dt}) = \gamma^{!}_{\rho^{\varepsilon}}(\pi_{*}L')1^{!}_{b} = \gamma^{!}_{\rho^{\varepsilon}}(\frac{\lambda}{n})\gamma^{!}_{\rho^{\varepsilon}}(\frac{\lambda+1}{n})\cdots\gamma^{!}_{\rho^{\varepsilon}}(\frac{\lambda+n-1}{n})1^{!}_{b}$, and we are done since ρ^{ε} is compatible with 5.1(iii).

(ii) Denote by E the set of functions γ that satisfy properties from (i). Let E^{0} be the set of functions $e(\lambda)$ which are invertible and holomorphic on the whole $\mathbb{C}$ and satisfy the relations (a) $e(\lambda+1) = e(\lambda)$; (b) $e(\frac{\lambda}{n})e(\frac{\lambda+1}{n})\cdots e(\frac{\lambda+n-1}{n}) = e(\lambda)$ for any positive integer n. Then E^{0} is a group with respect to multiplication, and E is an E^{0}-torsor.

Notice that the function $\mu(\lambda) := -\exp(-2\pi i\lambda)$ belongs to E^{0}, and $\{\lambda_{a}\}_{a\in\mathbb{Z}}$ is a $\mu^{\mathbb{Z}}$-torsor. Recall that $\Gamma(\lambda)$ is holomorphic and invertible for $\lambda \in \mathbb{C} \setminus \mathbb{Z}_{\leq 0}$, and satisfies the next relations: (a) $\Gamma(\lambda+1) = \lambda\Gamma(\lambda)$; (b) $\Gamma(\frac{\lambda}{n})\dots\Gamma(\frac{\lambda+n-1}{n}) = (2\pi)^{\frac{n-1}{2}}n^{\frac{1}{2}-\lambda}\Gamma(\lambda)$ for any positive integer n. This implies that γ_{a} belong to E. To prove the lemma, it remains to check that μ generates E^{0}.

Pick any $e \in E^{0}$; let a be the index of the holomorphic map $e : \mathbb{C}/\mathbb{Z} \to \mathbb{C}^{\times}$. Let us check that $e\mu^{a} \equiv 1$. Indeed, $e\mu^{a}$ has index 0, so $e\mu^{a}(\lambda) = \exp f(\lambda)$ for some holomorphic $f : \mathbb{C}/\mathbb{Z} \to \mathbb{C}$. Notice that for any $n \in \mathbb{Z}_{>0}$ the function $\lambda \mapsto f(\frac{\lambda}{n}) + f(\frac{\lambda+1}{n}) + \dots + f(\frac{\lambda+n-1}{n}) - f(\lambda)$ takes values in $2\pi i\mathbb{Z}$, hence constant. Consider the coefficients of the Laurent series $f(\lambda) = \Sigma b_{m}\exp(2\pi im\lambda)$. The above property implies that $(n-1)b_{\pm n} = 0$ for $n > 1$, i.e., $b_{m} = 0$ for $|m| > 1$. The case $n = 2$ shows that $b_{\pm 1} = 0$. Finally the fact that $(n-1)b_{0} \in 2\pi\mathbb{Z}$ for any $n > 0$ implies that $b_{0} \in 2\pi i\mathbb{Z}$, and we are done. $\square$

Corollary. *For ρ^{ε} as in (6.1.2) the isomorphism $\rho^{\varepsilon} : \mathcal{E}_{dR}(M)^{(1)}_{X\setminus T} \xrightarrow{\sim} \mathcal{E}_{B}(M)^{(1)}_{X\setminus T}$ equals the composition $\mathcal{E}_{dR}(M)^{(1)}_{X\setminus T} \xrightarrow{(2.6.1)} (\det M_{X\setminus T})^{\otimes-1} \xrightarrow{(4.7.3)} \mathcal{E}_{B}(M)^{(1)}_{X\setminus T}$ multiplied by $((2\pi)^{1/2}i)^{rk(M)}$. Replacing ρ^{ε} by $\rho^{\varepsilon}+a$ multiplies it by $(-1)^{rk(M)a}$.*

Proof Suppose M is smooth at b. Compatibility with 5.1(iii) implies, as in Remark in 5.3, that $\rho^{\varepsilon}_{b} : \mathcal{E}_{\mathrm{dR}}(M)^{(1)}_{b} \xrightarrow{\sim} \mathcal{E}_{\mathrm{B}}(M)^{(1)}_{b}$ does not depend on whether b is viewed as a point of T or not. The exact sequence $0 \to M \to j_{b*}j^{*}_{b}M \to i_{b*}M_{b} \to 0$ shows that $R\Gamma_{\mathrm{dR}\,b}(X, M) = M_{b}[-1]$, hence $\mathcal{E}_{?}(M)^{(1)}_{b} = \mathcal{E}_{?}(j_{b*}M)^{(1)} \otimes (\det M_{b})^{\otimes-1}$. The isomorphisms $\mathcal{E}_{?}(M)^{(1)}_{b} \xrightarrow{\sim} (\det M_{b})^{\otimes-1}$ come from trivializations of $\mathcal{E}_{?}(j_{b*}M)^{(1)}$, which are $\iota(M)_{t^{-1}dt}$ in the de Rham and $1^{!}_{b}$ in the Betti case. Since $\gamma^{!}_{\rho^{\varepsilon}}(M) = ((2\pi)^{1/2}i)^{\mathrm{rk}(M)}$ by the theorem, we are done. $\square$

The corollary together with the theorem completely determines $\rho^{\varepsilon}(M)$.

6.2   Here is another explicit formula for

$$\rho^{\varepsilon} : \mathcal{E}_{\mathrm{dR}}(M)_{(b,-t^{-1}dt)} \xrightarrow{\sim} \mathcal{E}_{\mathrm{B}}(M)_{(b,-t^{-1}dt)}.$$

Recall that $\mathcal{E}_{\mathrm{B}}(M)_{(b,-t^{-1}dt)} \xrightarrow{\sim} \det R\Gamma_{\mathrm{dR}}(X_{b}, M)$ by (4.7.1); here X_{b} is a small open disc at b. Let L be a $t\partial_{t}$-invariant b-lattice in M such that $\Lambda(L)$ does not contain non-positive integers, L_{ω} be the $\mathcal{O}$-submodule of ωM generated by $\nabla(L)$ (this is a b-lattice). Then the projection

$$\Gamma(X_{b}, dR(M)) \twoheadrightarrow C(L, L_{\omega}) \tag{6.2.1}$$

is a quasi-isomorphism. Together with isomorphism $r_{L,-t^{-1}dt} : \mathcal{E}_{\mathrm{dR}}(M)_{(b,-t^{-1}dt)} \xrightarrow{\sim} \det C(L, L_{\omega})$ from (2.5.6), it yields an identification $e(L) : \mathcal{E}_{\mathrm{dR}}(M)_{(b,-t^{-1}dt)} \xrightarrow{\sim} \mathcal{E}_{\mathrm{B}}(M)_{(b,-t^{-1}dt)}$. Thus $\rho^{\varepsilon}_{(b,-t^{-1}dt)} = \gamma^{*}_{\rho^{\varepsilon}}(L)e(L)$ for some $\gamma^{*}_{\rho^{\varepsilon}}(L) \in \mathbb{C}^{\times}$.

Proposition. *One has* $\gamma^*_{\rho^\varepsilon}(L) = \prod\limits_{\lambda \in \Lambda(L)} \gamma^*_{\rho^\varepsilon}(\lambda)$, *and for* ρ^ε *as in (6.1.2)*

$$\gamma^*_{\rho^\varepsilon}(\lambda) = (2\pi)^{-1/2} \exp(\pi i(\lambda - 1/2)) \Gamma(\lambda). \qquad (6.2.2)$$

Proof If $L' \supset L$ is another lattice as above, then $e(L')/e(L)$ is the determinant of the action of $-t\partial_t$ on tL'/tL, so the validity of the assertion does not depend on the choice of L. It is compatible with filtrations, and holds for M supported at b, so we can assume that M has rank 1 and is the $*$-extension at b. Thus $\Lambda(L) = \{\lambda\}$; by continuity, it suffices to consider the case of $\lambda \notin \mathbb{Z}$. Then the complexes in (6.2.1) are acyclic. The corresponding trivializations of $\mathcal{E}_{dR}(M)_{(b,-t^{-1}dt)}$ and $\mathcal{E}_B(M)_{(b,-t^{-1}dt)}$ are $\iota(L)_{-t^{-1}dt}$ from 6.1 and 1_b^* from 4.7; by construction, $e(L)(\iota(L)_{-t^{-1}dt}) = 1_b^*$.

By (2.11.3), the counterclockwise monodromy from $t^{-1}dt$ to $-t^{-1}dt$ sends $\iota(L)_{t^{-1}dt}$ to $\exp(\pi i(\lambda - 1/2))\iota(L)_{-t^{-1}dt}$. According to 4.8, the same monodromy sends $1_b^!$ to $(1 - \exp(-2\pi i\lambda))^{-1}1_b^*$. Since ρ^ε is horizontal, one has $\gamma^*_{\rho^\varepsilon}(\lambda) = (1 - \exp(-2\pi i\lambda))^{-1} \gamma^!_{\rho^\varepsilon}(\lambda) \exp(\pi i(1/2 - \lambda)) = \exp(\pi i(\lambda - 1/2))(1 - \exp(2\pi i\lambda))^{-1} \gamma^!_{\rho^\varepsilon}(\lambda)$, and we are done by (6.1.2). $\qquad\square$

Example. If M is smooth at b and $L = tM$, then isomorphism (6.2.1) is $\Gamma(X_b, M^\nabla) \xrightarrow{\sim} M_b$, $m \mapsto m_b$, and $\gamma^*_{\rho^\varepsilon}(tM) = ((2\pi)^{-\frac{1}{2}}i)^{\mathrm{rk}(M)}e(L)$.

Exercise. Deduce Euler's reflection formula $\Gamma(\lambda)\Gamma(1 - \lambda) = \pi \sin^{-1}(\pi\lambda)$ from (6.1.2), (6.2.2), the lemma in 2.7, and Exercise in 4.7.

6.3   Let us write down a formula for the factors $[\rho^\varepsilon_{(O,\nu)}]$ from 0.3.

Recall that we have X, M and ν defined over a subfield k of $\mathbb{C}$, and $B(M)$ is defined over a subfield k'. The finite set of singular points of M and ν is defined then over k; it is partitioned by $\mathrm{Aut}(\mathbb{C}/k)$-orbits. Let O be such an orbit. The $\mathbb{C}$-line $\mathcal{E}_{dR}(M)_{(O,\nu)} = \bigotimes\limits_{x \in O} \mathcal{E}_{dR}(M)_{(x,\nu)}$ is defined over k by §2,[29] and $\mathcal{E}_B(M)_{(O,\nu)} = \bigotimes\limits_{x \in O} \mathcal{E}_B(M)_{(x,\nu)}$ is defined over k' by §4. Computing $\rho^\varepsilon : \mathcal{E}_{dR}(M)_{(O,\nu)} \xrightarrow{\sim} \mathcal{E}_B(M)_{(O,\nu)}$, $\rho^\varepsilon \in E_{B/dR}$, in k- and k'-bases, we get a number whose class $[\rho^\varepsilon_{(O,\nu)}]$ in $\mathbb{C}^\times/k'^\times k^\times$ does not depend on the choice of the bases and the choice of ρ^ε in $E_{B/dR}$. Let us compute $[\rho^\varepsilon_{(O,\nu)}]$ explicitly assuming that M has regular singularities.

For $b \in O$ let $k_b \subset \mathbb{C}$ be its field of definition; let X_b be a small disc around b. Choose an auxiliary datum on the de Rham side: it is (t, L, u, v), where t is a parameter at b, L is a $t\partial_t$-invariant b-lattice in M, u is a non-zero vector in $\det L_b$, and v is a non-zero vector in $\det \mathcal{C}(L, L_\omega)$ (see 6.2); we assume that (t, L, u, v) are defined over k_b. Let Λ_b be the spectrum (with multiplicities) of $t\partial_t$ acting on the fiber L_x; we assume that Λ_b does not contain non-positive integers. An auxiliary datum on the Betti side is (ϕ, w), where ϕ is a non-zero horizontal section of $\det M$ over the half-disc $\mathrm{Re}(t) > 0$, which is defined over k' (with respect to the Betti k'-structure on the sheaf of horizontal sections), w is a non-zero vector in $\det R\Gamma_{dR}(X_b, M)$ defined over k'.

The data yield numbers: The leading term of ν at b is $\alpha_b t^{-\ell} dt$, $\alpha_b \in k_b^\times$; let $r_b \in k_b$ be the trace of $t\partial_t$ acting on L_b. Notice that $m_b := \exp(-2\pi r_b)$ is the monodromy of $\det M_b^\nabla$ around b, so $m_b \in k'^\times$. Then the section $t^{r_b}\phi$ on the half-disc extends

---

[29]The group $\mathrm{Aut}(\mathbb{C}/k)$ acts on $\mathcal{E}_{dR}(M)_{(O,\nu)}$ by transport of structure; its fixed points is the k-structure on $\mathcal{E}_{dR}(M)_{(O,\nu)}$.

to an invertible holomorphic section of $\det L$ on X_b; set $\beta_b := (t^{r_b}\phi)_b/u \in \mathbb{C}^\times$. Let $\delta_b \in \mathbb{C}^\times$ be the ratio of v and the image of w by the determinant of (6.2.1).

Let us compute the numbers α_b, β_b, δ_b and the spectrum Λ_b for each $b \in O$ using Galois-conjugate de Rham side data. Set $n := \mathrm{rk}(M)$.

Proposition. *One has*

$$[\rho^\varepsilon_{(O,\nu)}] = \prod_{b \in O} (2\pi)^{-\frac{n\ell}{2}} i^{n\frac{\ell(\ell-1)}{2}} m_b^{\frac{\ell-1}{2}} \alpha_b^{\frac{n\ell}{2}-r_b} \beta_b^{\ell-1} \delta_b \prod_{\lambda \in \Lambda_b} \Gamma(\lambda). \qquad (6.3.1)$$

Proof For the sake of clarity, we do the computation assuming that b is a k-point, leaving the general case to the reader.

As follows from (2.11.3), the validity of formula does not depend on α_b. Notice that the class of $\alpha_b^{-r_b} := \exp(-r_b \log(\alpha_b))$ in $\mathbb{C}^\times/k'^\times$ is well defined: adding $2\pi i$ to the logarithm multiplies the exponent by $m_b \in k'^\times$.

If $\ell = 1$ and $\alpha_b = -1$, then the formula follows from (6.2.2).

To finish the proof, it remains to check that

$$[\rho^\varepsilon_{(b,-t^{-\ell-1}dt)}] = (2\pi)^{-\frac{n}{2}} i^n \beta [\rho^\varepsilon_{(b,t^{-\ell}dt)}].$$

Consider a family of forms $\nu_x := t^{-\ell}(x-t)^{-1}dt$. Then $\mathcal{E}_{\mathrm{dR}}(M)_{(x,1_b,\nu_x)} = \mu^\nabla(t^{\ell-1}(t-x)L/L_\omega) = \det \mathcal{C}(L,L_\omega) \otimes \lambda(L/t^{\ell-1}(x-t)L)^{\otimes-1} = \mathcal{C}(L,L_\omega) \otimes \lambda(L/t^{\ell-1}L)^{\otimes-1} \otimes \lambda(t^{\ell-1}L/t^{\ell-1}(x-t)L)^{\otimes-1}$. We fix a non-zero l in $\lambda(L/t^{\ell-1}L)$ defined over k. Any local trivialization g of $\det L$ yields then a trivialization $e(g)_x := v \otimes l^{-1} \otimes t^{n(1-\ell)} g_x^{-1}$ of $\mathcal{E}_{\mathrm{dR}}(M)_{(x,1_b,\nu_x)}$; if g is defined over k, then so is $e(g)$.

The leading terms of ν_x at $t = 0$ and $t = x$ are $x^{-1}t^{-\ell}dt$ and $x^{-\ell}(x-t)^{-1}dt$. Applying (2.11.2) to the x-lattice $(x-t)L$ and (2.11.3) to the b-lattice $t^{\ell-1}L$, we see that $e(t^{r_b}\phi)_x = v \otimes (x^{-\frac{n(\ell-2)}{2}-r_b}\ell^{-1}) \otimes (x^{-\frac{n\ell}{2}}\phi^{-1})$ is a horizontal (with respect to x) section of $\mathcal{E}_{\mathrm{dR}}(M)_{(x,1_b,\nu_x)}$. Since the value at b of $\beta_b t^{n(1-\ell)}(t^{r_b}\phi)^{-1}$ is a generator of $\det(t^{\ell-1}L/t^\ell L)^{\otimes-1}$ defined over k, we see that $\beta_b e(t^{r_b}\phi)_b \in \mathcal{E}_{\mathrm{dR}}(M)_{(b,-t^{-\ell-1}dt)}$ is defined over k. If s a horizontal section of $\mathcal{E}_{\mathrm{B}}(M)_{(x,1_b,\nu_x)}$ over X_b defined over k', then $\rho^\varepsilon(e(t^{r_b}\phi))/s$ is a constant function. Its value at $x = 0$, i.e., at b, belongs to $\beta_b^{-1}[\rho^\varepsilon_{(b,-t^{-\ell-1}dt)}]$. By factorization and Example in 6.2, its value at $x = 1$ belongs to $[\rho^\varepsilon_{(b,t^{-\ell}dt)}](2\pi)^{-\frac{n}{2}}i^n$, and we are done. $\qquad \square$

Notation. a_ψ 2.9; $\mathcal{C}(U)$ 4.2; $\mathcal{C}(W,\mathcal{N})$ 4.4; $\mathcal{C}(L,L_\omega)$ 2.4; $\mathfrak{D}$, $\mathfrak{D}^\circ$, (D,c,ν_P), (D,c,ν) 1.1; $|D|$ 1.1; $dR(L,L_\omega)$ 2.7; $\mathcal{D}et_{P/S}(E)$, $\mathcal{D}et_{P/S}(E_1/E_2)$ 2.3; $\mathrm{Div}(X)$ 1.1; ${}^w\mathbb{E}$, ${}^w\mathbb{E}^0$ 5.1; $\mathbb{E}$, $\mathbb{E}^0$ 5.2; $\mathcal{E}$ 1.2; $\mathcal{E}^{(\ell)}$ 1.6; $E_{\mathrm{dR/B}}$ 5.4; e, e_L 2.5; $\mathcal{E}_B(M)$ 4.6; $\mathcal{E}_{\mathrm{dR}}(M)$ 2.5; $\mathcal{I}(U,\mathcal{N})$ 4.3; $G^\flat$ 2.2; $Hom_\mathcal{L}(J_1,J_2)$ 2.2; $j_{T*}M$ 2.4; $K(D)$, $K(D)^\times_{D,c}$ 1.1; $\mathcal{K}^{(\ell)}$ 1.1; $K^\times_{T_b}$ 1.1; $\mathcal{L}_?$, $\mathcal{L}_k$, $\mathcal{L}_O$, $\mathcal{L}_{\mathrm{dR}}$ 1.2; $\mathcal{L}^{\mathrm{inv}}_{\mathrm{dR}}(K^\times_{T_b})$ 1.11; $\mathcal{L}^0_{\mathrm{dR}}$ 1.7; $\mathcal{L}_{\mathrm{dR}}(X,T)$, $\mathcal{L}_{\mathrm{dR}}(X,T)^{(\ell)}$, $\mathcal{L}_{\mathrm{dR}}(X \setminus T)^{(\ell)}$ 1.6; $\mathcal{L}^\Phi_?$ 1.5; $\mathcal{L}^\natural_{\mathrm{dR}}(X,T;K)$ 1.11; $\mathcal{L}^\Phi_{\mathrm{dR}}(X,T)^{O\text{-triv}}$, $\mathcal{L}^\Phi_?(X,T;K)_T$ 1.13; $\mathcal{M}(\mathcal{C}(U))$ 4.1; $O^\times_{D,c}$ 1.1; $O^\times_{T_b}$ 1.1; $P_{D,c}$ 1.1; Rat 1.4; $r_{L,\nu}$ 2.5; $\mathcal{V}_{\mathrm{crys}}$ 1.2; $\mathcal{V}^{\mathrm{sm}}$ 1.3; τ_ψ 2.9; $\tau_\mathcal{N}$ 4.3; T, T^c_S 1.1; $\gamma^*_{\rho^\varepsilon}(L)$, $\gamma^*_{\rho^\varepsilon}(\lambda)$ 6.2; $\gamma^!_{\rho^\varepsilon}(L)$, $\gamma^!_{\rho^\varepsilon}(\lambda)$ 6.1; λ_P 2.3; $\lambda(F)$ 4.2; $\Lambda(L)$ 6.1; μ^∇_P 2.4; μ^ν_P 2.5; μ^ψ 2.9; $\mu^{\nabla/\kappa}_P$, $\mu^{\kappa/\nu}_P$ 2.10; $\pi_0(X)$, $\pi_0(\mathcal{L})$, $\pi_1(\mathcal{L})$ 1.1; ϕ_κ 2.10; ∇^ε 2.11; $\omega(X,T)$ 1.12; $\Omega^1(K^\times_T)^{\mathrm{inv}}$ 1.10; τ_ψ 2.9; $1^!_b$, 1^*_b 4.7; 2^T 1.1.

References

[A]   G. Anderson, *Local factorization of determinants of twisted DR cohomology groups*, Compositio Math. **83** (1992), no. 1, 69–105.

[B]   A. Beilinson, *Topological $\mathcal{E}$-factors*, Pure Appl. Math. Q. **3** (2007), no. 1, 357–391.

[BBE]   A. Beilinson, S. Bloch, H. Esnault, *$\mathcal{E}$-factors for Gauß-Manin determinants*, Moscow Mathematical Journal **2** (2002), no. 3, 477–532.

[BD]   A. Beilinson, V. Drinfeld, *Quantization of Hitchin's integrable system and Hecke eigensheaves*, http://www.math.uchicago.edu/~mitya/langlands.html

[BG]   A. Beilinson, D. Gaitsgory, *A corollary of the b-function lemma*, math.AG 0810.1504 (2008).

[Bj]   J.-E. Björk, *Analytic $\mathcal{D}$-modules and applications*, Mathematics and its Applications, Kluwer Academic Publishers, Dordrecht-Boston-London, 1993.

[BDE]   S. Bloch, P. Deligne, H. Esnault, *Periods for irregular connections on curves* (2005).

[BE]   S. Bloch, H. Esnault, *Homology for irregular connections*, J. Théor. Nombres Bordeaux **16** (2004), no. 2, 357–371.

[CC]   C. Contou-Carrère, *Jacobienne locale, groupe de bivecteurs de Witt universel, et symbole modéré*, C. R. Acad. Sci. Paris Sér. I Math. **318** (1994), no. 8, 743–746.

[Del]   P. Deligne, *Seminar at IHES, Spring 1984, handwritten notes by G. Laumon*, http://www.math.uchicago.edu/~mitya/langlands.html

[Den]   C. Deninger, *Local L-factors of motives and regularized determinants*, Invent. Math. **107** (1992), 135–150.

[Dr]   V. Drinfeld, *Infinite-dimensional vector bundles in algebraic geometry*, The Unity of Mathematics. In Honor of the 90th Birthday of I.M. Gelfand. Progress in Mathematics, vol. 244, Birkhäuser, Boston-Basel-Berlin, 2006, pp. 263–304.

[E]   H. Esnault, *Talk at the Tokyo conference "Ramification and vanishing cycles"*, 2007, http://www.ms.u-tokyo.ac.jp/~t-saito/conf/rv/rv.html

[G]   W. Goldman, *The complex-symplectic geometry of $SL(2,\mathbb{C})$-characters over surfaces*, Algebraic groups and arithmetic, Tata Inst. Fund. Res., Mumbai, 2004, pp. 375–407.

[Gr1]   A. Grothendieck, *Sur certains espaces de fonctions holomorphes, I,II*, J. Reine Angew. Math. **192** (1953), 35–64, 77–95.

[Gr2]   A. Grothendieck, *La théorie de Fredholm*, Bulletin de la S.M.F. **84** (1956), 319–384.

[I]   R. Ishimura, *Homomorphismes du faisceau des germes de fonctions holomorphes dans lui-même et opérateurs différentiels*, Memoirs of the Faculty of Science, Kyushu University **32** (1978), 301–312.

[L]   G. Laumon, *Transformation de Fourier, constantes d'equations fonctionelles et conjecture de Weil*, Publ. Math. IHES **65** (1987), 131–210.

[M]   B. Malgrange, *Équations différentielles à coefficients polynomiaux*, Progress in Mathematics, vol. 96, Birkhäuser, Boston, MA, 1991.

[LS]   F. Loeser, C. Sabbah, *Équations aux différences finies et déterminants d'intégrales de fonctions multiformes*, Comment. Math. Helv. **66** (1991), no. 3, 458–503.

[Me]   Z. Mebkhout, *Une équivalence de catégories*, Compositio Math. **51** (1984), no. 1, 51–62.

[P]   D. Patel, *Thesis*, University of Chicago (2008).

[PX1]   D. Pickrell, E. Xia, *Ergodicity of mapping class group actions on representation varieties, I. Closed surfaces*, Comment. Math. Helv. **77** (2002), 339–362

[PX2]   D. Pickrell, E. Xia, *Ergodicity of mapping class group actions on representation varieties, II. Surfaces with boundary*, Transformation Groups **8** (2003), no. 4, 397–402.

[PS]   A. Pressley, G. Segal, *Loop groups*, Oxford Mathematical Monographs, The Clarendon Press, Oxford University Press, New York, 1986.

[PSch]   F. Prosmans, J.-P. Schneiders, *A topological reconstruction theorem for $\mathcal{D}^\infty$-modules*, Duke Math J. **102** (2000), no. 1, 39–86.

[ST]   T. Saito, T. Terasoma, *Determinant of period integrals*, J. Amer. Math. Soc. **10**, (1997), no. 4, 865–937.

[SW]   G. Segal, G. Wilson, *Loop groups and equations of KdV type*, Publ. Math. IHES **61** (1985), 5–65.

[T]   T. Terasoma, *A product formula for period integrals*, Math. Ann. **298** (1994), 577–589.

Fields Institute Communications
Volume **56**, 2009

Hodge Cohomology of Invertible Sheaves

Hélène Esnault
Universität Duisburg-Essen, Mathematik
45117 Essen, Germany
esnault@uni-due.de

Arthur Ogus
Department of Mathematics
University of California
Berkeley, CA 94720-3840 USA
ogus@math.berkeley.edu

Abstract. Let X/k be a smooth proper scheme over an algebraically closed field and let L be an invertible sheaf on X with $L^n \cong \mathcal{O}_X$. Pink and Roessler have shown that if k has characteristic zero, then for any m the dimension of $\oplus_{a+b=m} H^b(X, L^i \otimes \Omega^a_{X/k})$ is independent of i if i is relatively prime to n. We discuss a motivic interpretation of their result and prove that the same holds if k has characteristic $p = n$, provided X/k is ordinary, has dimension $\leq p$ and lifts to the ring of second Witt vectors $W_2(k)$.

1 Introduction

Let k be an algebraically closed field and let X/k be a smooth projective connected k-scheme. Let L be an invertible sheaf on X, and for each integer m, let

$$H^m_{Hdg}(X/k, L) := \bigoplus_{a+b=m} H^b(X, L \otimes \Omega^a_{X/k}).$$

We wish to study how the dimensions of the k-vector spaces $H^m_{Hdg}(X/k, L)$ and $H^b(X, L \otimes \Omega^a_{X/k})$ vary with L. For example, if k has characteristic zero, Green and Lazarsfeld [4] proved that for given a, b, d, the subloci

$$\{L \in \mathrm{Pic}^0(X) : \dim H^b(X, \Omega^a_X \otimes L) \geq d\}$$

of $\mathrm{Pic}^0(X)$ are translates of abelian subvarieties, and Simpson [12] showed that they in fact are translates by torsions points. Both these papers use analytic methods, but Pink and Roessler [10] obtained the same results purely algebraically, using the

---

2000 *Mathematics Subject Classification.* Primary 14F17 14F20 14F30 14F40 .

We thank D. Roessler for explaining to us his and R. Pink's analytic proof of equation 1.1. We also thank the referee for very useful, accurate and friendly remarks which helped us to improve the exposition of this note.

technique of mod p reduction and the decomposition theorem of Deligne-Illusie. A key point of their proof is the fact that if $L^n \cong \mathcal{O}_X$ for some positive integer n, then for all integers i with $(i,n) = 1$ one has

$$\dim H^m_{Hdg}(X/k, L) = \dim H^m_{Hdg}(X/k, L^i) \tag{1.1}$$

([10, Proposition 3.5]). They conjecture that equation 1.1 remains true in characteristic $p > 0$ if X/k lifts to $W_2(k)$ and has dimension $\leq p$. The purpose of this note is to discuss a few aspects of this conjecture and some variants.

Our main result (see Theorem 3.6) says that the conjecture is true if $n = p$ and X is ordinary in the sense of Bloch-Kato [2, Definition 7.2]. We also explain in section 2 some motivic variants of (1) and, in particular in Proposition 2.2, a proof (due to Pink and Roessler) of the characteristic zero case of (1.1), using the language of Grothendieck Chow motives. See [7, 9.3] for a discussion of a related problem using similar techniques. We should remark that there are also some log versions of these questions, which we will not make explicit.

2 Motivic variants

Question 2.1 *Let X be a smooth projective connected variety defined over an algebraically closed field k. Let L be an invertible sheaf on X and n a positive integer such that $L^n \cong \mathcal{O}_X$. Is*

$$\dim H^m_{Hdg}(X/k, L^i) = \dim H^m_{Hdg}(X/k, L)$$

for every i relatively prime to n?

Let us explain how this question can be given a motivic interpretation. We refer to [11] for the definition of Grothendieck's Chow motives over a field k. In particular, objects are triples (Y, p, n) where Y is a smooth projective variety over k, p is an element $CH^{\dim(Y)}(Y \times_k Y) \otimes \mathbb{Q}$, the rational Chow group of $\dim(Y)$-cycles, which, as a correspondence, is an idempotent, and n is a natural number.

Let $\pi : Y \to X$ be a principal bundle under a k-group scheme μ, where X and Y are smooth and projective over k. Recall that this means that there is a k-group scheme action $\mu \times_k Y \to Y$ with the property that one has an isomorphism

$$(\xi, y) \mapsto (y, \xi y) : \mu \times_k Y \cong Y \times_X Y \subseteq Y \times_k Y.$$

Thus a point $\xi \in \mu(k)$ defines a closed subset Γ_ξ of $Y \times_k Y$, the graph of the endomorphism of Y defined by ξ. The map $\xi \mapsto \Gamma_\xi$ extends uniquely to a map of $\mathbb{Q}$-vector spaces

$$\Gamma : \mathbb{Q}[\mu(k)] \to CH^{\dim(Y)}(Y \times_k Y) \otimes \mathbb{Q}.$$

Here $\mathbb{Q}[\mu(k)]$ is the $\mathbb{Q}$-group algebra, so the product structure is induced by the product of k-roots of unity. We can think of $CH^{\dim(Y)}(Y \times_k Y) \otimes \mathbb{Q}$ as a $\mathbb{Q}$-algebra of correspondences acting on $CH^*(Y) \otimes \mathbb{Q}$, where for $\beta \in CH^s(Y) \otimes \mathbb{Q}, \gamma \in CH^{\dim(Y)}(Y \times_k Y) \otimes \mathbb{Q}$, one defines as usual

$$\gamma \cdot \beta := (p_2)_*(\gamma \cup p_1^* \beta).$$

Then the map Γ is easily seen to be compatible with composition, as on closed points $y \in Y$ one has $\Gamma_\xi(y) = \xi \cdot y$. In particular if $\xi \in \mathbb{Q}[\mu]$ is idempotent in the group ring $\mathbb{Q}[\mu(k)]$, then $\Gamma_\xi \cong Y \times \xi$ is idempotent as a correspondence. In this case we let Y_ξ be the Grothendieck Chow motive $(Y, \xi, 0)$.

Let L be an n-torsion invertible sheaf on smooth irreducible projective scheme X/k. Recall that the choice of an $\mathcal{O}_X$-isomorphism $L^n \xrightarrow{\alpha} \mathcal{O}_X$ defines an $\mathcal{O}_X$-algebra structure on

$$\mathcal{A} := \bigoplus_{i=0}^{n-1} L^i \qquad (2.1)$$

via the tensor product $L^i \times L^j \to L^i \otimes_{\mathcal{O}_X} L^j = L^{i+j}$ for $i+j < n$ and its composition with the isomorphism $L^i \times L^j \to L^i \otimes_{\mathcal{O}_X} L^j = L^{i+j} \xrightarrow{\alpha^{-1}} L^{i+j-n}$ for $0 \le i+j-n$. Then the corresponding X-scheme $\pi : Y := \mathrm{Spec}_X \mathcal{A} \to X$ is a torsor under the group scheme μ_n of nth roots of unity. Indeed, Zariski locally on X, $\mathcal{A} \cong \mathcal{O}_X[t]/(t^n - u)$ for a local unit u, the μ_n-action is defined by $\mathcal{A} \to \mathcal{A} \otimes \mathbb{Q}[\zeta]/(\zeta^n - 1)$, $t \mapsto t\zeta$, and the torsor structure is given by $\mathcal{A} \otimes \mathbb{Q}[\zeta]/(\zeta^n - 1) \cong \mathcal{A} \otimes_{\mathcal{O}_X} \mathcal{A}$, $(t, \zeta) \mapsto (t, t\zeta)$. This construction defines an equivalence between the category of pairs (L, α) and the category of μ_n-torsors over X. Assuming now that n is invertible in k, μ_n is étale, hence π is étale and Y is smooth and projective over k. Note that the character group $X_n := \mathrm{Hom}(\mu_n, \mathbf{G}_m)$ is cyclic of order n with a canonical generator (namely, the inclusion $\mu_n \to \mathbf{G}_m$). By construction, the direct sum decomposition (2.1) of $\mathcal{A}$ corresponds exactly to its eigenspace decomposition according to the characters of μ_n.

We can now apply the general construction of motives to this situation. Since μ_n is étale over the algebraically closed field k, it is completely determined by the finite group $\Gamma := \mu_n(k)$, which is cyclic of order n. The group algebra $\mathbb{Q}[\Gamma]$ is a finite separable algebra over $\mathbb{Q}$, hence is a product of fields:

$$\mathbb{Q}[\Gamma] = \prod E_e.$$

Here $E_e = \mathbb{Q}[T]/(\Phi_e(T)) = \mathbb{Q}(\xi_e)$, where e is a divisor of n, $\Phi_e(T)$ is the cyclotomic polynomial, and ξ_e is a primitive eth root of unity. There is an (indecomposable) idempotent e corresponding to each of these fields, and for each e we find a Chow motive Y_e.

The indecomposable idempotents of $\mathbb{Q}[\Gamma]$ can also be thought of as points of the spectrum T of $\mathbb{Q}[\Gamma]$. If K is a sufficiently large extension of $\mathbb{Q}$, then

$$T(K) = \mathrm{Hom}_{\mathrm{Alg}}(\mathbb{Q}[\Gamma], K) = \mathrm{Hom}_{\mathrm{Gr}}(\Gamma, K^*), \qquad (2.2)$$

$$\text{and } K \otimes \mathbb{Q}[\Gamma] \cong K[\Gamma] \cong K^{T(K)}. \qquad (2.3)$$

Thus $T(K)$ can be identified with the character group X_n of Γ, and is canonically isomorphic to $\mathbf{Z}/n\mathbf{Z}$, with canonical generator the inclusion $\Gamma \subseteq k$. Suppose that $K/\mathbb{Q}$ is Galois. Then $\mathrm{Gal}(K/\mathbb{Q})$ acts on $T(K)$, and the points of T correspond to the $\mathrm{Gal}(K/\mathbb{Q})$-orbits. By the theory of cyclotomic extensions of $\mathbb{Q}$, this action factors through a surjective map

$$\mathrm{Gal}(K/\mathbb{Q}) \to (\mathbf{Z}/n\mathbf{Z})^*$$

and the usual action of $(\mathbf{Z}/n\mathbf{Z})^*$ on $\mathbf{Z}/n\mathbf{Z}$ by multiplication. Thus the orbits correspond precisely to the divisors d of n; we shall associate to each orbit S the index d of the subgroup of $\mathbf{Z}/n\mathbf{Z}$ generated by any element of S. (Note that in fact the image of d in $\mathbf{Z}/n\mathbf{Z}$ belongs to S.) We shall thus identify the indecomposable idempotents of $\mathbb{Q}[\Gamma]$ and the divisors of n.

Let us suppose that $k = \mathbf{C}$. Then we can consider the Betti cohomologies of X and Y, and in particular the group algebra $\mathbb{Q}[\Gamma]$ operates on $H^m(Y, \mathbb{Q})$. We can thus view $H^m(Y, \mathbb{Q})$ as a $\mathbb{Q}[\Gamma]$-module, which corresponds to a coherent sheaf

$\tilde{H}^m(Y, \mathbb{Q})$ on T. If e is an idempotent of $\mathbb{Q}[\Gamma]$, then $H^m(Y_e, \mathbb{Q})$ is the image of the action of e on $H^m(Y, \mathbb{Q})$, or equivalently, it is the stalk of the sheaf $\tilde{H}^m(Y, \mathbb{Q})$ at the point of T corresponding to e, or equivalently, it is $H^m(Y, \mathbb{Q}) \otimes E_e$ where the tensor product is taken over $\mathbb{Q}[\Gamma]$. If K is a sufficiently large field as above, then equation (2.3) induces an isomorphism of K-vectors spaces:

$$H^m(Y_e, \mathbb{Q}) \otimes_{\mathbb{Q}} K \cong \bigoplus \{H^m(Y, K)_t : t \in T^e(K)\},$$

where here $T^e(K)$ means the set of points of $T(K)$ in the Galois orbit corresponding to e, and $H^m(Y, K)_t$ means the t-eigenspace of the action of Γ on $H^m(Y, \mathbb{Q}) \otimes_{\mathbb{Q}} K$. The de Rham and Hodge cohomologies of Y_e are defined in the same way: they are the images of the actions of the idempotent e acting on the k-vector spaces $H_{DR}(Y/k)$ and $H_{Hdg}(Y/k)$.

The following result is due to Pink and Roessler. Their article [10] contains a proof using reduction modulo p techniques and the results of [3]; the following analytic argument is based on oral communications with them.

Proposition 2.2 *The answer to Question 2.1 is affirmative if k is a field of characteristic zero.*

Proof As both sides of the equality in Question 2.1 satisfy base change with respect to field extensions, we may assume that $k = \mathbb{C}$. Let $i \to t_i$ denote the isomorphism $\mathbb{Z}/n\mathbb{Z} \cong T(\mathbb{C})$. For each divisor e of n there is a corresponding idempotent e of $\mathbb{Q}[\Gamma] \subseteq K[\Gamma]$, the sum over all i such that $t_i \in T^e(\mathbb{C})$. Consider the Hodge cohomology of the motive Y_e:

$$
\begin{aligned}
H^m_{Hdg}(Y_e/\mathbb{C}) := H^m_{Hdg}(Y/\mathbb{C}) \otimes_{\mathbb{Q}[\Gamma]} E_e &\cong H^m_{Hdg}(Y/\mathbb{C}) \otimes_{\mathbb{C}[\Gamma]} (\mathbb{C} \otimes E_e). \\
&\cong \bigoplus \{H^m_{Hdg}(Y/\mathbb{C})_i : i \in T^e(k)\}.
\end{aligned}
$$

Since $\pi\colon Y \to X$ is finite and étale,

$$
\begin{aligned}
H^b(Y, \Omega^a_{Y/\mathbb{C}}) \cong H^b(X, \pi_* \pi^* \Omega^a_{X/\mathbb{C}}) &\cong H^b(X, \Omega^a_{X/\mathbb{C}} \otimes \pi_* \mathcal{O}_Y) \\
&\cong \bigoplus \{H^b(X, \Omega^a_{X/\mathbb{C}} \otimes L^i) : i \in \mathbb{Z}/n\mathbb{Z}\}.
\end{aligned}
$$

Thus

$$H^m_{Hdg}(Y/\mathbb{C}) \cong \bigoplus \{H^m_{Hdg}(X, L^i) : i \in \mathbb{Z}/n\mathbb{Z}\},$$

and hence from the explicit description of the action of μ_n on $\mathcal{A}$ above it follows that

$$H^m_{Hdg}(Y_e/\mathbb{C}) = \bigoplus \{H^m_{Hdg}(X, L^i) : i \in T_e(\mathbb{C})\}.$$

The Hodge decomposition theorem for Y provides us with an isomorphism:

$$H^m_{Hdg}(Y/\mathbb{C}) \cong \mathbb{C} \otimes H^m(Y, \mathbb{Q}),$$

compatible with the action of $\mathbb{Q}[\Gamma]$. This gives us, for each idempotent e, an isomorphism of $\mathbb{C} \otimes E_e$-modules.

$$H^m_{Hdg}(Y_e/\mathbb{C}) \cong \mathbb{C} \otimes H^m(Y_e, \mathbb{Q}).$$

The action on $\mathbb{C} \otimes H^m(Y_e, \mathbb{Q})$ on the right just comes from the action of E_e on $H^m(Y_e, \mathbb{Q})$ by extension of scalars. Since E_e is a field, $H^m(Y_e, \mathbb{Q})$ is free as an E_e-module, and hence the $\mathbb{C} \otimes E_e$-module $H^m_{Hdg}(Y_e/\mathbb{C})$ is also free. It follows that its rank is the same at all the points $t \in T^e(\mathbb{C})$, affirming Question 2.1. $\square$

Let us now formulate an analog of Question 1 for the ℓ-adic and crystalline realizations of the motive Y_e in characteristic p.

Question 2.3 *Suppose that k is an algebraically closed field of characteristic p and $(n,p) = 1$. Let ℓ be a prime different from p, let e be a divisor of n, and let E_e be the corresponding factor of $\mathbb{Q}[\Gamma]$. Is it true that each $H^m(Y_e, \mathbb{Q}_\ell)$ is a free $\mathbb{Q}_\ell \otimes E_e$-module? And is it true that $H^m_{cris}(Y_e/W) \otimes \mathbb{Q}$ is a free $W \otimes E_e$-module, where $W := W(k)$?*

If K is an extension of $\mathbb{Q}_\ell$ (resp. of $W(k)$) which contains a primitive nth root of unity, then as above we have eigenspace decompositions:

$$K \otimes H^m(Y_{\acute{e}t}, \mathbb{Q}_\ell) \;\cong\; \bigoplus \{H^m(Y_{\acute{e}t}, K)_t : t \in T(K)\} \qquad (2.4)$$

$$K \otimes H^m(Y_{cris}/W(k)) \;\cong\; \bigoplus \{H^m(Y_{cris}, K)_t : t \in T(K)\}, \qquad (2.5)$$

and this question asks whether the K-dimension of the t-eigenspace is constant over the orbits $T_e(K) \subseteq T(K)$. We show in the sequel that the question has a positive answer.

Suppose first that X/k lifts to characteristic zero, *i.e.*, that there exists a complete discrete valuation ring V with residue field k and fraction field of characteristic zero and a smooth proper $\tilde{X}/V$ whose special fiber is X/k. Let X_m be the closed subscheme of $\tilde{X}$ defined by π^{m+1}, where π is a uniformizing parameter of V. Choose a trivialization α of L^n. It follows from Theorem 18.1.2 of [6] that the étale μ_n-torsor Y on X corresponding to (L, α) lifts to X_m, uniquely up to a unique isomorphism, and hence that the same is true for (L, α). This fact can also be seen by chasing the exact sequences of cohomology corresponding to the following commutative diagram of exact sequences in the étale topology.

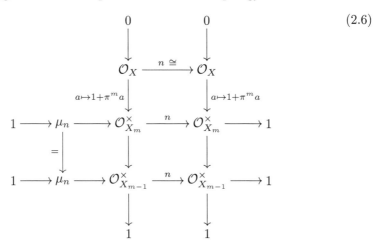

$$(2.6)$$

By Grothendieck's fundamental theorem for proper morphisms, it follows that (L, α) and Y lift to $(\tilde{L}, \tilde{\alpha})$ and $\tilde{Y}$ on $\tilde{X}$. Then by the étale to Betti and Betti to crystalline comparison theorems, we see that under the lifting assumption, the answer to Question 2.3 is affirmative.

In fact, the lifting hypothesis is superfluous, but this takes a bit more work.

Proposition 2.4 *The answer to Question 2.3 is affirmative.*

Proof It is trivially true that $H^m(Y_e, \mathbb{Q}_\ell)$ is free over $\mathbb{Q}_\ell \otimes E_e$ if $\mathbb{Q}_\ell \otimes E_e$ is a field. If $(\ell, n) = 1$, this is the case if and only if $(\mathbf{Z}/e\mathbf{Z})^*$ is cyclic and generated by ℓ. More generally, assuming ℓ is relatively prime to n, there is a decomposition of $\mathbb{Q}_\ell[\Gamma]$ into a product of fields $\mathbb{Q}_\ell[\Gamma] \cong \prod E_{\ell, e}$, where now e ranges over the orbits of $\mathbf{Z}/n\mathbf{Z}$ under the action of the cyclic subgroup of $(\mathbf{Z}/n\mathbf{Z})^*$ generated by ℓ. This is indeed the unramified lift of the decomposition of $A = \mathbb{F}_\ell[\Gamma]$ into a product of finite extensions of $\mathbb{F}_\ell$, corresponding to the orbits of Frobenius on the geometric points of A. This shows at least that the dimension of $H^m(Y, K)_t$ in (2.4) is, as a function of t, constant over the ℓ-orbits.

For the general statement, let K be an algebraically closed field containing $\mathbb{Q}_\ell$ for all primes $\ell \neq p$, and containing $W(k)$. For $\ell \neq p$ let $V_\ell := H^m(Y_{\acute{e}t}, \mathbb{Q}_\ell) \otimes_{\mathbb{Q}_\ell} K$, and let $V_p := H^m(Y_{cris}, W(k)) \otimes_{W(k)} K$. Then each V_ℓ is a finite-dimensional representation of Γ, and the isomorphisms (2.4) and (2.5) are just its decomposition as a direct sum of irreducible representations:

$$V_\ell \cong \bigoplus \{n_{\ell,i} V_i : i \in \mathbf{Z}/n\mathbf{Z}\},$$

where $V_i = K$, with $\gamma \in \Gamma$ acting by multiplication by γ^i. By [8, Theorem 2.2)] (and [1], [5] and [9] for the existence of cycle classes in crystalline cohomology) the trace of any $\gamma \in \Gamma$ acting on V_ℓ is an integer independent of ℓ, including $\ell = p$. Since Γ is a finite group, it follows from the independence of characters that for each i, $n_i := n_{\ell,i}$ is independent of ℓ. We saw above that $n_{\ell,\ell i} = n_{\ell,i}$ if $(\ell, n) = 1$ and $\ell \neq p$, so that in fact $n_{\ell i} = n_i$ for all $\ell \neq p$ with $(\ell, n) = 1$. Since the group $(\mathbf{Z}/n\mathbf{Z})^*$ is generated by all such ℓ, it follows that n_i is indeed constant over the ℓ-orbits. $\square$

What does this tell us about Question 2.1? If $(p, n) = 1$ and k is algebraically closed, $W[\Gamma]$ is still semisimple, and can be written canonically as a product of copies of W, indexed by $i \in T(W) \cong \mathbf{Z}/n\mathbf{Z}$. For every $t \in T(W) \cong T(k)$, we have an injective base change map from crystalline to de Rham cohomology:

$$k \otimes H^m(Y/W)_t \to H^m(Y/k)_t.$$

Question 2.5 *In the above situation, is $H^q(Y/W)$ torsion free when $(p, n) = 1$?*

If the answer is yes, then the maps $k \otimes H^m(Y/W)_t \to H^m(Y/k)_t$ are isomorphisms, and this means that we can compute the dimensions of the de Rham eigenspaces from the ℓ-adic ones. Assuming also that the Hodge to de Rham spectral sequence of Y/k degenerates at E_1, this should give an affirmative answer to Question 1. Note that if X/k lifts mod p^2, then Y/k lifts mod p^2 as well, and if the dimension is less than or equal to p, the E_1-degeneration is true by [3].

Of course, there is no reason for Question 2.5 to have an affirmative answer in general. Is there a reasonable hypothesis on X which guarantees it? For example, is it true if the crystalline cohomology of X/W is torsion free?

3 The p-torsion case in characteristic p

Let us assume from now on that k is a perfect field of characteristic $p > 0$. In this case we can reduce question 2.1 to a question about connections, using the following construction of [3]. First let us recall some standard notations. Let X' be the pull back of X via the Frobenius of k, let $\pi: X' \to X$ be the projection, and let $F: X \to X'$ and $F_X: X \to X$ be the relative and absolute Frobenius morphisms.

Then $F_X^* L = L^p = F^* L'$, where $L' := \pi^* L$. Then $L^p = F^{-1}L' \otimes_{\mathcal{O}_{X'}} \mathcal{O}_X$ is endowed with the Frobenius descent connection $1 \otimes d$, *i.e.* the unique connection spanned by its flat sections L'. In general, for a given integrable connection (E, ∇), we set

$$H_{DR}^i(X, (E, \nabla)) = \mathbb{H}^i(X/k, (\Omega_{X/k}^{\cdot} \otimes E, \nabla)),$$

and we use again the notation

$$H_{Hdg}^i(X/k, L) = \bigoplus_{a+b=i} H^b(X, \Omega_{X/k}^a \otimes L)$$

and write h_{DR}^i and h_{Hdg}^i for the respective dimensions of these spaces.

Proposition 3.1 *Let L be an invertible sheaf on a smooth proper scheme X over k and let ∇ be the Frobenius descent connection on L^p. Suppose that X/k lifts to $W_2(k)$ and has dimension at most p. Then for every natural number m,*

$$h_{DR}^m(X/k, (L^p, \nabla)) = h_{Hdg}^m(X/k, L).$$

Corollary 3.2 *Under the assumptions of Proposition 3.1, if $L^p \cong \mathcal{O}_X$ and $\omega := \nabla(1)$, then for any integer a,*

$$h_{Hdg}^m(X/k, L^a) = h_{DR}^m(X/k, (\mathcal{O}_X, d + a\omega)).$$

Remark 3.3 If p divides a, this just means the degeneration of the Hodge to de Rham spectral sequence for $(\mathcal{O}_X, d)$.

Proof Let $Hdg_{X'/k}^{\cdot}$ denote the Hodge complex of X'/k, *i.e.*, the direct sum $\oplus_i \Omega_{X'/k}^i[-i]$. Recall from [3] that the lifting yields an isomorphism in the bounded derived category of $\mathcal{O}_{X'}$-modules:

$$Hdg_{X'/k}^{\cdot} \cong F_*(\Omega_{X/k}^{\cdot}, d).$$

Tensoring this isomorphism with $L' := \pi^* L$ and using the projection formula for F, we find an isomorphism

$$Hdg_{X'/k}^{\cdot} \otimes L' \cong F_*(\Omega_{X/k}^{\cdot} \otimes L^p, \nabla).$$

Hence

$$H_{Hdg}^m(X/k, L) \xrightarrow{F_k^* \ \cong} H_{Hdg}^m(X'/k, L') \xleftarrow{F_* \ \cong} H_{DR}^m(X, (L^p, \nabla)).$$

This proves the proposition. If $L^p = \mathcal{O}_X$, the corresponding Frobenius descent connection ∇ on $\mathcal{O}_X$ is determined by $\omega_L := \nabla(1)$. It follows from the tensor product rule for connections that $\omega_{L^a} = a\omega_L$ for any integer a. $\qquad\square$

The corollary suggests the following question.

Question 3.4 *Let ω be a closed one-form on X and let c be a unit of k. Is the dimension of $H_{DR}^m(X, (\mathcal{O}_X, d + c\omega))$ independent of c?*

Remark 3.5 Some properness is necessary, since the p-curvature of $d_\omega := d + \omega$ can change from zero to non-zero as one multiplies by an invertible constant. If the p-curvature is non-zero, then the sheaf $\mathcal{H}^0(\Omega_{X/k}^{\cdot}, d_\omega)$ vanishes, and hence so does $H^0(X, (\Omega_{X/k}^{\cdot}, d_\omega))$. If the p-curvature vanishes, then $\mathcal{H}^0(\Omega_{X/k}^{\cdot}, d_\omega)$ is an invertible sheaf L, which can have nontrivial sections if X is allowed to shrink. However, since by definition, $L \subset \mathcal{O}_X$, it can have a global section on a proper X only if $L = \mathcal{O}_X$.

We can answer Question 3.4 under a strong hypothesis.

Theorem 3.6 *Suppose that X/k is smooth, proper, and ordinary in the sense of Bloch and Kato [2, Definition 7.2]: $H^i(X, B^j_{X/k}) = 0$ for all i, j, where*

$$B^j_{X/k} := Im\left(d \colon \Omega^{j-1}_{X/k} \to \Omega^j_{X/k}\right).$$

Then the answer to question 3.4 is affirmative. Hence if moreover X/k lifts to $W_2(k)$, has dimension at most p, and if $n = p$, the answer to Question 2.1 is also affirmative.

We begin with the following lemmas.

Lemma 3.7 *Let ω be a closed one-form on X, and let*

$$d_\omega := d + \omega \wedge \quad : \quad \Omega^{\cdot}_{X/k} \to \Omega^{\cdot+1}_{X/k}.$$

Then the standard exterior derivative induces a morphism of complexes:

$$(\Omega^{\cdot}_{X/k}, d_\omega) \xrightarrow{\ \delta\ } (\Omega^{\cdot}_{X/k}, d_\omega)[1].$$

Proof If α is a section of $\Omega^q_{X/k}$,

$$
\begin{aligned}
dd_\omega(\alpha) &= d(d\alpha + \omega \wedge \alpha) \\
&= dd\alpha + d\omega \wedge \alpha - \omega \wedge d\alpha \\
&= -\omega \wedge d\alpha.
\end{aligned}
$$

Since the sign of the differential of the complex $(\Omega^{\cdot}_{X/k}, d_\omega)[1]$ is the negative of the sign of the differential of $(\Omega^{\cdot}_{X/k}, d_\omega)$,

$$
\begin{aligned}
d_\omega d(\alpha) &= -(d + \omega\wedge)(d\alpha) \\
&= -\omega \wedge d\alpha
\end{aligned}
$$

$\square$

Lemma 3.8 *Let $Z^{\cdot} := \ker(d) \subseteq (\Omega^{\cdot}_{X/k}, d_\omega)$ and $B^{\cdot} := Im(d)[-1] \subseteq (\Omega^{\cdot}_{X/k}, d_\omega)$. Then for any $a \in k^*$, multiplication by a^i in degree i induces isomorphisms*

$$(Z^{\cdot}, d_\omega) \xrightarrow{\ \lambda_a\ } (Z^{\cdot}, d_{a\omega})$$

$$(B^{\cdot}, d_\omega) \xrightarrow{\ \lambda_a\ } (B^{\cdot}, d_{a\omega}).$$

Proof It is clear that the boundary map d_ω on $Z^{\cdot}$ and on $B^{\cdot}$ is just wedge product with ω. $\square$

Proof of Theorem 3.6 The morphism δ of Lemma 3.7 induces an exact sequence:

$$0 \to (Z^{\cdot}, d_\omega) \to (\Omega^{\cdot}_{X/k}, d_\omega) \xrightarrow{\ \delta\ } (B^{\cdot}, d_\omega)[1] \to 0. \qquad (3.1)$$

As X/k is ordinary, the E_1 term of the first spectral sequence for $(B^{\cdot}, d_\omega)$ is $E^{i,j}_1 = H^j(X, B^i) = 0$, and it follows that the hypercohomology of $(B^{\cdot}, d_\omega)$ vanishes, for every ω. Hence the natural map $H^q(Z^{\cdot}, d_\omega) \to H^q(\Omega^{\cdot}_{X/k}, d_\omega)$ is an isomorphism. Since the dimension of $H^q(Z^{\cdot}, d_\omega)$ is unchanged when ω is multiplied by a unit of k, the same is true of $H^q(\Omega^{\cdot}_{X/k}, d_\omega)$. This completes the proof of Theorem 3.6. $\square$

Remark 3.9 A simple Riemann-Roch computation shows that on curves, question 1 has a positive answer with no additional assumptions. Indeed, if L is a nontrivial torsion sheaf, then its degree is zero and it has no global sections. It follows that $h^1(L) = g - 1$. Since the same is true for L^{-1}, $h^0(L \otimes \Omega_X^1) = h^1(L^{-1}) = g - 1$, and $h^1(L \otimes \Omega_X^1) = h^0(L^{-1}) = 0$.

Remark 3.10 In the absence of the ordinarity hypothesis, one can ask if the rank of the boundary map

$$\partial_\omega \colon H^{q+1}(B^{\cdot}, \omega\wedge) \to H^{q+1}(Z^{\cdot}, \omega\wedge)$$

of (3.1) changes if ω is multiplied by a unit of k. To analyze this question, let

$$c_\omega \colon (B^{\cdot}, \omega\wedge) \to (Z^{\cdot}, \omega\wedge)$$

be the morphism in the derived category $D(X', \mathcal{O}_{X'})$ defined by the exact sequence (3.1), so that ∂_ω can be identified with $H^{q-1}(c_\omega)$. Similarly, the exact sequence

$$0 \to (Z^{\cdot}, \omega\wedge) \to (\Omega^{\cdot}, \omega\wedge) \to (B^{\cdot}, \omega\wedge)[1] \to 0$$

defines a morphism

$$a_\omega \colon (B^{\cdot}, \omega\wedge) \to (Z^{\cdot}, \omega\wedge)$$

in $D(X', \mathcal{O}_{X'})$ as well. There is also an inclusion morphism:

$$b_\omega \colon (B^{\cdot}, \omega\wedge) \to (Z^{\cdot}, \omega\wedge).$$

Then it is not difficult to check that $c_\omega = a_\omega + b_\omega$. If $a \in k^*$, we have isomorphisms of complexes

$$\lambda_a \colon (Z^{\cdot}, \omega\wedge) \to (Z^{\cdot}, a\omega\wedge), \qquad \lambda_a \colon (B^{\cdot}, \omega\wedge) \to (B^{\cdot}, a\omega\wedge).$$

Using these as identifications, one can check that $c_{a\omega} = a^{-1}a_\omega + b_\omega$. This would suggest a negative answer to Question 3.4, but we do not have an example.

References

[1] Berthelot, P.: *Cohomologie Cristalline des Schémas de Caractéristique $p > 0$*, Lecture Notes in Math. No. 407 (1974), Springer-Verlag.

[2] Bloch, S.; Kato, K.: p-adic étale cohomology, Inst. Hautes Études Sci. Publ. Math. No. **63** (1986).

[3] Deligne, P.; Illusie, L.: Relèvements modulo p^2 et décomposition du complexe de de Rham, Invent. Math. **89** (1987), no. 2, 247–270.

[4] Green, M.; Lazarsfeld, R.: Higher obstructions to deforming cohomology groups of line bundles, J. Am. Math. Soc. **4** (1991), no 1, 87–103.

[5] Gros, M.: Classes de Chern et classes de cycles en cohomologie de Hodge-Witt logarithmiques, Bull. SMF, Mémoire **21** (1985).

[6] Grothendieck, A.; Dieudonné, J.: Éléments de géométrie algébrique: étude locale des schémas et des morphismes des schémas. *Publ. Math. de l'I.H.E.S.*, 32, 1964.

[7] Katz, N.: *Rigid Local Systems*. Number 139 in Annals of Mathematics Studies. Princeton University Press, Princeton, New Jersey, 1996.

[8] Katz, N.; Messing, W.: Some consequences of the Riemann hypothesis for varieties over finite fields, Invent. Math. **23** (1974), 73–77.

[9] Petrequin, D.: Classes de Chern et classes de cycles en cohomologie rigide, Bull. Soc. Math. France **131** (2003), no 1, 59-121.

[10] Pink, R.; Roessler, D.: A conjecture of Beauville and Catanese revisited, Math.Ann. **330** (2004), no 2, 293–308.

[11] Scholl, A.: Classical motives. Motives (Seattle, WA, 1991), 163–187, Proc. Sympos. Pure Math., **55**, Part 1, Amer. Math. Soc., Providence, RI, 1994.

[12] Simpson, C.: Subspaces of moduli spaces of rank one local systems, Ann. Sc. Éc. Norm. Sup. (4) **26** (1993), no 3, 361–401.

Fields Institute Communications
Volume **56**, 2009

Arithmetic Intersection Theory on Deligne-Mumford Stacks

Henri Gillet

Department of Mathematics, Statistics, and Computer Science
University of Illinois at Chicago
322 Science and Engineering Offices (M/C 249)
851 S. Morgan Street
Chicago, IL 60607-7045 USA
gillet@uic.edu

To Spencer Bloch

Abstract. In this paper the arithmetic Chow groups and their product structure are extended from the category of regular arithmetic varieties to regular Deligne-Mumford stacks proper over a general arithmetic ring. The method used also gives another construction of the product on the usual Chow groups of a regular Deligne-Mumford stack.

Introduction

Because of the importance of moduli stacks in arithmetic geometry, it is natural to ask whether the arithmetic intersection theory introduced in [13] can be extended to stacks. Indeed arithmetic intersection numbers on stacks and moduli spaces have been studied by a number of authors; see [25] and [4], for example.

Recall that the arithmetic Chow theory of *op. cit.* has the following properties.

1. $X \mapsto \widehat{\mathrm{CH}}^*(X)$ is a contravariant functor from the category of schemes which are regular flat and projective over $S = \mathrm{Spec}(\mathbb{Z})$, to graded abelian groups.

2. $\widehat{\mathrm{CH}}^*(X)_{\mathbb{Q}}$ has a functorial graded ring structure.

3. $\widehat{\mathrm{CH}}^1(X) \simeq \widehat{\mathrm{Pic}}(X)$, the group of isomorphism classes of Hermitian line bundles on X. (A Hermitian line bundle (L, h) on X, is a line bundle L on X together with a choice of a C^∞ Hermitian metric h on the associated holomorphic line bundle $L(\mathbb{C})$ over the complex manifold $X(\mathbb{C})$.)

4. Each class in $\widehat{\mathrm{CH}}^p(X)$ is represented by a pair (ζ, g_ζ) with $\zeta = \sum_i [Z_i]$ a codimension p algebraic cycle on X and g_ζ a "Green current" for ζ, *i.e.* a

2000 *Mathematics Subject Classification.* Primary 14G40, 14A20; Secondary 19E08, 14C17.
Supported by NSF grant DMS-0500762, the Clay Mathematics Institute, and the Fields Institute.

current of degree $(p-1, p-1)$ on $X(\mathbb{C})$ such that

$$dd^c(g_\zeta) + \delta_\zeta$$

is a C^∞ (p,p)-form. Here δ_ζ is the (p,p)-current $\sum_i \delta_{Z_i(\mathbb{C})}$ where $\delta_{Z_i(\mathbb{C})}$ is the current of integration associated to the analytic subspace $Z_i(\mathbb{C}) \subset X(\mathbb{C})$.

5. There is an exact sequence, for each $p \geq 0$:

$$\mathrm{CH}^{p,p-1}(X) \xrightarrow{\rho} H_{\mathcal{D}}^{2p-1}(X_\mathbb{R}, \mathbb{R}(p))$$
$$\rightarrow \widehat{\mathrm{CH}}^p(X) \rightarrow \mathrm{CH}^p(X) \oplus Z^{p,p}(X_\mathbb{R})) \rightarrow \mathrm{H}_{\mathcal{D}}^{2p}(X_\mathbb{R}, \mathbb{R}(p)).$$

Here the $X_\mathbb{R}$ indicates that we are taking real forms on $X(\mathbb{C})$ on which the anti-holomorphic involution induced by complex conjugation acts by $(-1)^p$ on (p,p)-forms.

6. There is a theory of Chern classes for Hermitian vector bundles

$$(E, \| \ \|) \mapsto \widehat{C}_p(E, \| \ \|).$$

Extending the definition of the Chow groups to stacks is straightforward. A cycle on a stack is an element of the free abelian group on the set of integral substacks, and rational equivalence is defined similarly. (See [10].) To define the arithmetic Chow groups, we must associate Green currents to cycles. First we show that for a smooth separated stack $\mathfrak{X}$ over $\mathbb{C}$ the sheaves of differential forms are acyclic, and that the groups $H^q(\mathfrak{X}(\mathbb{C}), \Omega^p)$ are computed by the global Dolbeault complexes. If $\mathfrak{X}$ is proper, one can show that the Hodge spectral sequence degenerates, and that the cohomology $H^q(\mathfrak{X}(\mathbb{C}), \mathbb{C})$ has a Hodge decomposition "in the strong sense". In particular the $\partial\bar{\partial}$-lemma holds. This allows one to give a definition of the arithmetic Chow groups analogous to that of [13] and [6], though it does not give the product structure.

Before discussing how to define the product on the arithmetic Chow groups of stacks, let us briefly review how the classical Chow groups and their product structure are defined for stacks. In the 1980's two different approaches to intersection theory on stacks over fields were introduced; the first in [10] was via Bloch's formula. This approach has not been applied to the arithmetic Chow groups of schemes, never mind stacks, in part because Bloch's formula depends on Gersten's conjecture, which is not known for general regular schemes. The other construction of intersection theory on stacks was by Vistoli [32], using "Fulton style" intersection theory, and in particular operational Chow groups. While Fulton's operational Chow groups make sense for regular schemes, it is not clear how to construct the arithmetic Chow groups of schemes (or stacks), and their product structure, operationally. The problem with both of these approaches is that the product on the arithmetic Chow groups depends on the $*$- product for the Green current of two cycles. which is only defined when the cycles intersect properly, and thus the moving lemma plays a key role in the construction of the intersection product on the arithmetic Chow groups. Note however that combining the method of Hu [20] with Kresch's approach in [24] to the construction of the intersection product on stacks over fields, via deformation to the normal cone, might provide a way around this problem.

In this paper, we shall use a construction that is a variant on the operational method. It was first observed by Kimura [22] (see also [3]) that if X is a possibly singular variety proper over a field of characteristic zero then its operational Chow groups can be computed using hypercovers. It follows from this result that for a proper variety X over a field of characteristic zero, the operational Chow groups of X are isomorphic to the inverse limit of $\mathrm{CH}^*(Y)$ over the category all surjective morphisms $Y \to X$ with Y smooth and projective. This suggests using a similar construction for the arithmetic Chow groups of stacks.

Suppose for a moment that we have extended the functor $\widehat{\mathrm{CH}}^*$ to the category of separated stacks over a fixed base S. Then for each $p : V \to \mathfrak{X}$, with p proper and surjective, and V a regular quasi-projective variety over S we will have a natural homomorphism $p^* : \widehat{\mathrm{CH}}^*(\mathfrak{X}) \to \widehat{\mathrm{CH}}^*(V)$ and hence a homomorphism

$$\widehat{\mathrm{CH}}^*(\mathfrak{X}) \to \varprojlim_{p : V \to \mathfrak{X}} \widehat{\mathrm{CH}}^*(V).$$

Since we already have well defined functorial products on the groups $\widehat{\mathrm{CH}}^*(V)$, it follows that $\varprojlim \widehat{\mathrm{CH}}^*(V)$ has a natural product structure, which is contravariant with respect to $\mathfrak{X}$. A Hermitian vector bundle $\overline{E} = (E, h)$ on $\mathfrak{X}$ has Chern classes in $\varprojlim \widehat{\mathrm{CH}}^*(V)$, since the bundle pulls back to any V over $\mathfrak{X}$. Note that a similar construction appears in [4], where they consider towers of Shimura varieties with level structures, rather than the underlying stack.

Now the key point is that, even though we do not, *a priori*, have products and pull-backs on $\widehat{\mathrm{CH}}^*(\mathfrak{X})$, we have:

Theorem *If $\mathfrak{X}$ is regular Deligne-Mumford stack which is flat and proper over a base $S = \mathrm{Spec}(\mathcal{O}_K)$ for $\mathcal{O}_K$ the ring of integers in a number field K, there is a canonical isomorphism*

$$\varprojlim_{p : V \to \mathfrak{X}} \widehat{\mathrm{CH}}^*(V)_{\mathbb{Q}} \to \widehat{\mathrm{CH}}^*(\mathfrak{X})_{\mathbb{Q}}.$$

The idea of the proof is to show that the appropriate variants of this statement are true for differential forms, and also for both the usual Chow groups (tensored with $\mathbb{Q}$) and for cohomology.

Corollary *There is a product structure on $\widehat{\mathrm{CH}}^*(\mathfrak{X})_{\mathbb{Q}}$ which is functorial in $\mathfrak{X}$, and the theory of Chern classes for Hermitian vector bundles on arithmetic varieties extends to Hermitian vector bundles on stacks over S.*

Following preliminaries on the Dolbeault cohomology of stacks in section 1, in section 2 we discuss the G-theory and K-theory of stacks. We show that the isomorphism (due to Grothendieck) between the graded vector space associated to the γ-filtration on K_0 and the Chow groups of a regular scheme also holds for regular Deligne-Mumford stacks. This gives yet another construction of the product on the Chow groups, with rational coefficients, of a stack. We then go on to show that the motivic weight complex of a regular *proper* Deligne-Mumford stack over S is (up to homotopy) concentrated in degree zero. Thus to each regular proper Deligne-Mumford stack we can associate a pure motive. This extends a result of Toen for varieties over perfect fields. We then prove the main theorem in section 3. Finally in section 5 we consider the case of non-proper stacks.

Throughout the paper we shall fix a base S which is the spectrum of the ring of integers in a number field or more generally of an arithmetic ring in the sense of [13]. In particular a variety (over S) will be an integral scheme which is separated and of finite type over S.

I would like to thank the referee for a very careful reading of the manuscript and for many valuable comments and questions.

1 Dolbeault cohomology of stacks

Let $\mathfrak{X}$ be a regular Deligne-Mumford stack of finite type over the field of complex numbers $\mathbb{C}$. Let $\mathfrak{X}(\mathbb{C})$ be the associated smooth stack in the category of complex analytic spaces.

Recall that $\mathfrak{X}$ is a category cofibered in groupoids over the category of algebraic spaces; since $\mathfrak{X}$ is a Deligne-Mumford stack, we will, equivalently, view $\mathfrak{X}$ as cofibered over the category of schemes over $\mathbb{C}$. A 1-morphism from a scheme U to $\mathfrak{X}$ is really an object in the fiber of $\mathfrak{X}$ over U. A morphism between two objects in the category $\mathfrak{X}$ is étale (respectively surjective) if the morphism of the associated schemes is étale (respectively surjective), and a morphism $p : U \to \mathfrak{X}$ from a scheme to $\mathfrak{X}$ is étale (respectively surjective) if for every morphism $f : X \to \mathfrak{X}$ with domain a scheme, the projection $f \times_{\mathfrak{X}} p \to f$ is étale (respectively surjective).

Following the discussion in section 12.1 of [26] and definition 4.10 of [8], we can take the étale site of $\mathfrak{X}$ to consist of all schemes $p : U \to \mathfrak{X}$ étale over $\mathfrak{X}$, with covering families consisting of those families of morphisms in the category $\mathfrak{X}$ for which the associated family of morphisms of schemes is a covering family for the étale topology.

Because every Deligne-Mumford stack admits an étale cover $\pi : U \to \mathfrak{X}$ by a scheme, to give a sheaf of sets F on the étale site of $\mathfrak{X}$ is equivalent to giving a sheaf F_U together with an isomorphism between the two pull-backs of F_U to $U \times_{\mathfrak{X}} U$ satisfying the cocycle condition of [26] 12.2.1.

Similarly, to give a sheaf F on the stack $\mathfrak{X}(\mathbb{C})$ over the category of analytic spaces, is equivalent to giving a sheaf F_U for any étale cover $\pi : U \to \mathfrak{X}$ in the classical topology on $U(\mathbb{C})$ together with an isomorphism between the two pull-backs of F_U to $U(\mathbb{C}) \times_{\mathfrak{X}(\mathbb{C})} U(\mathbb{C})$ satisfying the cocycle condition.

Note that if $f : U \to V$ is an étale map between complex analytic manifolds, then we have *isomorphisms*, for all (p, q):

$$f^* : f^{-1}\mathcal{A}_U^{p,q} \to \mathcal{A}_V^{p,q}$$

where $\mathcal{A}_V^{p,q}$ is the usual sheaf of (p, q)-forms. We therefore have the sheaf $\mathcal{A}_{\mathfrak{X}}^{p,q}$ of differential forms of type (p, q) on the stack $\mathfrak{X}(\mathbb{C})$, together with the ∂ and $\bar{\partial}$ operators. Notice that if $p : U \to \mathfrak{X}$ is étale, then the group of automorphisms of p over U acts *trivially* on $\mathcal{A}_U^{p,q}$.

The total de Rham complex of $\mathfrak{X}$ is a resolution of the constant sheaf $\mathbb{C}$, since this can be checked locally in the étale topology, and similarly the complexes $(\mathcal{A}_{\mathfrak{X}}^{p,*}, \bar{\partial})$ are resolutions of the sheaves of holomorphic p-forms $\Omega_{\mathfrak{X}}^p$. We have the usual Hodge spectral sequence:

$$E_1^{p,q}(\mathfrak{X}) = H^q(\mathfrak{X}, \Omega^p) \Rightarrow H^{p+q}(\mathfrak{X}, \mathbb{C}) \,.$$

If U is a complex manifold, we also have for each (p, q), the sheaf $\mathcal{D}_U^{p,q}$ of (p, q)-forms with distribution coefficients; if $V \subset U$, $\mathcal{D}_U^{p,q}(V)$ is the bornological dual $D^{p,q}(V)$ of the Frechet space $A_c^{n-p,n-q}(V)$ of compactly supported $(n-p, n-q)$-forms on V, where $n = \dim(V)$. If $p : V \to V'$ is étale, we have a push forward map $p_* : A_c^{n-p,n-q}(V) \to A_c^{n-p,n-q}(V')$ which is a continuous map of Frechet spaces inducing a pull back map $p^* : D^{p,q}(V') \to D^{p,q}(V)$, thus for $\mathfrak{X}$ a Deligne-Mumford stack over $\mathbb{C}$ we get a sheaf $\mathcal{D}_{\mathfrak{X}}^{p,q}$ of (p, q)-forms with distribution coefficients on

$\mathfrak{X}(\mathbb{C})$. There is a natural inclusion $\mathcal{A}_{\mathfrak{X}}^{*,*} \subset \mathcal{D}_{\mathfrak{X}}^{*,*}$, the operators ∂ and $\overline{\partial}$ extend to $\mathcal{D}_{\mathfrak{X}}^{*,*}$, and the inclusion is a quasi-isomorphism of double complexes. It follows that the Hodge spectral sequence also arises as the cohomology spectral sequence of the filtered complex associated to the double complex of sheaves $\mathcal{D}_{\mathfrak{X}}^{*,*}$ with the Hodge filtration.

Note that in general on a smooth Artin or Deligne-Mumford stack over the complex numbers, the sheaves $\mathcal{A}_{\mathfrak{X}}^{*,*}$ and $\mathcal{D}_{\mathfrak{X}}^{*,*}$ need not be acyclic. For example this is not the case for $B\mathbb{G}L_{n,\mathbb{C}}$, nor for the affine line with the origin doubled. However we have:

Proposition 1.1 *If $\mathfrak{X}$ is a smooth separated Deligne-Mumford stack over $\mathbb{C}$, then the sheaves $\mathcal{A}_{\mathfrak{X}}^{*,*}$ and $\mathcal{D}_{\mathfrak{X}}^{*,*}$ are acyclic.*

Proof We know ([26] 19.1, [9] Ch. I. section 8, and [21]) that $\mathfrak{X}$ has a "coarse space" X which is a separated algebraic space, and for which the map $\pi : \mathfrak{X} \to X$ is finite and induces a bijection on (isomorphism classes of) geometric points. Furthermore, given an étale cover $p : P \to \mathfrak{X}$, a closed point $\xi : \mathrm{Spec}(\mathbb{C}) \to \mathfrak{X}$ and a lifting $x : \mathrm{Spec}(\mathbb{C}) \to P$ of ξ, the map π^* induces an isomorphism between $\mathcal{O}_{X,\pi(\xi)}^h$ and the invariants $\mathcal{O}_{P,x}^h/G$ of the action of the (finite) group of automorphisms of ξ on the Henselization of the local ring of P at x (see [9] Ch. I. 8.2.1, and [26], 6.2.1). Since $\mathfrak{X}$ is *separated*, it is straightforward to check that this isomorphism extends to an open neighborhood U of x in the analytic topology, giving an isomorphism $[U/G] \to \pi^{-1}(V)$ where $V \subset X$ is an open neighborhood of $\pi(\xi)$. (See II.5.3 of [31], which is based on 2.8 of [32].) Since $[U/G] = \pi^{-1}(V)$ is a finite étale groupoid, a standard transfer argument shows that the restrictions of the sheaves $\mathcal{A}_{\mathfrak{X}}^{*,*}$ and $\mathcal{D}_{\mathfrak{X}}^{*,*}$ to $[U/G]$ are acyclic. Now take a cover of the analytic space $X(\mathbb{C})$ by open sets V such that $\pi^{-1}(V)$ is a finite étale groupoid, and using a partition of unity subordinate to this cover, the proof finishes by a standard argument. $\square$

We immediately obtain:

Corollary 1.2 *Let $\mathfrak{X}$ be a Deligne-Mumford stack smooth and separated over the complex numbers $\mathbb{C}$. Since the sheaves $\mathcal{D}_{\mathfrak{X}}^{*,*}$ and $\mathcal{A}_{\mathfrak{X}}^{*,*}$ are acyclic, the groups $H^*(\mathfrak{X}, \mathbb{C})$ and $H^*(\mathfrak{X}, \Omega^*)$ are computed by the global de Rham and Dolbeault complexes of $\mathfrak{X}(\mathbb{C})$ respectively.*

If $f : \mathfrak{X} \to \mathfrak{Y}$ is a proper, representable, morphism of relative dimension d between regular stacks over $\mathbb{C}$, there is a push forward map:

$$f_* : f_*(\mathcal{D}_{\mathfrak{X}}^{p,q}) \to \mathcal{D}_{\mathfrak{Y}}^{p-d,q-d}$$

which is defined as follows. If $p : U \to \mathfrak{Y}$ is an étale morphism from a scheme to $\mathfrak{Y}$, let $\pi_U : \mathfrak{X} \times_{\mathfrak{Y}} U \to U$ be the induced proper morphism of schemes. Suppose that $T \in D^{p,q}((\mathfrak{X} \times_{\mathfrak{Y}} U)(\mathbb{C}))$; then if $\phi \in A_c^{n-p,n-q}(U)$, where n is the dimension of $\mathfrak{Y}$, is a compactly supported form, we have that $\pi_U^*(\phi)$ is a compactly supported form on $(\mathfrak{X} \times_{\mathfrak{Y}} U)(\mathbb{C})$, and hence we can define $\pi_*(T)$ to be the current in $D^{p-d,q-d}(U(\mathbb{C}))$ such that $\pi_*(T)(\phi) := T(\pi_U^*(\phi))$.

Lemma 1.3 *Let $f : \mathfrak{X} \to \mathfrak{Y}$ be a proper, representable, morphism between smooth Deligne-Mumford stacks over the complex numbers $\mathbb{C}$. Suppose that f is generically finite of degree d. Then the composition of the maps*

$$\mathcal{A}_{\mathfrak{Y}}^{p,q} \xrightarrow{f^*} f_*\mathcal{A}_{\mathfrak{X}}^{p,q} \to f_*\mathcal{D}_{\mathfrak{X}}^{p,q} \xrightarrow{f_*} \mathcal{D}_{\mathfrak{Y}}^{p,q}$$

is d times the natural inclusion $\mathcal{A}_{\mathfrak{Y}}^{p,q} \to \mathcal{D}_{\mathfrak{Y}}^{p,q}$.

Proof Suppose that $p : U \to \mathfrak{Y}$ is étale, with U a scheme, and that $\alpha \in A^{p,q}(U(\mathbb{C}))$. Without loss of generality, we can suppose that U is irreducible, so that $f_U : f^{-1}(U) \to U$ has a well defined degree. Since $f_U^*(\alpha)$ is a smooth form $f_{U,*} f_U^*(\alpha)$ is a current which is represented by a locally L^1 form, and hence is determined by its value on the complement of any subset of $U(\mathbb{C})$ with measure zero. But outside a set of measure zero on $U(\mathbb{C})$, the map of complex manifolds $f_U : f^{-1}(U)(\mathbb{C}) \to U(\mathbb{C})$ is a finite étale cover, and hence $f_{U,*} f_U^*(\alpha) = d\alpha$, where d is the degree of π. $\square$

Since the inclusions $\mathcal{A}_{\mathfrak{X}}^{p,*} \subset \mathcal{D}_{\mathfrak{X}}^{p,*}$ induce quasi-isomorphisms on the sheaves of Dolbeault complexes, we get:

Corollary 1.4 *Let* $f : \mathfrak{X} \to \mathfrak{Y}$ *be a proper, representable, generically finite morphism between smooth Deligne-Mumford stacks over the complex numbers* $\mathbb{C}$. *Then we have split injections:*

$$f^* : H^*(\mathfrak{Y}, \mathbb{C}) \to H^*(\mathfrak{X}, \mathbb{C})$$

and

$$f^* : H^*(\mathfrak{Y}, \Omega_{\mathfrak{Y}}^*) \to H^*(\mathfrak{X}, \Omega_{\mathfrak{X}}^*)$$

compatible with the Hodge spectral sequence.

Proposition 1.5 *Let* $\mathfrak{X}$ *be a Deligne-Mumford stack smooth and proper over the complex numbers* $\mathbb{C}$. *The Hodge spectral sequence for* $\mathfrak{X}(\mathbb{C})$ *degenerates at* E_1, *and the cohomology groups* $H^*(\mathfrak{X}, \mathbb{C})$ *admit a Hodge decomposition "in the strong sense" (see* [27] *Definition 2.24).*

Proof Since $\mathfrak{X}$ is proper over $\mathbb{C}$ it is in particular separated. Hence by Chow's lemma [26] 16.6.1, we know that there exists a proper surjective and generically finite map $\pi : X \to \mathfrak{X}$ with X smooth and proper over $\mathbb{C}$. Since the Hodge spectral sequence for $X(\mathbb{C})$ degenerates, by corollary 1.4 we know that the same is true for $\mathfrak{X}(\mathbb{C})$. Furthermore since the $\mathcal{A}_{\mathfrak{X}}^{*,*}$ are acyclic by proposition 1.1, the pull back map is induced by the map of complexes: $f^* : A^{*,*}(\mathfrak{X}(\mathbb{C})) \to A^{*,*}(X(\mathbb{C}))$, and is invariant under complex conjugation. Hence for $p + q = n$, $F^p H^n(\mathfrak{X}(\mathbb{C}), \mathbb{C}) \cap \overline{F^{q+1} H^n(\mathfrak{X}(\mathbb{C}), \mathbb{C})} = \{0\}$. $\square$

The proposition and the fact that the Hodge cohomology of $\mathfrak{X}(\mathbb{C})$ is computed by the Dolbeault complexes $A^{*,*}(\mathfrak{X}(\mathbb{C}))$ give us:

Corollary 1.6 *The* $\partial\bar{\partial}$-*lemma holds for* $\mathfrak{X}(\mathbb{C})$.

Proof See [27] 2.27 and 2.28. $\square$

This proof is a variation on an argument for complex manifolds, by which one deduces a strong Hodge decomposition and the $\partial\bar{\partial}$-lemma for Moishezon manifolds (see, for example, section 9 of Demailly's article in [2]).

We shall need the following lemma later.

Lemma 1.7 *If* $\mathfrak{X}$ *is a smooth separated Deligne-Mumford stack over* $\mathbb{C}$, *the natural map:*

$$A^{*,*}(\mathfrak{X}) \to \varprojlim_{p : X \to \mathfrak{X}} A^{*,*}(X)$$

where the inverse limit is over all proper surjective maps with domain a smooth variety, is an isomorphism.

Proof From lemma 1.3 we know the map is injective. We therefore need only show that it is surjective. Suppose then that $(p : X \to \mathfrak{X}) \mapsto \alpha_p$ is an element in the inverse limit.

First note that by Zariski's main theorem ([26], 16.5) any étale map $\pi : U \to \mathfrak{X}$ from a smooth variety to $\mathfrak{X}$ extends to a finite map $V \to \mathfrak{X}$, and hence by resolution of singularities to a proper surjective map $\tilde{\pi} : \tilde{V} \to \mathfrak{X}$. Thus, restricting $\alpha_{\tilde{\pi}}$, we get a form on U. A priori this depends depends on the choice of the factorization of π through $\tilde{\pi}$. However given two such factorizations, $U \subset \tilde{V}_1 \to \mathfrak{X}$ and $U \subset \tilde{V}_2 \to \mathfrak{X}$ we can factor through the fiber product $\tilde{V}_1 \times_{\mathfrak{X}} \tilde{V}_2$ and applying resolution of singularities again we see that the form is indeed independent of the factorization. Writing α_π for this form, by a slight variation of the the preceding argument, we also see that α_π is contravariant with respect to π. Hence we get an element in the inverse limit of $A^{*,*}(U)$ over all étale maps $U \to \mathfrak{X}$, i.e. an element $\alpha \in A^{*,*}(\mathfrak{X})$.

Now we want to show that for any $p : X \to \mathfrak{X}$, $\alpha_p = p^*(\alpha)$. Let $\pi : P \to \mathfrak{X}$ be étale and surjective with P a smooth scheme over $\mathbb{C}$. Let $i : P \to \bar{P}, \bar{\pi} : \bar{P} \to \mathfrak{X}$ be a factorization of π through a finite map with P dense in $\bar{P}$. Since P is dense in $\bar{P}$ $\alpha_{\bar{\pi}} = \bar{\pi}^*\alpha$, with $\bar{P}$ smooth over $\mathbb{C}$. Now by resolution of singularities, we know that there exists a proper morphism $Q \to \bar{P} \times_{\mathfrak{X}} X$, with Q smooth over $\mathbb{C}$ such that the induced maps $f : Q \to \bar{P}, g : Q \to X$ are surjective. Now $\alpha_Q = (f \cdot \bar{\pi})^*(\alpha)$. It follows that $g^*(\alpha_p) = f^*(\alpha_{\bar{\pi}}) = (g \cdot p)^*\alpha = g^*(p^*(\alpha))$. However, since g is a surjective map between smooth varieties over $\mathbb{C}$, g^* is injective, and hence $\alpha_p = p^*(\alpha)$. $\square$

Given a codimension p cycle $\zeta = \sum_i [Z_i]$ on a stack $\mathfrak{X}$, we know, using the local to global spectral sequence for cohomology (in the étale topology) with supports in $|\zeta|$, that the associated current δ_ζ represents the cycle class $[\zeta] \in H^{2p}(\mathfrak{X}, \mathbb{R}(p))$. Hence as in [13], if we choose an arbitrary C^∞ (p,p)-form ω representing this class, then by the $\partial\bar{\partial}$-lemma, we know that there is a current $g \in D^{(p-1,p-1)}(\mathfrak{X}(\mathbb{C}))$ such that

$$dd^c(g) + \delta_\zeta = \omega.$$

We can compute the real Deligne Cohomology of a Deligne-Mumford stack which is proper and smooth over $\mathbb{C}$ using the Deligne complexes of Burgos associated to the Dolbeault complex of $\mathfrak{X}(\mathbb{C})$ (see [5]). However, we do not know how to extend the results of Burgos to non-proper stacks, and therefore we do not know how to give a cohomological construction of Green currents analogous to that of *op. cit.*

2 Chow groups and the K-theory of stacks

2.1 Chow groups of Deligne-Mumford stacks. Recall the definition of cycles on, and Chow groups of, stacks from [10].

Definition 2.1 1. If $\mathfrak{X}$ is a stack, the group of codimension p cycles $Z^p(\mathfrak{X})$ is the free abelian group on the set of codimension p reduced irreducible substacks of $\mathfrak{X}$. This is isomorphic to the group of codimension p cycles on the coarse space $|\mathfrak{X}|$.

2. If $\mathfrak{W}$ is a integral stack, write $\mathbf{k}(\mathfrak{W})$ for its function field. This may be defined as étale H^0 of the sheaf of total quotient rings on $\mathfrak{W}$; and if $f \in \mathbf{k}(\mathfrak{W})$, we may define $\mathrm{div}(f)$ locally in the étale topology. Therefore, if $\mathfrak{W} \subset \mathfrak{X}$ is a codimension p integral substack, and $f \in \mathbf{k}(\mathfrak{W})$, we have $\mathrm{div}(f) \in Z^{p+1}(\mathfrak{X})$.

3. We define $\mathrm{CH}^p(\mathfrak{X})$ to be the quotient of $Z^p(\mathfrak{X})$ by the subgroup consisting of divisors of rational functions on integral subschemes of codimension $p-1$.

4. We define $\mathrm{CH}^{p,p-1}(\mathfrak{X})$ as in section 3.3.5 of [13], to be the homology of the complex

$$\bigoplus_{x \in \mathfrak{X}^{(p-2)}} K_2(\mathbf{k}(x)) \to \bigoplus_{x \in \mathfrak{X}^{(p-1)}} \mathbf{k}(x)^* \to Z^p(\mathfrak{X})$$

at the middle term. These groups are isomorphic to the groups of section 4.7 of [10], though there they are written $\mathrm{CH}^{p-1,p}(\mathfrak{X})$; this follows from the argument of the proof of theorem 6.8 of *op. cit.*

Note that this definition does not require that the Chow groups have rational coefficients. However once we pass to the étale site of $\mathfrak{X}$, we will need to tensor with $\mathbb{Q}$.

2.2 G-theory of stacks. Let us review the basic facts about the G-theory of Deligne-Mumford stacks, following section 4.2 of [17], where details and proofs of the following may be found.

Let $\mathbf{G}$ be the presheaf of spectra on the étale site induced by the functor $\mathbf{G}$. We shall take this functor with rational coefficients, *i.e.*, we take the usual $\mathbf{G}$-theory functor to the category of symmetric spectra, and take the smash product with the Eilenberg-Maclane spectrum $H\mathbb{Q}$. It is a (by now) elementary observation, first made by Thomason (*c.f.* [29]) that if X is a scheme, the natural map on *rational* G-theory:

$$\mathbf{G}(X) \to \mathbf{H}_{\text{ét}}(X, \mathbf{G})$$

is a weak equivalence. Hence we can extend rational G-theory from schemes to stacks by defining:

Definition 2.2 If $\mathfrak{X}$ is a stack, we define the G-theory spectrum $\mathbf{G}(\mathfrak{X})$ to be $\mathbf{H}_{\text{ét}}(\mathfrak{X}, \mathbf{G})$.

Notice that it follows from the definition, and Thomason's result, that if $\pi : V. \to \mathfrak{X}$ is an étale hypercover of a Deligne-Mumford stack, the natural map:

$$\mathbf{G}(\mathfrak{X}) \to \mathrm{holim}_i(\mathbf{G}(V_i))$$

is a weak equivalence.

It follows immediately that if $f : \mathfrak{Y} \to \mathfrak{X}$ is a *representable* proper morphism of stacks, then there is a natural map $f_* : \mathbf{G}(\mathfrak{Y}) \to \mathbf{G}(\mathfrak{X})$ (this was already observed in [10]). Similarly, if f is a *flat* representable morphism, there is a natural pullback map $f^* : \mathbf{G}(\mathfrak{X}) \to \mathbf{G}(\mathfrak{Y})$. If $f : \mathfrak{Y} \to \mathfrak{X}$ is a closed substack, with complement $j : \mathfrak{U} \to \mathfrak{X}$, it is then straightforward to check that Quillen's localization theorem extends to stacks, *i.e.*,

$$\mathbf{G}(\mathfrak{Y}) \xrightarrow{i_*} \mathbf{G}(\mathfrak{X}) \xrightarrow{j^*} \mathbf{G}(\mathfrak{U})$$

is a fibration sequence.

In order to define a pushforward for non-representable morphisms, in [17], we replace stacks by simplicial varieties, as follows. First observe, that using the strict covariance of our model for G-theory, we can extend the functor $\mathbf{G}$ from schemes to simplicial schemes with proper face maps.

Let $p : X. \to \mathfrak{X}$ be a proper morphism to a stack from a simplicial variety with proper face maps. We construct, in *op. cit.*, a map $p_* : \mathbf{G}(X.) \to \mathbf{G}(\mathfrak{X})$, which is well defined in the rational stable homotopy category and is compatible with composition, in the sense that if $f. : Y. \to X.$ is a map of simplicial varieties, we have that

$$(p \cdot f)_* = p_* \cdot f_* : \mathbf{G}(Y.) \to \mathbf{G}(\mathfrak{X})$$

in the rational stable homotopy category. This construction gives an extension of $\mathbf{G}$-theory from simplicial varieties to stacks, because we have:

Lemma 2.3 *Suppose that $p : X. \to \mathfrak{X}$ is a proper hypercover. Then $p_* : \mathbf{G}(X.) \to \mathbf{G}(\mathfrak{X})$ is a weak equivalence.*

From which we then get:

Theorem 2.4 *Let $f : \mathfrak{X} \to \mathfrak{Y}$ be a proper, not necessarily representable, morphism of stacks. Then there exists a canonical map (in the homotopy category)*

$$\mathbf{G}(\mathfrak{X}) \to \mathbf{G}(\mathfrak{Y})$$

with the property that for any commutative square:

$$
\begin{array}{ccc}
X. & \xrightarrow{\tilde{f}} & Y. \\
{\scriptstyle p.}\downarrow & & \downarrow{\scriptstyle q.} \\
\mathfrak{X} & \xrightarrow{f} & \mathfrak{Y}
\end{array}
$$

in which $p.$ and $q.$ are proper morphisms with domains simplicial schemes with proper face maps, we have a commutative square in the stable homotopy category:

$$
\begin{array}{ccc}
\mathbf{G}(X.) & \xrightarrow{\tilde{f}_*} & \mathbf{G}(Y.) \\
{\scriptstyle \mathbf{G}(p.)}\downarrow & & \downarrow{\scriptstyle \mathbf{G}(q.)} \\
\mathbf{G}(\mathfrak{X}) & \xrightarrow{\mathbf{G}(f)} & \mathbf{G}(\mathfrak{Y})
\end{array}
$$

It is shown in [17] that the natural (in general non-representable) map from a separated Deligne-Mumford stack to its coarse space induces an isomorphism on $\mathbf{G}$-theory.

Lemma 2.5 *Let $\mathfrak{X}$ be a Deligne-Mumford stack, and suppose that $\pi : X \to \mathfrak{X}$ is a proper and surjective map, with X a variety. Then $\pi_* : G_0(X) \to G_0(\mathfrak{X})$ is surjective. (Recall that we are using $\mathbb{Q}$ coefficients.)*

Proof First, observe that if ξ is a generic point of $\mathfrak{X}$, the stack ξ is equivalent to a finite étale groupoid acting on a field F, and $G_*(\xi)$ is a direct summand of $G_*(F)$ (see [10]). Hence if $F \subset E$ is any finite extension of F, the direct image, or transfer, map $G_*(E) \to G_*(\xi)$ is surjective, since we are using G-theory with rational coefficients). Hence if $f : X \to \xi$ is a proper morphism from a scheme to ξ, it also induces a surjective map on G-theory. The lemma now follows by localization and noetherian induction. □

2.3 K-theory of stacks. Just as we defined the G-theory of a stack as the hypercohomology of the G-theory sheaf, we can consider the presheaf $\mathbf{K}$ of K-theory spectra which associates to X the K-theory spectrum (again with rational coefficients) of the category of locally free sheaves on the big Zariski site, and then define:

$$\mathbf{K}(\mathfrak{X}) := \mathbf{H}_{\text{ét}}(\mathfrak{X}, \mathbf{K}).$$

This gives a contravariant functor from stacks to ring spectra. The tensor product between locally free and coherent sheaves induces a pairing

$$\mathbf{K}(\mathfrak{X}) \wedge \mathbf{G}(\mathfrak{X}) \to \mathbf{G}(\mathfrak{X}).$$

Lemma 2.6 *If* $f : \mathfrak{X} \to \mathfrak{Y}$ *is a proper representable morphism between stacks, the projection formula holds, i.e., the following diagram is commutative in the stable homotopy category*

$$
\begin{array}{ccccc}
\mathbf{K}(\mathfrak{Y}) \wedge \mathbf{G}(\mathfrak{X}) & \xrightarrow{1 \wedge f_*} & \mathbf{K}(\mathfrak{Y}) \wedge \mathbf{G}(\mathfrak{Y}) & \longrightarrow & \mathbf{G}(\mathfrak{Y}) \\
{\scriptstyle f^* \wedge 1} \downarrow & & & & \uparrow {\scriptstyle f_*} \\
\mathbf{K}(\mathfrak{X}) \wedge \mathbf{G}(\mathfrak{X}) & \longrightarrow & \mathbf{G}(\mathfrak{X}) & = & \mathbf{G}(\mathfrak{X})
\end{array}
$$

Proof Since f is representable, if $\pi : V. \to \mathfrak{Y}$ is any étale hypercover, $W. = V. \times_{\mathfrak{Y}} \mathfrak{X} \to \mathfrak{X}$ is also an étale hypercover, and so the result follows from the same statement for the morphisms $W_i \to V_i$ of schemes. For schemes the assertion follows from projection formula for the tensor product of locally free and flasque quasi-coherent sheaves, which is true up to canonical isomorphism. □

Combining lemma 2.5 with the previous lemma, we get:

Corollary 2.7 *If* $\pi : X \to \mathfrak{Y}$ *is a proper surjective morphism from a regular variety to a regular Deligne-Mumford stack, then* $\pi^* : K_0(\mathfrak{Y}) \to K_0(X)$ *is a split injection.*

2.4 Operations on the K-theory of stacks. Before constructing an intersection product on the arithmetic Chow groups of a stack, we must first construct a product on the ordinary Chow groups of stacks over an arithmetic ring. As in [1] and [13], we shall replace the Chow groups of $\mathfrak{X}$ by the graded group associated to the γ-filtration on $K_0(\mathfrak{X})$. However there are some technicalities involved in doing this.

Recall that if $\mathfrak{X}$ is a Deligne-Mumford stack, then the G-theory, with rational coefficients of $\mathfrak{X}$ is defined to be

$$\mathbf{G}(\mathfrak{X}) := \mathbf{H}_{et}(\mathfrak{X}, \mathbf{G})$$

where $\mathbf{G}$ is the presheaf of G-theory spaces in the étale topology of $\mathfrak{X}$ associated to the Quillen K-theory of coherent sheaves of $\mathcal{O}_{\mathfrak{X}}$-modules, and similarly the K-theory of $\mathfrak{X}$ is defined to be

$$\mathbf{K}(\mathfrak{X}) := \mathbf{H}_{et}(\mathfrak{X}, \mathbf{K})$$

where $\mathbf{K}$ is the presheaf of K-theory spaces associated to the K-theory of locally free sheaves. In order to define the γ-filtration we need to have λ-operations on the sheaf $\mathbf{K}$. While the method of [16], applied to any étale hypercover $V. \to \mathfrak{X}$ by a simplicial regular scheme, should lead to a construction of λ-operations, it is not clear that the simplicial scheme $V.$ is K-coherent, in the terminology of *op. cit.* However if we take as a model for K-theory the presheaf of simplicial sets associated to the G-construction of [11], and the variations on the G-construction described in [18] applied to the sheaf of categories $\mathcal{P}_{\mathfrak{X}}$ associated to the category of locally free sheaves of $\mathcal{O}_{\mathfrak{X}}$-modules on the étale site of $\mathfrak{X}$, then we have, following *op. cit*, maps of simplicial sheaves

$$\lambda^k : \mathrm{Sub}_k G(\mathcal{P}_{\mathfrak{X}}) \to G^{(k)}(\mathcal{P}_{\mathfrak{X}})$$

with both the domain and the target canonically weakly equivalent to $G(\mathcal{P}_{\mathfrak{X}})$. It is shown in *op. cit.* that if R is a commutative ring, then these maps induce the same maps on K-theory as those of [19], [23] and [28].

Theorem 2.8 *Let* $\mathfrak{X}$ *be a regular Deligne-Mumford stack. Then the γ-filtration on $K_0(\mathfrak{X})$ has finite length and there are isomorphisms:*

$$\mathrm{Gr}^p_\gamma(K_0(\mathfrak{X})) \simeq \mathrm{Gr}^p(K_0(\mathfrak{X})) \simeq \mathrm{CH}^p(\mathfrak{X})_{\mathbb{Q}}.$$

Here Gr^* is the filtration by codimension of supports. Furthermore, if $\mathfrak{Y} \subset \mathfrak{X}$ is a closed substack of pure codimension e, then:

$$Gr^*_\gamma K_0^{\mathfrak{Y}}(\mathfrak{X}) \simeq \mathrm{CH}^{*-e}(\mathfrak{Y})_{\mathbb{Q}}.$$

Proof The coniveau spectral sequence for $\mathbf{K}(\mathfrak{X})$ has

$$E_1^{p,q} = \bigoplus_{\xi \in \mathfrak{X}^{(p)}} K_{q-p}(\xi)$$

where the direct sum is over punctual substacks of $\mathfrak{X}$. As in [28], consider the action of the Adams operations on the coniveau spectral sequence (the Adams operations are linear combinations of the lambda operations). If $\xi \in \mathfrak{X}$ is a point, let $i : \xi \to \mathfrak{X}$ be the inclusion, and let $i^!$ be the "sections with support" functor. Quillen's localization theorem (applied locally in the étale topology) implies that we have a natural weak equivalence $Ri^! \mathbf{K}_{\mathfrak{X}} \simeq \mathbf{K}_\xi$. Following the argument in [28], this isomorphism shifts the weights of the Adams operations by the codimension of ξ. We have isomorphisms $K_*(\xi) \simeq K_*(\mathbf{k}(\xi))$ (remembering that we are using $\mathbb{Q}$-coefficients), and just as in *op. cit* we get that the coniveau spectral sequence is partially degenerate. Using the fact that $E_2^{p,p} \simeq \mathrm{CH}^p(\mathfrak{X})_{\mathbb{Q}}$, we then have the isomorphism of the theorem. The assertion about K-theory with supports follows from the same argument. □

The product structure on K-theory is compatible with the Adams operations (see *op. cit* and [16]), and so the groups $\mathrm{Gr}^p_\gamma(K_0(\mathfrak{X}))$ form a contravariant functor from stacks to commutative graded rings. Via the isomorphism of the theorem, this gives the Chow groups of stacks the same structure. We know from the appendix to exp. 0 of SGA6, [1], that if X is a regular noetherian scheme, then the isomorphism $Gr^*_\gamma K_0(X) \simeq \mathrm{CH}^*(X)_{\mathbb{Q}}$ is compatible with the product on the Chow groups whenever it is defined. Furthermore by the argument of [12], applied on an étale presentation of $\mathfrak{X}$, we know that if $\mathfrak{Y} \subset \mathfrak{X}$ and $\mathfrak{Z} \subset \mathfrak{X}$ are integral substacks of codimension p and q respectively, intersecting properly, the intersection product $[\mathfrak{Y}].[\mathfrak{Z}]$ computed using the isomorphism $Gr^p_\gamma K_0^{\mathfrak{Y}}(\mathfrak{X}) \simeq \mathrm{CH}^0(\mathfrak{Y})_{\mathbb{Q}}$ and $Gr^q_\gamma K_0^{\mathfrak{Z}}(\mathfrak{X}) \simeq \mathrm{CH}^0(\mathfrak{Z})_{\mathbb{Q}}$, agrees with the product computed using Serre's intersection multiplicities.

Thus we have given another construction of the product structure on the Chow groups in the case of stacks over a field.

2.5 Motives of stacks. Let $\mathfrak{X}$ be a proper Deligne-Mumford stack over S. Recall from [17] that there exist proper hypercovers $\pi : X. \to \mathfrak{X}$ with the X_i regular. (We refer to these as non-singular proper hypercovers). Following *op. cit.*, there is a covariant functor h from the category of regular projective schemes over S to the category of K_0-motives with rational coefficients. Applying this functor to a proper hypercover $\pi : X. \to \mathfrak{X}$ we get a chain complex of motives

$h(X.)$. It is proved in *op. cit.* that if we have two non-singular proper hypercovers $\pi_i : (X.)_i \to \mathfrak{X}$ for $i = 1, 2$, and a map $f : (X.)_1 \to (X.)_2$ of hypercovers (*i.e.*, $f \circ \pi_2 = \pi_1$) then the induced map of chain complexes of motives $h((X.)_1) \to h((X.)_2)$ is a homotopy equivalence, and then that given any two nonsingular hypercovers $\pi_i : (X.)_i \to \mathfrak{X}$, irrespective of whether there is a map between them, there is, nonetheless, a canonical isomorphism in the homotopy category of chain complexes quasi-isomorphic to bounded complexes of K_0-motives: $h((X.)_1) \to h((X.)_2)$. It follows upon applying the functor $K_0(\ )_{\mathbb{Q}}$, that we get, for each nonsingular proper hypercover $\pi : X. \to \mathfrak{X}$ a cochain complex $K_0(X.)_{\mathbb{Q}}$ which is independent, up to homotopy equivalence, of the choice of hypercover.

Theorem 2.9 *Suppose that the Deligne-Mumford stack $\mathfrak{X}$, flat and proper over S, is regular. Then for any non-singular proper hypercover $\pi : X. \to \mathfrak{X}$, any regular variety Y and any $p \geq 0$, the augmentation*

$$\pi_* : (i \mapsto G_p(X_i \times_S Y)) \to G_p(\mathfrak{X} \times_S Y)$$

is a quasi-isomorphism. It follows by the argument of [17] that the associated complex of motives $h(X.)$ is exact in positive degrees.

Proof As in [15], we want to construct by induction K_0-correspondences $h_i : X_i \to X_{i+1}$ for $i \geq -1$, (where we set $X_{-1} = \mathfrak{X}$) which provide a contracting homotopy of the complex. The slight complication here is that we need to start the induction with a stack. We could extend the theory of correspondences and motives to stacks, as was done in [30] for stacks over fields. However we shall do something more ad-hoc. By lemma 2.5, we know that there is a class $\eta \in G_0(X_0)$ such that $\pi_*(\eta) = [\mathcal{O}_{\mathfrak{X}}]$. It follows from the projection formula that if $\alpha \in K_p(\mathfrak{X})$, then $\pi_*(\pi^*(\alpha)\gamma) = \alpha$, and furthermore, since we assume that $\mathfrak{X}$ is flat over S, for any scheme Y over S, if $\alpha \in K_0(\mathfrak{X} \times_S Y)$, and $\pi_Y : X_0 \times_S Y \to \mathfrak{X} \times_S Y$ is the induced map, and $p : X_0 \times_S Y \to X_0$ the projection, we have:

$$(\pi_Y)_*((\pi_Y)^*(\alpha)(\pi_Y)^*(\eta)) = \alpha.$$

Thus we can view the class η as determining a correspondence $[\eta]$ from $\mathfrak{X}$ to X_0, which induces a splitting of the maps $\pi_* : G_*(X_0) \to G_*(\mathfrak{X})$.

Consider the diagram:

$$
\begin{array}{ccc}
X \times_{\mathfrak{X}} X & \xrightarrow{\ q\ } & X \\
{\scriptstyle p}\big\downarrow & & \big\downarrow{\scriptstyle \pi} \\
X & \xrightarrow{\ \pi\ } & \mathfrak{X}
\end{array}
$$

If $\mathcal{F}$ is a coherent sheaf on X, then by the projection formula:

$$\pi^*\pi_*([\mathcal{F}]) = p_*(q^*(\alpha))[\mathcal{O}_X{}^L \otimes_{\mathcal{O}_{\mathfrak{X}}} \mathcal{O}_X],$$

and so $[\eta] \cdot \pi_*$ is induced by the correspondence $\phi_*(\theta)q'^*(\eta) \in G_0(X \times_S X)$ where $\theta = [\mathcal{O}_X{}^L \otimes_{\mathcal{O}_{\mathfrak{X}}} \mathcal{O}_X]$ where $\phi : X \times_{\mathfrak{X}} X \to X \times_S X$ is the natural map, and $q' : X \times_S X \to X$ is projection onto the second factor. It is straightforward to check that $\phi_*(\theta)q'^*(\eta)$ is a projector in the group of K-correspondences from X to itself, and that $(1_X \times \pi)_*(\phi_*(\theta)q'^*(\eta)) = [\mathcal{O}_{\Gamma_\pi}] \in G_0(X \times_S \mathfrak{X})$. The proof now follows the argument in [15].   □

Corollary 2.10 *The weight complex of a regular Deligne-Mumford stack proper over S is isomorphic, in the homotopy category of (homological) motives to a direct summand of the motive of a regular projective variety over S.*

Remark 2.11 For stacks over perfect fields follows immediately from (indeed is essentially equivalent to) the conclusion of Toen in [30].

3 Arithmetic Chow groups of proper stacks

Given a regular Deligne-Mumford stack proper over S, we define its arithmetic Chow groups in exactly the same fashion as in [13] and [6].

Definition 3.1 We define $\widehat{Z}^p(\mathfrak{X})$ to be the group consisting of pairs (ζ, g) where ζ is cycle on $\mathfrak{X}$, and $g \in D^{p-1,p-1}/(Im(\partial) + Im(\bar{\partial}))(\mathfrak{X}_\mathbb{R})$ is a green current for ζ in the sense of section 1.

Here $\mathfrak{X}_\mathbb{R}$ indicates that we are taking forms, or currents, on $\mathfrak{X}(\mathbb{C})$ on which the complex conjugation F_∞ on the base arithmetic ring acts by the rule of [13] and [6].

If $\eta \in \mathfrak{X}$ is a point, the Zariski closure of which has codimension p in $\mathfrak{X}$, and $f \in \mathbf{k}(\eta)$, then f pulls back to a rational function $\tilde{f}$ on the inverse image of the Zariski closure of η on any étale cover $U \to \mathfrak{X}$, and hence determines a $(p-1, p-1)$-current $-\log(|\tilde{f}|^2)$ on $U(\mathbb{C})$. Since this construction is local in the analytic, and hence in the étale topologies, we get a current $-\log(|f|^2)$ on $\mathfrak{X}(\mathbb{C})$, and we define $\widehat{CH}^p(\mathfrak{X})$ to be the quotient of $\widehat{Z}^p(\mathfrak{X})$ by the subgroup generated by cycles of the form $(\mathrm{div}(f), -\log(|f|^2))$.

The definition above makes sense for stacks which are not proper over S. However for such a stack it will give "naive" $\widehat{CH}$ groups as in [13] rather than the groups of [6]. As remarked in section 1, we do not know a definition of a forms "with logarithmic growth" on a stack over $\mathbb{C}$ which is not proper, and so we do not know how to modify the definition above to give the "non-naive" groups.

The proof of the following proposition is exactly the same as in the case of arithmetic schemes.

Proposition 3.2 *If $\mathfrak{X}$ is proper over S, There is an exact sequence*

$$\mathrm{CH}^{p,p-1}(\mathfrak{X}) \to H_\mathcal{D}^{2p-1}(\mathfrak{X}_\mathbb{R}, \mathbb{R}(p))$$
$$\to \widehat{CH}^p(\mathfrak{X}) \to \mathrm{CH}^p(\mathfrak{X}) \oplus Z^{p,p}(\mathfrak{X}_\mathbb{R}) \to \mathrm{H}^{2p}(\mathfrak{X}_\mathbb{R}, \mathbb{R}(p)). \quad (1)$$

On the other hand, if $\mathfrak{X}$ is proper over S, it follows from theorem 2.9 and the discussion in section 1, in particular lemma 1.7 that we have an exact sequence:

$$\mathrm{CH}^{p,p-1}(\mathfrak{X})_\mathbb{Q} \to H_\mathcal{D}^{2p-1}(\mathfrak{X}_\mathbb{R}, \mathbb{R}(p))$$
$$\to \varprojlim_{f:X \to \mathfrak{X}} \widehat{CH}^p(X)_\mathbb{Q} \to \mathrm{CH}^p(\mathfrak{X})_\mathbb{Q} \oplus Z^{p,p}(\mathfrak{X}_\mathbb{R}) \to \mathrm{H}^{2p}(\mathfrak{X}_\mathbb{R}, \mathbb{R}(p)) \quad (2)$$

where the inverse limit is over all proper surjective maps to $f : X \to \mathfrak{X}$ with nonsingular domain.

Theorem 3.3 *There is a canonical isomorphism*

$$\varprojlim_{f:X \to \mathfrak{X}} \widehat{CH}^p(X)_\mathbb{Q} \to \widehat{CH}^p(\mathfrak{X})_\mathbb{Q}.$$

Proof Since all but one term of the exact sequences 1 and 2 coincide, we have simply to produce a map between the sequences.

In general if we are given a generically finite map $f : X \to \mathfrak{X}$, then we have push forward maps on cycles and currents which give a push forward from f_* :

$\widehat{\mathrm{CH}}^p(X) \to \widetilde{\mathrm{CH}^p(\mathfrak{X})}$, where $\widetilde{\mathrm{CH}^p(\mathfrak{X})}$ is defined in exactly the same way as $\widehat{\mathrm{CH}}^p(\mathfrak{X})$ *except* that we do not require the form $dd^c(g) + \delta_\zeta = \omega$ be C^∞. It is straight forward to check that this push forward map gives a map from the first exact sequence of proposition 3.2 for X to the corresponding exact sequence for $\mathfrak{X}$ in which $\widehat{\mathrm{CH}}^p(\mathfrak{X})$ is replaced by $\widetilde{\mathrm{CH}^p(\mathfrak{X})}$ and $Z^{p,p}(\mathfrak{X})$ is replaced by the space of closed (p,p)-currents.

Suppose now that $\alpha \in \varprojlim_{f:X\to\mathfrak{X}} \widehat{\mathrm{CH}}^q(X)_{\mathbb{Q}}$. Then for each $f : X \to \mathfrak{X}$, we have a class α_f which, by lemma 1.7, lies in the subgroup of $\widehat{\mathrm{CH}}^p(X)_{\mathbb{Q}}$ mapping to $f^* Z^{(p,p)}(\mathfrak{X}(\mathbb{C}))$. Therefore if $f : X \to \mathfrak{X}$ is generically finite, since any $(p-1, p-1)$-current on X pushes forward to a $(p-1, p-1)$-current on $\mathfrak{X}$, we can push forward $\frac{1}{\deg(f)} \alpha_p$ to a class $\frac{1}{\deg(f)} f_*(\alpha_f)$ in $\widehat{\mathrm{CH}}^p(\mathfrak{X})_{\mathbb{Q}}$.

Given generically finite maps $f : X \to \mathfrak{X}$ and $f' : X' \to \mathfrak{X}$, we can dominate both maps by a map $f'' : X'' \to \mathfrak{X}$ with X'' again regular and f'' generically finite. Now applying the projection formula, we get that $\frac{1}{\deg(f)} f_*(\alpha_f)$ is independent of f. Furthermore we now have a map from the first exact sequence of proposition 3.2 for X to the corresponding exact sequence for $\mathfrak{X}$. □

Since any map between stacks can be dominated by a map between non-singular hypercovers, $\widehat{\mathrm{CH}}^*(\mathfrak{X})_{\mathbb{Q}}$ becomes a contravariant functor from stacks over S to graded $\mathbb{Q}$-algebras.

If $\overline{E} = (E, \| \ \|)$ is a Hermitian vector bundle on $\mathfrak{X}$, then $\overline{E}$ pulls back by any proper surjective map $p : X \to \mathfrak{X}$ to a Hermitian vector bundle on X, and the Chern classes $\widehat{C}_p(\overline{E})$ of [14] define a class in $\varprojlim_{p:X\to\mathfrak{X}} \widehat{\mathrm{CH}}^p(X)$, and therefore in $\widehat{\mathrm{CH}}^p(\mathfrak{X})_{\mathbb{Q}}$, satisfying the properties of *op. cit.*

4 Arithmetic Chow groups of non-proper stacks

The construction of the previous section assumed that the stack $\mathfrak{X}$ was proper over S. Nonetheless we can make the following

Definition 4.1 If $\mathfrak{X}$ is a regular, separated, Deligne-Mumford stack over S, we define the groups $\widehat{\mathrm{CH}}^*(\mathfrak{X})_{\mathbb{Q}}$ to be the inverse limit of $\widehat{\mathrm{CH}}^*(X)_{\mathbb{Q}}$ over all proper surjective maps $\pi : X \to \mathfrak{X}$ with X nonsingular.

Note that this is equivalent to the inverse limit over the category of non-singular proper hypercovers $\pi : V. \to \mathfrak{X}$:

$$\widehat{\mathrm{CH}}^p(\mathfrak{X})_{\mathbb{Q}} := \varprojlim_{\pi:V.\to\mathfrak{X}} \mathrm{H}^0(i \mapsto \widehat{\mathrm{CH}}^p(V_i)_{\mathbb{Q}}).$$

To approach the basic properties of these groups, rather than working with the category of motives over S, we shall introduce the category of (homological) motives over $\mathfrak{X}$.

4.1 Motives over a stack. Let $\mathfrak{X}$ be a separated, regular Deligne-Mumford stack over $\mathrm{Spec}(\mathbb{Z})$. The category $\mathcal{V}ar_{\mathfrak{X}}$ of varieties over $\mathfrak{X}$ has finite products and fibred products. If $\alpha : X \to \mathfrak{X}$, $\beta : Y \to \mathfrak{X}$ and $\gamma : Z \to \mathfrak{X}$ are objects, and $f : (X, \alpha) \to (Z, \gamma)$ and $g : (Y, \beta) \to (Z, \gamma)$ are morphisms, in $\mathcal{V}ar_{\mathfrak{X}}$ we have that the fibred product $X \times_Z Y$ is the same whether we view $f : X \to Z$ and $g : Y \to Z$

as morphisms, in $\mathcal{V}ar_{\mathfrak{X}}$, or as morphisms in $\mathcal{V}ar_S$. In particular if $f : X \to \mathfrak{X}$, $g : Y \to \mathfrak{X}$, $h : Z \to \mathfrak{X}$ are objects in $\mathcal{V}ar_{\mathfrak{X}}$, we have that

$$X \times_{\mathfrak{X}} Y \times_{\mathfrak{X}} Z \simeq (X \times_{\mathfrak{X}} Y) \times_Y (Y \times_{\mathfrak{X}}) Z.$$

Following the method of [17], we define the category of (homological) K_0-motives over $\mathfrak{X}$ to be the idempotent completion of the category with objects the regular varieties which are projective over $\mathfrak{X}$, and

$$\mathrm{KC}_{\mathfrak{X}}(X, Y) := G_0(X \times_{\mathfrak{X}} Y),$$

where G_0 is the Grothendieck group of coherent sheaves of $\mathcal{O}_{X \times_{\mathfrak{X}} Y}$-modules (which as always we take with $\mathbb{Q}$-coefficients). Note that $X \times_{\mathfrak{X}} Y$ is not in general regular; however since Y is regular, $\mathcal{O}_Y$ is of finite global tor-dimension, and hence there is a bilinear product, given X, Y and Z regular, projective, $\mathfrak{X}$-varieties :

$$* : G_0(X \times_{\mathfrak{X}} Y) \times G_0(Y \times_{\mathfrak{X}} Z) \to G_0(X \times_{\mathfrak{X}} Y \times_{\mathfrak{X}} Z)$$

$$([\mathfrak{F}], [\mathfrak{G}]) \mapsto \sum_{i \geq 0} (-1)^i [\mathcal{T}or_i^{\mathcal{O}_Y}(\mathfrak{F}, \mathfrak{G})].$$

Composing with the direct image map ($p_{XZ} : X \times_{\mathfrak{X}} Y \times_Z Z \to X \times_{\mathfrak{X}} Z$ being the natural projection):

$$(p_{XZ})_* : G_0(X \times_{\mathfrak{X}} Y \times_{\mathfrak{X}} Z) \to G_0(X \times_Y Z)$$

we get a bilinear pairing:

$$G_0(X \times_{\mathfrak{X}} Y) \times G_0(Y \times_{\mathfrak{X}} Z) \to G_0(X \times_{\mathfrak{X}} Z)$$

and hence:

$$\mathrm{KC}_{\mathfrak{X}}(X, Y) \times \mathrm{KC}_{\mathfrak{X}}(Y, Z) \to \mathrm{KC}_{\mathfrak{X}}(X, Z).$$

The proofs of the following theorems are straightforward, following the pattern in [17] and of the previous sections and so we omit them.

Lemma 4.2 *Given regular varieties X, Y, Z and W projective over $\mathfrak{X}$, and elements $\alpha \in \mathrm{KC}_{\mathfrak{X}}(X, Y)$, $\beta \in \mathrm{KC}_{\mathfrak{X}}(Y, Z)$, $\gamma \in \mathrm{KC}_{\mathfrak{X}}(Z, W)$ we have*

$$\gamma \circ (\beta \circ \alpha) = (\gamma \circ \beta) \circ \alpha.$$

Note that just as in the case of varieties over a scheme, if $\alpha : X \to \mathfrak{X}$, $\beta : Y \to \mathfrak{X}$ and $f : X \to Y$ is a morphism in $\mathcal{V}ar_{\mathfrak{X}}$, then the graph of f is a closed subscheme of $X \times_{\mathfrak{X}} Y$ which is isomorphic to X, and its structure sheaf defines a class $\Gamma(f)$ in $\mathrm{KC}_{\mathfrak{X}}(X, Y)$. and that it is straightforward to check:

Lemma 4.3

$$f \mapsto \Gamma(f)$$

defines a covariant functor $\Gamma : \mathcal{V}ar_{\mathfrak{X}} \to \mathrm{KC}_{\mathfrak{X}}$.

Proposition 4.4 *The functors from regular varieties projective over $\mathfrak{X}$ to rational vector spaces, G_i (Quillen K-theory of coherent sheaves, with rational coefficients), the associated graded for the γ-filtration on G_0, i.e $\mathrm{CH}_{\mathbb{Q}}^*$, and real Deligne cohomology $X \mapsto \mathrm{H}_{\mathcal{D}}^*(X(\mathbb{C}), \mathbb{R}(*))$ all factor through $\mathrm{KC}_{\mathfrak{X}}$.*

Theorem 4.5 *Suppose that $V. \to \mathfrak{X}$ is a non-singular proper hypercover of the stack $\mathfrak{X}$, then the complex $\Gamma(V.)$ is exact, except in degree zero, and $\mathrm{H}_0(\Gamma(V.))$ is a direct summand of $\Gamma(V_0)$. Furthermore, if $\pi : V. \to \mathfrak{X}$ and $\pi' : W. \to \mathfrak{X}$ are two non-singular proper hypercovers of $\mathfrak{X}$, and $f. : V. \to W.$ is a morphism over $\mathfrak{X}$, then the induced map $\mathrm{H}_0(\Gamma(V.)) \to \mathrm{H}_0(\Gamma(W.))$ is an isomorphism.*

Corollary 4.6 *Suppose that* $\mathfrak{X}$ *is a regular separated Deligne-Mumford stack over* S, *and* $\pi : V. \to \mathfrak{X}$ *is non-singular proper hypercover. There are natural isomorphisms, for all p and q:*

$$H_{\mathcal{D}}^p(\mathfrak{X}_{\mathbb{R}}, \mathbb{R}(q)) \xrightarrow{\sim} H^0(i \mapsto H_{\mathcal{D}}^p((V_i)_{\mathbb{R}}, \mathbb{R}(q))),$$

$$\pi^* : CH^p(\mathfrak{X})_{\mathbb{Q}} \xrightarrow{\sim} H^0(i \mapsto CH^p(V_i)_{\mathbb{Q}}),$$

$$\pi^* : CH^{p,1}(\mathfrak{X})_{\mathbb{Q}} \xrightarrow{\sim} H^0(i \mapsto CH^{p,1}(V_i)_{\mathbb{Q}}).$$

Furthermore, these isomorphisms are compatible with both the cycle class maps

$$\gamma : CH^p(-)_{\mathbb{Q}} \to H_{\mathcal{D}}^{2p}(-, \mathbb{R}(p)))$$

and the regulator:

$$\rho : CH^{p,1}(-)_{\mathbb{Q}} \to H_{\mathcal{D}}^{2p-1}(-, \mathbb{R}(p))).$$

Theorem 4.7 *The groups* $\widehat{CH}^*(\mathfrak{X})_{\mathbb{Q}}$ *are contravariant with respect to* $\mathfrak{X}$ *and have a natural product structure. There are exact sequences:*

$$CH^{p,p-1}(\mathfrak{X})_{\mathbb{Q}} \to H_{\mathcal{D}}^{2p-1}(\mathfrak{X}_{\mathbb{R}}, \mathbb{R}(p))$$

$$\to \widehat{CH}^p(\mathfrak{X})_{\mathbb{Q}} \to CH^p(\mathfrak{X})_{\mathbb{Q}} \oplus \widetilde{Z}_{\log}^{p,p}(\mathfrak{X}_{\mathbb{R}}) \to H_{\mathcal{D}}^{2p}(\mathfrak{X}(\mathbb{C}), \mathbb{R}(p)).$$

Here $\widetilde{Z}_{\log}^{p,p}(\mathfrak{X}_{\mathbb{R}})$ *is the direct limit, over all proper surjective maps* $X \to \mathfrak{X}$ *of the groups* $Z_{\log}^{p,p}(X_{\mathbb{R}})$ *of [5] and [6].*

Unfortunately, I do not know whether the complex $\widetilde{A}_{\log}^{*,*}(\mathfrak{X}(\mathbb{C}))$, which is the direct limit over all proper surjective maps $X \to \mathfrak{X}$ of the logarithmic Dolbeault complexes $A_{\log}^{*,*}(X(\mathbb{C}))$ of [6], will compute the Dolbeault cohomology of $\mathfrak{X}(\mathbb{C})$, and thus the question of constructing a nice complex computing the real Deligne cohomology of $\mathfrak{X}(\mathbb{C})$ is open.

References

[1] P. Berthelot, A. Grothendieck, and L. Illusie, *Séminaire de géométrie algébrique: Théorie des intersections et théoréme de Riemann-Roch*, Lecture Notes in Math. 225, Springer-Veflag (1971).

[2] José Bertin, Jean-Pierre Demailly, Luc Illusie and Chris Peters, *Introduction to Hodge theory*, SMF/AMS Texts and Monographs, **8**, Translated from the 1996 French original by James Lewis and Chris Peters, American Mathematical Society, Providence, RI, (2002).

[3] S. Bloch, H. Gillet, and C. Soulé *Non-Archimedean Arakelov theory*, J. Algebraic Geom., **4**, (1995), 427–485.

[4] Jan H. Bruinier, José I. Burgos Gil, and Ulf Kühn, *Borcherds products and arithmetic intersection theory on Hilbert modular surfaces* Duke Math. J., **139**, (2007), 1-88.

[5] José I. Burgos Gil, *Green forms and their product*, Duke Math. J. **75** (1994), 529–574.

[6] José I. Burgos Gil, *Arithmetic Chow rings and Deligne-Beilinson cohomology*, J. Algebraic Geom. **6** (1997), 335–377.

[7] A.J. de Jong, *Families of curves and alterations*, Ann. Inst. Fourier (Grenoble) **47** (1997), 599-621.

[8] P. Deligne, D. Mumford, *The irreducibility of the space of curves of given genus*, Inst. Hautes Études Sci. Publ. Math., **36** (1969) 75–109.

[9] P. Deligne, M. Rapoport, *Les schémas de modules de courbes elliptiques*, In Modular functions of one variable, II (Proc. Internat. Summer School, Univ. Antwerp, Antwerp, 1972), . Lecture Notes in Math., **349** 143–316 Springer, Berlin, (1973).

[10] H. Gillet, *Intersection theory on algebraic stacks and Q-varieties*, J. Pure Appl. Algebra **34** (1984) 193–240.

[11] H. Gillet, D. Grayson, *The loop space of the Q-construction*. Illinois J. Math. **31** (1987), no. 4, 574–597.

[12] H. Gillet, C. Soulé, *Intersection theory using Adams operations*, Inventiones Mathematicae **90** (1987) 243–277.

[13] H. Gillet, C. Soulé, *Arithmetic intersection theory*, Inst. Hautes Études Sci. Publ. Math. **72** (1990), 93–174.

[14] H. Gillet, C. Soulé, *Characteristic classes for algebraic vector bundles with Hermitian metric. I, II*, Ann. of Math. **131** (1990), 163–238.

[15] H. Gillet, C. Soulé, *Descent, motives and K-theory* J. Reine Angew. Math. **478** (1996), 127–176.

[16] H. Gillet C. Soulé. *Filtrations on higher algebraic K-theory*, In Algebraic K-theory (Seattle, WA, 1997), pages 89–148. Amer. Math. Soc., Providence, RI, 1999

[17] H. Gillet, C. Soulé, *Motivic Weight Complexes for Arithmetic Varieties*, preprint 2008, http://arxiv.org/abs/0804.4853.

[18] D. Grayson, *Exterior power operations on higher K-theory*. K-Theory **3** (1989), no. 3, 247–260.

[19] H. Hiller, *λ-rings and algebraic K-theory* J. Pure Appl. Algebra **20** (1981), no. 3, 241–266.

[20] J. Hu, *Specialization of Green forms and arithmetic intersection theory*, 2003 preprint, available at http://www.institut.math.jussieu.fr/Arakelov/0002.

[21] Seán Keel, Shigefumi Mori, *Quotients by groupoids*, Ann. of Math. (2), **145**, (1997), 193–213.

[22] Shun-ichi Kimura, *Fractional intersection and bivariant theory*, Comm. Algebra, **20**, (1992), 285–302.

[23] Ch. Kratzer, *λ-structure en K-théorie algébrique*, Comment. Math. Helv. **55** (1980), no. 2, 233–254.

[24] Andrew Kresch, *Canonical rational equivalence of intersections of divisors*, Invent. Math., **136**, 1999, 483–496.

[25] Stephen S. Kudla, Michael Rapoport, and Tonghai Yang, *Modular forms and special cycles on Shimura curves*, Annals of Mathematics Studies, **161**, Princeton University Press, Princeton, NJ, (2006).

[26] G. Laumon, L. Moret-Bailly, *Champs algébriques*, Ergebnisse der Mathematik und ihrer Grenzgebiete. 3. Folge. **39**, Springer Verlag, Berlin, (2000)

[27] Chris A. M. Peters, Joseph H. M. Steenbrink, *Mixed Hodge structures*, Ergebnisse der Mathematik und ihrer Grenzgebiete. 3. Folge. **52**, Springer-Verlag, Berlin, (2008).

[28] C. Soulé, *Opérations en K-théorie algébrique*, Canadian J. Math. **37** (1985), 488–550.

[29] THOMASON, R. W. Algebraic K-theory and étale cohomology. *Ann. Sci. École Norm. Sup.* (4) **18** (1985), no. 3, 437–552.

[30] B. Toen, *On motives for Deligne-Mumford stacks*. Internat. Math. Res. Notices, **17**,(2000), 909–928.

[31] Theo van den Bogaart, *Links between cohomology and arithmetic* Thesis, University of Leiden, (2008) https://www.openaccess.leidenuniv.nl/dspace/handle/1887/12928

[32] A. Vistoli, *Intersection theory on algebraic stacks and on their moduli spaces*, Invent. Math. **97** (1989) 613–670.

Fields Institute Communications
Volume **56**, 2009

Notes on the Biextension of Chow Groups

Sergey Gorchinskiy
Steklov Mathematical Institute
8, Gubkina str.
Moscow, Russia, 119991
gorchins@mi.ras.ru

To Spencer Bloch, with respect and admiration

Abstract. The paper discusses four approaches to the biextension of Chow groups and their equivalences. These are the following: an explicit construction given by S. Bloch, a construction in terms of the Poincaré biextension of dual intermediate Jacobians, a construction in terms of K-cohomology, and a construction in terms of determinant of cohomology of coherent sheaves. A new approach to J. Franke's Chow categories is given. An explicit formula for the Weil pairing of algebraic cycles is obtained.

1 Introduction

One of the questions about algebraic cycles is the following: what can be associated in a bilinear way to a pair of algebraic cycles (Z, W) of codimensions p and q, respectively, on a smooth projective variety X of dimension d over a field k with $p + q = d + 1$, i.e., what is an analogue of the linking number for algebraic cycles? This question arose naturally from some of the approaches to the intersection index of arithmetic cycles on arithmetic schemes, i.e., to the height pairing for algebraic cycles.

For homologically trivial cycles, a "linking invariant" was constructed by S. Bloch in [5] and [7], and independently by A. Beilinson in [3]. It turns out that for an *arbitrary* ground field k, this invariant is no longer a number, but it is a k^*-torsor, i.e., a set with a free transitive action of the group k^*. More precisely, in [7] a biextension P of $(CH^p(X)_{\text{hom}}, CH^q(X)_{\text{hom}})$ by k^* is constructed, where $CH^p(X)_{\text{hom}}$ is the group of codimension p homologically trivial algebraic cycles on X up to rational equivalence. This means that there is a map of sets

$$\pi : P \to CH^p(X)_{\text{hom}} \times CH^q(X)_{\text{hom}},$$

2000 *Mathematics Subject Classification.* Primary 14C15, 14C35; Secondary 14C30, 13D15.

The author was partially supported by the grants RFBR 08-01-00095, Nsh-1987.2008.1, and INTAS 05-1000008-8118.

a free action of k^* on the set P such that π induces a bijection

$$P/k^* \cong CH^p(X)_{\text{hom}} \times CH^q(X)_{\text{hom}},$$

and for all elements $\alpha, \beta \in CH^p(X)_{\text{hom}}$, $\gamma, \delta \in CH^q(X)_{\text{hom}}$, there are fixed isomorphisms

$$P_{(\alpha,\gamma)} \otimes P_{(\beta,\gamma)} \cong P_{(\alpha+\beta,\gamma)},$$

$$P_{(\alpha,\gamma)} \otimes P_{(\alpha,\delta)} \cong P_{(\alpha,\gamma+\delta)}$$

such that certain compatibility axioms are satisfied (see [9]). Here the tensor product is taken in the category of k^*-torsors and $P_{(*,*)}$ denotes a fiber of P at $(*, *) \in CH^p(X)_{\text{hom}} \times CH^q(X)_{\text{hom}}$. In other words, the biextension P defines a bilinear pairing between Chow groups of homologically trivial cycles with value in the category of k^*-torsors.

The biextension P generalizes the Poincaré line bundle on the product of the Picard and Albanese varieties. Note that if the ground field k is a *number field*, then each embedding of k into its completion k_v induces a trivialization of the biextension $\log|P|_v$ and the collection of all these trivializations defines in a certain way the height pairing for algebraic cycles (see [5]).

On the other hand, the biextension P of $(CH^p(X)_{\text{hom}}, CH^q(X)_{\text{hom}})$ by k^* for $p + q = d + 1$ is an analogue of the intersection index $CH^p \times CH^q(X) \to \mathbb{Z}$ for $p + q = d$. There are several approaches to algebraic cycles, each of them giving its own definition of the intersection index. A natural question is to find analogous definitions for the biextension P. The main goal of the paper is to give a detailed answer to this question. Namely, we discuss four different constructions of biextensions of Chow groups and prove their equivalences.

Let us remind several approaches to the intersection index and mention the corresponding constructions of the biextension of Chow groups that will be given in the article.

The most explicit way to define the intersection index is to use the moving lemma and the definition of local multiplicities for proper intersections. Analogous to this is the explicit definition of the biextension P given in [7].

If $k = \mathbb{C}$, one can consider classes of algebraic cycles in Betti cohomology groups $H_B^{2p}(X(\mathbb{C}), \mathbb{Z})$ and then use the product between them and the push-forward map. Corresponding to this in [7] it was suggested that for $k = \mathbb{C}$, the biextension P should be equal to the pull-back via the Abel–Jacobi map of the Poincaré line bundle on the product of the corresponding dual intermediate Jacobians. This was partially proved in [22] by using the functorial properties of higher Chow groups and the regulator map to Deligne cohomology.

A different approach uses the Bloch–Quillen formula $CH^p(X) = H^p(X, \mathcal{K}_p)$, the product between cohomology groups of sheaves, the product between K-groups, and the push-forward for K-cohomology. Here $\mathcal{K}_p$ is the Zariski sheaf associated to the presheaf given by the formula $U \mapsto K_p(U)$ for an open subset $U \subset X$. A corresponding approach to the biextension uses the pairing between complexes

$$R\Gamma(X, \mathcal{K}_p) \times R\Gamma(X, \mathcal{K}_q) \to k^*[-d]$$

for $p + q = d + 1$.

Finally, one can associate with each cycle $Z = \sum_i n_i Z_i$ an element $[\mathcal{O}_Z] = \sum n_i [\mathcal{O}_{Z_i}] \in K_0(X)$ and to use the natural pairing on $K_0(X)$, i.e., to define the intersection index by the formula $\text{rk} R\Gamma(X, \mathcal{O}_Z \otimes_{\mathcal{O}_X}^L \mathcal{O}_W)$. Analogous to this one considers the determinant of cohomology $\det R\Gamma(X, \mathcal{O}_Z \otimes_{\mathcal{O}_X}^{L} \mathcal{O}_W)$ to get a biextension of Chow groups. This is a generalization of what was done for divisors on curves by P. Deligne in [11]. With this aim a new approach to J. Franke's Chow categories (see [12]) is developed. This approach uses a certain filtration "by codimension of support" on the Picard category of virtual coherent sheaves on a variety (see [11]).

One interprets the compatibility of the first definition of the intersection index with the second and the third one as the fact that the cycle maps

$$CH^p(X) \to H_B^{2p}(X(\mathbb{C}), \mathbb{Z}),$$

$$CH^p(X) \to H^{2p}(X, \mathcal{K}_p[-p])$$

commute with products and push-forwards. Note that the cycle maps are particular cases of canonical morphisms (regulators) from motivic cohomology to various cohomology theories. This approach explains quickly the comparison isomorphism between the corresponding biextensions. Besides this we give a more explicit and elementary proof of the comparison isomorphism in each case.

Question 1.1 What should be associated to a pair of cycles of codimensions p, q with $p + q = d + i$, $i \geq 2$? Presumably, when $i = 2$, one associates a $K_2(k)$-gerbe.

The paper has the following structure. Sections 2.1-2.3 contain the description of various geometric constructions that are necessary for definitions of biextensions of Chow groups. In particular, in Section 2.4 we recall several facts from [16] about adelic resolution for sheaves of K-groups.

In Sections 3.1 and 3.2 general algebraic constructions of biextensions are discussed. In particular, we introduce the notion of a bisubgroup and give an explicit construction of a biextension induced by a pairing between complexes. Though these constructions are elementary and general, the author could not find any reference for them. As an example to the above notions, in Section 3.3 we consider the definition of the Poincaré biextension of dual complex compact tori by $\mathbb{C}^*$.

Sections 4.1-4.4 contain the constructions of biextensions of Chow groups according to different approaches to algebraic cycles. In Section 4.1 we recall from [7] an explicit construction of the biextension of Chow groups. This biextension is interpreted in terms of the pairing between higher Chow complexes (Proposition 4.2), as was suggested to the author by S. Bloch. In Section 4.2 for the complex base field, we consider the pull-back of the Poincaré biextension of dual intermediate Jacobians (Proposition 4.3). Section 4.3 is devoted to the construction of the biextension in terms of K-cohomology groups and a pairing between sheaves of K-groups (Proposition 4.9). We give an explicit description of this biextension in terms of the adelic resolution for sheaves of K-groups introduced in [16]. In Section 4.4 we construct a filtration on the Picard category of virtual coherent sheaves on a variety (Definition 4.10) and we define the biextension of Chow groups in terms of the determinant of cohomology (Proposition 4.29). In each section we establish a canonical isomorphism of the constructed biextension with the explicit biextension from [7] described firstly. In Sections 4.2 and 4.3 we give both explicit proofs and the proofs that use properties of the corresponding regulator maps. Finally, Section 4.5 gives an explicit formula for the Weil pairing between torsion elements in $CH^*(X)_{\text{hom}}$. This can be considered as a generalization of the classical Weil's formula for divisors on a curve. Also, the equivalence of different constructions of biextensions of Chow groups implies the interpretation of the Weil pairing in terms of a certain Massey triple product.

The author is deeply grateful to A.A. Beilinson, S. Bloch, A.M. Levin, A.N. Parshin, C. Soulé, and V. Vologodsky for very stimulating discussions, and to the referee whose numerous suggestions helped improving the paper very much. In particular, the first proof of Proposition 4.9 was proposed by the referee.

The paper is dedicated to Spencer Bloch, whose work in mathematics is a bright leading light for the author and many others.

2 Preliminary results

2.1 Facts on higher Chow groups. All varieties below are defined over a fixed ground field k. For an equidimensional variety S, by $Z^p(S)$ denote the free abelian group generated by all codimension p irreducible subvarieties in S. For an element $Z = \sum n_i Z_i \in Z^p(S)$, by $|Z|$ denote the union of codimension p irreducible subvarieties Z_i in S such that $n_i \neq 0$. Let $S^{(p)}$ be the set of all codimension p schematic points on S.

Let us recall the definition of higher Chow groups (see [6]). We put

$$\Delta^n = \left\{ \sum_{i=0}^{n} t_i = 1 \right\} \subset \mathbb{A}^{n+1};$$

note that $\Delta^\bullet$ is a cosimplicial variety. In particular, for each $m \leq n$, there are several face maps $\Delta^m \to \Delta^n$. For an equidimensional variety X, by $Z^p(X, n)$ denote the free abelian group generated by codimension p irreducible subvarieties in $X \times \Delta^n$ that meet the subvariety $X \times \Delta^m \subset X \times \Delta^n$ properly for any face $\Delta^m \subset \Delta^n$. The simplicial group $Z^p(X, \bullet)$ defines a homological type complex; by definition, $CH^p(X, n) = H_n(Z^p(X, \bullet))$ is the *higher Chow group* of X. Note that $CH^p(X, 0) = CH^p(X)$ and $CH^1(X, 1) = k[X]^*$ for a regular variety X (see op.cit.). For a projective morphism of varieties $f : X \to Y$, there is a push-forward morphism of complexes $f_* : Z^p(X, \bullet) \to Z^{p+\dim(Y)-\dim(X)}(Y, \bullet)$.

For an equidimensional subvariety $S \subset X$, consider the subcomplex

$$Z_S^p(X, \bullet) \subset Z^p(X, \bullet)$$

generated by elements from $Z^p(X, n)$ whose support meets the subvariety $S \times \Delta^m \subset X \times \Delta^n$ properly for any face $\Delta^m \subset \Delta^n$. For a collection $\mathcal{S} = \{S_1 \ldots, S_r\}$ of equidimensional subvarieties in X, we put $Z_{\mathcal{S}}^p(X, \bullet) = \cap_{i=1}^{r} Z_{S_i}^p(X, \bullet)$. The following moving lemma is proven in Proposition 2.3.1 from [8] and in [21]:

Lemma 2.1 *Provided that X is smooth over k and either projective or affine, the inclusion $Z_{\mathcal{S}}^p(X, \bullet) \subset Z^p(X, \bullet)$ is a quasiisomorphism for any $\mathcal{S}$ as above.*

In particular, Lemma 2.1 allows to define the multiplication morphism

$$m \in \mathrm{Hom}_{D^-(\mathcal{A}b)}(Z^p(X, \bullet) \otimes_{\mathbb{Z}}^{L} Z^q(X, \bullet), Z^{p+q}(X, \bullet)).$$

Recall that a cycle $Z \in Z^p(X)$ is called *homologically trivial* if its class in the étale cohomology group $H_{\acute{e}t}^{2p}(X_{\overline{k}}, \mathbb{Z}_l(p))$ is zero for any prime $l \neq \mathrm{char}(k)$, where $\overline{k}$ is the algebraic closure of the field k. Note that when $\mathrm{char}(k) = 0$ the cycle Z is homologically trivial if and only if its class in the Betti cohomology group $H_B^{2p}(X_{\mathbb{C}}, \mathbb{Z})$ is zero after we choose any model of X defined over $\mathbb{C}$. Denote by $CH^p(X)_{\mathrm{hom}}$ the subgroup in $CH^p(X)$ generated by classes of homologically trivial cycles.

The following result is proved in [7], Lemma 1.

Lemma 2.2 *Let X be a smooth projective variety, $\pi : X \to \mathrm{Spec}(k)$ be the structure morphism. Suppose that a cycle $Z \in Z^p(X)$ is homologically trivial; then the natural homomorphism $CH^{d+1-p}(X,1) \stackrel{m(-\otimes Z)}{\longrightarrow} CH^{d+1}(X,1) \stackrel{\pi_*}{\longrightarrow} CH^1(k,1) = k^*$ is trivial.*

Remark 2.3 The proof of Lemma 2.2 uses the regulator map from higher Chow groups to Deligne cohomology if the characteristic is zero and to étale cohomology if the characteristic is positive.

Question 2.4 According to Grothendieck's standard conjectures, the statement of Lemma 2.2 (at least up to torsion in k^*) should be true if one replaces the homological triviality of the cycle W by the numerical one. Does there exist a purely algebraic proof of this fact that does not use Deligne cohomology or étale cohomology?

Remark 2.5 In Section 4.2 we give an analytic proof of Lemma 2.2 for complex varieties, which uses only general facts from the Hodge theory (see Lemma 4.8).

2.2 Facts on K_1-chains. Let X be an equidimensional variety over the ground field k. We put $G^p(X,n) = \bigoplus_{\eta \in X^{(p)}} K_n(k(\eta))$ (in this section we use these groups only for $n = 0, 1, 2$). Elements of the group $G^{p-1}(X,1)$ are called K_1-*chains*. There are natural homomorphisms $\mathrm{Tame} : G^{p-2}(X,2) \to G^{p-1}(X,1)$ and $\mathrm{div} : G^{p-1}(X,1) \to G^p(X,0) = Z^p(X)$. Note that $\mathrm{div} \circ \mathrm{Tame} = 0$. The subgroup $\mathrm{Im}(\mathrm{Tame}) \subset G^{p-1}(X,1)$ defines an equivalence on K_1-chains; we call this a K_2-*equivalence* on K_1-chains. For a K_1-chain $\{f_\eta\} \in G^{p-1}(X,1)$, by $\mathrm{Supp}(\{f_\eta\})$ denote the union of codimension $p-1$ irreducible subvarieties $\bar{\eta}$ in X such that $f_\eta \neq 1$.

Let S be an equidimensional subvariety; by $G_S^{p-1}(X,1)$ denote the group of K_1-chains $\{f_\eta\}$ such that for any $\eta \in X^{(p-1)}$, either $f_\eta = 1$, or the closure $\bar{\eta}$ and the support $|\mathrm{div}(f_\eta)|$ meet S properly. For a collection $S = \{S_1 \ldots, S_r\}$ of equidimensional subvarieties in X, we put $G_S^{p-1}(X,1) = \cap_{i=1}^r G_{S_i}^{p-1}(X,1)$.

Define the homomorphism $N : Z^p(X,1) \to G^{p-1}(X,1)$ as follows. Let Y be an irreducible subvariety in $X \times \Delta^1$ that meets properly both faces $X \times \{(0,1)\}$ and $X \times \{(1,0)\}$ (recall that $\Delta^1 = \{t_0 + t_1 = 1\} \subset \mathbb{A}^2$). By p_X and p_{Δ^1} denote the projections from $X \times \Delta^1$ to X and Δ^1, respectively. If the morphism $p_X : Y \to X$ is not generically finite onto its image, then we put $N(Y) = 0$. Otherwise, let $\eta \in X^{(p-1)}$ be the generic point of $p_X(Y)$; we put $f_\eta = (p_X)_*(p_{\Delta^1}^*(t_1/t_0)) \in k(\eta)^*$ and $N(Y) = f_\eta \in G^{p-1}(X,1)$. We extend the homomorphism N to $Z^p(X,1)$ by linearity.

Conversely, given a point $\eta \in X^{(p-1)}$ and a rational function $f_\eta \in k(\eta)^*$ such that $f_\eta \neq 1$, let $\Gamma(f_\eta)$ be the closure of the graph of the rational map $(\frac{1}{1+f_\eta}, \frac{f_\eta}{1+f_\eta})$: $\bar{\eta} \dashrightarrow \Delta^1 \subset \mathbb{A}^2$. This defines the map of sets $\Gamma : G^{p-1}(X,1) \to Z^p(X,1)$ such that $N \circ \Gamma$ is the identity. For an element $Y \in Z^p(X,1)$, we have $\mathrm{div}(N(Y)) = d(Y) \in Z^p(X)$, where d denotes the differential in the complex $Z^p(X,\bullet)$.

Given an equidimensional subvariety $S \subset X$, it is easy to check that $\Gamma(G_S^{p-1}(X,1)) \subset Z_S^p(X,1)$ and $N(Z_S^p(X,1)) \subset G_S^{p-1}(X,1)$.

Lemma 2.6 *Suppose that X is smooth over k and either projective or affine. Let $S = \{S_1, \ldots, S_r\}$ be a collection of equidimensional closed subvarieties in X,*

$\{f_\eta\} \in G^{p-1}(X, 1)$ be a K_1-chain such that the support $|\mathrm{div}(\{f_\eta\})|$ meets S_i properly for all i, $1 \leq i \leq r$; then there exists a K_1-chain $\{g_\eta\} \in G_{\mathcal{S}}^{p-1}(X, 1)$ such that $\{g_\eta\}$ is K_2-equivalent to $\{f_\eta\}$.

Proof Denote by d the differential in the complex $Z^p(X, \bullet)$. Since $\mathrm{div}(\{f_\eta\}) \in Z_{\mathcal{S}}^p(X, 0)$, by Lemma 2.1, there exists an element $Y' \in Z_{\mathcal{S}}^p(X, 1)$ such that $d(Y') = \mathrm{div}(\{f_\eta\}) = d(\Gamma(\{f_\eta\}))$. Again by Lemma 2.1, there exists an element $Y'' \in Z_{\mathcal{S}}^p(X, 1)$ such that $d(Y'') = 0$ and $Y'' + Y' - \Gamma(\{f_\eta\}) = d(\widetilde{Y})$ for some $\widetilde{Y} \in Z^p(X, 2)$.

Recall that $(N \circ d)(Z^p(X, 2)) \subset \mathrm{Im}(\mathrm{Tame}) \subset G^{p-1}(X, 1)$, see [23], Remark on p. 13 for more details. Therefore, the K_1-chain $\{g_\eta\} = N(Y'' + Y') \in G_{\mathcal{S}}^{p-1}(X, 1)$ is K_2-equivalent to $\{f_\eta\}$. □

Corollary 2.7 In notations from Lemma 2.6 let $Z = \mathrm{div}(\{f_\eta\})$ and suppose that $\mathrm{codim}_Z(Z \cap S_i) \geq n_i$ for all i, $1 \leq i \leq r$ (in particular, $n_i \leq \mathrm{codim}_X(S_i)$); then there exists a K_1-chain $\{g_\eta\} \in G^{p-1}(X, 1)$ such that $\{g_\eta\}$ is K_2-equivalent to $\{f_\eta\}$, $\mathrm{codim}_Y(Y \cap S_i) \geq n_i$ for all i, $1 \leq i \leq r$, where $Y = \mathrm{Supp}(\{g_\eta\})$, and for each $\eta \in Z^{p-1}(X)$, we have $\mathrm{codim}_{\mathrm{div}(g_\eta)}(\mathrm{div}(g_\eta) \cap S_i) \geq n_i$ for all i, $1 \leq i \leq r$.

Proof We claim that for each i, $1 \leq i \leq r$, there exists an equidimensional subvariety $S_i' \subset X$ of codimension n_i such that $S_i' \supset S_i$ and Z meets S_i' properly. By Lemma 2.6, this immediately implies the needed statement.

For each i, $1 \leq i \leq r$, we prove the existence of S_i' by induction on n_i. Suppose that $n_i = 1$. Then there exists an effective reduced divisor $H \subset X$ such that $H \supset S_i$ and H meets Z properly: to construct such divisor we have to choose a closed point on each irreducible component of Z outside of S_i and take an arbitrary H that does not contain any of these points and such that $H \supset S_i$.

Now let us do the induction step from $n_i - 1$ to n_i. Let $\widetilde{S}_i \subset X$ be an equidimensional subvariety that satisfies the needed condition for $n_i - 1$. For each irreducible component of $\widetilde{S}_i$ choose a closed point on it outside of S_i. Also, for each irreducible component of $\widetilde{S}_i \cap Z$ choose a closed point on it outside of S_i. Thus we get a finite set T of closed points in X outside of S_i. Let H be an effective reduced divisor on X such that $H \supset S_i$ and $H \cap T = \emptyset$; then we put $S_i' = \widetilde{S}_i \cap H$. □

Let W be a codimension q cycle on X, Y be an irreducible subvariety of codimension $d - q$ in X that meets $|W|$ properly, and let f be a rational function on Y such that $\mathrm{div}(f)$ does not intersect with $|W|$. We put $f(Y \cap W) = \prod_{x \in Y \cap |W|} \mathrm{Nm}_{k(x)/k}(f^{(Y,W;x)}(x))$, where $(Y, W; x)$ is the intersection index of Y with the cycle W at a point $x \in Y \cap |W|$.

Lemma 2.8 Let X be a smooth projective variety over k and let $p + q = d + 1$; consider cycles $Z \in Z^p(X)$, $W \in Z^q(X)$, and K_1-chains $\{f_\eta\} \in G_{|W|}^{p-1}(X, 1)$, $\{g_\xi\} \in G_{|Z|}^{q-1}(X, 1)$ such that $\mathrm{div}(\{f_\eta\}) = Z$, $\mathrm{div}(\{g_\xi\}) = W$. Then we have $\prod_\eta f_\eta(\overline{\eta} \cap W) = \prod_\xi g_\xi(Z \cap \overline{\xi})$.

Proof Note that the left hand side depends only in the K_2-equivalence class of the K_1-chain $\{f_\eta\}$. Therefore by Lemma 2.6, we may assume that $\{f_\eta\} \in G_{\mathcal{S}}^{p-1}(X, 1)$, where $\mathcal{S} = \{\mathrm{Supp}(\{g_\xi\}), \cup_\xi |\mathrm{div}(g_\xi)|\}$. For each pair $\eta \in \mathrm{Supp}(\{f_\eta\})$, $\xi \in \mathrm{Supp}(\{g_\xi\})$, let $C_{\eta\xi}^\alpha$ be an irreducible component of the intersection $\overline{\eta} \cap \overline{\xi}$ and let $n_{\eta\xi}^\alpha$ be the intersection index of the subvarieties $\overline{\eta}$ and $\overline{\xi}$ at the irreducible curve $C_{\eta\xi}^\alpha$.

By condition, for all η, ξ, α as above, the restrictions $f_{\eta\xi}^{\alpha} = f_{\eta}|_{C_{\eta\xi}^{\alpha}}$ and $g_{\eta\xi}^{\alpha} = g_{\xi}|_{C_{\eta\xi}^{\alpha}}$ are well defined as rational functions on the irreducible curve $C_{\eta\xi}^{\alpha}$. It follows that $\prod_{\eta} f_{\eta}(\overline{\eta} \cap W) = \prod_{\eta,\xi,\alpha} f_{\eta\xi}^{\alpha}(\mathrm{div}(g_{\eta\xi}^{\alpha}))^{n_{\eta\xi}^{\alpha}}$; thus we conclude by the classical Weil reciprocity law for curves. $\qquad\square$

Remark 2.9 The same reasoning as in the proof of Lemma 2.8 is explained in a slightly different language in the proof of Proposition 3 from [7].

Lemma 2.10 *Let X be a smooth projective variety over k. Suppose that a cycle $W \in Z^q(X)$ is homologically trivial; then for any K_1-chain $\{f_{\eta}\} \in G_{|W|}^{d-q}(X,1)$ with $\mathrm{div}(\{f_{\eta}\}) = 0$, we have $\prod_{\eta} f_{\eta}(\overline{\eta} \cap W) = 1$.*

Proof By condition, $\Gamma(\{f_{\eta}\}) \in Z_{|W|}^{d-q}(X,1)$. Keeping in mind the explicit formula for product in higher Chow groups for cycles in general position (see [6]), we see that $((\pi_*) \circ m)(\Gamma(\{f_{\eta}\}) \otimes W) \in k^*$ is well defined and coincides with $\prod_{\eta} f_{\eta}(\overline{\eta} \cap W)$. Hence we conclude by Lemma 2.2. $\qquad\square$

Remark 2.11 If $q = d$, then Lemma 2.10 is trivial. An elementary proof of Lemma 2.10 for the case $q = 1$ can be found in [16].

2.3 Facts on the Abel–Jacobi map. Notions and results of this section are used in Section 4.2.

Let X be a complex smooth variety of dimension d. By A_X^n denote the group of complex valued smooth differential forms on X of degree n. Let $F^p A_X^n$ be the subgroup in A_X^n that consists of all differential forms with at least p "dz_i". If X is projective, then the classical Hodge theory implies $H^n(F^p A_X^{\bullet}) = F^p H^n(X, \mathbb{C})$, where we consider the Hodge filtration in the right hand side.

Let S be a closed subvariety in a smooth projective variety X; then the notation $\eta \in F_{\log}^p A_{X\setminus S}^n$ means that there exists a smooth projective variety X' together with a birational morphism $f : X' \to X$ such that $D = f^{-1}(S)$ is a normal crossing divisor on X', f induces an isomorphism $X'\setminus f^{-1}(S) \to X\setminus S$, and $f^*\eta \in F^p A_{X'}^n\langle D\rangle$, where $A_{X'}^n\langle D\rangle$ is the group of complex valued smooth differential forms on $X'\setminus D$ of degree n with logarithmic singularities along D. Recall that any class in $F^p H^n(X\setminus S, \mathbb{C})$ can be represented by a closed form $\eta \in F_{\log}^p A_{X\setminus S}^n$ (see [10]).

In what follows the variety X is supposed to be projective. By $H^*(X, \mathbb{Z})$ we often mean the image of this group in $H^*(X, \mathbb{C})$. Recall that the p-th *intermediate Jacobian* $J^{2p-1}(X)$ of X is the compact complex torus given by the formula

$$J^{2p-1}(X) = H^{2p-1}(X, \mathbb{C})/(H^{2p-1}(X, \mathbb{Z}) + F^p H^{2p-1}(X, \mathbb{C}))$$
$$= F^{d-p+1} H^{2d-2p+1}(X, \mathbb{C})^* / H_{2d-2p+1}(X, \mathbb{Z}).$$

Let Z be a homologically trivial algebraic cycle on X of codimension p. Then there exists a differentiable singular chain Γ of dimension $2d - 2p + 1$ with $\partial\Gamma = Z$, where ∂ denotes the differential in the complex of singular chains on X. Consider a closed differential form $\omega \in F^{d-p+1} A_X^{2d-2p+1}$. It can be easily checked that the integral $\int_{\Gamma} \omega$ depends only on the cohomology class $[\omega] \in F^{d-p+1} H^{2d-2p+1}(X, \mathbb{C}) = H^{2d-2p+1}(F^{d-p+1} A_X^{\bullet})$ of ω. Thus the assignment

$$Z \mapsto \{[\omega] \mapsto \textstyle\int_{\Gamma} \omega\}$$

defines a homomorphism $AJ : Z^p(X)_{\mathrm{hom}} \to J^{2p-1}(X)$, which is called the *Abel–Jacobi map*.

We give a slightly different description of the Abel–Jacobi map. Recall that there is an exact sequence of integral mixed Hodge structures:

$$0 \to H^{2p-1}(X)(p) \to H^{2p-1}(X\backslash|Z|)(p) \overset{\partial_Z}{\to} H_{2d-2p}(|Z|) \to H^{2p}(X)(p).$$

Recall that $H_{2d-2p}(|Z|) = \oplus_i \mathbb{Z}(0)$, where the sum is taken over all irreducible components in $|Z| = \cup_i Z_i$. Thus the cycle Z defines an element $[Z] \in F^0 H_{2d-2p}(|Z|, \mathbb{C})$ with a trivial image in the group $H^{2p}(X, \mathbb{Z}(p))$. Hence there exists a closed differential form $\eta \in F^p_{\log} A^{2p-1}_{X\backslash|Z|}$ such that $\partial_Z([(2\pi i)^p \eta]) = [Z]$. The difference $PD[\Gamma] - [\eta]$ defines a unique element in the group $H^{2p-1}(X, \mathbb{C})$, where $PD : H_*(X, |Z|; \mathbb{Z}) \to H^{2d-*}(X\backslash|Z|, \mathbb{Z})$ is the canonical isomorphism induced by Poincaré duality.

Lemma 2.12 *The image of $PD[\Gamma] - [\eta] \in H^{2p-1}(X, \mathbb{C})$ in the intermediate Jacobian $J^{2p-1}(X)$ is equal to the image of Z under the Abel–Jacobi map.*

Proof Consider a closed differential form $\omega \in F^{d-p+1} A^{2d-2p+1}_X$. Since $\dim(Z) = d-p$, the form ω also defines the class $[\omega] \in F^{d-p+1} H^{2d-2p+1}(X, |Z|; \mathbb{C})$. Denote by

$$(\cdot, \cdot) : H^*(X\backslash Z, \mathbb{C}) \times H^{2d-*}(X, |Z|; \mathbb{C}) \to \mathbb{C}$$

the natural pairing. Then we have $(PD[\Gamma] - [\eta], [\omega]) = (PD[\Gamma], [\omega]) = \int_\Gamma \omega$; this proves the needed result. $\square$

Remark 2.13 It follows from Lemma 2.12 that $AJ(Z) = 0$ if and only if there exists an element $\alpha \in F^p H^{2p-1}(X\backslash|Z|, \mathbb{C}) \cap H^{2p-1}(X\backslash|Z|, \mathbb{Z}(p))$ such that $\partial_Z(\alpha) = [Z]$.

Example 2.14 Suppose that $X = \mathbb{P}^1$, $Z = \{0\} - \{\infty\}$. Let z be a coordinate on $\mathbb{P}^1$ and Γ be a smooth generic path on $\mathbb{P}^1$ such that $\partial\Gamma = \{0\} - \{\infty\}$; then we have $\alpha = [\frac{dz}{z}] = 2\pi i PD[\Gamma] \in F^1 H^1(X\backslash|Z|, \mathbb{C}) \cap H^1(X\backslash|Z|, \mathbb{Z}(1))$ and $\partial_Z(\alpha) = Z$.

Lemma 2.15 *Suppose that the cycle Z is rationally trivial; then $AJ(Z) = 0$.*

Proof 1 By linearity, it is enough to consider the case when $Z = \text{div}(f)$, $f \in \mathbb{C}(Y)^*$, $Y \subset X$ is an irreducible subvariety of codimension $p - 1$. Let $\widetilde{Y}$ be the closure of the graph of the rational function $f : Y \dashrightarrow \mathbb{P}^1$ and let $p : \widetilde{Y} \to X$ be the natural map. In [17] it was shown that the following map is holomorphic:

$$\varphi : \mathbb{P}^1 \to J^{2p-1}(X), z \mapsto AJ(p_* f^{-1}(\{z\} - \{\infty\}));$$

therefore φ is constant and $AJ(Z) = \varphi(0) = \{0\}$.

Proof 2 Let α be as in Example 2.14; then by Lemma 2.17 with $X_1 = \mathbb{P}^1$, $X_2 = X$, $C = \widetilde{Y}$, $Z_1 = \{0, \infty\}$, $W_2 = \emptyset$, we have $[\widetilde{Y}]^* \alpha \in F^p H^{2p-1}(X\backslash|Z|, \mathbb{C}) \cap H^{2p-1}(X\backslash|Z|, \mathbb{Z}(p))$ and $\partial_Z([\widetilde{Y}]^* \alpha) = Z$; thus we conclude by Remark 2.13. $\square$

Remark 2.16 If there is a differentiable triangulation of the closed subset $p(f^{-1}(\gamma)) \subset X$, then we have a well defined class $[p(f^{-1}(\gamma))] \in H_{2d-2p+1}(X, |Z|; \mathbb{Z})$ and $\alpha = (2\pi i)^p PD[p(f^{-1}(\gamma))]$.

In particular, we see that the Abel–Jacobi map factors through Chow groups. In what follows we consider the induced map $AJ : CH^p(X)_{\text{hom}} \to J^{2p-1}(X)$.

In the second proof of Lemma 2.15 we have used the following simple fact. Let X_1 and X_2 be two complex smooth projective varieties of dimensions d_1 and d_2, respectively. Suppose that $C \subset X_1 \times X_2$, $Z_1 \subset X_1$, and $W_2 \subset X_2$ are closed subvarieties; we put $Z_2 = \pi_2(\pi_1^{-1}(Z_1) \cap C)$, $W_1 = \pi_1(\pi_2^{-1}(W_2) \cap C)$, where $\pi_i : X_1 \times X_2 \to X_i$, $i = 1, 2$ denote the natural projections.

Lemma 2.17   (i) *Let c be the codimension of C in $X_1 \times X_2$; then there is a natural morphism of integral mixed Hodge structures*

$$[C]^* : H^*(X_1 \backslash Z_1, W_1) \to H^{*+2c-2d_1}(X_2 \backslash Z_2, W_2)(c - d_1).$$

(ii) *For $i = 1, 2$, let p_i be the codimension of Z_i in X_i. Suppose that C meets $\pi_1^{-1}(Z_1)$ properly, the intersection $\pi_1^{-1}(Z_1) \cap \pi_2^{-1}(W_2) \cap C$ is empty, and $p_1 + c - d_1 = p_2$; then $Z_1 \cap W_1 = Z_2 \cap W_2 = \emptyset$ and the following diagram commutes:*

$$
\begin{array}{ccc}
H^{2p_1-1}(X_1\backslash Z_1, W_1)(p_1) & \overset{\partial_{Z_1}}{\to} & H_{2d_1-2p_1}(Z_1) = Z^0(Z_1) \otimes \mathbb{Z}(0) \\
\downarrow [C]^* & & \downarrow \pi_2(C \cap \pi_1^{-1}(\cdot)) \\
H^{2p_2-1}(X_2\backslash Z_2, W_2)(p_2) & \overset{\partial_{Z_2}}{\to} & H_{2d_2-2p_2}(Z_2) = Z^0(Z_2) \otimes \mathbb{Z}(0),
\end{array}
$$

where the right vertical arrow is defined via the corresponding natural homomorphisms of groups of algebraic cycles.

Proof The needed morphism $[C]^*$ is the composition of the following natural morphisms

$$
\begin{array}{rcl}
H^*(X_1\backslash Z_1, W_1) & \to & H^*((X_1 \times X_2)\backslash\pi_1^{-1}(Z_1), \pi_2^{-1}(W_2) \cap C) \\
& \overset{\cap[C]}{\to} & H^{*+2c}_{C\backslash\pi_1^{-1}(Z_1)}((X_1 \times X_2)\backslash\pi_1^{-1}(Z_1), \pi_2^{-1}(W_2) \cap C)(c) \\
& = & H_{2d_1+2d_2-*-2c}(C\backslash\pi_2^{-1}(W_2), \pi_1^{-1}(Z_1) \cap C) \\
& = & H^{*+2c}_{C\backslash\pi_1^{-1}(Z_1)}((X_1 \times X_2)\backslash(\pi_1^{-1}(Z_1) \cap C), \pi_2^{-1}(W_2))(c) \\
& \to & H^{*+2c}((X_1 \times X_2)\backslash(\pi_1^{-1}(Z_1) \cap Y), \pi_2^{-1}(W_2))(c) \\
& \to & H^{*+2c-2d_1}(X_2\backslash Z_2, W_2)(c - d_1),
\end{array}
$$

where the first morphism is the natural pull-back map, the second one is multiplication by the fundamental class $[C] \in H^{2c}_C(X_1 \times X_2)(c)$, the equalities in the middle follow from the excision property, and the last morphism is the push-forward map. The second assertion follows from the commutativity of the following diagram:

$$
\begin{array}{ccc}
H^*((X_1 \times X_2)\backslash(\pi_1^{-1}(Z_1) \cap C), \pi_2^{-1}(W_2)) & \to & H_{2d_1+2d_2-1-*}(\pi_1^{-1}(Z_1) \cap C)(-c - p_1) \\
\downarrow & & \downarrow \\
H^{*-2d_1}(X_2\backslash Z_2, W_2)(-d_1) & \to & H_{2d_1+2d_2-1-*}(Z_2)(-c - p_1).
\end{array}
$$

$\square$

2.4 Facts on K-adeles. Notions and results of this section are used in Section 4.3.

Let X be an equidimensional variety over the ground field k. Let $\mathcal{K}_n$ be the sheaf on X associated to the presheaf given by the formula $U \mapsto K_n(U)$, where $K_n(-)$ is the Quillen K-group and U is an open subset in X. Zariski cohomology groups of the sheaves $\mathcal{K}_n$ are called K-*cohomology groups*. When it will be necessary for us to point out the underline variety, we will use notation $\mathcal{K}_n^X$ for the defined above sheaf $\mathcal{K}_n$ on X.

Recall that in notations from Section 2.2, for all integers $n \geq 1, p \geq 0$, there are natural homomorphisms $d : G^p(X, n) \to G^{p+1}(X, n-1)$ such that $d^2 = 0$. Thus for each $n \geq 0$, there is a complex $Gers(X, n)^\bullet$, where $Gers(X, n)^p = G^p(X, n - p)$; this complex is called the *Gersten complex*. Note that the homomorphisms $d : G^{p-1}(X, 1) \to G^p(X, 0)$ and $d : G^{p-2}(X, 2) \to G^{p-1}(X, 1)$ coincide with the homomorphisms div and Tame, respectively. For a projective morphism of varieties

$f : X \to Y$, there is a push-forward morphism of complexes $f_* : Gers(X, n)^\bullet \to Gers(Y, n + \dim(Y) - \dim(X))^\bullet[\dim(Y) - \dim(X)]$.

In what follows we suppose that X is *smooth* over the field k. By results of Quillen (see [27]), for each $n \geq 0$, there is a canonical isomorphism between the classes of the complexes $Gers(X, n)^\bullet$ and $R\Gamma(X, \mathcal{K}_n)$ in the derived category $D^b(\mathcal{A}b)$. In particular, there is a canonical isomorphism $H^p(X, \mathcal{K}_p) \cong CH^p(X)$ for all $p \geq 0$. The last statement is often called the Bloch–Quillen formula.

There is a canonical product between the sheaves of K-groups, induced by the product in K-groups themselves. However, the Gersten complex *is not multiplicative*, i.e., there is no a product between Gersten complexes that would correspond to the product between sheaves $\mathcal{K}_n$: otherwise there would exist an intersection theory for algebraic cycles without taking them modulo rational equivalence. Explicitly, there is no a morphism of complexes

$$Gers(X, m)^\bullet \otimes_{\mathbb{Z}} Gers(X, n)^\bullet \to Gers(X, m + n)^\bullet$$

that would correspond to the natural product between cohomology groups

$$H^\bullet(X, \mathcal{K}_m) \otimes H^\bullet(X, \mathcal{K}_n) \to H^\bullet(X, \mathcal{K}_{m+n}).$$

Therefore if one would like to work explicitly with the pairing of objects in the derived category

$$R\Gamma(X, \mathcal{K}_m) \otimes_{\mathbb{Z}}^{L} R\Gamma(X, \mathcal{K}_n) \to R\Gamma(X, \mathcal{K}_{m+n}),$$

then a natural way would be to use a different resolution rather than the Gersten complex. There are general multiplicative resolutions of sheaves, for example a Godement resolution, but it does not see the Bloch–Quillen isomorphism. In particular, there is no explicit quasiisomorphism between the Godement and the Gersten complexes.

In [16] the author proposed another way to construct resolutions for a certain class of abelian sheaves on smooth algebraic varieties, namely, the *adelic resolution*. This class of sheaves includes the sheaves $\mathcal{K}_n$. It is *multiplicative* and there is an *explicit quasiisomorphism* from the adelic resolution to the Gersten complex.

Remark 2.18 Analogous adelic resolutions for coherent sheaves on algebraic varieties have been first introduced by A. N. Parshin (see [26]) in the two-dimensional case, and then developed by A. A. Beilinson (see [1]) and A. Huber (see [19]) in the higher-dimensional case.

Remark 2.19 When the paper was finished, the author discovered that a similar but more general construction of a resolution for sheaves on algebraic varieties was independently done in [4, Section 4.2.2] by A. A. Beilinson and V. Vologodsky.

Let us briefly recall several notions and facts from [16]. A *non-degenerate flag of length p* on X is a sequence of schematic points $\eta_0 \dots \eta_p$ such that $\eta_{i+1} \in \bar{\eta}_i$ and $\eta_{i+1} \neq \eta_i$ for all i, $0 \leq i \leq p - 1$. For $n, p \geq 0$, there are *adelic groups*

$$\mathbf{A}(X, \mathcal{K}_n)^p \subset \prod_{\eta_0 \dots \eta_p} K_n(\mathcal{O}_{X, \eta_0}),$$

where the product is taken over all non-degenerate flags of length p, $\mathcal{O}_{X, \eta_0}$ is the local ring of the scheme X at a point η_0, and the subgroup $\mathbf{A}(X, \mathcal{K}_n)^p$ is defined by certain explicit conditions concerning "singularities" of elements in K-groups. Elements of the adelic groups are called K-*adeles* or just adeles. Explicitly, an adele

$f \in \mathbf{A}(X, \mathcal{K}_n)^p$ is a collection $f = \{f_{\eta_0 \dots \eta_p}\}$ of elements $f_{\eta_0 \dots \eta_p} \in K_n(\mathcal{O}_{X, \eta_0})$ that satisfies certain conditions.

Example 2.20 If X is a smooth curve over the ground field k, then

$$\mathbf{A}(X, \mathcal{K}_n)^0 = K_n(k(X)) \times \prod_{x \in X} K_n(\mathcal{O}_{X,x}),$$

where the product is taken over all closed points $x \in X$, and an adele $f \in \mathbf{A}(X, \mathcal{K}_n)^1$ is a collection $f = \{f_{Xx}\}$, $f_{Xx} \in K_n(k(X))$, such that $f_{Xx} \in K_n(\mathcal{O}_{X,x})$ for almost all $x \in X$ (this is the restricted product condition in this case). The apparent similarity with classical adele and idele groups explains the name of the notion.

There is a differential

$$d : \mathbf{A}(X, \mathcal{K}_n)^p \to \mathbf{A}(X, \mathcal{K}_n)^{p+1}$$

defined by the formula

$$(df)_{\eta_0 \dots \eta_p} = \sum_{i=0}^{p} (-1)^i f_{\eta_0 \dots \hat{\eta}_i \dots \eta_p},$$

where the hat over η_i means that we omit a point η_i. It can be easily seen that $d^2 = 0$, so one gets an *adelic complex* $\mathbf{A}(X, \mathcal{K}_n)^\bullet$. There is a canonical morphism of complexes $\nu_X : \mathbf{A}(X, \mathcal{K}_n)^\bullet \to Gers(X, n)^\bullet$. There is also an adelic complex of flabby sheaves $\underline{\mathbf{A}}(X, \mathcal{K}_n)^\bullet$ given by the formula $\underline{\mathbf{A}}(X, \mathcal{K}_n)^p(U) = \mathbf{A}(U, \mathcal{K}_n)^p$ and a natural morphism of complexes of sheaves $\mathcal{K}_n[0] \to \underline{\mathbf{A}}(X, \mathcal{K}_n)^\bullet$.

In what follows we suppose that the ground field k is *infinite and perfect*.

Lemma 2.21 ([16, Theorem 3.34]) *The complex of sheaves $\underline{\mathbf{A}}(X, \mathcal{K}_n)^\bullet$ is a flabby resolution for the sheaf $\mathcal{K}_n$. The morphism ν_X is a quasiisomorphism; in particular, this induces a canonical isomorphism between the classes of the complexes $\mathbf{A}(X, \mathcal{K}_n)^\bullet$ and $R\Gamma(X, \mathcal{K}_n)$ in the derived category $D^b(\mathcal{A}b)$.*

In particular, there is a canonical isomorphism

$$H^p(X, \mathcal{K}_n) = H^p(\mathbf{A}(X, \mathcal{K}_n)^\bullet).$$

The main advantages of adelic complexes are the contravariancy and the multiplicativity properties.

Lemma 2.22 ([16, Remark 2.12])

(i) *Given a morphism $f : X \to Y$ of smooth varieties over k, for each $n \geq 0$, there is a morphism of complexes*

$$f^* : \mathbf{A}(Y, \mathcal{K}_n^Y)^\bullet \to \mathbf{A}(X, \mathcal{K}_n^X)^\bullet;$$

this morphism agrees with the natural morphism

$$\mathrm{Hom}_{D^b(\mathcal{A}b)}(R\Gamma(Y, f_* \mathcal{K}_n^X), R\Gamma(X, \mathcal{K}_n^X))$$

and the morphism $\mathcal{K}_n^Y \to f_* \mathcal{K}_n^X$ of sheaves on Y.

(ii) *For all $p, q \geq 0$, there is a morphism of complexes*

$$m : \mathbf{A}(X, \mathcal{K}_p)^\bullet \otimes \mathbf{A}(X, \mathcal{K}_q)^\bullet \to \mathbf{A}(X, \mathcal{K}_{p+q})^\bullet;$$

this morphism agrees with the multiplication morphism

$$m \in \mathrm{Hom}_{D^b(\mathcal{A}b)}(R\Gamma(X, \mathcal{K}_p) \otimes_{\mathbb{Z}}^L R\Gamma(X, \mathcal{K}_q), R\Gamma(X, \mathcal{K}_{p+q})).$$

In what follows we recollect some technical notions and facts that are used in calculations with elements of the adelic complex. The idea is to associate with each cocycle in the Gersten complex, an explicit cocycle in the adelic resolution with the same class in K-cohomology. The adelic cocycle should be good enough so that it would be easy to calculate its product with other (good) adelic cocycles. This allows to analyze explicitly the interrelation between the product on complexes $R\Gamma(X, \mathcal{K}_n)$ and the Bloch–Quillen isomorphism.

For any equidimensional subvariety $Z \subset X$ of codimension p in X, there is a notion of a *patching system* $\{Z_r^{1,2}\}$, $1 \leq r \leq p-1$ for Z on X, where Z_r^1 and Z_r^2 are equidimensional subvarieties in X of codimension r such that the system $\{Z_r^{1,2}\}$ satisfies certain properties; in particular, we have:

(i) the varieties Z_r^1 and Z_r^2 have no common irreducible components for all $r, 1 \leq r \leq p-1$;

(ii) the variety Z is contained in both varieties Z_{p-1}^1 and Z_{p-1}^2, and the variety $Z_r^1 \cup Z_r^2$ is contained in both varieties Z_{r-1}^1 and Z_{r-1}^2 for all $r, 2 \leq r \leq p-1$.

Remark 2.23 What we call here a patching system is what is called in [16] *a patching system with the freedom degree at least zero.*

Lemma 2.24 *([16, Remark 3.32])*

(i) *Suppose that $Z \subset X$ is an equidimensional subvariety of codimension p in X; then there exists a patching system $\{Z_r^{1,2}\}$, $1 \leq r \leq p-1$ for Z on X such that each irreducible component of Z_{p-1}^1 and Z_{p-1}^2 contains some irreducible component of Z and for any r, $1 \leq r \leq p-2$, each irreducible component of Z_r^1 and Z_r^2 contains some irreducible component of $Z_{r+1}^1 \cup Z_{r+1}^2$;*

(ii) *given an equidimensional subvariety $W \subset X$ that meets Z properly, one can require in addition that no irreducible component of $W \cap Z_r^1$ is contained in Z_r^2 for all r, $1 \leq r \leq p-1$.*

Remark 2.25 If an equidimensional subvariety $W \subset X$ meets Z properly and the patching system $\{Z_r^{1,2}\}$ satisfies the condition (i) from Lemma 2.24, then W meets Z_r^i properly for $i = 1, 2$ and all $r, 1 \leq r \leq p-1$.

Suppose that $\{f_\eta\} \in Gers(X, n)^p$ is a cocycle in the Gersten complex and let Z be the support of $\{f_\eta\}$. Given a patching system $Z_r^{1,2}$, $1 \leq r \leq p-1$ for Z on X, there is a notion of a *good cocycle* $[\{f_\eta\}] \in \mathbf{A}(X, \mathcal{K}_n)^p$ with respect to the patching system $\{Z_r^{1,2}\}$. In particular, we have $d[\{f_\eta\}] = 0$, $\nu_X[\{f_\eta\}] = \{f_\eta\}$, and $i_U^*[\{f_\eta\}] = 0 \in \mathbf{A}(U, \mathcal{K}_n)^p$, where $i_U : U = X \backslash Z \hookrightarrow X$ is the open embedding. Thus the good cocycle $[\{f_\eta\}]$ is a cocycle in the adelic complex $\mathbf{A}(X, \mathcal{K}_n)^\bullet$, which represents the cohomology class in $H^p(X, \mathcal{K}_n)$ of the Gersten cocycle $\{f_\eta\}$. In addition, $[\{f_\eta\}]$ satisfies certain properties that allow to consider explicitly its products with other cocycles.

Lemma 2.26 *([16, Claim 3.47])* *Let $\{f_\eta\}$, Z, and $\{Z_r^{1,2}\}$, $1 \leq r \leq p-1$ be as above; then there exists a good cocycle for $\{f_\eta\}$ with respect to the patching system $Z_r^{1,2}$.*

Consider a cycle $Z \in Z^p(X)$; suppose that a K_1-chain $\{f_\eta\} \in G^{p-1}(X, 1)$ is such that $\mathrm{div}(\{f_\eta\}) = Z$ and a K-adele $[Z] \in \mathbf{A}(X, \mathcal{K}_p)^p$ is such that $d[Z] = 0$, $\nu_X([Z]) = Z$, and $i_U^*[Z] = 0 \in \mathbf{A}(U, \mathcal{K}_p)^p$, where $i_U : U = X \backslash |Z| \hookrightarrow X$ is the open embedding.

Lemma 2.27 ([16, Lemma 3.48]) *In the above notations, let* $Y = \mathrm{Supp}(\{f_\eta\})$ *and let* $\{Y_r^{1,2}\}$ *be a patching system for* Y *on* X; *then there exists a* K-*adele* $[\{f_\eta\}] \in \mathbf{A}(X, \mathcal{K}_p)^{p-1}$ *such that* $d[\{f_\eta\}] = [Z]$, $\nu_X[\{f_\eta\}] = \{f_\eta\}$, *and* $i_U^*[\{f_\eta\}] \in \mathbf{A}(U, \mathcal{K}_p)^{p-1}$ *is a good cocycle with respect to the restriction of the patching system* $\{Y_r^{1,2}\}$ *to* U.

For a cycle $Z \in Z^p(X)$, let $\{Z_r^{1,2}\}$, $1 \le r \le p-1$ be a patching system for $|Z|$ on X and $[Z] \in \mathbf{A}(X, \mathcal{K}_p)^p$ be a good cocycle for Z with respect to the patching system $\{Z_r^{1,2}\}$. Given a K_1-chain $\{f_z\} \in G^p(X, 1)$ with support on $|Z|$ and $\mathrm{div}(\{f_z\}) = 0$, let $[\{f_z\}] \in \mathbf{A}(X, p+1)^p$ be a good cocycle for $\{f_z\}$ with respect to the patching system $\{Z_r^{1,2}\}$. Let $\{W_s^{1,2}\}$, $1 \le s \le q-1$, $[W]$, and $[\{g_w\}] \in \mathbf{A}(X, q+1)^q$ be the analogous objects for a cycle $W \in Z^q(X)$ and a K_1-chain $\{g_w\} \in G^q(X, 1)$ with support on $|W|$ and $\mathrm{div}(\{g_w\}) = 0$.

Lemma 2.28 ([16, Theorem 4.22]) *In the above notations, suppose that* $p+q = d$, $|Z|$ *meets* $|W|$ *properly, the patching systems* $\{Z_r^{1,2}\}$ *and* $\{W_s^{1,2}\}$ *satisfy the condition* (i) *from Lemma 2.24, and that the patching system* $\{W_s^{1,2}\}$ *satisfies the condition* (ii) *from Lemma 2.24 with respect to the subvariety* $|Z|$. *Then we have*

$$\nu_X(m([\{f_z\}] \otimes [W])) = (-1)^{(p+1)q}\{(\prod_{z \in Z^{(0)}} f_z^{(\overline{z}, W; x)}(x))_x\} \in G^d(X, 1),$$

$$\nu_X(m([Z] \otimes [\{g_w\}])) = (-1)^{pq}\{(\prod_{w \in W^{(0)}} g_w^{(Z, \overline{w}; x)}(x))_x\} \in G^d(X, 1),$$

where $(\overline{z}, W; x)$ *is the intersection index of the subvariety* $\overline{z}$ *with the cycle* W *at a point* $x \in X^{(d)}$ *(the same for* $(Z, \overline{w}; x)$).

2.5 Facts on determinant of cohomology and Picard categories. Notions and results of this section are used in Section 4.4.

Let X be a smooth projective variety over a field k. Given two coherent sheaves $\mathcal{F}$ and $\mathcal{G}$ on X, one has a well-defined k^*-torsor $\det R\Gamma(X, \mathcal{F} \otimes_{\mathcal{O}_X}^L \mathcal{G})\backslash\{0\}$, where $\det(V^\bullet) = \otimes_{i \in \mathbb{Z}} \det^{(-1)^i} H^i(V^\bullet)$ for a bounded complex of vector spaces $V^\bullet$. We will need to study a behavior of this k^*-torsor with respect to exact sequences of coherent sheaves. With this aim it is more convenient to consider a $\mathbb{Z}$-graded k^*-torsor

$$\langle \mathcal{F}, \mathcal{G} \rangle = (\mathrm{rk}R\Gamma(X, \mathcal{F} \otimes_{\mathcal{O}_X}^L \mathcal{G}), \det R\Gamma(X, \mathcal{F} \otimes_{\mathcal{O}_X}^L \mathcal{G})\backslash\{0\})$$

and to use the construction of virtual coherent sheaves and virtual vector spaces.

Recall that for any exact category $\mathcal{C}$, P. Deligne has defined in [11] the category of virtual objects $V\mathcal{C}$ together with a functor $\gamma : \mathcal{C}_{iso} \to V\mathcal{C}$ that has a certain universal property, where $\mathcal{C}_{iso}$ is the category with the same objects as $\mathcal{C}$ and with morphisms being all isomorphisms in $\mathcal{C}$. The category $V\mathcal{C}$ is a *Picard category*: it is non-empty, every morphism in $V\mathcal{C}$ is invertible, there is a functor $+ : V\mathcal{C} \times V\mathcal{C} \to V\mathcal{C}$ such that it satisfies some compatible associativity and commutativity constraints and such that for any object L in $V\mathcal{C}$, the functor $(\cdot + L) : V\mathcal{C} \to V\mathcal{C}$ is an autoequivalence of $V\mathcal{C}$ (see op.cit.)[1]. In particular, this implies the existence of a unit object 0 in any Picard category.

Explicitly, the objects of $V\mathcal{C}$ are based loops on the H-space $BQ\mathcal{C}$ and for any two loops γ_1, γ_2 on $BQ\mathcal{C}$, the morphisms in $\mathrm{Hom}_{V\mathcal{C}}(\gamma_1, \gamma_2)$ are the homotopy

---

[1]In op.cit. this notion is called a *commutative* Picard category; since we do not consider non-commutative Picard categories, we use a shorter terminology.

classes of homotopies from γ_1 to γ_2. For an object E in $\mathcal{C}$, the object $\gamma(E)$ in $V\mathcal{C}$ is the canonical based loop on $BQ\mathcal{C}$ associated with E. By $[E]$ denote the class of E in the group $K_0(\mathcal{C})$. Note that $[E] = [E']$ if and only if there is a morphism between $\gamma(E)$ and $\gamma(E')$ in $V\mathcal{C}$. Moreover, an exact sequence in $\mathcal{C}$

$$0 \to E' \to E \to E'' \to 0$$

defines in a canonical way an isomorphism in $V\mathcal{C}$

$$\gamma(E') + \gamma(E'') \cong \gamma(E).$$

Furthermore, there are canonical isomorphisms $\pi_i(V\mathcal{C}) \cong K_i(\mathcal{C})$ for $i = 0, 1$ and $\pi_i(V\mathcal{C}) = 0$ for $i > 1$.

Let us explain in which sense the functor $\gamma : \mathcal{C}_{iso} \to V\mathcal{C}$ is universal. Given a Picard category $\mathcal{P}$, consider the category $\mathrm{Det}(\mathcal{C}, \mathcal{P})$ of *determinant functors*, i.e., a category of pairs (δ, D), where $\delta : \mathcal{C}_{iso} \to \mathcal{P}$ is a functor and a D is a functorial isomorphism

$$D(\Sigma) : \delta(E') + \delta(E'') \to \delta(E)$$

for each exact sequence

$$\Sigma : 0 \to E' \to E \to E'' \to 0$$

such that the pair (δ, D) is compatible with zero objects, associativity and commutativity (see op.cit., 4.3). Note that $\mathrm{Det}(\mathcal{C}, \mathcal{P})$ has a natural structure of a Picard category.

For Picard categories $\mathcal{P}$ and $\mathcal{Q}$, denote by $\mathrm{Fun}^+(\mathcal{P}, \mathcal{Q})$ the category of symmetric monoidal functors $F : \mathcal{P} \to \mathcal{Q}$. Morphisms in $\mathrm{Fun}^+(\mathcal{P}, \mathcal{Q})$ are monoidal morphisms between monoidal functors. Denote by $0 : \mathcal{P} \to \mathcal{Q}$ a functor that sends every object of $\mathcal{P}$ to the unit object 0 in $\mathcal{Q}$ and sends every morphism to the identity. Note that $\mathrm{Fun}^+(\mathcal{P}, \mathcal{Q})$ has a natural structure of a Picard category. The universality of $V\mathcal{C}$ is expressed by the following statement (see op.cit., 4.4).

Lemma 2.29 *For any Picard category $\mathcal{P}$ and an exact category $\mathcal{C}$, the composition with the functor $\gamma : \mathcal{C} \to V\mathcal{C}$ defines an equivalence of Picard categories*

$$\mathrm{Fun}^+(V\mathcal{C}, \mathcal{P}) \to \mathrm{Det}(\mathcal{C}, \mathcal{P}).$$

By universality of $V\mathcal{C}$, the smallest Picard subcategory in $V\mathcal{C}$ containing $\gamma(\mathcal{C}_{iso})$ is equivalent to the whole category $V\mathcal{C}$.

Example 2.30 Let us describe explicitly the category $V\mathcal{M}_k$, where $\mathcal{M}_k$ is the exact category of finite-dimensional vector spaces over a field k (see op.cit., 4.1 and 4.13). Objects of $V\mathcal{M}_k$ are $\mathbb{Z}$-graded k^*-torsors, i.e, pairs (l, L), where $l \in \mathbb{Z}$ and L is a k^*-torsor. For any objects (l, L) and (m, M) in $V\mathcal{M}_k$, we have $\mathrm{Hom}_{V\mathcal{M}_k}((l, L), (m, M)) = 0$ if $l \neq m$, and $\mathrm{Hom}_{V\mathcal{M}_k}((l, L), (m, M)) = \mathrm{Hom}_{k^*}(L, M)$ if $l = m$. The functor $(\mathcal{M}_k)_{iso} \to V\mathcal{M}_k$ sends a vector space V over k to the pair $(\mathrm{rk}_k(V), \det_k(V) \backslash \{0\})$. We have

$$(l, L) + (m, M) = (l + m, L \otimes M)$$

and the commutativity constraint

$$(l, L) + (m, M) \cong (m, M) + (l, L)$$

is given by the formula $u \otimes v \mapsto (-1)^{lm} v \otimes u$, where $u \in L$, $v \in M$.

Given exact categories $\mathcal{C}$, $\mathcal{C}'$, $\mathcal{E}$, and a biexact functor $\mathcal{C} \times \mathcal{C}' \to \mathcal{E}$, one has a corresponding functor $G : V\mathcal{C} \times V\mathcal{C}' \to V\mathcal{E}$, which is *distributive* with respect to addition in categories of virtual objects: for a fixed object L in $V\mathcal{C}$ or M in $V\mathcal{C}'$, the functor $G(L, M)$ is a symmetric monoidal functor and the choices of a fixed argument are compatible with each other (see op.cit., 4.11). A distributive functor is an analog for Picard categories of what is a biextension for abelian groups (see Section 3.1).

Now let $\mathcal{M}_X$, $\mathcal{P}_X$, and $\mathcal{P}'_X$ denote the categories of coherent sheaves on X, vector bundles on X, and vector bundles E on X such that $H^i(X, E) = 0$ for $i > 0$, respectively. The natural symmetric monoidal functors $V\mathcal{P}_X \to V\mathcal{M}_X$ and $V\mathcal{P}'_X \to V\mathcal{P}_X$ are equivalences of categories, see op.cit., 4.12. A choice of corresponding finite resolutions for all objects gives inverse functors to these functors. Thus, there is a distributive functor

$$\langle \cdot, \cdot \rangle : V\mathcal{M}_X \times V\mathcal{M}_X \cong V\mathcal{P}_X \times V\mathcal{P}_X \to V\mathcal{P}_X \cong V\mathcal{P}'_X \to V\mathcal{M}_k.$$

There is a canonical isomorphism of the composition $\langle \cdot, \cdot \rangle \circ \gamma : (\mathcal{M}_X)_{iso} \times (\mathcal{M}_X)_{iso} \to V\mathcal{M}_k$ and the previously defined functor

$$\langle \mathcal{F}, \mathcal{G} \rangle = (\mathrm{rk}R\Gamma(X, \mathcal{F} \otimes^L_{\mathcal{O}_X} \mathcal{G}), \det R\Gamma(X, \mathcal{F} \otimes^L_{\mathcal{O}_X} \mathcal{G}) \backslash \{0\})$$

By Lemma 2.29, the last condition defines the distributive functor $\langle \cdot, \cdot \rangle$ up to a unique isomorphism.

In Section 4.4 we will need some more generalities about Picard categories that we describe below. For Picard categories $\mathcal{P}_1$ and $\mathcal{P}_2$, denote by $\mathrm{Distr}(\mathcal{P}_1, \mathcal{P}_2; \mathcal{Q})$ the category of distributive functors $G : \mathcal{P}_1 \times \mathcal{P}_2 \to \mathcal{Q}$. Morphisms between G and G' in $\mathrm{Distr}(\mathcal{P}_1, \mathcal{P}_2; \mathcal{Q})$ are morphisms between functors such that for a fixed object L in $\mathcal{P}_1$ or M in $\mathcal{P}_2$, the corresponding morphism of monoidal functors $G(L, M) \to G'(L, M)$ is monoidal. Denote by $0 : \mathcal{P}_1 \times \mathcal{P}_2 \to \mathcal{Q}$ a functor that sends every object of $\mathcal{P}_1 \times \mathcal{P}_2$ to the unit object 0 in $\mathcal{Q}$ and sends every morphism to the identity. As above, the category $\mathrm{Distr}(\mathcal{P}_1, \mathcal{P}_2; \mathcal{Q})$ has a natural structure of a Picard category.

We will use quotients of Picard categories.

Definition 2.31 Given a functor F in $\mathrm{Fun}^+(\mathcal{P}', \mathcal{P})$, a *quotient* $\mathcal{P}/\mathcal{P}'$ is the following category: objects of $\mathcal{P}/\mathcal{P}'$ are the same as in $\mathcal{P}$, and morphisms are defined by the formula

$$\mathrm{Hom}_{\mathcal{P}/\mathcal{P}'}(L, M) = \mathrm{colim}_{K \in Ob(\mathcal{P}')} \mathrm{Hom}_{\mathcal{P}}(L, M + F(K)),$$

i.e., we take the colimit of the functor $\mathcal{P} \to Sets$, $K \mapsto \mathrm{Hom}_{\mathcal{P}}(L, M + F(K))$. The composition of morphisms $f \in \mathrm{Hom}_{\mathcal{P}/\mathcal{P}'}(L, M)$ and $g \in \mathrm{Hom}_{\mathcal{P}/\mathcal{P}'}(M, N)$ is defined as follows: represent f and g by elements $\widetilde{f} \in \mathrm{Hom}_{\mathcal{P}}(L, M + F(K_1))$ and $\widetilde{g} \in \mathrm{Hom}_{\mathcal{P}}(M, N + F(K_2))$, respectively, and take the composition

$$(\widetilde{g} + id_{F(K_1)}) \circ \widetilde{f} \in \mathrm{Hom}_{\mathcal{P}}(L, N + F(K_1) + F(K_2)) = \mathrm{Hom}_{\mathcal{P}}(L, N + F(K_1 + K_2)).$$

Remark 2.32

(i) Taking a representative $K \in Ob(\mathcal{P}')$ for each class in $\pi_0(\mathcal{P}')$, we get a decomposition

$$\mathrm{Hom}_{\mathcal{P}/\mathcal{P}'}(L, M) = \coprod_{[K] \in \pi_0(\mathcal{P}')} \mathrm{Hom}_{\mathcal{P}}(L, M + F(K))/\pi_1(\mathcal{P}'),$$

where $\pi_1(\mathcal{P}')$ acts on $\mathrm{Hom}_{\mathcal{P}}(L, M + F(K))$ via the second summand and the identification $\pi_1(\mathcal{P}') = \mathrm{Hom}_{\mathcal{P}'}(K, K)$.

(ii) A more natural way to define a quotient of Picard categories would be to consider 2-Picard categories instead of taking quotients of sets of morphisms (or, more generally, to consider homotopy quotients of ∞-groupoids); the above construction is a truncation to the first two layers in the Postnikov tower of the "right" quotient.

Let us consider a descent property for quotients of Picard categories.

For a symmetric monoidal functor $F : \mathcal{P}' \to \mathcal{P}$ and a Picard category $\mathcal{Q}$, consider a category $\mathrm{Fun}_F^+(\mathcal{P}, \mathcal{Q})$, whose objects are pairs (G, Ψ), where G is an object in $\mathrm{Fun}^+(\mathcal{P}, \mathcal{Q})$ and $\Psi : G \circ F \to 0$ is an isomorphism in $\mathrm{Fun}^+(\mathcal{P}', \mathcal{Q})$. Morphisms in $\mathrm{Fun}_F^+(\mathcal{P}, \mathcal{Q})$ are defined in a natural way. Note that $\mathrm{Fun}_F^+(\mathcal{P}, \mathcal{Q})$ is a Picard category.

Analogously, for two symmetric monoidal functors of Picard categories $F_1 : \mathcal{P}_1' \to \mathcal{P}_1$, $F_2 : \mathcal{P}_2' \to \mathcal{P}_2$, and a Picard category $\mathcal{Q}$, consider a category

$$\mathrm{Distr}_{(F_1, F_2)}(\mathcal{P}_1, \mathcal{P}_2; \mathcal{Q}),$$

whose objects are triples (G, Ψ_1, Ψ_2), where G is an object in $\mathrm{Distr}(\mathcal{P}_1, \mathcal{P}_2; \mathcal{Q})$, $\Psi_i : G \circ (F_i \times \mathrm{Id}) \to 0$, $i = 1, 2$, are isomorphisms in $\mathrm{Distr}(\mathcal{P}_i', \mathcal{P}_{3-i}; \mathcal{Q})$ such that

$$\Psi_1 \circ (\mathrm{Id} \times F_2) = \Psi_2 \circ (F_1 \times \mathrm{Id}) \in \mathrm{Hom}(G \circ (F_1 \times F_2), 0),$$

where morphisms are taken in the category $\mathrm{Distr}(\mathcal{P}_1', \mathcal{P}_2'; \mathcal{Q})$. Morphisms in

$$\mathrm{Distr}_{(F_1, F_2)}(\mathcal{P}_1, \mathcal{P}_2; \mathcal{Q})$$

are defined in a natural way. Note that $\mathrm{Distr}_{(F_1, F_2)}(\mathcal{P}_1, \mathcal{P}_2; \mathcal{Q})$ is a Picard category.

Lemma 2.33

(i) *The category $\mathcal{P}/\mathcal{P}'$ inherits a canonical Picard category structure such that the natural functor $P : \mathcal{P} \to \mathcal{P}/\mathcal{P}'$ is a symmetric monoidal functor.*

(ii) *There is an exact sequence of abelian groups*

$$\pi_1(\mathcal{P}') \to \pi_1(\mathcal{P}) \to \pi_1(\mathcal{P}/\mathcal{P}') \to \pi_0(\mathcal{P}) \to \pi_0(\mathcal{P}') \to \pi_0(\mathcal{P}/\mathcal{P}') \to 0.$$

(iii) *The composition of functors with P defines an isomorphism of Picard categories*

$$\mathrm{Fun}^+(\mathcal{P}/\mathcal{P}', \mathcal{Q}) \to \mathrm{Fun}_F^+(\mathcal{P}, \mathcal{Q}).$$

(iv) *The composition of functors with $P_1 \times P_2$ defines an isomorphism of Picard categories*

$$\mathrm{Distr}(\mathcal{P}_1/\mathcal{P}_1', \mathcal{P}_2/\mathcal{P}_2'; \mathcal{Q}) \to \mathrm{Fun}_{(F_1 \times F_2)}^+(\mathcal{P}_1, \mathcal{P}_2; \mathcal{Q}).$$

Proof (i) Let L, M be objects in $\mathcal{P}/\mathcal{P}'$; by definition, the sum $L + M$ in $\mathcal{P}/\mathcal{P}'$ is equal the sum of L and M as objects of the Picard category $\mathcal{P}$. The commutativity and associativity constraints are also induced by that in the category $\mathcal{P}$. The check of the needed properties is straightforward.

(ii) Direct checking that uses an explicit description of morphisms given in Remark 2.32(i).

(iii) Let $H : \mathcal{P}/\mathcal{P}' \to \mathcal{Q}$ be a symmetric monoidal functor and let K be an object in $\mathcal{P}'$. There is a canonical morphism $F(K) \to 0$ in $\mathcal{P}/\mathcal{P}'$ given by the morphism $F(K) \to F(K)$ in $\mathcal{P}$; this induces a canonical morphism $(H \circ P \circ F)(K) \to 0$ in $\mathcal{Q}$ and an isomorphism of symmetric monoidal functors $\Psi : H \circ P \circ F \to 0$. Thus the assignment $H \mapsto (H \circ P, \Psi)$ defines defines a functor $\mathrm{Fun}^+(\mathcal{P}/\mathcal{P}', \mathcal{Q}) \to \mathrm{Fun}_F^+(\mathcal{P}, \mathcal{Q})$.

Let (G, Ψ) be an object in $\mathrm{Fun}_F^+(\mathcal{P}, \mathcal{Q})$. We put $H(L) = G(L)$ for any object L in $\mathcal{P}/\mathcal{P}'$. Let $f : L \to M$ be a morphism in $\mathcal{P}/\mathcal{P}'$ represented by a morphism $\widetilde{f} : L \to M + F(K)$ in $\mathcal{P}$. We put $H(f)$ to be the composition

$$H(L) = G(L) \xrightarrow{G(\widetilde{f})} G(M) + (G \circ F)(K) \xrightarrow{id_{G(M)} + \Psi(K)} G(M) = H(M).$$

This defines a symmetric monoidal functor $H : \mathcal{P}/\mathcal{P}' \to \mathcal{Q}$. The assignment $(G, \Psi) \mapsto H$ defines an inverse functor $\mathrm{Fun}_F^+(\mathcal{P}, \mathcal{Q}) \to \mathrm{Fun}^+(\mathcal{P}/\mathcal{P}', \mathcal{Q})$ to the functor constructed above.

(iv) The proof is completely analogous to the proof of (iii). $\square$

For an abelian group A, denote by $\mathrm{Picard}(A)$ a Picard category whose objects are elements of A and whose only morphisms are identities.

Example 2.34

(i) Let $\mathcal{P}$ be a Picard category and let $0_\mathcal{P}$ be a full subcategory in $\mathcal{P}$, whose only object is 0; then the natural functor $\mathcal{P}/0_\mathcal{P} \to \mathrm{Picard}(\pi_0(\mathcal{P}))$ is an equivalence of Picard categories.

(ii) Let $\mathcal{A}'$ be an abelian Serre subcategory in an abelian category $\mathcal{A}$; then the natural functor $V\mathcal{A}/V\mathcal{A}' \to V(\mathcal{A}/\mathcal{A}')$ is an equivalence of Picard categories. This follows immediately from the 5-lemma and Lemma 2.33 (ii).

For an abelian group N, denote by N_{tors} the Picard category of N-torsors. Note that for abelian groups A, B, a biextension of (A, B) by N is the same as a distributive functor $\mathrm{Picard}(A) \times \mathrm{Picard}(B) \to N_{tors}$. Let $\mathcal{P}_1$, $\mathcal{P}_2$ be Picard categories, P be a biextension of $(\pi_0(\mathcal{P}_1), \pi_0(\mathcal{P}_2))$ by N. Then P naturally defines a distributive functor $G_P : \mathcal{P}_1 \times \mathcal{P}_2 \to N_{tors}$.

Combining Lemma 2.33(iv) and Example 2.34(i), we get the next statement.

Lemma 2.35 *A distributive functor $G : \mathcal{P}_1 \times \mathcal{P}_2 \to N_{tors}$ is isomorphic to the functor G_P for some biextension P of $(\pi_0(\mathcal{P}_1), \pi_0(\mathcal{P}_2))$ by N if and only if $G_*(\pi_0(\mathcal{P}_1) \times \pi_1(\mathcal{P}_2)) = G_*(\pi_1(\mathcal{P}_1) \times \pi_0(\mathcal{P}_2)) = 0 \in N = \pi_1(N_{tors})$.*

3 General facts on biextensions

3.1 Quotient biextension. For all groups below, we write the groups law in the additive manner. For an abelian group A and a natural number $l \in \mathbb{Z}$, let A_l denote the l-torsion in A.

The notion of a biextension was first introduced in [24]; see more details on biextensions in [9] and [28]. Recall that the set of isomorphism classes of biextensions of (A, B) by N is canonically bijective with the group $\mathrm{Ext}^1_{\mathcal{A}b}(A \otimes_{\mathbb{Z}}^L B, N)$ for any abelian groups A, B, and N, where $\mathcal{A}b$ is the category of all abelian groups.

Let us describe one explicit construction of biextensions. As before, let A and B be abelian groups.

Definition 3.1 A subset $T \subset A \times B$ is a *bisubgroup* if for all elements (a, b), (a, b'), and (a', b) in T, the elements $(a + a', b)$ and $(a, b + b')$ belong to T. For a bisubgroup $T \subset A \times B$ and an abelian group N, a *bilinear map* $\psi : T \to N$ is a map of sets such that for all elements (a, b), (a, b'), and (a', b) in T, we have $\psi(a, b) + \psi(a', b) = \psi(a + a', b)$ and $\psi(a, b) + \psi(a, b') = \psi(a, b + b')$.

Suppose that $T \subset A \times B$ is a bisubgroup and $\varphi_A : A \to A'$, $\varphi_B : B \to B'$ are surjective group homomorphisms such that $(\varphi_A \times \varphi_B)(T) = A' \times B'$. Consider the set

$$S = T \cap (\mathrm{Ker}(\varphi_A) \times B \cup A \times \mathrm{Ker}(\varphi_B)) \subset A \times B;$$

clearly, S is a bisubgroup in $A \times B$. Given a bilinear map $\psi : S \to N$, let us define an equivalence relation on $N \times T$ as the transitive closure of the isomorphisms

$$N \times \{(a, b)\} \overset{\psi(a', b)}{\longrightarrow} N \times \{(a + a', b)\},$$

$$N \times \{(a, b)\} \overset{\psi(a, b')}{\longrightarrow} N \times \{(a, b + b')\}$$

for all $(a, b) \in T$ and $(a', b), (a, b') \in S$. It is easy to check that the quotient P_ψ of $N \times T$ by this equivalence relation has a natural structure of a biextension of (A', B') by N.

Remark 3.2 Suppose that we are given two bisubgroups $T_2 \subset T_1$ in $A \times B$ satisfying the above conditions; if the restriction of ψ_1 to S_2 is equal to ψ_2, then the biextensions P_{ψ_1} and P_{ψ_2} are canonically isomorphic.

Remark 3.3 Suppose that we are given two bilinear maps $\psi_1, \psi_2 : S \to N$. Let $\phi : T \to N$ be a bilinear map that satisfies $\psi_1 = \phi|_S + \psi_2$; then the multiplication by ϕ defines an isomorphism of the biextensions P_{ψ_2} and P_{ψ_1} of (A', B') by N.

Remark 3.4 There is a natural generalization of the construction mentioned above. Suppose that we are given a biextension Q of (A, B) by N and surjective homomorphisms $\varphi_A : A \to A'$, $\varphi_B : B \to B'$. Consider the bisubgroup $S = \mathrm{Ker}(\varphi_A) \times B \cup A \times \mathrm{Ker}(\varphi_B)$ in $A \times B$; a bilinear map (in the natural sense) $\psi : S \to Q$ is called a *trivialization of the biextension Q over the bisubgroup S*. The choice of the trivialization $\psi : S \to Q$ canonically defines a biextension P of (A', B') by N such that the biextension $(\varphi_A \times \varphi_B)^* P$ is canonically isomorphic to the biextension Q (one should use the analogous construction to the one described above).

Now let us recall the definition of a Weil pairing associated to a biextension.

Definition 3.5 Consider a biextension P of (A, B) by N and a natural number $l \in \mathbb{N}$, $l \geq 1$; for elements $a \in A_l$, $b \in B_l$, their *Weil pairing* $\phi_l(a, b) \in N_l$ is the obstruction to a commutativity of the diagram

$$
\begin{array}{ccc}
P_{(a, lb)} & \overset{\alpha}{\longrightarrow} & P_{(a, b)}^{\wedge l} \\
\uparrow \beta & & \uparrow \gamma \\
P_{(0, 0)} & \overset{\delta}{\longrightarrow} & P_{(la, b)},
\end{array}
$$

where arrows are canonical isomorphisms of N-torsors given by the biextension structure on P: $\phi_l(a, b) = \gamma \circ \delta - \alpha \circ \beta$.

It is easily checked that the Weil pairing defines a bilinear map $\phi_l : A_l \times B_l \to N_l$.

Remark 3.6 There is an equivalent definition of the Weil pairing: an isomorphism class of a biextension is defined by a morphism $A \otimes_{\mathbb{Z}}^L B[-1] \to N$ in $D^b(\mathcal{A}b)$, and the corresponding Weil pairing is obtained by taking the composition

$$\oplus_l (A_l \otimes B_l) \to \mathrm{Tor}_1^{\mathbb{Z}}(A, B) \to A \otimes_{\mathbb{Z}}^L B[-1] \to N.$$

Lemma 3.7 *Let* $\varphi_A : A \to A'$, $\varphi_B : B \to B'$, $T \subset A \times B$, *and* $\psi : S \to N$ *be as above. Consider a pair* $(a,b) \in T$ *such that* $\varphi_A(la) = 0$ *and* $\varphi_B(lb) = 0$; *then we have*

$$\phi_l(\varphi_A(a), \varphi_B(b)) = \psi(la, b) - \psi(a, lb),$$

where ϕ_l *denotes the Weil pairing associated to the biextension* P *of* (A', B') *by* N *induced by the bilinear map* $\psi : S \to N$.

Proof By construction the pull-back of the biextension P with respect to the map $T \to A' \times B'$ is canonically trivial. Furthermore, the pull-back of the diagram that defines the Weil pairing $\phi_l(\varphi(a), \varphi(b))$ is equal to the following diagram:

$$
\begin{array}{ccc}
N \times \{(a, lb)\} & \xrightarrow{\ id_N\ } & N^{\wedge l} \times \{(a,b)\} \\
\uparrow +\psi(a, lb) & & \uparrow id_N \\
N \times \{(0,0)\} & \xrightarrow{\ +\psi(la,b)\ } & N \times \{(la, b)\}.
\end{array}
$$

This concludes the proof. $\qquad\qquad\qquad\qquad\qquad\qquad\qquad\qquad\qquad\qquad$ $\square$

3.2 Biextensions and pairings between complexes. We describe a way to construct a biextension starting from a pairing between complexes. Suppose that $A^\bullet$ and $B^\bullet$ are two bounded from above complexes of abelian groups, N is an abelian group, and that we are given a pairing $\phi \in \mathrm{Hom}_{D^-(\mathcal{A}b)}(A^\bullet \otimes_{\mathbb{Z}}^L B^\bullet, N)$. We fix an integer p. Let $H^p(A^\bullet)' \subset H^p(A^\bullet)$ be the annulator of $H^{-p}(B^\bullet)$ with respect to the induced pairing $\phi : H^p(A^\bullet) \times H^{-p}(B^\bullet) \to N$. Analogously, we define the subgroup $H^{1-p}(B^\bullet)' \subset H^{1-p}(B^\bullet)$. Let $\tau'_{\leq p}A^\bullet \subset A^\bullet$ be a subcomplex such that $(\tau'_{\leq p}A^\bullet)^n = 0$ if $n > p$, $(\tau'_{\leq p}A^\bullet)^n = A^n$ if $n < p$, and $(\tau'_{\leq p}A^\bullet)^p = \mathrm{Ker}(d_A^p)'$, where $d_A^p : A^p \to A^{p+1}$ is the differential in the complex $A^\bullet$ and $\mathrm{Ker}(d_A^p)'$ is the group of cocycles that map to $H^p(A^\bullet)'$. Note that the operation $\tau'_{\leq p}$ is well defined for complexes up to quasiisomorphisms, i.e., the class of the complex $\tau'_{\leq p}A^\bullet$ in the derived category $D^-(\mathcal{A}b)$ depends only on the classes of the complexes $A^\bullet$, $B^\bullet$ and the morphism ϕ. Analogously, we define the subcomplex $\tau'_{\leq(d+1-p)}B^\bullet \subset B^\bullet$. The restriction of ϕ defines an element $\phi' \in \mathrm{Ext}^1_{D^-(\mathcal{A}b)}(\tau'_{\leq p}A^\bullet \otimes_{\mathbb{Z}}^L \tau'_{\leq(1-p)}B^\bullet, N[-1])$. It can be easily checked that ϕ' passes in a unique way through the morphism $\tau'_{\leq p}A^\bullet \otimes_{\mathbb{Z}}^L \tau'_{\leq(1-p)}B^\bullet \to H^p(A^\bullet)'[-p] \otimes_{\mathbb{Z}}^L H^{1-p}(B^\bullet)'[p-1]$. Thus we get a canonical element

$$
\begin{aligned}
\phi' \ \in\ & \mathrm{Hom}_{D^-(\mathcal{A}b)}(H^p(A^\bullet)'[-p] \otimes_{\mathbb{Z}}^L H^{1-p}(B^\bullet)'[1-p], N) \\
& = \mathrm{Ext}^1_{\mathcal{A}b}(H^p(A^\bullet)' \otimes_{\mathbb{Z}}^L H^{1-p}(B^\bullet)', N),
\end{aligned}
$$

i.e., a biextension P_ϕ of $(H^p(A^\bullet)', H^{1-p}(B^\bullet)')$ by N.

We construct this biextension explicitly for the case when ϕ is induced by a true morphism of complexes $\phi : A^\bullet \otimes B^\bullet \to N$ (this can be always obtained by taking projective resolutions). Let $A = \mathrm{Ker}(d_A^p)'$, $B = \mathrm{Ker}(d_B^{1-p})'$, $T = A \times B$, $\phi_A : \mathrm{Ker}(d_A^p)' \to H^p(A^\bullet)'$, $\phi_B : \mathrm{Ker}(d_B^{1-p})' \to H^{1-p}(B^\bullet)'$, $S = \mathrm{Im}(d_A^{p-1}) \times \mathrm{Ker}(d_B^{1-p})' \cup \mathrm{Ker}(d_A^p)' \times \mathrm{Im}(d_B^{-p})$. We define a bilinear map $\psi : S \to N$ as follows: $\psi(d_A^{p-1}(a^{p-1}), b^{1-p}) = \phi(a^{p-1} \otimes b^{1-p})$ and $\psi(a^p, d_B^{-p}(b^{-p})) = (-1)^p \phi(a^p \otimes b^{-p})$. It is readily seen that this does not depend on the choices of a^{p-1} and b^{-p}, and that $\psi(d_A^{p-1}(a^{p-1}), d_B^{-p}(b^{-p}))$ is well defined (the reason is that ϕ is a morphism of complexes). The application of the construction from Section 3.1 gives a biextension P_ψ of $(H^p(A^\bullet)', H^{1-p}(B^\bullet)')$ by N; we have $P_\phi = P_\psi$.

Remark 3.8 There is an equivalent construction of the biextension P_ϕ in terms of Picard categories (see Section 2.5). For a two-term complex $C^\bullet = \{C^{-1} \xrightarrow{d} C^0\}$, let $\mathrm{Picard}(C^\bullet)$ be the following Picard category: objects in $\mathrm{Picard}(C^\bullet)$ are elements in the group C^0 and morphisms from $c \in C^0$ to $c' \in C^0$ are elements $\widetilde{c} \in C^{-1}$ such that $d\widetilde{c} = c' - c$. A morphism of complexes ϕ defines a morphism of complexes

$$\tau_{\geq(p-1)}\tau_{\leq p}A^\bullet \times \tau_{\geq -p}\tau_{\leq(1-p)}A^\bullet \to N$$

that defines a distributive functor

$$F : \mathrm{Picard}(\tau_{\geq(p-1)}\tau_{\leq p}A^\bullet) \times \mathrm{Picard}(\tau_{\geq -p}\tau_{\leq(1-p)}A^\bullet) \to BN,$$

where BN is a Picard category with one object 0 and with $\pi_1(BN) = N$. The functor F is defined as follows: for morphisms $\widetilde{a} : a \to a'$ and $\widetilde{b} : b \to b'$, we have $F(\widetilde{a}, \widetilde{b}) = \phi(\widetilde{a} \otimes b') + \phi(a \otimes \widetilde{b})$. Consider the restriction of F to the full Picard subcategories that consist of objects whose cohomology classes belong to $H^p(A^\bullet)'$ and $H^{1-p}(B^\bullet)'$, respectively. By Lemma 2.35, we get a biextension P_ϕ of $(H^p(A^\bullet)', H^{1-p}(B^\bullet)')$ by N.

Remark 3.9 In what follows there will be given shifted pairings

$$p \in \mathrm{Hom}_{D-(\mathcal{A}b)}(A^\bullet \otimes_{\mathbb{Z}}^L B^\bullet, N[m]), \quad m \in \mathbb{Z};$$

this gives a biextension of $(H^p(A^\bullet), H^{-m+1-p}(B^\bullet))$ by N.

Example 3.10 Consider a unital DG-algebra $A^\bullet$ over $\mathbb{Z}$, an abelian group N, and an integer p. Suppose that $A^i = 0$ for $i > d$. Given a homomorphism $H^d(A^\bullet) \to N$, by $H^p(A^\bullet)' \subset H^p(A^\bullet)$ denote the annulator of the group $H^{d-p}(A^\bullet)$ with respect to the induced pairing $H^p(A^\bullet) \times H^{d-p}(A^\bullet) \to N$. Analogously, define the subgroup $H^{d+1-p}(A^\bullet)' \subset H^{d+1-p}(A^\bullet)$. By the above construction, this defines a biextension P of $(H^p(A^\bullet)', H^{d+1-p}(A^\bullet)')$ by N. The associated Weil pairing $\phi_l : H^p(A^\bullet)'_l \times H^{d+1-p}(A^\bullet)'_l \to N_l$ has the following interpretation as a Massey triple product. For the triple $a \in H^p(A^\bullet)'_l$, $l \in H^0(A^\bullet)$, and $b \in H^{d+1-p}(A^\bullet)'_l$, there is a well defined Massey triple product

$$m_3(a, l, b) \in H^d(A^\bullet)/(a \cdot H^{d-p}(A^\bullet) + H^{p-1}(A^\bullet) \cdot b).$$

By condition, the image $\overline{m}_3(a, l, b) \in N_l$ of $m_3(a, l, b)$ with respect to the map $H^d(A^\bullet) \to N$ is well defined. By Lemma 3.7, we have $\phi_l(a, b) = \overline{m}_3(a, l, b)$.

Example 3.11 Let Y be a Noetherian scheme of finite dimension d; then any sheaf of abelian groups $\mathcal{F}$ on Y has no non-trivial cohomology groups with numbers greater than d and there is a natural morphism $R\Gamma(Y, \mathcal{F}) \to H^d(Y, \mathcal{F})[-d]$ in $D^b(\mathcal{A}b)$. Let $\mathcal{F}, \mathcal{G}$ be sheaves of abelian groups on Y, N be an abelian group, and $p \geq 0$ be an integer. Given a homomorphism $H^d(Y, \mathcal{F} \otimes \mathcal{G}) \to N$, let $H^p(Y, \mathcal{F})'$ be the annulator of $H^{d-p}(Y, \mathcal{G})$ with respect to the natural pairing $H^p(Y, \mathcal{F}) \times H^{d-p}(Y, \mathcal{G}) \to N$. Analogously, we defined the subgroup

$$H^{d+1-p}(Y, \mathcal{G})' \subset H^{d+1-p}(Y, \mathcal{G}).$$

Then the multiplication morphism

$$m \in \mathrm{Hom}_{D^b(\mathcal{A}b)}(R\Gamma(Y, \mathcal{F}) \otimes_{\mathbb{Z}}^L R\Gamma(Y, \mathcal{G}), R\Gamma(Y, \mathcal{F} \otimes \mathcal{G}))$$

and the morphism $R\Gamma(Y, \mathcal{F} \otimes \mathcal{G}) \to N[-d]$ lead to a biextension of

$$(H^p(Y, \mathcal{F})', H^{d+1-p}(Y, \mathcal{G})')$$

by N.

3.3 Poincaré biextension. Consider an example to the construction described above. Let $\mathcal{H}$ be the abelian category of integral mixed Hodge structures. For an object H in $\mathcal{H}$, by $H_{\mathbb{Z}}$ denote the underlying finitely generated abelian group, by $W_*H_{\mathbb{Q}}$ denote the increasing weight filtration, and by $F^*H_{\mathbb{C}}$ denote the decreasing Hodge filtration, where $H_{\mathbb{C}} = H_{\mathbb{Z}} \otimes_{\mathbb{Z}} \mathbb{C}$. Recall that if $H_{\mathbb{Z}}$ is torsion-free, then the complex $R\mathrm{Hom}_{\mathcal{H}}(\mathbb{Z}(0), H)$ is canonically quasiisomorphic to the complex

$$B(H)^{\bullet} : 0 \to W_0 H_{\mathbb{Z}} \oplus (F^0 \cap W_0) H_{\mathbb{C}} \to W_0 H_{\mathbb{C}} \to 0,$$

where the differential is given by $d(\gamma, \eta) = \gamma - \eta$ for $\gamma \in W_0 H_{\mathbb{Z}}$, $\eta \in (F^0 \cap W_0)H_{\mathbb{C}}$, and $W_0 H_{\mathbb{Z}} = H_{\mathbb{Z}} \cap W_0 H_{\mathbb{Q}}$ (see [2]). Moreover, for any two integral mixed Hodge structures H and H' with $H_{\mathbb{Z}}$ and $H'_{\mathbb{Z}}$ torsion-free, the natural pairing $R\mathrm{Hom}_{\mathcal{H}}(\mathbb{Z}(0), H) \otimes_{\mathbb{Z}}^{L} R\mathrm{Hom}_{\mathcal{H}}(\mathbb{Z}(0), H') \to R\mathrm{Hom}_{\mathcal{H}}(\mathbb{Z}(0), H \otimes H')$ corresponds to the morphism

$$m : B(H)^{\bullet} \otimes_{\mathbb{Z}} B(H')^{\bullet} \to B(H \otimes H')^{\bullet}$$

given by the formula $m(\varphi \otimes \eta') = \varphi \otimes \eta' \in (H \otimes H')_{\mathbb{C}}$, $m(\gamma \otimes \varphi') = \gamma \otimes \varphi' \in (H \otimes H')_{\mathbb{C}}$, $m(\gamma \otimes \gamma') = \gamma \otimes \gamma' \in W_0(H \otimes H')_{\mathbb{Z}}$, $m(\eta \otimes \eta') = \eta \otimes \eta' \in F^0(H \otimes H')_{\mathbb{C}}$, and $m = 0$ otherwise, where $\gamma \in W_0 H_{\mathbb{Z}}$, $\gamma' \in W_0 H'_{\mathbb{Z}}$, $\eta \in F^0 H_{\mathbb{C}}$, $\eta' \in F^0 H'_{\mathbb{C}}$, $\varphi \in H_{\mathbb{C}}$, $\varphi' \in H'_{\mathbb{C}}$.

Let H be a pure Hodge structure of weight -1, and let $H^{\vee} = \mathrm{Hom}(H, \mathbb{Z}(1))$ be the corresponding internal Hom in $\mathcal{H}$. By $\langle \cdot, \cdot \rangle : H_{\mathbb{C}} \otimes_{\mathbb{C}} H_{\mathbb{C}}^{\vee} \to \mathbb{C}^*$ denote the composition of the natural map $H_{\mathbb{C}} \otimes_{\mathbb{C}} H_{\mathbb{C}}^{\vee} \to \mathbb{C}$ and the exponential map $\mathbb{C} \to \mathbb{C}/\mathbb{Z}(1) = \mathbb{C}^*$. We put $J(H) = \mathrm{Ext}_{\mathcal{H}}^1(\mathbb{Z}(0), H) = H_{\mathbb{C}}/(H_{\mathbb{Z}} + F^0 H_{\mathbb{C}})$. This group has a natural structure of a compact complex torus; we call it the *Jacobian of H*. Note that the Jacobian $J(H^{\vee})$ is naturally isomorphic to the dual compact complex torus $J(H)^{\vee}$.

By Section 3.2, the natural pairing

$$\begin{aligned} R\mathrm{Hom}(\mathbb{Z}(0), H) \quad &\otimes_{\mathbb{Z}}^{L} \quad R\mathrm{Hom}(\mathbb{Z}(0), H^{\vee}) \to R\mathrm{Hom}(\mathbb{Z}(0), H \otimes H^{\vee}) \\ &\to \quad R\mathrm{Hom}(\mathbb{Z}(0), \mathbb{Z}(1)) \to \mathbb{C}^*[-1] \end{aligned}$$

defines a biextension P of $(J(H), J(H^{\vee}))$ by $\mathbb{C}^*$. The given above explicit description of the pairing m shows that the biextension P is induced by the bilinear map $\psi : S \to \mathbb{C}^*$ (see Section 3.1), where

$$S = ((H_{\mathbb{Z}} + F^0 H_{\mathbb{C}}) \times H_{\mathbb{C}}^{\vee}) \cup (H_{\mathbb{C}} \times (H_{\mathbb{Z}}^{\vee} + F^0 H_{\mathbb{C}}^{\vee}))$$

is a bisubgroup in $H_{\mathbb{C}} \times H_{\mathbb{C}}^{\vee}$, $\psi(\gamma + \eta, \varphi^{\vee}) = \langle \gamma, \varphi^{\vee} \rangle$, and $\psi(\varphi, \gamma^{\vee} + \eta^{\vee}) = \langle \varphi, \eta^{\vee} \rangle$ for all $\gamma \in H_{\mathbb{Z}}$, $\gamma^{\vee} \in H_{\mathbb{Z}}^{\vee}$, $\eta \in F^0 H_{\mathbb{C}}$, $\eta^{\vee} \in F^0 H_{\mathbb{C}}^{\vee}$, $\varphi \in H_{\mathbb{C}}$, $\varphi^{\vee} \in H_{\mathbb{C}}^{\vee}$.

Remark 3.12 The given above explicit construction of the biextension P shows that it coincides with the biextension constructed in [18], Section 3.2. In particular, by Lemma 3.2.5 from op.cit., P is canonically isomorphic to the Poincaré line bundle over $J(H) \times J(H)^{\vee}$.

By Remark 3.12, it makes sense to call P the *Poincaré biextension*.

Remark 3.13 It follows from [18], Section 3.2 that the fiber of the biextension P over a pair $(e, f) \in J(H) \times J(H^{\vee}) = \mathrm{Ext}_{\mathcal{H}}^1(\mathbb{Z}(0), H) \times \mathrm{Ext}_{\mathcal{H}}^1(H, \mathbb{Z}(1))$ is canonically bijective with the set of isomorphism classes of integral mixed Hodge structures V whose weight graded quotients are identified with $\mathbb{Z}(0)$, H, $\mathbb{Z}(1)$ and such that $[V/W_{-2}V] = e \in \mathrm{Ext}_{\mathcal{H}}^1(\mathbb{Z}(0), H)$, $[W_{-1}V] = f \in \mathrm{Ext}_{\mathcal{H}}^1(H, \mathbb{Z}(1))$.

Remark 3.14 There is a canonical trivialization of the biextension $\log |P|$ of $(J(H), J(H^{\vee}))$ by $\mathbb{R}$; this trivialization is given by the bilinear map $H_{\mathbb{C}} \times H_{\mathbb{C}}^{\vee} \to \mathbb{R}$, $(\varphi, \varphi^{\vee}) \mapsto \log |\langle r, \varphi^{\vee} \rangle|$, where $\varphi = r + \eta$, $r \in H_{\mathbb{R}}$, $\eta \in F^0 H$.

4 Biextensions over Chow groups

4.1 Explicit construction. The construction described below was first given in [7]. We use notions and notations from Section 2.2. Let X be a smooth projective variety of dimension d over k. For an integer $p \geq 0$, by $Z^p(X)'$ denote the subgroup in $Z^p(X)$ that consists of cycles Z such that for any K_1-chain $\{f_\eta\} \in G_{|Z|}^{d-p}(X, 1)$ with $\mathrm{div}(\{f_\eta\}) = 0$, we have $\prod_\eta f_\eta(\overline{\eta} \cap Z) = 1$. Let $CH^p(X)'$ be the image of $Z^p(X)'$ in $CH^p(X)$; by Lemma 2.10, we have $CH^p(X)_{\mathrm{hom}} \subset CH^p(X)'$.

Remark 4.1 If the group k^* has an element of infinite order, then it is easy to see that any cycle $Z \in Z^p(X)'$ is numerically trivial.

Consider integers $p, q \geq 0$ such that $p + q = d + 1$. Let $T \subset Z^p(X)' \times Z^q(X)'$ be the bisubgroup that consists of pairs (Z, W) such that $|Z| \cap |W| = \emptyset$. By the classical moving lemma, we see that the map $T \to CH^p(X)' \times CH^q(X)'$ is surjective. Let $S \subset T$ be the corresponding bisubgroup as defined in Section 3.1. Define a bilinear map $\psi : S \to k^*$ as follows. Suppose that Z is rationally trivial; then by Lemma 2.6, there exists an element $\{f_\eta\} \in G_{|W|}^{p-1}(X, 1)$ such that $\mathrm{div}(\{f_\eta\}) = Z$. We put $\psi(Z, W) = \prod_\eta f_\eta(\overline{\eta} \cap W)$. By condition, this element from k^* does not depend on the choice of $\{f_\eta\}$. Similarly, we put $\psi(Z, W) = \prod_\xi g_\xi(Z \cap \overline{\xi})$ if $W = \mathrm{div}(\{g_\xi\})$, $\{g_\xi\} \in G_{|Z|}^{q-1}(X, 1)$. By Lemma 2.8, $\psi(\mathrm{div}(\{f_\eta\}), \mathrm{div}(\{g_\xi\}))$ is well defined. The application of the construction from Section 3.1 yields a biextension P_E of $(CH^p(X)', CH^q(X)')$ by k^*.

The following interpretation of the biextension P_E from Section 4.1 in terms of pairings between higher Chow complexes was suggested to the author by S. Bloch.

We use notations and notions from Section 2.1. As above, let X be a smooth projective variety of dimension d over k and let the integers $p, q \geq 0$ be such that $p + q = d + 1$. There is a push-forward morphism of complexes $\pi_* : Z^{d+1}(X, \bullet) \to Z^1(\mathrm{Spec}(k), \bullet)$, where $\pi : X \to \mathrm{Spec}(k)$ is the structure map. Further, there is a morphism of complexes $Z^1(\mathrm{Spec}(k), \bullet) \to k^*[-1]$ (which is actually a quasiisomorphism). Taking the composition of these morphisms with the multiplication morphism

$$m \in \mathrm{Hom}_{D^-(\mathcal{A}b)}(Z^p(X, \bullet) \otimes_{\mathbb{Z}}^L Z^q(X, \bullet), Z^{d+1}(X, \bullet)),$$

we get an element

$$\phi \in \mathrm{Hom}_{D^-(\mathcal{A}b)}(Z^p(X, \bullet) \otimes_{\mathbb{Z}}^L Z^q(X, \bullet), k^*[-1]).$$

Note that by Lemma 2.1, in notations from Section 3.2, we have $H_0(Z^p(X, \bullet))' = CH^p(X)'$ and $H_0(Z^q(X, \bullet))' = CH^q(X)'$. Hence the construction from Section 3.2 gives a biextension P_{HC} of $(CH^p(X)', CH^q(X)')$ by k^*.

Proposition 4.2 *The biextension P_{HC} is canonically isomorphic to the biextension P_E.*

Proof Recall that the multiplication morphism m is given by the composition

$$Z^p(X, \bullet) \otimes Z^q(X, \bullet) \overset{ext}{\to} Z^{d+1}(X \times X, \bullet) \hookleftarrow Z_D^{d+1}(X \times X, \bullet) \overset{D^*}{\to} Z^{d+1}(X, \bullet),$$

where $D \subset X \times X$ is the diagonal. Let $A_0 \subset Z^p(X) \otimes Z^q(X)$ be the subgroup generated by elements $Z \otimes W$ such that $|Z| \cap |W|$, and let

$$A_1 \subset Z^p(X, 1) \otimes Z^q(X, 0) \oplus Z^p(X, 0) \otimes Z^q(X, 1) = (Z^p(X, \bullet) \otimes Z^q(X, \bullet))_1$$

be the subgroup generated by elements $(\alpha \otimes W, Z \otimes \beta)$ such that $\alpha \in Z^p_{|W|}(X, 1)$, $\beta \in Z^q_{|Z|}(X, 1)$; we have $ext(A_i) \subset Z^{d+1}_D(X \times X, i)$ for $i = 0, 1$. Since all terms of the complex $Z^p(X, \bullet) \otimes Z^q(Z, \bullet)$ are free $\mathbb{Z}$-modules and the groups A_i are direct summands in the groups $(Z^p(X, \bullet) \otimes Z^q(X, \bullet))_i$ for $i = 0, 1$, there exists a true morphism of complexes $ext' : Z^p(X, \bullet) \otimes Z^q(X, \bullet) \to Z^{p+q}_D(X \times X, \bullet)$ such that ext' is equivalent to ext in the derived category $D^-(\mathcal{A}b)$ and ext' coincides with ext on A_i, $i = 0, 1$.

On the other hand, for a cycle $Z \in Z^p(X)$ and a K_1-chain $\{g_\xi\} \in G^{q-1}_{|Z|}(X, 1)$, we have $\beta = \Gamma(\{g_\xi\}) \in Z^q_{|Z|}(X, 1)$ and $(\pi_* \circ D^* \circ ext)(Z \otimes \beta)) = \prod_\xi g_\xi(Z \cap \bar{\xi})$ (see Section 2.2).

Therefore the needed statement follows from Remark 3.2 applied to the bisubgroup $T \subset Z^p(X) \times Z^q(X)$ together with the bilinear map ψ from Section 4.1 and the bisubgroup in $Z^p(X) \times Z^q(X)$ together the bilinear map induced by the morphism of complexes $\phi = \pi_* \circ D^* \circ ext'$ as described in Section 3.2. $\square$

4.2 Intermediate Jacobians construction.

The main result of this section was partially proved in [22] by using the functorial properties of higher Chow groups and the regulator map to Deligne cohomology. We give a more elementary proof that uses only the basic Hodge theory.

We use notions and notations from Sections 3.3 and 2.3. Let X be a complex smooth projective variety of dimension d. Suppose that p and q are natural numbers such that $p + q = d + 1$. Multiplication by $(2\pi i)^p$ induces the isomorphism $J^{2p-1}(X) \to J(H^{2p-1}(X)(p))$. Since $H^{2p-1}(X)(p)^\vee = H^{2q-1}(X)(q)$, by Section 2.3, there is a Poincaré biextension P_{IJ} of $(J^{2p-1}(X), J^{2q-1}(X))$ by $\mathbb{C}^*$.

Proposition 4.3 *The pull-back $AJ^*(P_{IJ})$ is canonically isomorphic to the restriction of the biextension P_E constructed in section 4.1 to subgroups $CH^p(X)_{\mathrm{hom}} \subset CH^p(X)'$.*

Proof 1 First, let us give a short proof that uses functorial properties of the regulator map from the higher Chow complex to the Deligne complex.

Let $\mathbb{Z}_D(p)^\bullet = \mathrm{cone}(\mathbb{Z}(p) \oplus F^p\Omega^\bullet_X \to \Omega^\bullet_X)[-1]$ be the Deligne complex, where all sheaves are considered in the analytic topology. There are a multiplication morphism

$$R\Gamma_{an}(X(\mathbb{C}), \mathbb{Z}_D(p)^\bullet) \otimes^L_{\mathbb{Z}} R\Gamma_{an}(X(\mathbb{C}), \mathbb{Z}_D(q)^\bullet) \to R\Gamma_{an}(X(\mathbb{C}), \mathbb{Z}_D(p+q)^\bullet)$$

and a push-forward morphism

$$R\Gamma_{an}(X(\mathbb{C}), \mathbb{Z}_D(d+1)^\bullet) \to \mathbb{C}^*[-2d-1]$$

in $D^b(\mathcal{A}b)$. The application of the construction from Section 3.2 to the arising pairing of complexes

$$R\Gamma_{an}(X(\mathbb{C}), \mathbb{Z}_D(p)^\bullet) \otimes^L_{\mathbb{Z}} R\Gamma_{an}(X(\mathbb{C}), \mathbb{Z}_D(q)^\bullet) \to \mathbb{C}^*[-2d-1]$$

for $p + q = d + 1$ yields a biextension of $(J^{2p-1}(X), J^{2q-1}(X))$ by $\mathbb{C}^*$, since $J^{2p-1}(X) \subset H^{2p}_{an}(X(\mathbb{C}), \mathbb{Z}_D(p)^\bullet)'$ and $J^{2q-1}(X) \subset H^{2q}_{an}(X(\mathbb{C}), \mathbb{Z}_D(q)^\bullet)'$.

Furthermore, there is a canonical morphism

$$\epsilon : \mathrm{RHom}_{D^b(\mathcal{H})}(\mathbb{Z}(0), RH^\bullet(X)(p)) \to R\Gamma_{an}(X(\mathbb{C}), \mathbb{Z}_D(p)^\bullet),$$

where $RH^\bullet(X)$ is the "derived" Hodge structure of X. The morphism ϵ induces isomorphisms on cohomology groups in degrees less or equal to $2p$ (this follows from the explicit formula for the complex $B(H)^\bullet$ given in Section 3.3). Moreover, the morphism ϵ commutes with the multiplication and the push-forward. Note that the multiplication morphism on the left hand side is defined as the composition

$$\mathrm{RHom}_{D^b(\mathcal{H})}(\mathbb{Z}(0), RH^\bullet(X)(p)) \otimes^L_{\mathbb{Z}} \mathrm{RHom}_{D^b(\mathcal{H})}(\mathbb{Z}(0), RH^\bullet(X)(q))$$
$$\to \mathrm{RHom}_{D^b(\mathcal{H})}(\mathbb{Z}(0), RH^\bullet(X) \otimes RH^\bullet(X)(p+q))$$
$$\to \mathrm{RHom}_{D^b(\mathcal{H})}(\mathbb{Z}(0), RH^\bullet(X)(p+q))$$

and the push-forward is defined as the composition

$$\mathrm{RHom}_{D^b(\mathcal{H})}(\mathbb{Z}(0), RH^\bullet(X)(d+1))$$
$$\to \quad \mathrm{RHom}_{D^b(\mathcal{H})}(\mathbb{Z}(0), H^{2d}(X)(d+1)[-2d])$$
$$= \quad \mathrm{RHom}_{D^b(\mathcal{H})}(\mathbb{Z}(0), \mathbb{Z}(1))[-2d] = \mathbb{C}^*[-2d-1].$$

Therefore, the biextension of $(J^{2p-1}(X), J^{2q-1}(X))$ by $\mathbb{C}^*$ defined via the pairing of complexes on the left hand side is canonically isomorphic to the previous one.

The multiplicativity of the spectral sequence

$$\mathrm{Ext}^i_{\mathcal{H}}(\mathbb{Z}(0), H^j(X)(p)) \Rightarrow \mathrm{Hom}^{i+j}_{D^b(\mathcal{H})}(\mathbb{Z}(0), RH^\bullet(X)(p))$$

shows that the last biextension is canonically isomorphic to the Poincaré biextension defined in Section 3.3.

Now consider the cohomological type complex $Z^p(X)^\bullet = Z^p(X, 2p - \bullet)$ built out of the higher Chow complex (see Section 2.1). In [20] it is constructed an explicit morphism

$$\rho : Z^p(X)^\bullet \to R\Gamma_{an}(X(\mathbb{C}), \mathbb{Z}_D(p)^\bullet)$$

of objects in the derived category $D^-(\mathcal{A}b)$, which is actually the regulator map. It follows from op.cit. that the regulator map ρ commutes with the corresponding multiplication morphisms for higher Chow complexes and Deligne complexes. By Proposition 4.2, this implies immediately the needed result. $\square$

Question 4.4 Does there exist a morphism of DG-algebras $\rho : A^\bullet_M \to A^\bullet_D$ such that a DG-algebra $A^\bullet_M$ is quasiisomorphic to the DG-algebra $\bigoplus_{p \geq 0} R\Gamma_{Zar}(X, \mathbb{Z}(p)^\bullet)$, where $\mathbb{Z}(p)^\bullet$ is the Suslin complex, and a DG-algebra $A^\bullet_D$ is quasiisomorphic to the DG-algebra $\bigoplus_{p \geq 0} R\Gamma_{an}(X, \mathbb{Z}_D(p)^\bullet)$? The identification of higher Chow groups with motivic cohomology $CH^p(X, n) = H^{2p-n}_M(X, \mathbb{Z}(p))$ shows that a positive answer to the question would be implied by the existence of a DG-realization functor from the DG-category of Voevodsky motives (see [4]) to the DG-category associated with integral mixed Hodge structures. It was pointed out to the author by V. Vologodsky that the existence of this DG-realization functor follows immediately from op.cit. Does there exist an explicit construction of the DG-algebras $A^\bullet_M$, $A^\bullet_D$, and the morphism ρ?

Proof 2 Let $\widetilde{Z}^p(X)_{\mathrm{hom}}$ be the group that consists of all triples $\widetilde{Z} = (Z, \Gamma_Z, \eta_Z)$, where $Z \in Z^p(X)_{\mathrm{hom}}$, Γ_Z is a differentiable singular chain on X such that $\partial \Gamma_Z = Z$, and $\eta_Z \in F^p_{\log} A^{2p-1}_{X \setminus |Z|}$ is such that $\partial_Z([(2\pi i)^p \eta_Z]) = [Z]$. It follows from Section 2.3 that the natural map $\widetilde{Z}^p(X)_{\mathrm{hom}} \to Z^p(X)_{\mathrm{hom}}$ is surjective. There is a homomorphism $\widetilde{Z}^p(X) \to H^{2p-1}(X, \mathbb{C})$ given by the formula $\widetilde{Z} \mapsto PD[\Gamma_Z] - [\eta_Z]$; its composition with the natural map $H^{2p-1}(X, \mathbb{C}) \to J^{2p-1}(X)$ is equal to the Abel–Jacobi map $\widetilde{Z}^p(X)_{\mathrm{hom}} \to J^{2p-1}(X)$ given by the formula $\widetilde{Z} \mapsto AJ(Z)$ (see section 2.3).

Let T be a bisubgroup in $\widetilde{Z}^p(X)_{\hom} \times \widetilde{Z}^q(X)_{\hom}$ that consists of all pairs of triples $(\widetilde{Z}, \widetilde{W}) = ((Z, \Gamma_Z, \eta_Z), (W, \Gamma_W, \eta_W))$ such that $|Z| \cap |W| = \emptyset$, Γ_Z does not meet $|W|$, and $|Z|$ does not meet Γ_W. Note that for any pair $(\widetilde{Z}, \widetilde{W}) \in T$, there are well defined classes $PD[\Gamma_Z], [\eta_Z] \in H^{2p-1}(X \backslash |Z|, |W|; \mathbb{C})$ and $PD[\Gamma_W], [\eta_W] \in H^{2q-1}(X \backslash |W|, |Z|; \mathbb{C})$. As before, denote by

$$(\cdot, \cdot) : H^*(X \backslash |Z|, |W|; \mathbb{C}) \times H^{2d-*}(X \backslash |W|, |X|; \mathbb{C}) \to \mathbb{C}$$

the natural pairing.

Let $S \subset T$ be a bisubgroup that consists of all pairs of triples $(\widetilde{Z}, \widetilde{W})$ such that Z or W is rationally trivial. Let $\psi_1 : S \to \mathbb{C}^*$ be the pull-back of the bilinear map constructed in Section 4.1 via the natural map $\widetilde{Z}^p(X)_{\hom} \times \widetilde{Z}^q(X)_{\hom} \to Z^p(X)_{\hom} \times Z^q(X)_{\hom}$. Let $\psi_2 : S \to \mathbb{C}^*$ be the pull-back of the bilinear constructed in Section 3.3 via the defined above map $\widetilde{Z}^p(X)_{\hom} \times \widetilde{Z}^q(X)_{\hom} \to H^{2p-1}(X, \mathbb{C}) \times H^{2q-1}(X, \mathbb{C})$. The construction from Section 3.1 applied to the maps ψ_1 and ψ_2 gives the biextensions P_E and P_{IJ}, respectively. Define a bilinear map $\phi : T \to \mathbb{C}^*$ by the formula $\phi(\widetilde{Z}, \widetilde{W}) = \exp(2\pi i \int_{\Gamma_Z} \eta_W)$. By Remark 3.3, it is enough to show that $\psi_1 = \phi|_S \cdot \psi_2$.

Consider a pair $(\widetilde{Z}, \widetilde{W}) \in S$. First, suppose that Z is rationally trivial. By linearity and Lemma 2.6, we may assume that $Z = \operatorname{div}(f)$, where $f \in \mathbb{C}^*(Y)$ and $Y \subset X$ is an irreducible subvariety of codimension $p - 1$ such that Y meets $|W|$ properly. Let $\widetilde{Y}$ be the closure of the graph of the rational function $f : Y \dashrightarrow \mathbb{P}^1$ and let $p : \widetilde{Y} \to X$ be the natural map. Let Γ be a smooth generic path on $\mathbb{P}^1$ such that $\partial \Gamma = \{0\} - \{\infty\}$ and Γ does not intersect with the finite set $f(p^{-1}(W))$; there is a cohomological class $PD[\Gamma] \in H^1(\mathbb{P}^1 \backslash \{0, \infty\}, f(p^{-1}(|W|)); \mathbb{Z})$. We put $\alpha_Z = (2\pi i)^{-p} [\widetilde{Y}]^* (2\pi i PD[\gamma]) \in H^{2p-1}(X \backslash |Z|, |W|; \mathbb{Z})$. By Lemma 2.17 with $X_1 = \mathbb{P}^1$, $X_2 = X$, $C = \widetilde{Y}$, $Z_1 = \{0, \infty\}$, and $W_2 = |W|$, we get $PD[\Gamma_Z] - \alpha_Z \in H^{2p-1}(X, |W|; \mathbb{Z})$ and $\alpha_Z - [\eta_Z] \in F^p H^{2p-1}(X, \mathbb{C})$ (note the element $\alpha_Z - [\eta_Z] \in H^{2p-1}(X, |W|; \mathbb{C})$ does not necessary belong to the subgroup $F^p H^{2p-1}(X, |W|; \mathbb{C})$). Since $PD[\Gamma_Z] - [\eta_Z] = (PD[\Gamma_Z] - \alpha_Z) + (\alpha_Z - [\eta_Z])$, it follows from the explicit construction given in Section 3.3 that

$$\begin{aligned} \psi_2(\widetilde{Z}, \widetilde{W}) &= \exp(2\pi i (PD[\Gamma_Z] - \alpha_Z, PD[\Gamma_W] - [\eta_W])) \\ &= \exp(-2\pi i \int_{\Gamma_Z} \eta_W + 2\pi i (\alpha_Z, [\eta_W])). \end{aligned}$$

In addition, combining Lemma 4.5(i) and Lemma 2.17 with $X_1 = X$, $X_2 = \mathbb{P}^1$, $C = \widetilde{Y}$, $Z_1 = |W|$, and $W_2 = \{0, \infty\}$, we see that $\exp(2\pi i (\alpha_Z, [\eta_W])) = f(Y \cap W) = \psi_1(\widetilde{Z}, \widetilde{W})$.

Now suppose that W is rationally trivial. As before, by linearity and Lemma 2.6, we may assume that $W = \operatorname{div}(g)$, where $W \in \mathbb{C}^*(V)$ and $V \subset X$ is an irreducible subvariety of codimension $p - 1$ such that $|Z|$ meets V properly. Let $\widetilde{V}$ be the closure of the graph of the rational function $g : V \dashrightarrow \mathbb{P}^1$ and let $p : \widetilde{V} \to X$ be the natural map. Note that the differential form $\frac{dz}{z}$ defines a cohomological class $[\frac{dz}{z}] \in F^1 H^1(\mathbb{P}^1 \backslash \{0, \infty\}, g(p^{-1}(|Z|)); \mathbb{C})$. We put $\alpha_W = (2\pi i)^{-q} [\widetilde{V}]^* [\frac{dz}{z}] \in F^q H^{2q-1}(X \backslash |W|, |Z|; \mathbb{C})$. By Lemma 2.17 with $X_1 = \mathbb{P}^1$, $X_2 = X$, $C = \widetilde{V}$, $Z_1 = \{0, \infty\}$, and $W_2 = |Z|$, we get $\alpha_W - [\eta_W] \in F^q H^{2q-1}(X, |Z|; \mathbb{Z})$ and $PD[\Gamma_W] - \alpha_W \in H^{2q-1}(X, \mathbb{Z})$ (note that the element $PD[\Gamma_W] - \alpha_W \in H^{2q-1}(X, |Z|; \mathbb{C})$ does not necessary belong to the subgroup $H^{2q-1}(X, |Z|; \mathbb{Z})$). Since $PD[\Gamma_W] - [\eta_W] =$

$(PD[\Gamma_W] - \alpha_W) + (\alpha_W - [\eta_W])$, it follows from the explicit construction given in Section 3.3 that

$$\begin{aligned}\psi_2(\widetilde{Z}, \widetilde{W}) &= \exp(2\pi i(PD[\Gamma_Z] - [\eta_Z], \alpha_W - [\eta_W])) \\ &= \exp(2\pi i(PD[\Gamma_Z], \alpha_W) - 2\pi i \int_{\Gamma_Z} \eta_W).\end{aligned}$$

In addition, combining Lemma 4.5(ii) and Lemma 2.17 with $X_1 = X$, $X_2 = \mathbb{P}^1$, $C = \widetilde{Y}$, $Z_1 = |Z|$, and $W_2 = \{0, \infty\}$, we see that $\exp(2\pi i(PD[\Gamma_Z], \alpha_W)) = g(Z \cap Y) = \psi_1(\widetilde{Z}, \widetilde{W})$. This concludes the proof. $\qquad\qquad\square$

During the proof of Proposition 4.3 we have used the following simple fact.

Lemma 4.5

(i) *Let η be a meromorphic 1-form on $\mathbb{P}^1$ with poles of order at most one (i.e., η is a differential of the third kind), Γ be a smooth generic path on $\mathbb{P}^1$ such that $\partial\Gamma = \{0\} - \{\infty\}$ and Γ does not contain any pole of η; if $\mathrm{res}(2\pi i \eta) = \sum_i n_i\{z_i\}$ for some integers n_i, then we have $\exp(2\pi i \int_\Gamma \eta) = \prod_i z_i^{n_i}$.*

(ii) *Let z be a coordinate on $\mathbb{P}^1$, Γ be a differentiable singular 1-chain on $\mathbb{P}^1$ that does not intersect with the set $\{0, \infty\}$; if $\partial\Gamma = \sum_i n_i\{z_i\}$, then we have $\exp(\int_\Gamma \frac{dz}{z}) = \prod_i z_i^{n_i}$.*

Proof (i) Let $\log z$ be a branch of logarithm on $\mathbb{P}^1 \backslash \Gamma$, T_ϵ be a tubular neighborhood of Γ with radius ϵ, and let $B_{i,\epsilon}$ be disks around $\{z_i\}$ with radius ϵ; we put $X_\epsilon = \mathbb{P}^1 \backslash (\cup_i B_{i,\epsilon} \cup T_\epsilon)$. Then we have $0 = \lim_{\epsilon \to 0} \int_{X_\epsilon} d((\log z)\eta) = \sum_i n_i \log z_i - 2\pi i \int_\Gamma \eta$; this concludes the proof.

(ii) One may assume that Γ does not contain any loop around $\{0\}$ or $\{\infty\}$; hence there exists a smooth path γ on $\mathbb{P}^1$ such that $\partial\gamma = \{0\} - \{\infty\}$ and γ does not intersect Γ. Let $\log z$ be a branch of logarithm on $\mathbb{P}^1 \backslash \gamma$. Then we have $\int_\Gamma \frac{dz}{z} = \int_\Gamma d\log z = \sum_i n_i \log z_i$; this concludes the proof. $\qquad\square$

Remark 4.6 For a cycle $Z \in Z^p(X)_{\mathrm{hom}}$, consider an exact sequence of integral mixed Hodge structures

$$0 \to H^{2p-1}(X)(p) \to H^{2p-1}(X \backslash |Z|)(p) \to H_{2d-2p}(|Z|) \to H^{2p}(X)(p).$$

Its restriction to $[Z]_{\mathbb{Z}} = \mathbb{Z}(0) \subset H_{2d-2p}(|Z|)$ defines a short exact sequence

$$0 \to H^{2p-1}(X)(p) \to E_Z \to \mathbb{Z}(0) \to 0.$$

Analogously, for a cycle $W \in Z^q(X)_{\mathrm{hom}}$, we get an exact sequence

$$0 \to H^{2q-1}(X)(q) \to E_W \to \mathbb{Z}(0) \to 0$$

and a dual exact sequence

$$0 \to \mathbb{Z}(1) \to E_W^\vee \to H^{2p-1}(X)(p) \to 0.$$

By Remark 3.13, we see that the fiber $P_{IJ}|_{(AJ(Z), AJ(W))}$ is canonically bijective with the set of isomorphism classes of all integral mixed Hodge structures V whose weight graded quotients are identified with $\mathbb{Z}(0)$, $H^{2p-1}(X)(p)$, $\mathbb{Z}(1)$ and such that

$$\begin{aligned}[V/W_{-2}V] &= [E_Z] \in \mathrm{Ext}^1_{\mathcal{H}}(\mathbb{Z}(0), H^{2p-1}(X)(p)), \\ [W_{-1}V] &= [E_W^\vee] \in \mathrm{Ext}^1_{\mathcal{H}}(H^{2p-1}(X)(p), \mathbb{Z}(1)).\end{aligned}$$

Remark 4.7 Under the assumptions of Corollary 4.32 suppose that $|Z| \cap |W| = \emptyset$. Then there are two canonical trivializations of the fiber of the biextension $\log|P_E|$ over (Z, W): the first one follows from the construction of P_E and the second one follows from Proposition 4.3 and Remark 3.14. Let $h(Z, W) \in \mathbb{R}$ be the quotient of these two trivializations. Consider closed forms $\eta_Z^r \in F_{\log}^p A_{X \backslash |Z|}^{2p-1}$ and $\eta_W^r \in F_{\log}^q A_{X \backslash |W|}^{2q-1}$ such that $\partial_Z((2\pi i)^p \eta_Z^r) = [Z]$, $\partial_W((2\pi i)^q \eta_W^r) = [W]$, η_Z^r has real periods on $X \backslash |Z|$, and η_W^r has real periods on $X \backslash |W|$ (it is easy to see that such forms always exist). In notations from the proof of Proposition 4.3 the isomorphism $P_E \to P_{IJ}$ is given by the multiplication by $\exp(2\pi i \int_{\Gamma_Z} \eta_W)$. On the other hand, the trivialization from Remark 3.14 is given by the section $\mathrm{Re}(2\pi i(PD[\Gamma_Z] - [\eta_Z^r], PD[\Gamma_W] - [\eta_W])) = \mathrm{Re}(2\pi i \int_{\Gamma_Z}(\eta_W^r - \eta_W))$. Therefore, $h(Z, W) = \mathrm{Re}(2\pi i \int_{\Gamma_Z} \eta_W^r)$. This number is called the *archimedean height pairing* of Z and W.

Finally, using Lemma 4.5 we give an explicit analytic proof of Lemma 2.10 for complex varieties.

Lemma 4.8 *Let X be a complex smooth projective variety of dimension d, $W \in Z^q(X)_{\mathrm{hom}}$ be a homologically trivial cycle, $\{Y_i\}$ be a finite collection of irreducible subvarieties of codimension $d-q$ in X, and let $f_i \in \mathbb{C}(Y_i)^*$ be a collection of rational functions such that $\sum_i \mathrm{div}(f_i) = 0$ and for any i, the support $|W|$ meets Y_i properly and $|W| \cap |\mathrm{div}(f_i)| = \emptyset$. Then we have $f(Y \cap W) = \prod_i f_i(Y_i \cap W) = 1$; here for each i, we put $f_i(Y_i \cap W) = \prod_{x \in Y_i \cap |W|} f_i^{(Y_i, W; x)}(x)$, where $(Y_i, W; x)$ is the intersection index of Y_i and W at a point $x \in Y_i \cap |W|$.*

Proof Since W is homologically trivial, by Section 2.3, there exists a closed differential form $\eta \in F_{\log}^q A_{X \backslash |W|}^{2q-1}$ such that $\partial_W([(2\pi i)^q \eta]) = [W]$, i.e., for a small $(2q-1)$-sphere S_j around an irreducible component W_j of W, we have $\int_{S_j} \eta = m_j$, where $W = \sum m_j W_j$.

For each i, let $\widetilde{Y}_i$ be the closure of the graph of the rational function $f_i : Y_i \dashrightarrow \mathbb{P}^1$ and let $p_i : \widetilde{Y}_i \to X$ be the natural map. Let γ be a smooth generic path on $\mathbb{P}^1$ such that $\partial\gamma = \{0\} - \{\infty\}$ and γ does not intersect with the finite set $\Sigma = \cup_i p_i(f_i^{-1}(|W|))$. There is a cohomological class $PD[\gamma] \in H^1(\mathbb{P}^1 \backslash \{0, \infty\}, \Sigma; \mathbb{Z})$; we put $\alpha_i = (2\pi i)^{-(d-q+1)}[\widetilde{Y}_i]^*(2\pi i PD[\gamma]) \in H^{2d-2q+1}(X \backslash |\mathrm{div}(f_i)|, |W|; \mathbb{Z})$. Let Γ_i be a differentiable representative for $PD(\alpha_i) \in H_{2q-1}(X \backslash |W|, |\mathrm{div}(f_i)|; \mathbb{Z})$ (see Remark 2.16). In particular, Γ_i is a differentiable singular $(2q-1)$-chain on $X \backslash |W|$ with boundary $\mathrm{div}(f_i)$.

By Lemma 4.5 and Lemma 2.17 with $X_1 = \mathbb{P}^1$, $X_2 = X$, $C = \widetilde{Y}_i$, $Z_1 = \{0, \infty\}$, and $W_2 = |W|$, we get $f_i(Y_i \cap W) = \exp(2\pi i(\alpha_i, \eta)) = \exp(2\pi i \int_{\Gamma_i} \eta)$. Therefore, $f(Y \cap W) = \exp(2\pi i \int_\Gamma \eta)$, where $\Gamma = \sum_i \Gamma_i$. Since $\sum_i \mathrm{div}(f_i) = 0$, the chain Γ on $X \backslash |W|$ has no boundary.

Note that the image of the class α_i under the natural homomorphism

$$H^{2d-2q+1}(X \backslash |\mathrm{div}(f_i)|, |W|; \mathbb{C}) \to H^{2d-2q+1}(X \backslash |\mathrm{div}(f_i)|, \mathbb{C})$$

belongs to the subgroup $F^{d-q+1} H^{2d-2q+1}(X \backslash |\mathrm{div}(f_i)|, \mathbb{C})$, i.e, for any closed form $\omega \in F^q A_X^{2q-1}$, we have $\int_{\Gamma_i} \omega = 0$. It follows that the chain Γ is homologous to zero on X. Therefore, Γ is homologous on $X \backslash |W|$ to an integral linear combination of small $(2q-1)$-spheres around irreducible components of the subvariety $|W|$; hence $\int_\Gamma \eta \in \mathbb{Z}$ and this concludes the proof. $\square$

4.3 K-cohomology construction. We use notions and notations from Section 2.4. Let X be a smooth projective variety of dimension d over a field k, and let $p, q \geq 0$ be such that $p+q = d+1$. There is a product morphism $m : \mathcal{K}_p \otimes \mathcal{K}_q \to \mathcal{K}_{d+1}$ and a push-forward map $\pi_* : H^d(X, \mathcal{K}_{d+1}) \to k^*$, where $\pi : X \to \mathrm{Spec}(k)$ is the structure map. By Example 3.11, for any integer $p \geq 0$, we get a biextension of $(H^p(X, \mathcal{K}_p)', H^q(X, \mathcal{K}_q)')$ by k^*, where $H^p(X, \mathcal{K}_p)' \subset H^p(X, \mathcal{K}_p)$ is the annulator of the group $H^{q-1}(X, \mathcal{K}_q)$ with respect to the pairing $H^p(X, \mathcal{K}_p) \times H^{q-1}(X, \mathcal{K}_q) \to k^*$ and the analogous is true for the subgroup $H^q(X, \mathcal{K}_q)' \subset H^q(X, \mathcal{K}_q)$.

Consider a cycle $W \in Z^q(X)$ and a K_1-chain $\{f_\eta\} \in G^{d-q}_{|W|}(X, 1)$ such that $\mathrm{div}(\{f_\eta\}) = 0$. By $\overline{W}$ and $\overline{\{f_\eta\}}$ denote the classes of W and $\{f_\eta\}$ in the corresponding K-cohomology groups. It follows directly from Lemma 2.28 that $\pi_*(m(\overline{\{f_\eta\}} \otimes \overline{W})) = \prod_\eta f_\eta(\overline{\eta} \cap W)$. Hence by Lemma 2.10, the identification $CH^p(X) = H^p(X, \mathcal{K}_p)$ induces the equality $CH^p(X)' = H^p(X, \mathcal{K}_p)'$; thus we get a biextension P_{KC} of $(CH^p(X)', CH^{d+1-1}(X)')$ by k^*.

Proposition 4.9 *The biextension P_{KC} is canonically isomorphic up to the sign $(-1)^{pq}$ to the biextension P_E.*

Proof 1 First, we give a short proof that uses a regulator map from higher Chow groups to K-cohomology. The author is very grateful to the referee who suggested this proof.

Consider the cohomological type higher Chow complex

$$Z^p(X)^\bullet = Z^p(X, 2p - \bullet)$$

and the complex of Zariski flabby sheaves $\underline{Z}^p(X)^\bullet$ defined by the formula $\underline{Z}^p(X)^\bullet(U) = Z^p(U)^\bullet$ for an open subset $U \subset X$. Let $\mathcal{H}^i$ be the cohomology sheaves of the complex $\underline{Z}^p(X)^\bullet$. It follows from [6] combined with [25] and [29] that $\mathcal{H}^i = 0$ for $i > p$ and that the sheaf $\mathcal{H}^p$ has the following flabby resolution $\underline{Gers}^M(X, p)^\bullet$: $\underline{Gers}^M(X, p)^\bullet(U) = Gers^M(U, p)^\bullet$ for an open subset $U \subset X$, where

$$Gers^M(U, p)^l = \sum_{\eta \in X^{(l)}} K^M_{p-l}(k(\eta))$$

and K^M denotes the Milnor K-groups. Combining the canonical homomorphism from Milnor K-groups to Quillen K-groups of fields and the fact that the Gersten complex is a resolution of the sheaf $\mathcal{K}_p$, we get a morphism

$$R\Gamma(X, Z^p(X)^\bullet) \to R\Gamma(X, \mathcal{K}_p[-p]).$$

Moreover, it follows from [25] and [29] that this morphism commutes with the multiplication morphisms. By Proposition 4.2, we get the needed result. $\square$

Proof 2 Let us give an explicit proof in the case when the ground field is infinite and perfect. We use adelic complexes (see Section 2.4 and [16]). The multiplicative structure on the adelic complexes defines the morphism of complexes $\phi : \mathbf{A}(X, p)^\bullet \otimes \mathbf{A}(X, q)^\bullet \to k^*[-d]$, which agrees with the natural morphism $\mathrm{Hom}_{D^b(\mathcal{A}b)}(R\Gamma(X, \mathcal{K}_p) \otimes^L_{\mathbb{Z}} R\Gamma(X, \mathcal{K}_q), k^*[-d])$ that defines the biextension P_{KC}.

By d denote the differential in the adelic complexes $\mathbf{A}(X, n)^\bullet$. In notations from Section 3.2, let $T \subset \mathrm{Ker}(d^p)' \times \mathrm{Ker}(d^q)'$ be the bisubgroup that consists of all pairs $(\sum_i [Z_i], \sum_j [W_j])$ with $[Z_i] \in \mathbf{A}(X, \mathcal{K}_p)^p$, $[W_j] \in \mathbf{A}(X, q)^q$ such that for all i, j, we have:

(i) the K-adeles $[Z_i] \in \mathbf{A}(X, \mathcal{K}_p)^p$ and $[W_j] \in \mathbf{A}(X, q)^q$ are good cocycles with respect to some patching systems $\{(Z_i)_r^{1,2}\}$ and $\{(W_j)_s^{1,2}\}$ for cycles $Z_i \in Z^p(X)'$ and $W_j \in Z^q(X)'$ on X, respectively;

(ii) the patching systems $\{(Z_i)_r^{1,2}\}$ and $\{(W_j)_s^{1,2}\}$ satisfy the condition (i) from Lemma 2.24 and the patching system $\{(W_j)_s^{1,2}\}$ satisfies the condition (ii) from Lemma 2.24 with respect to the subvariety $|Z_i| \subset X$;

(iii) the support $|Z_i|$ meets the support $|W_j|$ properly.

In particular, for all j, we have $\operatorname{codim}_Z(Z \cap ((W_j)_s^1 \cap (W_j)_s^2)) \geq s + 1$ for all s, $1 \leq s \leq q - 1$, where $Z = |\sum_i Z_i| \in Z^p(X)$.

Combining the classical moving lemma, Lemma 2.24, and Lemma 2.26, we see that the natural map $T \to H^p(X, \mathcal{K}_p)' \times H^q(X, \mathcal{K}_q)' = CH^p(X)' \times CH^q(X)'$ is surjective.

Suppose that $(\sum_i [Z_i], \sum_j [W_j]) \in T$ and the class of the cycle $Z = \sum_i Z_i$ in $H^p(X, \mathcal{K}_p) = CH^p(X)$ is trivial. By Lemma 2.6 and Corollary 2.7, there exists a K_1-chain $\{f_\eta\} \in G^{p-1}(X, 1)$ such that $\operatorname{div}(\{f_\eta\}) = Z$, the support $Y = \operatorname{Supp}(\{f_\eta\})$ meets $W = \sum_j W_j \in Z^q(X)$ properly and for all j, we have $\operatorname{codim}_Y(Y \cap ((W_j)_s^1 \cap (W_j)_s^2))) \geq s + 1$ for all s, $1 \leq s \leq q - 1$. It follows that for all j, the patching system $\{(W_j)_s^{1,2}\}$ satisfies the condition (ii) from Lemma 2.24 with respect to the subvariety $Y \subset X$. Combining Lemma 2.24, Lemma 2.27, and Lemma 2.28, we see that there exists a K-adele $[\{f_\eta\}] \in \mathbf{A}(X, \mathcal{K}_p)^{p-1}$ such that $d^{p-1}[\{f_\eta\}] = \sum_i [Z_i]$ and $\pi_*(m([\{f_\eta\}] \otimes \sum_j [W_j])) = (-1)^{pq} \prod_\eta f_\eta(\overline{\eta} \cap W)$.

Suppose that $(\sum_i [Z_i], \sum_j [W_j]) \in T$ and the class of the cycle $W = \sum_j W_j$ in $H^q(X, \mathcal{K}_q) = CH^q(X)$ is trivial. By Lemma 2.6, there exists a K_1-chain $\{g_\xi\} \in G^{p-1}(X, 1)$ such that $\operatorname{div}(\{g_\xi\}) = Z$ and the support $Y = \operatorname{Supp}(\{g_\xi\})$ meets $Z = \sum_i Z_i \in Z^p(X)$ properly. Combining Lemma 2.24, Lemma 2.27, and Lemma 2.28, we see that there exists a K-adele $[\{g_\xi\}] \in \mathbf{A}(X, \mathcal{K}_q)^{q-1}$ such that $d^{q-1}[\{g_\xi\}] = \sum_j [W_j]$ and $(-1)^p \pi_*(m([Z] \otimes [\{g_\xi\}])) = (-1)^{pq} \prod_\xi g_\xi(Z \cap \overline{\xi})$.

Therefore by Remark 3.2 applied to $T \subset \operatorname{Ker}(d^p)' \times \operatorname{Ker}(d^q)'$ and ψ induced by $\phi = \pi_* \circ m$, we get the needed result.

Therefore the needed statement follows from Remark 3.2 applied to the bisubgroup $T \subset \operatorname{Ker}(d^p)' \times \operatorname{Ker}(d^q)'$ and the bigger bisubgroup in $\operatorname{Ker}(d^p)' \times \operatorname{Ker}(d^q)'$ together with the bilinear map defined by the morphism of complexes ϕ as shown in Section 3.2. $\qquad\square$

4.4 Determinant of cohomology construction. In Section 2.5 we defined a "determinant of cohomology" distributive functor

$$\langle \cdot, \cdot \rangle : V\mathcal{M}_X \times V\mathcal{M}_X \to V\mathcal{M}_k.$$

Our goal is to descend this distributive functor to a biextension of Chow groups. The strategy is as follows. First, we define a filtration $C^p V\mathcal{M}_X$ on the category $V\mathcal{M}_X$, which is a kind of a filtration by codimension of support. The successive quotients of this filtration are isomorphic to certain Picard categories $\widetilde{CH}_X^p$ that are related to Chow groups. Then we show a homotopy invariance of the categories $C^p V\mathcal{M}_X$ and construct a specialization map for them. As usual, this allows to defined a contravariant structure on the categories $C^p V\mathcal{M}_X$ by using deformation to the normal cone. Next, the exterior product between the categories $C^p V\mathcal{M}_X$

together with the pull-back along diagonal allows one to define for a smooth variety X a collection of distributive functors

$$C^p V\mathcal{M}_X \times C^q V\mathcal{M}_X \to C^{p+q} V\mathcal{M}_X,$$

compatible with the distributive functor

$$V\mathcal{M}_X \times V\mathcal{M}_X \to V\mathcal{M}_X$$

defined by the derived tensor product of coherent sheaves. Since all constructions for the categories $C^p V\mathcal{M}_X$ are compatible for different p, we get the analogous constructions for the categories $\widetilde{CH}_X^p$ (contravariancy and a distributive functor). Taking the push-forward functor $C^{d+1} V\mathcal{M}_X \to V\mathcal{M}_X \to V\mathcal{M}_k$ for a smooth projective variety X, we get a distributive functor

$$\langle \cdot, \cdot \rangle_{pq} : \widetilde{CH}_X^p \times \widetilde{CH}_X^q \to V\mathcal{M}_k$$

for $p + q = d + 1$. Finally, applying Lemma 2.35, we get a biextension P_{DC} of $(CH^p(X)', CH^q(X)')$ by k^*. The fiber of this biextension at the classes of algebraic cycles $Z = \sum_i m_i Z_i$ and $W = \sum_j n_j W_j$ is canonically isomorphic to the k^*-torsor

$$\det(\sum_{i,j} m_i n_j R\Gamma(X, \mathcal{O}_{Z_i} \otimes^L_{\mathcal{O}_X} \mathcal{O}_{W_j})) \backslash \{0\}.$$

Also, we prove that the biextension P_{DC} is canonically isomorphic to the biextension P_E constructed in Section 4.1.

Let us follow this plan. For any variety X, and $p \geq 0$ denote by $\mathcal{M}_X^p$ the exact category of sheaves on X whose support codimension is at least p. By definition, put $\mathcal{M}_X^p = \mathcal{M}_X$ for $p < 0$. For $p \geq q$, there are natural functors $V\mathcal{M}_X^p \to V\mathcal{M}_X$ and $V\mathcal{M}_X^p \to V\mathcal{M}_X^q$.

Definition 4.10 For $p \geq 0$, let $C^p V\mathcal{M}_X$ be the following Picard category: objects in $C^p V\mathcal{M}_X$ are objects in the category $V\mathcal{M}_X^p$ and morphisms are defined by the formula

$$\mathrm{Hom}_{C^p V\mathcal{M}_X}(L, M) = \mathrm{Im}(\mathrm{Hom}_{V\mathcal{M}_X^{p-1}}(L, M) \to \mathrm{Hom}_{V\mathcal{M}_X^{p-2}}(L, M))$$

(more precisely, we consider images of objects L and M with respect to the corresponding functors from $V\mathcal{M}_X^p$); a monoidal structure on $C^p V\mathcal{M}_X$ is naturally defined by monoidal structures on the categories $V\mathcal{M}_X^*$.

By definition, we have $C^0 V\mathcal{M}_X = V\mathcal{M}_X$. Note that for $p > d+1$, $C^p V\mathcal{M}_X = 0$ and $C^{d+1} V\mathcal{M}_X$ consists of one object 0 whose automorphisms group is equal to $\mathrm{Im}(K_1(\mathcal{M}_X^d) \to K_1(\mathcal{M}_X^{d-1}))$ (the last group equals $H^d(X, \mathcal{K}_{d+1})$ provided that X is smooth). For $p \geq q \geq 0$, there are natural functors $C^p V\mathcal{M}_X \to C^q V\mathcal{M}_X$. It is clear that

$$\pi_0(C^p V\mathcal{M}_X) = \mathrm{Im}(K_0(\mathcal{M}_X^p) \to K_0(\mathcal{M}_X^{p-1})),$$
$$\pi_1(C^p V\mathcal{M}_X) = \mathrm{Im}(K_1(\mathcal{M}_X^{p-1}) \to K_1(\mathcal{M}_X^{p-2})).$$

The next definition is the same as the definition given in [12, Section 2].

Definition 4.11 For $p \geq 1$, let $\widetilde{CH}_X^p$ be the following Picard category: objects in $\widetilde{CH}_X^p$ are elements in the group $Z^p(X)$ and morphisms are defined by the formula

$$\mathrm{Hom}_{\widetilde{CH}_X^p}(Z, W) = \{f \in G^{p-1}(X, 1) | \mathrm{div}(f) = W - Z\} / \mathrm{Tame},$$

where Tame denotes the K_2-equivalence on K_1-chains, i.e., the equivalence defined by the homomorphism

$$\text{Tame}: G^{p-2}(X,2) \to G^{p-1}(X,1);$$

a monoidal structure on $\widetilde{CH}^p_X$ is defined by taking sums of algebraic cycles.

It is clear that for $p \geq 0$, we have

$$\pi_0(\widetilde{CH}^p_X) = CH^p(X),$$

$$\pi_1(\widetilde{CH}^p_X) = H^{p-1}(\text{Gers}(X,p)^\bullet) = \text{Ker}(\text{div})/\text{Im}(\text{Tame}).$$

Remark 4.12 In notations from Remark 3.8, we have

$$\widetilde{CH}^p_X = \text{Picard}(\tau_{\geq(p-1)}\text{Gers}(X,p)^\bullet).$$

Lemma 4.13 *For any $p \geq 0$, there is a canonical equivalence of Picard categories*

$$C^p V\mathcal{M}_X/C^{p-1}V\mathcal{M}_X \to \widetilde{CH}^p_X.$$

Proof First, let us construct a functor $F^p : C^p V\mathcal{M}_X \to \widetilde{CH}^p_X$. By Example 2.34 (ii), for any $p \geq 0$, there is an equivalence of Picard categories $H^p : V\mathcal{M}^p_X/V\mathcal{M}^{p-1}_X \to \oplus_{\eta \in X^{(p)}} V\mathcal{M}_{k(\eta)}$. In notation from Section 2.5, this defines a functor $\widetilde{H}^p : V\mathcal{M}^p_X \to \text{Picard}(Z^p(X))$. We put $F^p(L) = \widetilde{H}^p(L) \in Z^p(X)$ for any object L in $C^p V\mathcal{M}_X$. Let $f : L \to M$ be a morphism in $C^p V\mathcal{M}_X$ and let $\widetilde{f} : L \to M$ be a corresponding morphism in the category $V\mathcal{M}^{p-1}_X$ (note that $\widetilde{f}$ is not uniquely defined). There are canonical isomorphisms $H^{p-1}(L) \to 0$, $H^{p-1}(M) \to 0$, therefore $H^{p-1}(\widetilde{f})$ defines a canonical element in the group $\pi_1(\oplus_{\eta \in X^{(p-1)}} V\mathcal{M}_{k(\eta)}) = G^{p-1}(X,1)$ that we denote again by $H^{p-1}(\widetilde{f})$. It is easy to check that $\text{div}(H^{p-1}(\widetilde{f})) = F^p(M) - F^p(L)$. We put $F^p(f) = [H^{p-1}(\widetilde{f})]$, where brackets denote the class of a K_1-chain modulo K_2-equivalence. The exact sequence

$$G^{p-2}(X,2) \to K_1(\mathcal{M}^{p-1}_X) \to K_1(\mathcal{M}^{p-2}_X)$$

shows that $[F^{p-1}(\widetilde{f})]$ does not depend on the choice of $\widetilde{f}$.

Next, the exact sequences

$$K_0(\mathcal{M}^{p+1}_X) \to K_0(\mathcal{M}^p_X) \to Z^p(X),$$
$$K_1(\mathcal{M}^p_X) \to K_1(\mathcal{M}^{p-1}_X) \to G^{p-1}(X,1)$$

show that the composition $C^{p+1}V\mathcal{M}_X \to C^p V\mathcal{M}_X \to \widetilde{CH}^p_X$ is canonically trivial. By Lemma 2.33(iii), we get a well-defined functor

$$C^p V\mathcal{M}_X/C^{p-1}V\mathcal{M}_X \to \widetilde{CH}^p_X.$$

Finally, combining Lemma 2.33(ii) with the explicit description of π_i for all involved categories, we see that the last functor is an isomorphism on the groups π_i, $i = 0, 1$. This gives the needed result. $\square$

Remark 4.14 Let us construct explicitly an inverse functor $(F^p)^{-1}$ to the equivalence $F^p : C^p V\mathcal{M}_X/C^{p+1}V\mathcal{M}_X \to \widetilde{CH}^p_X$. For each cycle $Z = \sum_i m_i Z_i$, we put $(F^p)^{-1}(Z) = \sum_i m_i \gamma(\mathcal{O}_{Z_i})$, where we choose an order on the set of summands in the last expression. For a codimension $p-1$ irreducible subvariety $Y \subset X$ and a rational function $f \in k(Y)^*$, let $\widetilde{Y}$ be the closure of the graph of the rational function $f : Y \dashrightarrow \mathbb{P}^1$ and let $\pi : \widetilde{Y} \to Y \hookrightarrow X$ be the natural map. Denote by D_0

and D_∞ Cartier divisors of zeroes and poles of f on $\widetilde{Y}$, respectively. Denote by Z_0 and Z_∞ the positive and the negative part of the cycle $\mathrm{div}(f)$ on X, respectively. The exact sequences

$$0 \to \mathcal{O}_{\widetilde{Y}}(-D_\infty) \to \mathcal{O}_{\widetilde{Y}} \to \mathcal{O}_{D_\infty} \to 0,$$

$$0 \to \mathcal{O}_{\widetilde{Y}}(-D_0) \to \mathcal{O}_{\widetilde{Y}} \to \mathcal{O}_{D_0} \to 0,$$

and the isomorphism $f : \mathcal{O}_{\widetilde{Y}}(-D_\infty) \to \mathcal{O}_{\widetilde{Y}}(-D_0)$ define an element in the set

$$\mathrm{Hom}_{V\mathcal{M}_X^{p-1}}\left(\sum_i (-1)^i \gamma(R^i \pi_* \mathcal{O}_{D_\infty}), \sum_i (-1)^i \gamma(R^i \pi_* \mathcal{O}_{D_0})\right),$$

that in turn defines an element

$$(F^p)^{-1}(f_\eta) \in \mathrm{Hom}_{V\mathcal{M}_X^{p-1}}((F^p)^{-1}(Z_\infty), (F^p)^{-1}(Z_0) + L)$$

up to morphisms in the category $V\mathcal{M}_X^p$, where L is a well-defined object in $V\mathcal{M}_X^{p+1}$. For each morphism in the category $\widetilde{CH}_X^p$, we choose its representation as the composition of morphisms defined by f for some codimension $p-1$ irreducible subvarieties $Y \subset X$. This allows to extend $(F^p)^{-1}$ to all morphisms in the category $\widetilde{CH}_X^p$.

Now let us describe some functorial properties of the categories $C^p V\mathcal{M}_X$.

Definition 4.15 For a variety X, let $\{\mathcal{P}_X^p\}$, $p \geq 0$, be the collection of categories $\{V\mathcal{M}_X^p\}$ or $\{C^p \mathcal{M}_X\}$ (the same choice for all varieties), $G_X^{pq} : \mathcal{P}^p \to \mathcal{P}^q$ be the natural functors, and let S and T be two varieties. A *collection of compatible functors* from $\{\mathcal{P}_S^p\}$ to $\{\mathcal{P}_T^p\}$ is a pair $(\{F^p\}, \{\Psi^{pq}\})$, where $F^p : \mathcal{P}_S^p \to \mathcal{P}_T^p$, $p \geq 0$, are symmetric monoidal functors and $\Psi^{pq} : G_T^{pq} \circ F^p \to F^q \circ G_S^{pq}$ are isomorphisms in the category $\mathrm{Fun}^+(\mathcal{P}_S^p, \mathcal{P}_T^q)$ such that the following condition is satisfied: for all $p \geq q \geq r \geq 0$, we have $G_S^{pq}(\Psi^{qr}) \circ G_T^{qr}(\Psi^{pq}) = \Psi^{pr}$.

The proof of the next result is straightforward.

Lemma 4.16 *For varieties S and T, a collection of compatible functors from $\{V\mathcal{M}_S^p\}$ to $\{V\mathcal{M}_T^p\}$ defines in a canonical way a collection of compatible functors from $\{C^p \mathcal{M}_S\}$ to $\{C^p \mathcal{M}_S\}$.*

Example 4.17 Let $f : S \to T$ be a flat morphism of schemes; then there is a collection of compatible functors $f^* : V\mathcal{M}_T^p \to V\mathcal{M}_S^p$. By Lemma 4.16, this gives a collection of compatible functors $f^* : C^p \mathcal{M}_T \to C^p \mathcal{M}_S$.

Now let us prove homotopy invariance of the categories $C^p V\mathcal{M}_X$. Note that there is no homotopy invariance for the categories $V\mathcal{M}_X^p$. We use the following straightforward result on truncations of spectral sequences:

Lemma 4.18

(i) *Consider a spectral sequence E_r^{ij}, $r \geq 1$, a natural number s, and an integer $l \in \mathbb{Z}$. Then there is a unique spectral sequence $(t_s^l E)_r^{ij}$, $r \geq s$, such that $(t_s^l E)_s^{ij} = E_s^{ij}$ if $i \geq l$, $(t_s^l E)_s^{ij} = 0$ if $i < l$, and there is a morphism of spectral sequences $(t_s^l E)_r^{ij} \to E_r^{ij}$, $r \geq s$.*

(ii) *In the above notations, let s_1, s_2 be two natural numbers; then we have $(t_l^{s_1} E)_r^{ij} = (t_l^{s_2} E)_r^{ij}$ for all $r \geq s = \max\{s_1, s_2\}$ and $i \geq l + s - 1$.*

Lemma 4.19 *Consider a vector bundle $f : N \to S$; then the natural functors $f^* : C^p V\mathcal{M}_S \to C^p V\mathcal{M}_N$ are equivalencies of Picard categories.*

Proof We follow the proof of [14, Lemma 81]. It is enough to show that f^* induces isomorphisms on π_0 and π_1, i.e., it is enough to show that the natural homomorphisms

$$f^* : \operatorname{Im}(K_n(\mathcal{M}_S^p) \to K_n(\mathcal{M}_S^{p-1})) \to \operatorname{Im}(K_n(\mathcal{M}_N^p) \to K_n(\mathcal{M}_N^{p-1}))$$

are isomorphisms for all $p \geq 0$, $n \geq 0$.

For any variety T, consider Quillen spectral sequence $E_r^{ij}(T)$, $r \geq 1$, that converges to $K_{-i-j}(\mathcal{M}_T)$. In notations from Lemma 4.18, the filtration of abelian categories $\mathcal{M}_T^{p-1} \supset \mathcal{M}_T^p \supset \ldots$ defines the spectral sequence that is equal to the shift of the truncated Quillen spectral sequence $(t_1^{p-1}E)_r^{ij}(T)$, $r \geq 1$ and converges to $K_{-i-j}(\mathcal{M}_T^{p-1})$. By Lemma 4.18, $(t_1^{p-1}E)_r^{ij}(T) = (t_2^{p-1}E)_r^{ij}(T)$ for $i \geq p$. Therefore, there is a spectral sequence $(t_2^{p-1}E)_r^{ij}(T)$, $r \geq 2$ that converges to $\operatorname{Im}(K_{-i-j}(\mathcal{M}_T^p) \to K_{-i-j}(\mathcal{M}_T^{p-1}))$. Explicitly, we have $(t_2^{p-1}E)_2^{ij}(T) = H^i(Gers(T, -j)^\bullet)$.

The morphism f defines the morphism of spectral sequences $f^* : (t_2^{p-1}E)_r^{ij}(S) \to (t_2^{p-1}E)_r^{ij}(N)$, $r \geq 2$. Moreover, the homomorphism f^* is an isomorphism for $r = 2$ (see op.cit.), hence f^* is an isomorphism for all $r \geq 2$. This gives the needed result. $\qquad\square$

Corollary 4.20 *Taking inverse functors to the equivalences* $f^* : C^p V \mathcal{M}_V \to C^p V \mathcal{M}_N$, *we get a collection of compatible functors* $(f^*)^{-1}$ *from* $C^p V \mathcal{M}_N$ *to* $C^p V \mathcal{M}_S$.

Next we construct a specialization map for the categories $C^p V \mathcal{M}_X$.

Lemma 4.21 *Let* $j : D \subset Y$ *be a subscheme on a scheme given as a subscheme by the equation* $f = 0$, *where* f *is a regular function on* Y, *and let* $U = Y \backslash D$. *Then there exists a collection of compatible functors* $Sp^p : V\mathcal{M}_U^p \to V\mathcal{M}_D^p$ *such that the composition* $V\mathcal{M}_Y \to V\mathcal{M}_U \xrightarrow{Sp^0} V\mathcal{M}_D$ *is canonically isomorphic to* j^* *in the category of symmetric monoidal functors.*

Proof The composition $j^* \circ \gamma_Y : (\mathcal{M}_Y)_{iso} \to V\mathcal{M}_D$ is canonically isomorphic to the functor $\mathcal{F} \mapsto \gamma_D(\mathcal{G}_0) - \gamma_D(\mathcal{G}_{-1})$, where $\mathcal{G}_i$ are cohomology sheaves of the complex $\mathcal{F} \xrightarrow{f} \mathcal{F}$ placed in degrees -1 and 0. It follows that the composition $j^* \circ j_* \circ \gamma_D : (\mathcal{M}_D)_{iso} \to V\mathcal{M}_D$ is canonically isomorphic to 0 in the category $\operatorname{Det}(\mathcal{M}_D, V\mathcal{M}_D)$. By Lemma 2.29, this defines a canonical isomorphism $j^* \circ j_* \to 0$ in the category $\operatorname{Fun}^+(V\mathcal{M}_D, V\mathcal{M}_D)$. Combining Lemma 2.33 (iii) and Example 2.34 (ii), we get the functor $Sp : V\mathcal{M}_U \to V\mathcal{M}_D$, since there is a natural equivalence of abelian categories $\mathcal{M}_Y/\mathcal{M}_D \to \mathcal{M}_U$.

More explicitly, the functor $sp : V\mathcal{M}_U \to V\mathcal{M}_D$ corresponds via Lemma 2.29 to the following determinant functor: $\mathcal{F} \mapsto \gamma_D(j^*\widetilde{\mathcal{F}})$, where $\mathcal{F}$ is a coherent sheaf on U and $\widetilde{\mathcal{F}}$ is coherent sheaf on Y that restricts to $\mathcal{F}$ (one can easily check that two different collections of choices of $\widetilde{\mathcal{F}}$ for all $\mathcal{F}$ define canonically isomorphic determinant functors).

Take a closed subscheme $Z \subset U$ such that each component of Z has codimension at most p in U. Consider the closed subscheme $Z_D = \overline{Z} \times_X D$ in D, where $\overline{Z}$ is the closure of Z in X. We get a functor $Sp_Z : V\mathcal{M}_Z \to V\mathcal{M}_{Z_D} \to V\mathcal{M}_D^p$. Given a diagram $Z \xhookrightarrow{i} Z' \subset U$, we have a canonical isomorphism of symmetric monoidal functors $Sp_{Z'} \circ i_* \to Sp_Z$. Taking the limits over closed subschemes $Z \subset U$, we get the needed collection of compatible functors $Sp^p : V\mathcal{M}_U^p \to V\mathcal{M}_D^p$, $p \geq 0$.   □

Corollary 4.22 *Combining Lemma 4.21 and Lemma 4.16, we get a collection of compatible functors $Sp^p : C^p V \mathcal{M}_U \to C^p V \mathcal{M}_D$ in notations from Lemma 4.21.*

Now we show contravariancy for the categories $C^p V \mathcal{M}_X$. Let $i : S \subset T$ be a regular embedding of varieties; then there is a symmetric monoidal functor

$$i^* : V\mathcal{M}_T \cong V\mathcal{M}'_T \to V\mathcal{M}_S$$

where $\mathcal{M}'_T$ is a subcategory in $\mathcal{M}_T$ of $\mathcal{O}_S$-flat coherent sheaves. The composition of $\gamma_T : (\mathcal{M}_T)_{iso} \to V\mathcal{M}_T$ with i^* is canonically isomorphic to a functor that sends a coherent sheaf $\mathcal{F}$ on T to the object $\sum_i (-1)^i \gamma(\mathcal{T}or_i^{\mathcal{O}_T}(\mathcal{F}, \mathcal{O}_S))$. By Lemma 2.29, the last condition defines the functor i^* up to a unique isomorphism.

Proposition 4.23 *For each $p \geq 0$, there exists a collection of compatible functors $i_p^* : C^p V \mathcal{M}_T \to C^p V \mathcal{M}_S$ such that $i_0^* = i^*$.*

Proof We use deformation to the normal cone (see [13]). Let M be the blow-up of $S \times \{\infty\}$ in $T \times \mathbb{P}^1$ and let $M^0 = M \backslash \mathbb{P}(N)$, where N denotes the normal bundle to S in T. Then we have $M^0|_{\mathbb{A}^1} \cong T \times \mathbb{A}^1$ and $M^0|_{\{\infty\}} = N$, where $\mathbb{A}^1 = \mathbb{P}^1 \backslash \{\infty\}$. Combining Corollary 4.20 and Corollary 4.22, we get a compatible collection of functors i_p^* as the composition

$$C^p V \mathcal{M}_T \xrightarrow{pr_T^*} C^p V \mathcal{M}_{T \times \mathbb{A}^1} \xrightarrow{Sp^p} C^p V \mathcal{M}_N \xrightarrow{(pr_S^*)^{-1}} C^p V \mathcal{M}_S,$$

where $pr_T : T \times \mathbb{A}^1 \to T$, $pr_S : N \to S$ are natural projections, and $(pr_S^*)^{-1}$ is an inverse to the equivalence $pr_S^* : C^p V \mathcal{M}_S \to C^p V \mathcal{M}_N$.   □

Remark 4.24 Let $\mathcal{M}_{T,S}^p$ be a full subcategory in $\mathcal{M}_T^p$ whose objects are coherent sheaves $\mathcal{F}$ from $\mathcal{M}_T^p$ whose support meets S properly. Then the composition of the natural functor $(\mathcal{M}_{T,S}^p)_{iso} \to C^p V \mathcal{M}_T$ with i_p^* is canonically isomorphic to the functor that sends $\mathcal{F}$ in $\mathcal{M}_{T,S}^p$ to the object $\sum_i (-1)^i \gamma(\mathcal{T}or_i^{\mathcal{O}_T}(\mathcal{F}, \mathcal{O}_S))$.

Remark 4.25 Applying Proposition 4.23 to the embedding of the graph of a morphism of varieties $f : X \to Y$ with smooth Y, we get a collection of compatible functors $f^* : C^p V \mathcal{M}_Y \to C^p \mathcal{M}_X$. Combining Lemma 4.13 and Lemma 2.33 (iii), we get a collection of symmetric monoidal pull-back functors $f^* : \widetilde{CH}_Y^p \to \widetilde{CH}_X^p$.

Remark 4.26 The pull-back functors for the categories $\widetilde{CH}_X^p$ are also constructed by J. Franke in [12] by different methods: the categories $\widetilde{CH}_X^p$ are interpreted as equivalent categories to categories of torsors under certain sheaves. Also, a biextension of Chow groups is constructed by the same procedure as below. It is expected that the pull-back functor constructed in op.cit. is canonically isomorphic to the pull-back functor constructed in Remark 4.25. By Proposition 4.31 below, this would imply that the biextension constructed in op.cit. is canonically isomorphic to the biextension P_E.

Remark 4.27 By Lemma 2.21, for a smooth variety X over an infinite perfect field k, there is a canonical equivalence

$$\mathrm{Picard}(\tau_{\geq(p-1)}\tau_{\leq p}\mathbf{A}(X, \mathcal{K}_p)^\bullet) \to \widetilde{CH}_X^p.$$

Moreover, the Picard categories on the left hand side are canonically contravariant. This gives another way to define a contravariant structure on the Chow categories (note that the Godement resolution for a sheaf $\mathcal{K}_p$ is not enough to do this).

For three varieties R, S, T, a *collection of compatible distributive functors* from $C^p V \mathcal{M}_R \times C^q V \mathcal{M}_S$ to $C^{p+q} V \mathcal{M}_T$, $p, q \geq 0$ is defined analogously to the collection of compatible symmetric tensor functors.

Corollary 4.28 *Let X be a smooth variety. There exists a collection of compatible distributive functors*

$$C^p V \mathcal{M}_X \times C^q V \mathcal{M}_X \to C^{p+q} V \mathcal{M}_X$$

such that for $p = q = 0$, the composition of the corresponding functor with the functor $(\mathcal{M}_X)_{iso} \times (\mathcal{M}_X)_{iso} \to V \mathcal{M}_X \times V \mathcal{M}_X$ is canonically isomorphic to the functor $(\mathcal{F}, \mathcal{G}) \mapsto \sum_i (-1)^i \gamma(\mathcal{T}or_i^{\mathcal{O}_X}(\mathcal{F}, \mathcal{G}))$.

Proof The exterior product of sheaves defines an collection of compatible distributive functors

$$C^p V \mathcal{M}_X \times C^q V \mathcal{M}_X \to C^{p+q} \mathcal{M}_{X \times X}.$$

Applying Proposition 4.23 to the diagonal embedding $X \subset X \times X$, we get the needed statement. $\square$

For algebraic cycles $Z = \sum_i m_i Z_i$, $W = \sum_j n_j W_j$, we put

$$\langle Z, W \rangle = \det(\sum_{i,j} m_i n_j R\Gamma(X, \mathcal{O}_{Z_i} \otimes^L_{\mathcal{O}_X} \mathcal{O}_{W_j})) \backslash \{0\}.)$$

Proposition 4.29 *Let X be a smooth projective variety of dimension d. Then for $p + q = d + 1$, there is a biextension P_{DC} of $(CH^p(X)', CH^q(X)')$ such that the fiber $P_{DC}|_{(Z,W)}$ is canonically isomorphic to the k^*-torsor $\langle Z, W \rangle$.*

Proof By Lemma 2.33 (iv), the collection of compatible distributive functors from Corollary 4.28 defines a collection of distributive functors

$$\widetilde{CH}^p_X \times \widetilde{CH}^q_X \to \widetilde{CH}^{p+q}_X.$$

for all $p, q \geq 0$. If $p + q = d + 1$, then we get a distributive functor

$$\langle \cdot, \cdot \rangle_{pq} : \widetilde{CH}^p_X \times \widetilde{CH}^q_X \to \widetilde{CH}^{d+1}_X = C^{d+1} V \mathcal{M}_X \to V \mathcal{M}_X \to V \mathcal{M}_k.$$

Moreover, by Remark 4.14, we have $\langle Z, W \rangle_{pq} = (0, \langle Z, W \rangle)$.

Consider full Picard subcategories $(\widetilde{CH}^p_X)' \subset \widetilde{CH}^p_X$ whose objects are elements $Z \in Z^p(X)'$ and the restriction of the distributive functor $\langle \cdot, \cdot \rangle$ to the product $(\widetilde{CH}^p_X)' \times (\widetilde{CH}^q_X)'$. Applying Lemma 2.35, we get the next result. $\square$

Finally, let us compare the biextension P_{DC} with the biextension P_E. With this aim we will need the following result.

Lemma 4.30 *Let X be a smooth projective variety of dimension d, $p+q = d+1$, let $Y \subset X$ be a codimension $p-1$ irreducible subvariety, $f \in k(Y)^*$, and let $W \subset X$ be a codimension q irreducible subvariety. Suppose that Y meets W properly and that $|\text{div}(f)| \cap W = \emptyset$. Then f defines a isomorphism $0 \to \text{div}(f)$ in $\widetilde{CH}^p_X$, hence a morphism $\sigma(f) : k^* = \langle 0, W \rangle \to \langle \text{div}(f), W \rangle = k^*$ such that $\sigma(f) = f(Y \cap W)$.*

Proof By Remark 4.14 and in its notations, the action of f in question is induced by the exact sequences

$$0 \to \mathcal{O}_{\widetilde{Y}}(-D_\infty) \to \mathcal{O}_{\widetilde{Y}} \to \mathcal{O}_{D_\infty} \to 0,$$

$$0 \to \mathcal{O}_{\widetilde{Y}}(-D_0) \to \mathcal{O}_{\widetilde{Y}} \to \mathcal{O}_{D_0} \to 0,$$

and the isomorphism $f : \mathcal{O}_{\widetilde{Y}}(-D_\infty) \to \mathcal{O}_{\widetilde{Y}}(-D_0)$. Moreover, $\pi : \widetilde{Y} \to Y$ is an isomorphism outside of $D_0 \cup D_\infty$, so $\sigma(f)$ is induced by the automorphism of the complex $\mathcal{O}_Y \otimes^L_{\mathcal{O}_X} \mathcal{O}_W$ given by multiplication on f. Applying the functor $\det R\Gamma(-)$, we get $f(Y \cap W)$. ☐

Proposition 4.31 *The biextension P_{DC} is canonically isomorphic to the biextension P_E.*

Proof Let $T \subset Z^p(X)' \times Z^q(X)'$ be the bisubgroup that consists of pair (Z, W) such that $|Z| \cap |W| = \emptyset$. Suppose that (Z, W), (Z', W) are in T and that Z is rationally equivalent to Z'. By Lemma 2.6, we may suppose that there is a K_1-chain $\{f_\eta\} \in G^{p-1}_{|W|}(X, 1)$ such that $\operatorname{div}(\{f_\eta\}) = Z' - Z$. By Lemma 4.30, we immediately get the needed result. ☐

Combining Proposition 4.31, Proposition 4.3, and Remark 4.6, we obtain the following statement.

Corollary 4.32 *Suppose that X is a complex smooth projective variety of dimension d, $Z \in Z^p(X)_{\mathrm{hom}}$, $W \in Z^q(X)_{\mathrm{hom}}$, $p + q = d + 1$; then there is a canonical isomorphism of the following $\mathbb{C}^*$-torsors:*

- $\det R\Gamma(X, \mathcal{O}_Z \otimes^L_{\mathcal{O}_X} \mathcal{O}_W) \backslash \{0\}$;
- *the set of isomorphism classes of all integral mixed Hodge structures V whose weight graded quotients are identified with $\mathbb{Z}(0)$, $H^{2p-1}(X)(p)$, $\mathbb{Z}(1)$ and such that $[V/W_{-2}V] = [E_Z] \in \mathrm{Ext}^1_{\mathcal{H}}(\mathbb{Z}(0), H^{2p-1}(X)(p))$, $[W_{-1}V] = [E_W^\vee] \in \mathrm{Ext}^1_{\mathcal{H}}(H^{2p-1}(X)(p), \mathbb{Z}(1))$.*

Question 4.33 Does there exist a direct proof of Corollary 4.32?

The present approach is a higher-dimensional generalization for the description of the Poincaré biextension on a smooth projective curve C in terms of determinant of cohomology of invertible sheaves, suggested by P. Deligne in [11]. Let us briefly recall this construction for the biextension $\mathcal{P}$ of $(\mathrm{Pic}^0(C), \mathrm{Pic}^0(C))$ by k^*. For any degree zero invertible sheaves $\mathcal{L}$ and $\mathcal{M}$ on C there is an equality

$$\mathcal{P}_{([\mathcal{L}],[\mathcal{M}])} = \langle \mathcal{L} - \mathcal{O}_C, \mathcal{M} - \mathcal{O}_C \rangle = \det R\Gamma(C, \mathcal{L} \otimes_{\mathcal{O}_C} \mathcal{M} - \mathcal{L} - \mathcal{M} + \mathcal{O}_C)) \backslash \{0\},$$

where $[\cdot]$ denotes the isomorphism class of an invertible sheaf. Since $\chi(C, \mathcal{L} \otimes_{\mathcal{O}_C} \mathcal{M}) = \chi(C, \mathcal{L}) = \chi(C, \mathcal{M}) = 1 - g$, this k^*-torsor is well defined on $\mathrm{Pic}^0(C) \times \mathrm{Pic}^0(C)$. Consider a homomorphism $p : \mathrm{Pic}^0(C) \times \mathrm{Div}^0(C) \to \mathrm{Pic}^0(C) \times \mathrm{Pic}^0(C)$ given by the formula $([\mathcal{L}], E) \mapsto ([\mathcal{L}], [\mathcal{O}_C(E)])$. Then there is an isomorphism of k^*-torsors $\varphi : p^*\mathcal{P} \cong \mathcal{P}'$, where $\mathcal{P}'$ is a biextension of $(\mathrm{Pic}^0(C), \mathrm{Div}^0(C))$ by k^* given by the formula $\mathcal{P}'_{([\mathcal{L}],E)} = (\bigotimes_{x \in C} \mathcal{L}|_x^{\otimes \mathrm{ord}_x(E)}) \backslash \{0\}$. Thus φ induces a biextension structure on $p^*\mathcal{P}$; it turns out that this structure descends to $\mathcal{P}$. Moreover, in op.cit. it is proved that the biextension $\mathcal{P}$ is canonically isomorphic to the Poincaré line bundle on $\mathrm{Pic}^0(X) \times \mathrm{Pic}^0(X)$ without the zero section.

Claim 4.34 *In the above notations, the biextensions $\mathcal{P}$ and P_{DC} are canonically isomorphic.*

Proof Consider a map $\pi : \mathrm{Div}^0(C) \times \mathrm{Div}^0(C) \to \mathrm{Pic}^0(C) \times \mathrm{Pic}^0(C)$ given by the formula $(D, E) \mapsto ([\mathcal{O}_C(D), \mathcal{O}_C(E)])$. There are canonical isomorphisms of k^*-torsors

$$\pi^*\mathcal{P} \cong (\bigotimes_{x \in C} \mathcal{O}_C(D)|_x^{\otimes \mathrm{ord}_x(E)}) \backslash \{0\} \cong (\bigotimes_{x \in C} \mathcal{O}_C(x)|_x^{\otimes (\mathrm{ord}_x(D) + \mathrm{ord}_x(E))}) \backslash \{0\}$$

$$\cong \langle \sum_{x \in C} \mathrm{ord}_x(D)\mathcal{O}_x, \sum_{y \in C} \mathrm{ord}_y(E)\mathcal{O}_y \rangle.$$

Besides, these isomorphisms commute with the biextension structure. It remains to note that the rational equivalence on the first argument also commutes with the above isomorphisms (this is a particular case of Lemma 4.30). □

Remark 4.35 For a smooth projective curve, the biextension P_{KC} from Section 4.3 corresponds to a pairing between complexes given by a certain product of Hilbert tame symbols. Combining Proposition 4.9, Proposition 4.31, and Claim 4.34, we see that there is a connection between the tame symbol and the Poincaré biextension given in terms of determinant of cohomology of invertible sheaves. In [15] this connection is explained explicitly in terms of the central extension of ideles on a smooth projective curves constructed by Arbarello, de Concini, and Kac.

4.5 Consequences on the Weil pairing. Let X be a smooth projective variety of dimension d over a field k. As above, suppose that $p, q \geq 0$ are such that $p + q = d + 1$. Consider cycles $Z \in Z^p(X)'$ and $W \in Z^q(X)'$ (see Section 4.1) such that $lZ = \mathrm{div}(\{f_\eta\})$ and $lW = \mathrm{div}(\{g_\xi\})$ for an integer $l \in \mathbb{Z}$ and K_1-chains $\{f_\eta\} \in G_{|W|}^{p-1}(X, 1)$, $\{g_\xi\} \in G_{|Z|}^{q-1}(X, 1)$ (see Section 2.2). By $[Z]$ and $[W]$ denote the classes of the cycles Z and W in the groups $CH^p(X) = H^p(X, \mathcal{K}_p)$ and $CH^q(X) = H^q(X, \mathcal{K}_q)$, respectively. As explained in Example 3.10, the Massey triple product in K-cohomology

$$m_3([Z], l, [W]) \in H^d(X, \mathcal{K}_{d+1})/([Z] \cdot H^{q-1}(X, \mathcal{K}_q) + H^{p-1}(X, \mathcal{K}_p) \cdot [W])$$

has a well defined push-forward $\overline{m}_3([Z], l, [W])) \in \mu_l$ with respect to the map $\pi_* : H^d(X, \mathcal{K}_{d+1}) \to k^*$, where $\pi : X \to \mathrm{Spec}(k)$ is the structure morphism.

Combining Lemma 3.7, Example 3.10, Proposition 4.9, and Proposition 4.3, we get the following statements.

Corollary 4.36

(i) *In the above notations, we have*

$$\overline{m}_3(\alpha, l, \beta)^{(-1)^{pq}} = \prod_\eta f_\eta(\overline{\eta} \cap W) \cdot \prod_\xi g_\xi^{-1}(Z \cap \overline{\xi}) = \phi_l([Z], [W]),$$

where ϕ_l is the Weil pairing associated with the biextension P_E from Section 4.1.

(ii) *If $k = \mathbb{C}$ and $Z \in Z^p(X)_{\mathrm{hom}}$, $W \in Z^q(X)_{\mathrm{hom}}$, then we have*

$$\phi_l([Z], [W]) = \phi_l^{an}(AJ(Z), AJ(W)),$$

where ϕ_l^{an} is the Weil pairing between the l-torsion subgroups in the dual complex tori $J^{2p-1}(X)$ and $J^{2q-1}(X)$.

This is a generalization of the classical Weil's formula for divisors on a curve. The first part of Corollary 4.36 was proved in [16] without considering biextensions.

References

[1] Beilinson, A. A. *Residues and adeles*. Funct. Anal. And Appl. **14** (1980), 34–35.

[2] Beilinson, A. *Notes on absolute Hodge cohomology*. Contemporary Mathemetics **55** (1986), 35–68.

[3] Beilinson, A. *Height pairing between algebraic cycles*. Contemporary Mathemetics **67** (1987), 1–24.

[4] Beilinson, A., Vologodsky, V. *A DG guide to Voevodsky's motives*. GAFA, Geom. func. anal., **17** (2007), 1709–1787.

[5] Bloch, S. *Height pairing for algebraic cycles*. Journal of Pure and Applied Algebra **34** (1984), 119–145.

[6] Bloch, S. *Algebraic cycles and higher K-theory*. Advances in Mathematics **61** (1986), 267–304.

[7] Bloch, S. *Cycles and biextensions*. Contemporary Mathematics **83** (1989), 19–30.

[8] Bloch, S. http://www.math.uchicago.edu/~bloch/cubical_chow.pdf.

[9] Breen, L. *Fonctions thêta et théorème du cube*. Lecture Notes in Mathematics **980** (1983).

[10] Deligne, P. *Théorie de Hodge: II*. Publications Mathématiques de l'I.H.É.S. **40** (1971), 5–57.

[11] Deligne, P. *Le déterminant de la cohomologie*. Contemporary Mathematics **67** (1987), 93–177.

[12] Franke, J. *Chow categories*. Compositio Mathematica **76**:1,2 (1990), 101–162.

[13] Fulton, W. *Intersection Theory*. A Series of Modern Surveys in Mathematics, Springer Verlag (1984).

[14] Gillet, H. *K-theory and intersection theory*. Handbook of K-theory, **1** (2005), 236–293.

[15] Gorchinskiy, S. *Poincaré biextension and ideles on an algebraic curve*. Sbornik Mathematics **197**:1 (2006), 25–38; arXiv: math.AG/0511626v3.

[16] Gorchinskiy, S. *An adelic resolution for homology sheaves*. Izvestiya: Mathematics **72**:6 (2008), 133–202; arXiv:0705.2597.

[17] Griffiths, P. *Periods of integrals on algebraic manifolds, I, II*. American Jouranal of Mathematics **90** (1968), 568–626, 805–865.

[18] Hain, R. *Biextensions and heights associated to curves of odd genus*. Duke Math. Journal **61** (1990), 859–898.

[19] Huber, A. *On the Parshin–Beilinson adeles for schemes*. Abh. Math. Sem. Univ. Hamburg **61** (1991), 249–273.

[20] Kerr M., Lewis J. D., Müller-Stach, S. *The Abel–Jacobi map for higher Chow groups*. Compositio Math. **142** (2006), 374–396.

[21] Levine, M. *Mixed motives*. Math. Surveys and Monographs **57** (1998).

[22] Müller-Stach, S. $\mathbb{C}^*$-extensions of tori, higher Chow groups and applications to incidence equivalence relations for algebraic cycles. K-theory **9** (1995), 395–406.

[23] Müller-Stach, S. *Algebraic cycle complexes: basic properties*. The Arithmetic and Geometry of Algebraic cycles Banff 1998, Nato Sci. Series **548** (2000), 285-305.

[24] Mumford, D. *Biextensions of formal groups*. Proceedings of the Bombay Colloquium on Algebraic Geometry (1968).

[25] Nesterenko Yu. P., Suslin A. A., *Homology of the full linear group over a local ring, and Milnor's K-theory*. Math. USSR-Izv, **34** (1990), 121–145.

[26] Parshin, A. N. *On the arithmetic of two-dimensional schemes I. Repartitions and residues* Izv. Akad. Nauk SSSR **40**:4 (1976), 736–773.

[27] Quillen, D. *Higher Algebraic K-theory*. Lecture Notes in Mathematics **341** (1973), 85–147.

[28] *Groupes de Monodromie en Géométrie Algébrique (SGA 7)*. Lecture Note in Mathematics **288** (1972).

[29] Totaro B., *Milnor K-theory is the simplest part of algebraic K-theory*. K-theory **6** (1992), 177–189.

Fields Institute Communications
Volume **56**, 2009

Démonstration Géométrique du Théorème de Lang-Néron et Formules de Shioda-Tate

Bruno Kahn

Institut de Mathématiques de Jussieu
175–179 rue du Chevaleret
75013 Paris, France
kahn@math.jussieu.fr

To Spencer Bloch

Résumé. On donne une démonstration "sans hauteurs", dans l'esprit de [8], du théorème de Lang-Néron: si K/k est une extension de type fini régulière et A est une K-variété abélienne, le groupe $A(K)/\operatorname{Tr}_{K/k} A(k)$ est de type fini, où $\operatorname{Tr}_{K/k} A$ désigne la K/k-trace de A au sens de Chow. On en déduit des formules à la Shioda-Tate pour le rang de ce groupe.

Abstract. We give a proof without heights, in the spirit of [8], of the Lang-Néron theorem: if K/k is a regular extension of finite type and A is an abelian K-variety, the group $A(K)/\operatorname{Tr}_{K/k} A(k)$ is finitely generated, where $\operatorname{Tr}_{K/k} A$ denotes the K/k-trace of A in the sense of Chow. We derive Shioda-Tate–like formulas for the rank of this group.

Introduction

Soit K/k une extension de type fini régulière. Le foncteur d'extension des scalaires

$$\mathbf{Ab}(k) \to \mathbf{Ab}(K)$$

de la catégorie des variétés abéliennes sur k vers celle des variétés abéliennes sur K admet un adjoint à droite : la K/k-trace [8, app. A]. Nous noterons cet adjoint $\operatorname{Tr}_{K/k}$. Cette notion est due à Chow, mais semble avoir été anticipée par Néron [10], *cf.* remarque 1.1.

Nous aurons aussi besoin au §4 de l'adjoint à gauche du foncteur d'extension des scalaires : la K/k-image $\operatorname{Im}_{K/k}$ (*ibid.*).

Soit A une K-variété abélienne. On se propose de donner une démonstration "sans hauteurs" du théorème de Lang-Néron :

Théorème 1 ([9]) *Le groupe $A(K)/\operatorname{Tr}_{K/k} A(k)$ est de type fini.*

2000 *Mathematics Subject Classification.* Primary 11G10, 14C22, 14K30.

(La démonstration de Lang et Néron est exposée dans le langage des schémas dans [8, App. B] et dans [2, §7].)

La démonstration du théorème 1 est donnée au §1 : elle est dans l'esprit de [8], où je n'étais pas parvenu à l'obtenir dans ce style. Un sous-produit en est une formule à la Shioda-Tate [12, 13] pour le rang du groupe $A(K)/\operatorname{Tr}_{K/k} A(k)$ (corollaire 2.1). En fait, de nombreuses formules de ce genre peuvent être obtenues : ces variations sont expliquées au §3.

Remerciements. Je remercie Marc Hindry pour m'avoir signalé ses articles avec Amilcar Pacheco et Rania Wazir [6, 7] et Keiji Oguiso pour une correspondance relative à son article [11] : tous m'ont permis de replacer la présente démonstration dans le contexte de la formule de Shioda-Tate. Je remercie également le rapporteur pour ses remarques qui m'ont aidé à clarifier la rédaction.

Notation. Pour toute variété X lisse sur un corps k, on note $\operatorname{NS}(X)$ le groupe des cycles de codimension 1 modulo l'équivalence algébrique [4, 10.3] : si X est projective et k est algébriquement clos, c'est son groupe de Néron-Severi. Nous utiliserons :

Proposition 0.1 ([8, th. 3]) *Si k est algébriquement clos, $\operatorname{NS}(X)$ est un groupe de type fini pour toute k-variété lisse X.*

(Rappelons brièvement l'argument : on se réduit au cas projectif, en utilisant la résolution des singularités en caractéristique zéro ou le théorème des altérations de de Jong en toute caractérstique.)

1 Démonstration du théorème 1

Comme dans [8, app. B], on se ramène au cas où k est algébriquement clos. Il suffit de démontrer le théorème 1 pour la variété abélienne duale $\hat{A}$. Comme $\hat{A}(K) = \operatorname{Pic}^0(A)$, il suffit de démontrer la génération finie de $\operatorname{Pic}(A)/\operatorname{Tr}_{K/k} \hat{A}(k)$.

Soit X un modèle lisse de K/k choisi de telle sorte que A se prolonge en un schéma abélien $p : \mathcal{A} \to X$.

Lemme 1.1 *La suite*

$$0 \to \operatorname{Pic}(X) \xrightarrow{p^*} \operatorname{Pic}(\mathcal{A}) \xrightarrow{j^*} \operatorname{Pic}(A) \to 0$$

est exacte, où j est l'inclusion de A dans $\mathcal{A}$.

Démonstration On a un diagramme commutatif

$$
\begin{array}{ccccccccc}
\Gamma(A, \mathbb{G}_m) & \longrightarrow & \bigoplus\limits_{x \in (\mathcal{A}-A)^{(1)}} \mathbf{Z} & \longrightarrow & \operatorname{Pic}(\mathcal{A}) & \xrightarrow{j^*} & \operatorname{Pic}(A) & \longrightarrow & 0 \\
\wr \uparrow & & \wr \uparrow & & p^* \uparrow & & \uparrow & & \\
K^* & \longrightarrow & \bigoplus\limits_{x \in X^{(1)}} \mathbf{Z} & \longrightarrow & \operatorname{Pic}(X) & \longrightarrow & 0 & &
\end{array}
$$

de suites exactes de localisation, d'où la suite exacte désirée. $\qquad\square$

Lemme 1.2 *Soit $A^1(\mathcal{A}) \subset \operatorname{Pic}(\mathcal{A})$ le sous-groupe des cycles algébriquement équivalents à zéro. Alors $j^* A^1(\mathcal{A}) \subset \operatorname{Tr}_{K/k} \hat{A}(k)$, où $\hat{A} = \operatorname{Pic}^0_{A/K}$ est la variété abélienne duale de A.*

Démonstration Par définition, $A^1(\mathcal{A})$ est engendré via les correspondances algébriques par des jacobiennes de courbes, donc le lemme résulte de la définition de la K/k-trace.

(Voici quelques précisions demandées par le rapporteur : d'après [4, ex. 10.3.2], comme k est algébriquement clos, tout cycle algébriquement équivalent à zéro sur $\mathcal{A}$ est de la forme $\gamma_* Z$, où $\gamma \in \text{Pic}(C \times \mathcal{A})$ est une correspondance d'une courbe projective lisse C vers $\mathcal{A}$ et $Z \in \text{Pic}^0(C)$. Considérons le morphisme $j' : C_K \times_K A = C \times_k A \hookrightarrow C \times_k \mathcal{A}$: la correspondance $j'^* \gamma \in \text{Pic}(C_K \times_K A)$ définit un homomorphisme de foncteurs représentables sur les K-schémas affines lisses de type fini :

$$f : \text{Pic}^0_{C_K/K} \to \text{Pic}^0_{A/K}.$$

Par la propriété universelle de la K/k-trace, f provient d'un morphisme

$$\tilde{f} : \text{Pic}^0_{C/k} \to \text{Tr}_{K/k} \text{Pic}^0_{A/K},$$

donc $j^* \gamma_* Z \in \text{Tr}_{K/k} \text{Pic}^0_{A/K}(k)$.)   □

Les lemmes 1.1 et 1.2 montrent que j^* induit une surjection

$$\text{NS}(\mathcal{A}) \longrightarrow\!\!\!\!\!\rightarrow \text{Pic}(A)/\text{Tr}_{K/k} \hat{A}(k).$$

Le théorème 1 en résulte, grâce à la génération finie de $\text{NS}(\mathcal{A})$ (proposition 0.1).   □

Remarque 1.1 La démonstration ci-dessus déduit le théorème 1 de la génération finie du groupe de Néron-Severi, via la proposition 0.1. Lang et Néron, quant à eux, faisaient l'inverse dans [9]. Mais la génération finie du groupe de Néron-Severi peut s'obtenir en combinant un argument l-adique et la représentabilité de Pic^τ (*cf.* [8, §1, 4)]), donc l'argument n'est pas circulaire. (Naturellement cette génération finie a une démonstration transcendante facile en caractéristique zéro, *cf.* [8, §1].)

Il est d'ailleurs très intéressant de voir comment Néron procède dans [10], où il démontre que le groupe $\gamma(K)$ des points rationnels de la jacobienne d'une courbe C de genre g définie sur un corps K de type fini sur son sous-corps premier k_0 est de type fini. Le plus simple est de citer un extrait de l'introduction :

> *Nous allons étudier le cas général où, g étant quelconque, K est de caractéristique p quelconque et engendré sur son sous-corps premier par un nombre fini d'éléments. Nous montrerons que, dans ces conditions, le problème se rattache aux propriétés des diviseurs d'une certaine variété algébrique $\mathcal{C}$. Plus précisément, le groupe $\gamma(K)$ attaché à C est isomorphe à un sous-groupe du produit direct des deux groupes suivants : le groupe $\mathcal{G}/\mathcal{G}_a$ des classes de diviseurs algébriquement équivalents sur $\mathcal{C}$, et le groupe des points d'une variété abélienne Ω (voisine de la variété de Picard de $\mathcal{C}$) qui sont rationnels sur un certain corps de définition algébrique de Ω.*
>
> *D'après le théorème de Weil, le second de ces deux groupes est de type fini. Lorsque la caractéristique est nulle, on sait qu'il en est de même du premier, d'après un théorème de Severi; dans ce cas, l'application de ce théorème entraîne donc l'extension annoncée.*
>
> *Mais il était naturel de chercher à retrouver ce résultat au moyen de la méthode "de descente infinie", classique en Arithmétique et utilisée en particulier par Mordell et Weil dans la démonstration des*

*théorèmes cités plus haut. Le prolongement de cette méthode au cas
actuel est possible et conduit en fait à une démonstration du théorème
de Severi par voie algébrique et à une extension de ce théorème au
cas où la caractéristique est quelconque.*

En pratique, Néron considère au chapitre II, no 12 un modèle projectif normal
$\mathcal{M}$ de K/k_0 et un modèle projectif normal $\mathcal{C} \to \mathcal{M}$ de C/K. Il identifie $\gamma(K)$ à
un sous-quotient de ce que nous notons aujourd'hui $CH^1(\mathcal{C})$, puis décrit ce sous-
quotient comme extension d'un sous-quotient de $\mathrm{NS}(\mathcal{C})$ par les points rationnels
d'une certaine variété abélienne Ω définie sur une extension finie de k_0. Cette
variété Ω n'est autre que la K/k-trace de la jacobienne de C, où k est la fermeture
algébrique de k_0 dans k...

2 Un calcul de rang

On peut tirer un peu plus de renseignements de la démonstration ci-dessus :

Proposition 2.1 *On a $j^* A^1(\mathcal{A}) = \mathrm{Tr}_{K/k} \hat{A}(k)$ dans le lemme 1.1, et une suite
exacte scindée*

$$0 \to \mathrm{NS}(X) \xrightarrow{p^*} \mathrm{NS}(\mathcal{A}) \xrightarrow{j^*} \mathrm{Pic}(A)/\mathrm{Tr}_{K/k} \hat{A}(k) \to 0$$

où NS désigne le groupe des cycles de codimension 1 modulo l'équivalence algébrique.

Démonstration Le lemme du serpent appliqué au diagramme commutatif de
suites exactes

$$
\begin{array}{ccccccccc}
0 & \longrightarrow & A^1(\mathcal{A}) & \longrightarrow & \mathrm{Pic}(\mathcal{A}) & \longrightarrow & \mathrm{NS}(\mathcal{A}) & \longrightarrow & 0 \\
& & p_A^* \uparrow & & p^* \uparrow & & p_N^* \uparrow & & \\
0 & \longrightarrow & A^1(X) & \longrightarrow & \mathrm{Pic}(X) & \longrightarrow & \mathrm{NS}(X) & \longrightarrow & 0
\end{array}
$$

fournit, via le lemme 1.1, une suite exacte

$$\mathrm{Ker}\, p_N^* \to \mathrm{Coker}\, p_A^* \xrightarrow{\varphi} \mathrm{Pic}\, A \to \mathrm{Coker}\, p_N^* \to 0.$$

Par le lemme 1.2, $\varphi(\mathrm{Coker}\, p_A^*) \subset \mathrm{Tr}_{K/k} \hat{A}(k)$. D'après la proposition 0.1,
$\mathrm{Coker}\, p_N^*$ est de type fini. Le groupe $\mathrm{Tr}_{K/k} \hat{A}(k)/\varphi(\mathrm{Coker}\, p_A^*)$, divisible et de type
fini, est donc nul. On obtient donc un isomorphisme

$$\mathrm{Pic}(A)/\mathrm{Tr}_{K/k} \hat{A}(k) \xrightarrow{\sim} \mathrm{Coker}\, p_N^*.$$

Enfin, $\mathrm{Ker}\, p_N^* = 0$ car p a une section. $\qquad\square$

Corollaire 2.1 $\mathrm{rg}\left(\hat{A}(K)/\mathrm{Tr}_{K/k} \hat{A}(k)\right) = \rho(\mathcal{A}) - \rho(X) - \rho(\hat{A})$, *où $\rho(Y)$ désigne
le rang de $\mathrm{NS}(Y)$ pour toute variété lisse Y. (Noter que $\rho(\hat{A})$ est calculé "sur K".)*
$\qquad\square$

Remarque 2.1 La proposition 2.1 est parallèle à la proposition 2.2 de Hindry,
Pacheco et Wazir [7], qui donne (dans le cas d'une fibration de variétés propres et
lisses) une suite exacte de variétés de Picard. Mais les deux méthodes de démonstration sont différentes : celle de Hindry-Pacheco-Wazir repose sur une variante relative [7, lemme 2.1] du "théorème de l'indice de Hodge en codimension 1" (Segre-
Grothendieck, [5, ch. V, th. 1.9, p. 364]), alors que la présente méthode, reposant
sur le lemme 1.2, est inspirée d'une idée de Bloch [3, Lect. 1, lemme 1.3].

3 Variantes du corollaire 2.1

Dans ma grande inculture, je n'ai réalisé qu'après avoir obtenu le corollaire 2.1 qu'il était une variante des "formules de Shioda-Tate" [14, 12, 13, 6, 7, 11]. La méthode du §2 est assez flexible pour s'adapter à des situations et besoins divers, pour obtenir des variantes de ce corollaire. Voici quelques exemples :

3.1   On peut toujours remplacer X par un ouvert, de façon à supposer que $\rho(X) = 0$. Ou bien on peut se ramener au cas où X est un ouvert de $\mathbf{P}^n$, quitte à écrire K comme extension finie séparable d'un corps K_0 transcendant pur sur k et à remplacer A par sa restriction des scalaires à la Weil $R_{K/K_0}A$. (Dans ce cas, $\rho(X) = 1$ ou 0.)

3.2   a) Sous la résolution des singularités, on peut prendre une base X projective lisse, ainsi que l'espace total $\mathcal{A}$. Le calcul est alors différent : soit $U \subset X$ l'ouvert maximal de X au-dessus duquel p est lisse. Dans la démonstration du lemme 1.1, la seconde flèche verticale à partir de la gauche reste un isomorphisme au-dessus des points de U, mais les points de codimension 1 $P_1, \ldots, P_n$ de $X - U$ peuvent avoir des fibres réductibles. Soit m_i le nombre de composantes de codimension 1 dans $p^{-1}(\overline{\{P_i\}})$: on obtient alors un complexe

$$0 \to \mathrm{Pic}(X) \xrightarrow{p^*} \mathrm{Pic}(\mathcal{A}) \xrightarrow{j^*} \mathrm{Pic}(A) \to 0$$

exact sauf au milieu, où l'homologie est isomorphe à $\bigoplus_i \mathbf{Z}^{m_i}/\mathbf{Z}$.[1] Dans la démonstration du lemme 2.1, on ne peut plus invoquer le fait que p a une section. Toutefois, $A^1(X)$ et $A^1(\mathcal{A})$ ne sont autres que $\mathrm{Pic}^0(X)$ et $\mathrm{Pic}^0(\mathcal{A})$; de plus, la démonstration du lemme 1.2 montre que le morphisme φ intervenant dans la démonstration de la proposition 2.1 est induit par un morphisme de variétés abéliennes

$$\mathrm{Pic}^0_{\mathcal{A}/k} / \mathrm{Pic}^0_{X/k} \to \mathrm{Tr}_{K/k}\,\hat{A}.$$

Comme $\mathrm{Ker}\,p_N^*$ est de type fini, on en déduit que $\mathrm{Ker}\,\varphi$ *est fini*, d'où un complexe

$$0 \to \mathrm{NS}(X) \xrightarrow{p^*} \mathrm{NS}(\mathcal{A}) \xrightarrow{j^*} \mathrm{Pic}(A)/\mathrm{Tr}_{K/k}\,\hat{A}(k) \to 0$$

qui, modulo des groupes finis, est acyclique partout sauf en $\mathrm{NS}(\mathcal{A})$, où l'homologie est $\bigoplus_i \mathbf{Z}^{m_i}/\mathbf{Z}$. Ceci généralise la preuve de la formule de Shioda-Tate donnée par M. Hindry et A. Pacheco dans [6, §3], avec un argument un peu différent. On en déduit la formule

$$\mathrm{rg}\left(\hat{A}(K)/\mathrm{Tr}_{K/k}\,\hat{A}(k)\right) = \rho(\mathcal{A}) - \rho(X) - \rho(\hat{A}) - \sum_i (m_i - 1)$$

généralisant celle du corollaire 2.1.

b) Dans le cas ci-dessus, on pourrait demander que X et $\mathcal{A}$ soient seulement normaux, dans l'esprit de Néron (*cf.* remarque 1.1). Si X est tout de même lisse, on peut raisonner partiellement comme ci-dessus en remplaçant les groupes de Picard par des groupes de cycles de codimension 1 modulo l'équivalence rationnelle. Sans supposer X lisse, K. Oguiso a fait le même calcul dans un cas particulier [11, th. 1.1]. (Voir *loc. cit.*, th. 3.1 pour une jolie application.)

c) En généralisant encore, on peut considérer un morphisme propre surjectif $\mathcal{A} \to X$ de variétés lisses, à fibre générique A lisse et géométriquement irréductible (on ne

---

[1]On remarquera que les homomorphismes $\mathbf{Z} \to \mathbf{Z}^{m_i}$ sous-jacents à la somme directe peuvent faire intervenir des multiplicités : les groupes notés $\mathbf{Z}^{m_i}/\mathbf{Z}$ sont donc de rang $m_i - 1$, mais peuvent contenir un sous-groupe cyclique de torsion.

suppose plus que A est une variété abélienne). Les résultats sont les mêmes qu'en a) : on peut le voir en se ramenant au cas où X et $\mathcal{A}$ sont propres, grâce au théorème de de Jong. (Voir aussi [7].)

3.3 a) Si K est de degré de transcendance 1 sur k, on peut choisir pour X le modèle projectif lisse de K et prendre pour $\mathcal{A}$ le *modèle de Néron* de A. Comme p est alors lisse et a une section, les calculs du §1 et en particulier le corollaire 2.1 sont les mêmes. (Ici, $\rho(X) = 1$.)

b) Sous la même hypothèse, si A est la jacobienne d'une courbe C (cas auquel on peut toujours essentiellement se ramener en pratique), on peut remplacer $\mathcal{A}$ par une surface projective lisse minimale, fibrée sur X et de fibre générique C. C'est la situation de Hindry et Pacheco [6]. Les résultats sont les mêmes qu'en 3.2 a), *mutatis mutandis*.

4 Groupe de Lang-Néron et 1-motifs

Les calculs du §2 ont une interprétation éclairante en termes de 1-motifs, dans le cadre développé dans [1]. Supposons seulement k parfait ; soit $\mathcal{M}_1$ la catégorie des 1-motifs de Deligne sur k et soit $D^b(\mathcal{M}_1)$ sa catégorie dérivée au sens de [1, déf. 1.5.2]. Soient $\mathrm{LAlb}(X)$ et $\mathrm{LAlb}(\mathcal{A})$ les objets de $D^b(\mathcal{M}_1)$ associés à X et $\mathcal{A}$ par [1, déf. 8.1.1], et soit $\mathrm{LAlb}(\mathcal{A}/X)$ la fibre du morphisme $p_* : \mathrm{LAlb}(\mathcal{A}) \to \mathrm{LAlb}(X)$. D'après [1, cor. 10.2.3], on a, pour toute k-variété lisse Y :

$$\mathrm{L}_i\mathrm{Alb}(Y) = \begin{cases} [\mathbf{Z}[\pi_0(Y)] \to 0] & \text{si } i = 0 \\ [0 \to \mathcal{A}^0_{Y/k}] & \text{si } i = 1 \\ [0 \to \mathrm{NS}^*_{Y/k}] & \text{si } i = 2 \\ 0 & \text{sinon} \end{cases}$$

où $\pi_0(Y), \mathcal{A}^0_{Y/k}$ et $\mathrm{NS}^*_{Y/k}$ désignent respectivement l'ensemble des composantes connexes géométriques, la variété d'Albanese et le dual de Cartier du groupe de Néron-Severi (au sens ci-dessus) de Y ; les $\mathrm{L}_i\mathrm{Alb}(Y)$ sont calculés par rapport à une t-structure convenable sur $D^b(\mathcal{M}_1)$ ($\mathrm{L}_0\mathrm{Alb}(Y)$ et $\mathrm{L}_1\mathrm{Alb}(Y)$ sont des 1-motifs de Deligne, mais en général $\mathrm{L}_2\mathrm{Alb}(Y)$ est un 1-motif "avec cotorsion"). Le lemme 1.1, la proposition 2.1, [1, th. 10.3.2 b)] et un calcul facile de suite exacte donnent de plus

$$\mathrm{L}_i\mathrm{Alb}(\mathcal{A}/X) = \begin{cases} [0 \to \mathrm{Im}_{K/k} A] & \text{si } i = 1 \\ [0 \to M^*] & \text{si } i = 2 \\ 0 & \text{sinon} \end{cases}$$

où $\mathrm{Im}_{K/k} A$ est la K/k-image de A et $M(\bar{k}) := \mathrm{Pic}(A_{\bar{k}})/\mathrm{Tr}_{K/k}\hat{A}(\bar{k})$ est une extension

$$0 \to \hat{A}(\bar{k}K)/\mathrm{Tr}_{K/k}\hat{A}(\bar{k}) \to M(\bar{k}) \to \mathrm{NS}(A_{\bar{k}K}) \to 0$$

$\bar{k}$ étant une clôture algébrique de k. En particulier, $\mathrm{LAlb}(\mathcal{A}/X)$ ne dépend pas du choix de X.

Références

[1] L. Barbieri-Viale, B. Kahn *On the derived category of 1-motives*, prépublication, 2006.

[2] B. Conrad *Chow's K/k-image and K/k-trace, and the Lang-Néron theorem*, Enseign. Math. **52** (2006), 37–108.

[3] S. Bloch Lectures on algebraic cycles, Duke Univ. Math. Series IV, 1980.

[4] W. Fulton *Intersection theory*, Springer, 1984.

[5] R. Hartshorne *Algebraic geometry*, Springer, 1977.

[6] M. Hindry, A. Pacheco *Sur le rang des jacobiennes sur un corps de fonctions*, Bull. Soc. Math. France **133** (2005), 275–295.

[7] M. Hindry, A. Pacheco, R. Wazir *Fibrations et conjecture de Tate*, J. Number Theory **112** (2005), 345–368.

[8] B. Kahn *Sur le groupe des classes d'un schéma arithmétique* (avec un appendice de Marc Hindry), Bull. Soc. Math. France **134** (2006), 395–415.

[9] S. Lang, A. Néron. *Rational points of abelian varieties over function fields*, Amer. J. Math. **81** (1959), 95–118.

[10] A. Néron *Problèmes arithmétiques et géométriques rattachés à la notion de rang d'une courbe algébrique dans un corps*, Bull. SMF **80** (1952), 101–166.

[11] K. Oguiso *Shioda-Tate formula for an abelian fibred variety and applications*, prépublication, 2007, math.AG/0703245.

[12] T. Shioda *On elliptic modular surfaces*, J. Math. Soc. Japan **24** (1972), 20–59.

[13] T. Shioda *Mordell-Weil lattices for higher genus fibration over a curve*, in New Trends in Algebraic Geometry, London Math. Soc. Lect. Notes Series **264** (1999), 359–373.

[14] J. Tate *On the conjectures of Birch and Swinnerton-Dyer and a geometric analog*, Sém. Bourbaki **9** Exp. No. 306, 415–440, Soc. Math. France, Paris, 1995.

Fields Institute Communications
Volume **56**, 2009

Surjectivity of the Cycle Map for Chow Motives

Shun-ichi Kimura

Department of Mathematics
Hiroshima University
1-3-1 Kagamiyama
Higashi-Hiroshima, 739-8526 Japan
kimura@math.sci.hiroshima-u.ac.jp

Abstract. For a smooth projective variety X over $\mathbb{C}$, Uwe Jannsen proved that if the cycle map cl : $\mathrm{CH}^* X_{\mathbb{Q}} \to H^*(X, \mathbb{Q})$ is surjective, then it is actually bijective in [6]. In this paper, we generalize his result to Chow motives. We show that for a Chow motive $M = (X, \alpha, n)$ over a universal domain Ω, if its Chow group $\mathrm{CH}^* M_{\mathbb{Q}}$ is finitely generated as a vector space over $\mathbb{Q}$, then we can write $\alpha = \sum_i \gamma_i \times \delta_i$ with $\gamma_i, \delta_i \in \mathrm{CH}_* X$ satisfying $\gamma_i \cdot \delta_j = \begin{cases} 1 & (i = j), \\ 0 & (i \neq j). \end{cases}$ A possible application to the Hodge conjecture is discussed.

1 Introduction

In [2], Bloch and Srinivas introduced a fundamental technique to prove the degeneracy of the whole motive of X from the degeneracy of the 0-cycles. As an application, they proved that for a smooth algebraic variety X over $\mathbb{C}$, if its 0-cycles are supported on a 3 dimensional subscheme, then Hodge conjecture holds for codimension 2 cycles on X. In [6], Jannsen applied their technique to prove that for a smooth projective variety X over $\mathbb{C}$, the cycle map is surjective if it is injective. Esnault and Levine generalized Jannsen's result to Deligne cohomology in [4]. In particular, for a smooth algebraic variety X over $\mathbb{C}$, if its Chow group is representable, then Hodge conjecture holds for X.

The aim of this paper is to generalize Jannsen's result in a different direction: We prove that the cycle map is surjective if it is injective, for a Chow motive M. Our main theorem says the following.

Theorem 1.1 *(Theorem 3.7) Let $M = (X, \alpha, 0)$ be a Chow motive over a universal domain Ω such that $\mathrm{CH}_* M_{\mathbb{Q}}$ is finite dimensional as a $\mathbb{Q}$-vector space. Then we can write $\alpha = \sum \gamma_i \times \delta_i$ where $\gamma_i, \delta_i \in \mathrm{CH}_* X$ are algebraic cycles such*

2000 *Mathematics Subject Classification.* Primary 14C15; Secondary 14C30.
Partially supported by JSPS Core-to-Core Program, 18005 JAPAN.

that

$$\gamma_i \cdot \delta_j = \begin{cases} 1 & (i = j), \\ 0 & Otherwise. \end{cases}$$

In particular for such a motive, the cycle map is bijective, and all the cohomology classes of $H^*(M, \mathbb{Q})$ are algebraic. Our improvement from Jannsen's result is a small one, but one potentially big merit is that our method may be applicable to any varieties. In fact, if Hodge conjecture holds, then we can always find a Chow motive $M = (X, \alpha, 0)$ such that $CH^* M_{\mathbb{Q}}$ is finitely generated and $H^*(M, \mathbb{Q})$ contains all the Hodge cycles of X. Conversely using Theorem 1.1, we can recover Hodge conjecture for X from the existence of such a motive (Corollary 3.9). From our result, we can claim that if we understand the Chow groups deeply enough, then we can prove Hodge conjecture. In fact, if we understand the Chow group of X, in particular the homologically equivalent to 0 part, well enough, then we can write down the Chow motive $(X, \beta, 0)$ which controls that part of Chow group, then $M = (X, [\Delta_X] - \beta, 0)$ should satisfy the conditions above. Bad news is that we cannot prove Hodge conjecture by this method even for surfaces for now, because we have not understood the Chow groups of surfaces well enough, in particular the Albanese kernel.

The basic idea to prove our result is just to follow Bloch-Srinivas and Jannsen's line, but by introducing the notion of supported correspondences, we chase the supports of all the intermediate correspondences carefully, to obtain a finer result. In this way, we can argue everything in terms of algebraic cycles, without referring to any particular cohomology theory to obtain our main theorem.

Notation: In this paper, all Chow groups are with rational coefficients. We work in the category of algebraic schemes over a fixed universal domain Ω, unless otherwise stated. Varieties are irreducible and reduced algebraic schemes over a field.

2 Correspondence with support

Definition 2.1 Let X and Y be smooth complete varieties over Ω. A *correspondence of degree r* is an element of $CH^{d+r}(X \times Y)$, where $d = \dim Y$. When α is a correspondence from X to Y, we write as $\alpha : X \vdash Y$. When $\alpha : X \vdash Y$ and $\beta : Y \vdash Z$ are correspondences, we define their composition $\beta \circ \alpha : X \vdash Z$ to be $\pi_{XZ*}(\pi_{XY}^* \alpha \cdot \pi_{YZ}^* \beta) \in CH^*(X \times Z)$, where π_{XZ}, π_{XY} and π_{YZ} are the projections from $X \times Y \times Z$.

Remark 2.2 See [5, Chapter 16] for basic properties of correspondences. Correspondence can be regarded as a "generalized morphism", whose "graph" is the given element of the Chow group.

Definition 2.3 Let X and Y be smooth complete varieties, and $S \subseteq X$ and $T \subseteq Y$ be closed subschemes. A correspondence $\alpha : X \vdash Y$ is said to be *supported on $S \times T$ represented by $\tilde{\alpha}$* when $\tilde{\alpha} \in CH_*(S \times T)$ and $\alpha = i_* \tilde{\alpha}$ where $i : S \times T \hookrightarrow X \times Y$ is the closed immersion. When confusion is unlikely, we simply write $\tilde{\alpha} : S \vdash T$ to denote the correspondence $\alpha : X \vdash Y$ represented by $\tilde{\alpha}$. Notice that S and T may be singular and/or reducible.

Remark 2.4 When $\alpha : X \vdash Y$ is supported on $S \times T$ and $\beta : X \vdash Y$ is supported on $U \times V$, their sum $\alpha + \beta : X \vdash Y$ is supported on $(S \cup U) \times (T \cup V)$.

Remark 2.5 One can also define the notion of correspondence supported on $S \subset X \times Y$ for any closed subscheme S. In this paper, we don't need this generality.

Lemma 2.6 *Let X, Y and Z be smooth complete varieties, $S \subseteq X$ and $T, U \subseteq Y$ and $V \subseteq Z$ be closed subschemes, and $\alpha : X \vdash Y$ and $\beta : Y \vdash Z$ be correspondences supported on $S \times T$ and $U \times V$ represented by $\tilde{\alpha} : S \vdash T$ and $\tilde{\beta} : U \vdash V$ respectively. Then the composition $\beta \circ \alpha$ is supported on $S \times V$ represented by $\pi_{SV*}\left(\Delta_Y^!(\tilde{\alpha} \times \tilde{\beta})\right) \in \mathrm{CH}_*(S \times V)$ in the diagram below, where $(S \times T) \to T \to Y$ and $(U \times V) \to V \to Y$ are the projections composed with the closed immersions, and $\Delta_Y : Y \to Y \times Y$ is the diagonal morphism:*

$$
\begin{array}{ccc}
S \times V \xleftarrow{\ \pi_{SV}\ } S \times (T \times_Y U) \times V & \longrightarrow & (S \times T) \times (U \times V) \\
& \downarrow & \downarrow \\
Y & \xrightarrow{\ \ \Delta_Y\ \ } & Y \times Y
\end{array}
$$

The proof is straightforward, and left to the reader.

Definition 2.7 Using the notation as in Lemma 2.6, we define *the refined composition* of correspondences $\tilde{\alpha} : S \vdash T$ and $\tilde{\beta} : U \vdash V$ to be the correspondence represented by $\pi_{SV*}\left(\Delta_Y^!(\tilde{\alpha} \times \tilde{\beta})\right) : S \vdash V$ where $\pi_{SV} : S \times Y \times V \to S \times V$ is the projection.

Lemma 2.8 *Let X and Y be smooth complete varieties, $S \subseteq X$ and $T \subseteq Y$ be closed subschemes, $\alpha : X \vdash Y$ a correspondence supported on $S \times T$ represented by $\tilde{\alpha}$, and $\tilde{S} \to S$ and $\tilde{T} \to T$ be alterations from regular schemes $\tilde{S}$ and $\tilde{T}$ [3]. Then there exists a correspondence $A : \tilde{S} \vdash \tilde{T}$ such that $\tilde{\alpha} = [^t\Gamma_{\tilde{S} \to S}] \circ A \circ [\Gamma_{\tilde{T} \to T}] : S \vdash T$ as supported correspondences, where $\Gamma_{\tilde{S} \to S}$ and $\Gamma_{\tilde{T} \to T}$ are the graphs of the indicated morphisms.*

Proof As the morphism $\pi : \tilde{S} \times \tilde{T} \to S \times T$ is proper surjective, the pushforward $\pi_* : \mathrm{CH}_*(\tilde{S} \times \tilde{T}) \to \mathrm{CH}_*(S \times T)$ is also surjective with rational coefficients. Pick $A \in \mathrm{CH}_*(\tilde{S} \times \tilde{T})$ so that $\pi_* A = \tilde{\alpha}$. Now easy calculation shows that $\tilde{\alpha} = [\Gamma_{\tilde{T} \to T}] \circ A \circ [^t\Gamma_{\tilde{S} \to S}] : S \vdash T$, as supported correspondences. $\square$

Proposition 2.9 *Let X, Y and Z be smooth complete varieties, $S \subseteq X$ and $T, U \subseteq Y$ and $V \subseteq Z$ closed subschemes, and $\alpha : X \vdash Y$ (resp. $\beta : Y \vdash Z$) correspondence of degree 0, supported on $S \times T$ represented by $\tilde{\alpha}$ (resp. supported on $U \times V$, represented by $\tilde{\beta}$).*

(1) *When $\dim S + \dim V < \dim X$, then the refined composition $\tilde{\beta} \circ \tilde{\alpha}$ is 0.*

(2) *When $\dim T + \dim U < \dim Y$, then the refined composition $\tilde{\beta} \circ \tilde{\alpha}$ is 0.*

Proof For (1), the refined composition is an element of $\mathrm{CH}_{\dim X}(S \times V)$, which is the 0 group.

For (2), by Lemma 2.8, we can write $\alpha = [\Gamma_{\tilde{T} \to T}] \circ A$ and $\beta = B \circ [^t\Gamma_{\tilde{U} \to U}]$ as supported correspondences, where $\tilde{T} \to T$ and $\tilde{U} \to U$ are alterations from regular schemes, and $A : X \vdash \tilde{T}$ and $B : \tilde{U} \vdash Z$ are suitable correspondences supported on $S \times \tilde{T}$ and $\tilde{U} \times V$. Now it is enough to show that the composition $[^t\Gamma_{\tilde{U} \to Y}] \circ [\Gamma_{\tilde{T} \to Y}] : \tilde{T} \vdash \tilde{U}$ is 0. This composition is equal to $\Delta_Y^!\left([\tilde{T} \times \tilde{U}]\right)$ in the diagram below, which

lies in $\displaystyle\bigoplus_{i \le \dim T + \dim U - \dim Y}$ $\mathrm{CH}_i(\tilde{T} \times_Y \tilde{U})$, and as this Chow group is the 0 group, we
conclude $\tilde{\beta} \circ \tilde{\alpha} = 0$ under the condition (2).

$$
\begin{array}{ccc}
\tilde{T} \times_Y \tilde{U} & \longrightarrow & \tilde{T} \times \tilde{U} \\
\downarrow & & \downarrow \\
Y & \xrightarrow{\ \Delta_Y\ } & Y \times Y
\end{array}
$$

$\square$

3 Main result

Let us start this section by recalling the technique by Bloch-Srinivas [2].

Proposition 3.1 *When* $X = \varprojlim X_i$ *is the projective limit of Noetherian schemes with affine flat transition morphisms, we have* $\mathrm{CH}^p X = \varinjlim \mathrm{CH}^p X_i$.

Proof See [7, Section 7, Thm 5.4]. $\square$

Proposition 3.2 *Let* X *be a scheme over a field* k, $k \to K$ *is a field extension, and* $X_K := X \times_k K$ *be the base extension by the field extension, then the pull-back* $\mathrm{CH}^p X \to \mathrm{CH}^p X_K$ *is injective.*

Proof As X_K is the inverse limit of $X \times_k L$ with $K \supset L \supset k$ a middle field with $L \supset k$ finitely generated, by Proposition 3.1, we may assume that $K \supset k$ is finitely generated. Using the induction on the number of generators, we may assume that $K = k(\alpha)$ for some $\alpha \in K$. When α is algebraic over k, by usual norm argument, $\mathrm{CH}^p X$ injects into $\mathrm{CH}^p X_K$ with rational coefficients. When α is transcendental, X_K is the inverse limit of $X \times U$ with $U \subset \mathbb{A}^1$ open subscheme. Again by Proposition 3.1, it is enough to prove the injectivity for the pull-back $\mathrm{CH}^p X \to \mathrm{CH}^p(X \times U)$. Take a closed point $P \in U$, we have a Gysin map $\mathrm{CH}^p(X \times U) \to \mathrm{CH}^p(X \times P)$, and the composition $\mathrm{CH}^p X \to \mathrm{CH}^p(X \times U) \to \mathrm{CH}^p(X \times P)$ is injective by the algebraic extension case, so $\mathrm{CH}^p X \to \mathrm{CH}^p(X \times U)$ is also injective. $\square$

Lemma 3.3 *Let* $K \subset \Omega$ *be a subfield. When* X_K *is a d-dimensional smooth complete variety over* K, $S_K \subset X_K$ *a closed subscheme,* $\alpha_K \in \mathrm{CH}^d(X_K \times_K X_K)$ *an element of the Chow group. Define* $X := X_K \times_K \Omega$, $S := S_K \times_K \Omega$ *and define* $\alpha \in \mathrm{CH}^d(X \times_\Omega X)$ *to be the pull-back of* α_K. *We fix an irreducible component* $S_i \subset S_K$, *let* $Q(S_i)$ *be the function field of* S_i, *and assume that there is an inclusion of fields* $Q(S_i) \hookrightarrow \Omega$. *Then the following diagram commutes, where the top horizontal arrow is the composition with the correspondence* α_K, *the vertical arrows are the flat pull-backs by the base extension of* $\mathrm{Spec}\,\Omega \to S_K$, *and the bottom horizontal arrow is the action by* α.

$$
\begin{array}{ccc}
\mathrm{CH}^p(S_K \times_K X_K) & \xrightarrow{\ \alpha_K \circ\ } & \mathrm{CH}^p(S_K \times_K X_K) \\
\downarrow & & \downarrow \\
\mathrm{CH}^p(X) & \xrightarrow{\quad \alpha_* \quad} & \mathrm{CH}^p(X)
\end{array}
$$

Proof Horizontal arrows are the compositions of the exterior products with α_K, the Gysin map by $\Delta_{X_K} : X_K \to X_K \times_K X_K$, and the proper push-forward by the projections. All these operations are known to commute with flat pull-backs for algebraic schemes [5]. As the vertical arrows are the limits of the affine flat morphisms of algebraic schemes, the commutativity follows from Proposition 3.1. □

The following Proposition is the main technical result of this paper.

Proposition 3.4 *Let X be a d-dimensional smooth complete variety over Ω, m an integer with $0 \leq m < d$, $\alpha_m : X \vdash X$ be a correspondence of degree 0 supported on $S_m \times X$, represented by $\tilde{\alpha}_m : S_m \vdash X$ with S_m purely $d - m$ dimensional closed subscheme of X. We assume that $\tilde{\alpha}_m$ is an idempotent in the following sense; the refined composition $\tilde{\alpha}_m \circ \tilde{\alpha}_m$ equals $\tilde{\alpha}_m : S_m \vdash X$. Also we assume that the image of $\alpha_{m*} : \mathrm{CH}_* X \to \mathrm{CH}_* X$ is finitely generated as a vector space over $\mathbb{Q}$. Then there exist a purely 1-codimensional closed subscheme $S_{m+1} \subset S_m$, a purely m-dimensional closed subscheme $T_m \subset X$, correspondences $\alpha_{m+1} : X \vdash X$ supported on $S_{m+1} \times X$ represented by $\tilde{\alpha}_{m+1}$, and $\beta_m : X \vdash X$ supported on $S_m \times T_m$ represented by $\tilde{\beta}_m$ with the following properties:*

1. *$\alpha_m = \alpha_{m+1} + \beta_m$ as correspondences supported on $S_m \vdash X$ (cf, Remark 2.4).*
2. *The supported correspondence α_{m+1} is an idempotent in the following sense: The refined composition $\tilde{\alpha}_{m+1} \circ \tilde{\alpha}_{m+1} : S_{m+1} \vdash X$ equals $\tilde{\alpha}_{m+1}$.*
3. *The supported correspondence β_m is also an idempotent: The refined composition $\tilde{\beta}_m \circ \tilde{\beta}_m : S_m \vdash T_m$ equals $\tilde{\beta}_m$.*
4. *The supported correspondences α_{m+1} and β_m are orthogonal to each other in the following sense: the refined compositions $\tilde{\alpha}_{m+1} \circ \tilde{\beta}_m : S_m \vdash X$ and $\tilde{\beta}_m \circ \tilde{\alpha}_{m+1} : S_{m+1} \vdash T_m$ are 0.*

Proof As the image of $\alpha_{m*} : \mathrm{CH}_m X \to \mathrm{CH}_m X$ is finitely generated, we take a finite set of the generators and fix it, and let T_m be the union of the supports of the generators. Decompose S_m into irreducible components $S_m = \bigcup S_{m,i}$. Let $\mathbb{F}$ be the prime field in Ω, and $K \supset \mathbb{F}$ be a finitely generated field extension such that the schemes and correspondences $X, S_m, S_{m,i}, T_m$ and α_m are defined over K. By Proposition 3.2, $\alpha_{m,K} : S_{m,K} \vdash X_K$ is also idempotent. We denote the schemes and correspondences defined over K as $X_K, S_{m,K}, S_{m,i,K}, T_{m,K}$ and $\alpha_{M,K}$ respectively. Let $Q(S_{m,i,K})$ be the function field, then because Ω is a universal domain, the field inclusion $K \to \Omega$ can be extended to $Q(S_{m,i,K}) \to \Omega$. We fix this inclusion for each $Q(S_{m,i,k})$.

Consider the following diagram:

$$
\begin{array}{ccccc}
\mathrm{CH}^{d\text{-}m}(S_{m,K} \times_K X_K) & \xrightarrow{\;\alpha_{m,K} \circ\; -\;} & \mathrm{CH}^{d\text{-}m}(S_{m,K} \times_K X_K) & \longrightarrow & \mathrm{CH}^{d\text{-}m}(S_{m,K} \times_K (X_K \backslash T_{m,K})) \\
\Big\downarrow & & \Big\downarrow & & \Big\downarrow \\
 & & & & \underset{i}{\oplus}\, \mathrm{CH}^{d\text{-}m}(Q(S_{m,i,K}) \times_K (X_K \backslash T_{m,K})) \\
 & & & & \Big\uparrow \\
\underset{i}{\oplus}\, \mathrm{CH}^{d\text{-}m} X & \xrightarrow{\;\oplus\, \alpha_{m*}\;} & \underset{i}{\oplus}\, \mathrm{CH}^{d\text{-}m} X & \longrightarrow & \underset{i}{\oplus}\, \mathrm{CH}^{d\text{-}m}(X \backslash T_m)
\end{array}
$$

The left square commutes by Lemma 3.3. Also the right rectangle commutes, because all the homomorphisms are flat pull-backs, which are functorial.

Pick $\alpha_{m,K} \in \mathrm{CH}^{d-m}(S_{m,K} \times_K X_K)$ at the top left corner, and send it to $\bigoplus_i \mathrm{CH}^{d-m}(X \backslash T_m)$ at the bottom right corner. By the definition of T_m, any image of α_{m*} in $\mathrm{CH}^{d-m}(X)$ restricted to $X \backslash T_m$ is 0, hence the image of $\alpha_{m,K}$ via the bottom horizontal arrows is 0. By the commutativity of the diagram and the by the injectivity of the vertical arrow at the bottom right, the image of $\alpha_{m,K}$ at $\bigoplus_i \mathrm{CH}^{d-m}(Q(S_{m,i,K}) \times_K (X_K \backslash T_{m,K}))$ is already 0. As the $\alpha_{m,K}$ at the top left corner is sent to $\alpha_{m,K} \circ \alpha_{m,K} = \alpha_{m,K}$ at the top middle, we conclude that $\alpha_{m,K} \in \mathrm{CH}^{d-m}(S_{m,K} \times_K X_K)$ becomes 0 when it is restricted to $Q(S_{m,i,K}) \times_K (X_K \backslash T_{m,K})$.

Proposition 3.1 implies that there exists a dense open subscheme $U \subset S_{m,K}$ such that $\alpha_{m,K}$ is 0 at $\mathrm{CH}^{d-m}(U \times_K (X \backslash T_{m,K}))$. Let $S_{m+1,K} \subset S_{m,K}$ be a purely 1-codimensional subscheme of $S_{m,K}$ such that $S_{m+1,K} \supset S_{m,K} \backslash U$. Because $\alpha_{m,K}$ goes to 0 when it is restricted to the complement of $(S_{m+1,K} \times X_K) \cup (S_{m,K} \times T_{m,K})$, one can write $\alpha_{m,K} = \alpha'_{m+1,K} + \beta'_{m,K}$, where $\alpha'_{m+1,K}$ is supported on $S_{m+1,K} \times X_K$ represented by $\tilde{\alpha}'_{m+1,K}$, and $\beta'_{m,K}$ is supported on $S_{m,K} \times T_{m,K}$ represented by $\tilde{\beta}'_{m,K}$.

Let $\tilde{\beta}'_m := \tilde{\beta}'_{m,K} \times_K \Omega : S_m \vdash T_m$, and define $\tilde{\beta}_m : S_m \vdash T_m$ to be the refined composition $\tilde{\beta}'_m \circ \alpha_m : S_m \vdash T_m$. Define $\beta_m : X \vdash X$ to be the supported correspondence represented by $\tilde{\beta}_m$. We also define $\tilde{\alpha}'_{m+1} := \tilde{\alpha}'_{m+1,K} \times_K \Omega$, and define $\tilde{\alpha}_{m+1} : S_{m+1} \vdash X$ to be the refined composition $\alpha_m \circ \tilde{\alpha}'_{m+1}$ and $\alpha_{m+1} : X \vdash X$ to be the supported correspondence represented by $\tilde{\alpha}_{m+1}$. We also take β'_m and α'_{m+1} as the images of $\tilde{\beta}'_m$ and $\tilde{\alpha}'_{m+1}$ in $\mathrm{CH}_d(S_m \times X)$. We need to show that all these constructions satisfy the conditions (1)~(4).

By Proposition 2.9, the refined compositions $\tilde{\alpha}_{m+1} \circ \tilde{\beta}_m$ and $\tilde{\beta}_m \circ \tilde{\alpha}_{m+1}$ are 0, hence (4) is satisfied.

We prove that $\tilde{\alpha}_{m+1} : S_{m+1} \vdash X$ is idempotent. As $\alpha_m : S_m \vdash X$ is idempotent and $\tilde{\alpha}_{m+1}$ is defined to be $\alpha_m \circ \tilde{\alpha}'_{m+1}$, we have $\tilde{\alpha}_{m+1} = \alpha_m \circ \alpha_m \circ \tilde{\alpha}_{m+1}$. We substitute $\alpha_m = \tilde{\alpha}'_{m+1} + \tilde{\beta}'_m$ to obtain $\tilde{\alpha}_{m+1} = \alpha_m \circ (\tilde{\alpha}'_{m+1} + \tilde{\beta}'_m) \circ \tilde{\alpha}_{m+1}$. By Proposition 2.9, we have $\tilde{\beta}'_m \circ \tilde{\alpha}_{m+1} = 0$, hence $\tilde{\alpha}_{m+1} = \alpha_m \circ \tilde{\alpha}'_{m+1} \circ \tilde{\alpha}_{m+1} = \tilde{\alpha}_{m+1} \circ \tilde{\alpha}_{m+1}$, which shows (2).

For the idempotency of $\tilde{\beta}_m$, we can proceed similarly:

$$
\begin{aligned}
\tilde{\beta}_m &= \tilde{\beta}'_m \circ \alpha_m \\
&= \tilde{\beta}'_m \circ \alpha_m \circ \alpha_m \circ \alpha_m \\
&= \tilde{\beta}_m \circ (\tilde{\alpha}'_{m+1} + \tilde{\beta}'_m) \circ \alpha_m \\
&= \tilde{\beta}_m \circ \tilde{\beta}'_m \circ \alpha_m \\
&= \tilde{\beta}_m \circ \tilde{\beta}_m.
\end{aligned}
$$

(3) is proved.

As α_m is idempotent and $\alpha'_{m+1} \circ \beta'_m = \beta'_m \circ \alpha'_{m+1} = 0$ by Proposition 2.9, we have

$$
\begin{aligned}
\alpha_m &= \alpha_m \circ \alpha_m \\
&= (\alpha'_{m+1} + \beta'_m) \circ (\alpha'_{m+1} + \beta'_m) \\
&= \alpha'_{m+1} \circ \alpha'_{m+1} + \beta'_m \circ \beta'_m,
\end{aligned}
$$

where all the cycles are considered to be supported on $S_m \times X$. As $\alpha_{m+1} = (\alpha'_{m+1} + \beta'_m) \circ \alpha'_{m+1} = \alpha'_{m+1} \circ \alpha'_{m+1}$ and $\beta_m = \beta'_m \circ (\alpha'_{m+1} + \beta'_m) = \beta'_m \circ \beta'_m$

in $\mathrm{CH}_*(S_m \times X)$, we conclude $\alpha_{m+1} + \beta_m = \alpha_m$ as a supported correspondence $S_m \vdash X$. The condition (1) is proved. $\qquad \square$

Corollary 3.5 *In the situation of Proposition 3.4, the image of α_{m+1*} :* $\mathrm{CH}_*X \to \mathrm{CH}_*X$ *is finitely generated as a vector space over $\mathbb{Q}$.*

Proof The idempotent correspondence α_m decomposes into the sum of two orthogonal idempotents, α_{m+1} and β_m. Therefore, the Chow Motive $(X, \alpha_{m+1}, 0)$ is a direct summand of $(X, \alpha_m, 0)$, which implies that the image of $\alpha_{m+1*} : \mathrm{CH}_*X \to \mathrm{CH}_*X$ is a direct summand of the image of $\alpha_{m*} : \mathrm{CH}_*X \to \mathrm{CH}_*X$, a finite dimensional $\mathbb{Q}$-vector space, hence it is also finite dimensional. $\qquad \square$

We need a small lemma to prove our main result.

Lemma 3.6 *Let s and t be positive integers, and assume that C is an $s \times t$ matrix, A a $t \times s$ matrix, both $\mathbb{Q}$ coefficients, such that $CAC = C$. When $r = \mathrm{rk}\, C$, then there are $1 \times s$ matrices (namely row vectors) $v_1, v_2, \ldots, v_r$ and $t \times 1$ matrices (column vectors) $w_1, \ldots, w_r$ such that $C = \displaystyle\sum_{i=1}^{r} w_i v_i$ and $v_i A w_j = \begin{cases} (1) & (i = j), \\ (0) & (i \neq j). \end{cases}$*

Proof Let $v_1, \ldots, v_r$ be a basis of the vector space spanned by the row vectors of C. Then as $CAC = C$, we have $\mathrm{rk}\, CA = r$, and hence $v_1 A, v_2 A, \ldots, v_r A$ is a basis of the vector space spanned by the row vectors of CA. As $\mathrm{rk}\, (CA)C = r$, one can find the basis $\{w_1, w_2, \ldots, w_r\}$ of the vector space spanned by the columns of C so that $(v_i A) w_j = \begin{cases} (1) & (i = j), \\ (0) & (i \neq j). \end{cases}$

We claim that $C = \displaystyle\sum_{i=1}^{r} w_i v_i$. When we write the k-th column vector of C as $\sum \alpha_{jk} w_j$, namely if

$$C = \left(\sum_j \alpha_{j1} w_j, \sum_j \alpha_{j2} w_j, \ldots, \sum_j \alpha_{jt} w_j \right),$$

then we have $(v_i A) C = (\alpha_{i1}, \alpha_{i2}, \ldots, \alpha_{it})$, hence we obtain

$$\left(\sum_{i=1}^{r} w_i v_i \right) AC = \sum_{i=1}^{r} w_i (\alpha_{i1}, \alpha_{i2}, \ldots, \alpha_{it}) = C.$$

Hence we have $(C - \sum w_i v_i) AC = C - C = 0$. As the rows of $(C - \sum w_i v_i)$ is contained in the vector space spanned by the vectors $v_1, \ldots, v_r$, and for this vector space, the right multiplication by AC acts as injection because $\mathrm{rk}\, C(AC) = \mathrm{rk}\, C$, we conclude $C = \displaystyle\sum_{i=1}^{r} w_i v_i$ as required. $\qquad \square$

Now we are ready to prove our main result.

Theorem 3.7 *Let $M = (X, \alpha, 0)$ be a Chow motive such that CH_*M is finite dimensional as a $\mathbb{Q}$-vector space. Then we can write $\alpha = \sum \gamma_i \times \delta_i$ where $\gamma_i, \delta_i \in \mathrm{CH}_*X$ are algebraic cycles such that*

$$\gamma_i \cdot \delta_j = \begin{cases} 1 & (i = j), \\ 0 & Otherwise. \end{cases}$$

Proof Starting by $\alpha_0 := \alpha$, we can apply Proposition 3.4 to find β_m and α_{m+1} from α_m inductively. We obtain the decomposition $\alpha = \beta_0 + \beta_1 + \cdots + \beta_d$ where $d = \dim X$, and each correspondence $\beta_m : X \vdash X$, supported on $S_m \times T_m$ represented by $\tilde{\beta}_m$, is idempotent in the sense that the refined composition $\tilde{\beta}_m \circ \tilde{\beta}_m = \tilde{\beta}_m : S_m \vdash T_m$. As $\dim S_m \times T_m = d$, there is no non-trivial rational equivalence in $\mathrm{CH}^0(S_m \times T_m) \simeq Z^0(S_m \times T_m)$. Let us decompose S_m and T_m into the irreducible components as $S_m = \bigcup_{i=1}^{s} S_{m,i}$ and $T_m = \bigcup_{j=1}^{t} T_{m,j}$. We write the correspondence $\tilde{\beta}_m$ as $\tilde{\beta}_m = \sum c_{i,j}[S_{m,i} \times T_{m,j}]$ and take the matrix C to be $C = (c_{i,j})$. Also take the matrix $A = (a_{k,\ell})$ with $a_{k,\ell} = \deg([T_{m,k}] \cdot [S_{m,\ell}])$, the intersection numbers over X. Then the idempotency $\tilde{\beta}_m \circ \tilde{\beta}_m = \tilde{\beta}_m$ interprets as $CAC = C$, hence by Lemma 3.6, we can find vectors $v_1, \ldots, v_r$ and $w_1, \ldots, w_r$ so that $C = \sum w_i \times v_i$ and $v_i A w_j = \begin{cases} (1) & (i = j), \\ (0) & (i \neq j). \end{cases}$ Write $v_i = (e_{i,1}, e_{i,2}, \ldots, e_{i,t})$, and define $\delta_i = \sum_{i=1}^{r} e_{i,j}[T_{m,j}]$. Also write $w_j = {}^t(f_{1,j}, f_{2,j}, \ldots, f_{s,j})$ and define $\gamma_j = \sum_{j=1}^{r} f_{i,j}[S_{m,i}]$, then we have $\tilde{\beta}_m = \sum \gamma_i \times \delta_i$ and $\gamma_i \cdot \delta_j = \begin{cases} 1 & (i = j), \\ 0 & (i \neq j). \end{cases}$ We have written down only $\tilde{\beta}_m$ (hence β_m) in the required form, but it implies that the sum α also has the same form, because if γ_i and δ_j come from different m, then their intersection number is 0 by the dimension reason. $\square$

Remark 3.8 When κ is a coefficient field with characteristic 0, if one can show that $\mathrm{CH}^* M \otimes \kappa$ is finitely generated κ vector space, then it implies that $\mathrm{CH}^* M$ is finitely generated $\mathbb{Q}$ vector space (the assumption of Theorem 3.7), because $\dim_\kappa \mathrm{CH}^* M \otimes \kappa = \dim_\mathbb{Q} \mathrm{CH}^* M$. In particular, if the cycle map to a cohomology theory with coefficient field κ is injective, then we can conclude $M = (X, \sum \gamma_i \times \delta_i, *)$ by Theorem 3.7.

Corollary 3.9 *Let X be a smooth projective variety over $\mathbb{C}$. Assume that there exists a Chow motive $M = (X, \alpha, 0)$ such that*

(1) *$\mathrm{CH}^* M$ is a finitely generated as $\mathbb{Q}$-vector space.*
(2) *The cohomology group $H^*(M, \mathbb{C})$ contains all the Hodge cycles.*

Then Hodge conjecture holds for X. Conversely if Hodge conjecture holds for X, then one can find the Chow motive M which satisfies the conditions (1) and (2) above.

Proof By the condition (2), M contains all the Hodge cycles of X. By (1), $\mathrm{CH}^* M = \mathrm{CH}^* M$ is a finitely generated as a $\mathbb{Q}$-vector space, hence by Theorem 3.7, we can write $\alpha = \sum \gamma_i \times \delta_i$ with $\gamma_i \cdot \delta_j = \begin{cases} 1 & (i = j), \\ 0 & (i \neq j). \end{cases}$ Then we have $\mathrm{CH}^*(M) \simeq \bigoplus \mathbb{Q} \cdot [\delta_i] \simeq H^*(M, \mathbb{Q})$, hence the cycle map of M is bijective. In particular, all the Hodge cycles are algebraic.

For the converse, notice that in $H^*(X, \mathbb{Q})$, one can find a basis $\{\gamma_i\}$ of the $\mathbb{Q}$ vector space of Hodge cycles, and its dual basis $\{\delta_i\}$ which also consists of Hodge cycles. When Hodge conjecture holds, then all these cycles are algebraic, and it is enough to choose α to be $\sum \gamma_i \times \delta_i$. $\square$

Remark 3.10 Corollary 3.9 implies that if we understand the Chow group, in particular the homologically trivial part, deeply enough, then we can prove Hodge conjecture. In fact then, we should be able to construct a motive $M^\perp = (X, \beta, 0)$ so that $\mathrm{CH}^* M^\perp$ contains all the elements of homologically trivial elements in $\mathrm{CH}^* X$, and $H^*(M^\perp, \mathbb{Q})$ contains no Hodge cycles. Then we can take $\alpha := [\Delta_X] - \beta$.

Remark 3.11 If we try to prove Hodge conjecture for a surface X along the line of Remark 3.10, we have to construct a correspondence $\beta : X \vdash X$ such that the image of $\beta_* : \mathrm{CH}_0 X \to \mathrm{CH}_0 X$ contains all the degree 0 part, including the Albanese Kernel, and the image of $\beta_* : H^2(X, \mathbb{Q}) \to H^2(X, \mathbb{Q})$ contains no Hodge cycles. For now, our understanding of the Albanese Kernel is not enough to admit us this approach.

Example 3.12 Let $X = \mathbb{P}^1 \times E$ where E is a smooth elliptic curve. Fix points $P \in \mathbb{P}^1$ and $Q \in E$. Let $p_1(E) := [\Delta_E] - [Q \times E] - [E \times Q]$ and take the Chow motive $ch^1(E) := (E, p_1(E), 0)$, then the Chow motive $M^\perp := ch(\mathbb{P}^1) \otimes ch^1(E)$ contains all the "non-finitely-generated part" of the Chow group of X, with no Hodge cycles in $H^*(M^\perp)$. The Chow motive

$$M := (\mathbb{P}^1 \times E, [\Delta_{\mathbb{P}^1 \times E}] - [\Delta_{\mathbb{P}^1}] \times p_1(E), 0)$$

satisfies the conditions for Corollary 3.9, which verifies the Hodge conjecture for X. In this case, the projector of M is

$$[P \times Q] \times [\mathbb{P}^1 \times E] + [P \times E] \times [\mathbb{P}^1 \times Q] + [\mathbb{P}^1 \times Q] \times [P \times E] + [\mathbb{P}^1 \times E] \times [P \times Q].$$

References

[1] S. BLOCH, *Lectures on Algebraic Cycles* Duke Univ. 1980

[2] S. BLOCH, V. SRINIVAS, Remarks on correspondences and algebraic cycles, Amer, J. of Math. 105 (1983), 1235–1253

[3] A. J. DE JONG, Smoothness, semi-stability and alterations, Publ. Math. IHES 83 (1996) 51–93

[4] H. ESNAULT, M. LEVINE, Surjectivity of cycle maps, in *Journées de Géométrie Algébrique d'Orsay, Juillet 1992*, Astérisque 218 (1993), 203-226

[5] W. FULTON, *Intersection Theory.* Springer (1984)

[6] U. JANNSEN, Motivic Sheaves and Filtrations on Chow Groups in *Motives*, (Seattle, WA, 1991) Proceedings of Symposia in Pure Mathematics Vol. 55 Part 1 AMS (1994), 245–302

[7] D. QUILLEN, Higher algebraic K-theory. I. In *Algebraic K-theory, I: Higher K-theories*, Lecture Notes in Math., Vol. 341. Springer (1973) 85 – 147

Fields Institute Communications
Volume **56**, 2009

On Codimension Two Subvarieties in Hypersurfaces

N. Mohan Kumar
Department of Mathematics
Washington University in St. Louis
St. Louis, Missouri, 63130 USA
kumar@wustl.edu
http://www.math.wustl.edu/~kumar

A. P. Rao
Department of Mathematics
University of Missouri-St. Louis
St. Louis, Missouri 63121 USA
rao@arch.umsl.edu

G. V. Ravindra
Department of Mathematics
Indian Institute of Science
Bangalore 560012, India
ravindra@math.iisc.ernet.in

Dedicated to Spencer Bloch

Abstract. We show that for a smooth hypersurface $X \subset \mathbb{P}^n$ of degree at least 2, there exist arithmetically Cohen-Macaulay (ACM) codimension two subvarieties $Y \subset X$ which are not an intersection $X \cap S$ for a codimension two subvariety $S \subset \mathbb{P}^n$. We also show there exist $Y \subset X$ as above for which the normal bundle sequence for the inclusion $Y \subset X \subset \mathbb{P}^n$ does not split.

1 Introduction

In this note, we revisit some questions of Griffiths and Harris from 1985 [GH]:

Questions (Griffiths and Harris) Let $X \subset \mathbb{P}^4$ be a general hypersurface of degree $d \geq 6$ and $C \subset X$ be a curve.

1. Is the degree of C a multiple of d?
2. Is $C = X \cap S$ for some surface $S \subset \mathbb{P}^4$?

2000 *Mathematics Subject Classification.* Primary 14F05.
Key words and phrases. Arithmetically Cohen-Macaulay subvarieties, ACM vector bundles.
We thank the referee for pointing out some relevant references.

The motivation for these questions comes from trying to extend the Noether-Lefschetz theorem for surfaces to threefolds. Recall that the Noether-Lefschetz theorem states that if X is a very general surface of degree $d \geq 4$ in $\mathbb{P}^3$, then $\mathrm{Pic}(X) = \mathbb{Z}$, and hence every curve C on X is the complete intersection of X and another surface S.

C. Voisin very soon [Vo] proved that the second question had a negative answer by constructing counterexamples on any smooth hypersurface of degree at least 2. She also considered a third question:

Question With the same terminology and when C is smooth:

3. Does the exact sequence of normal bundles associated to the inclusions $C \subset X \subset \mathbb{P}^4$:

$$0 \to N_{C/X} \to N_{C/\mathbb{P}^4} \to \mathcal{O}_C(d) \to 0$$

split?

Her counterexamples provided a negative answer to this question as well. The first question, the Degree Conjecture of Griffiths-Harris, is still open. Strong evidence for this conjecture was provided by some elementary but ingenious examples of Kollár ([BCC], Trento examples). In particular he shows that if $\gcd(d, 6) = 1$ and $d \geq 4$ and X is a very general hypersurface of degree d^2 in $\mathbb{P}^4$, then every curve on X has degree a multiple of d. In the same vein, van Geemen shows that if $d > 1$ is an odd number and X is a very general hypersurface of degree $54d$, then every curve on X has degree a multiple of $3d$.

The main result of this note is the existence of a large class of counterexamples which subsumes Voisin's counterexamples and places them in the context of arithmetically Cohen-Macaulay (ACM) vector bundles on X. It is well known that ACM bundles which are not sums of line bundles can be found on any hypersurface of degree at least 2 [BGS], and for such a bundle, say of rank r, on X, ACM subvarieties of codimension two can be created on X by considering the dependency locus of $r - 1$ general sections. These subvarieties fail to satisfy Questions 2 and 3. We will be working on hypersurfaces in $\mathbb{P}^n$ for any $n \geq 4$ and our constructions of ACM subvarieties may not give smooth ones. Hence in Question 3, we will consider the splitting of the conormal sheaf sequence instead.

2 Main results

Let $X \subset \mathbb{P}^n$ be a smooth hypersurface of degree $d \geq 2$ and let $Y \subset X$ be a codimension 2 subscheme. Recall that Y is said to be an arithmetically Cohen-Macaulay (ACM) subscheme of X if $\mathrm{H}^i(X, I_{Y/X}(\nu)) = 0$ for $0 < i \leq \dim Y$ and for any $\nu \in \mathbb{Z}$. Similarly, a vector bundle E on X is said to be ACM if $\mathrm{H}^i(X, E(\nu)) = 0$ for $i \neq 0, \dim X$ and for any $\nu \in \mathbb{Z}$.

Let $\mathcal{F}$ be a coherent sheaf on X and let $s_i \in \mathrm{H}^0(\mathcal{F}(m_i))$ for $1 \leq i \leq k$, be generators for the $\oplus_{\nu \in \mathbb{Z}} \mathrm{H}^0(\mathcal{O}_X(\nu))$−graded module $\oplus_{\nu \in \mathbb{Z}} \mathrm{H}^0(\mathcal{F}(\nu))$. These sections give a surjection $\oplus_{i=1}^k \mathcal{O}_X(-m_i) \twoheadrightarrow \mathcal{F}$ which induces a surjection of global section $\oplus_{i=1}^k \mathrm{H}^0(\mathcal{O}_X(\nu - m_i)) \twoheadrightarrow \mathrm{H}^0(\mathcal{F}(\nu))$ for any $\nu \in \mathbb{Z}$.

Applying this to the ideal sheaf $I_{Y/X}$ of an ACM subscheme of codimension 2 in X, we obtain the short exact sequence

$$0 \to G \to \oplus_{i=1}^k \mathcal{O}_X(-m_i) \to I_{Y/X} \to 0,$$

where G is some ACM sheaf on X of rank $k-1$. Since Y is ACM as a subscheme of X, it is also ACM as a subscheme of $\mathbb{P}^n$. In particular, Y is locally Cohen-Macaulay. Hence G is a vector bundle by the Auslander-Buchsbaum Theorem (see [Mat] page 155). We will loosely say that G is associated to Y.

Conversely, the following Bertini type theorem which goes back to arguments of Kleiman in [Kl] (see also [Ban]) shows that given an ACM bundle G on X, we can use G to construct ACM subvarieties Y of codimension 2 in X:

Proposition 2.1 *(Kleiman). Given a bundle G of rank $k-1$ on X, a general map $G \to \oplus_{i=1}^k \mathcal{O}_X(m_i)$ for sufficiently large m_i will determine the ideal sheaf (up to twist) of a subvariety Y of codimension 2 in X with a resolution of sheaves:*

$$0 \to G \to \oplus_{i=1}^k \mathcal{O}_X(m_i) \to I_{Y/X}(m) \to 0.$$

Since the conclusion of Question 2 implies that of Question 3, we will look at just Question 3, in the conormal sheaf version.

Let X be a hypersurface of degree d in $\mathbb{P}^n$ defined by the equation $f = 0$. Let X_2 be the thickening of X defined by $f^2 = 0$ in $\mathbb{P}^n$. Given a subvariety Y of codimension 2 in X, let $I_{Y/\mathbb{P}}$ (resp. $I_{Y/X}$) denote the ideal sheaf of $Y \subset \mathbb{P}^n$ (resp. $Y \subset X$). The conormal sheaf sequence is

$$0 \to \mathcal{O}_Y(-d) \to I_{Y/\mathbb{P}}/I_{Y/\mathbb{P}}^2 \to I_{Y/X}/I_{Y/X}^2 \to 0. \qquad (2.1)$$

Lemma 2.2 *For the inclusion $Y \subset X \subset \mathbb{P}^n$, if the sequence of conormal sheaves (2.1) splits, then there exists a subscheme $Y_2 \subset X_2$ containing Y such that*

$$I_{Y_2/X_2}(-d) \xrightarrow{f} I_{Y_2/X_2} \to I_{Y/X} \to 0$$

is exact. Furthermore, $f I_{Y_2/X_2}(-d) = I_{Y/X}(-d)$.

Proof Suppose sequence (2.1) splits: then we have a surjection

$$I_{Y/\mathbb{P}} \twoheadrightarrow I_{Y/\mathbb{P}}/I_{Y/\mathbb{P}}^2 \twoheadrightarrow \mathcal{O}_Y(-d)$$

where the first map is the natural quotient map and the second is the splitting map for the sequence. The kernel of this composition defines a scheme Y_2 in $\mathbb{P}^n$. Since this kernel $I_{Y_2/\mathbb{P}}$ contains $I_{Y/\mathbb{P}}^2$ and hence f^2, it is clear that $Y \subset Y_2 \subset X_2$.

The splitting of (2.1) also means that $f \in I_{Y/\mathbb{P}}(d)$ maps to $1 \in \mathcal{O}_Y$. We get the commutative diagram:

$$
\begin{array}{ccccccccc}
 & & & & & & 0 & & \\
 & & & & & & \uparrow & & \\
0 & \to & I_{Y_2/\mathbb{P}} & \to & I_{Y/\mathbb{P}} & \to & \mathcal{O}_Y(-d) & \to & 0 \\
 & & \uparrow f^2 & & \uparrow f & & \uparrow & & \\
0 & \to & \mathcal{O}_\mathbb{P}(-2d) & \xrightarrow{f} & \mathcal{O}_\mathbb{P}(-d) & \to & \mathcal{O}_X(-d) & \to & 0 \\
 & & \uparrow & & \uparrow & & & & \\
 & & 0 & & 0 & & & &
\end{array}
$$

This induces

$$0 \to I_{Y/X}(-d) \to I_{Y_2/X_2} \to I_{Y/X} \to 0.$$

In particular, note that $I_{Y/X}(-d)$ is the image of the multiplication map $f : I_{Y_2/X_2}(-d) \to I_{Y_2/X_2}$. $\qquad\square$

Now assume that Y is an ACM subvariety on X of codimension 2. The ideal sheaf of Y in X has a resolution

$$0 \to G \to \oplus_{i=1}^{k} \mathcal{O}_X(-m_i) \to I_{Y/X} \to 0,$$

for some ACM bundle G on X associated to Y.

Lemma 2.3 *Suppose the conditions of the previous lemma hold, and in addition Y is an ACM subvariety. Then there is an extension of the ACM bundle G (associated to Y) on X to a bundle $\mathcal{G}$ on X_2. ie. there is a vector bundle $\mathcal{G}$ on X_2 such that the multiplication map $f : \mathcal{G}(-d) \to \mathcal{G}$ induces the exact sequence $0 \to G(-d) \to \mathcal{G} \to G \to 0$.*

Proof Since Y is ACM, $H^1(I_{Y/X}(-d + \nu)) = 0, \forall \nu$, hence in the sequence stated in the previous lemma, the right hand map is surjective on the level of sections. Therefore, the map $\oplus_{i=1}^{k} \mathcal{O}_X(-m_i) \to I_{Y/X}$ can be lifted to a map $\oplus_{i=1}^{k} \mathcal{O}_{X_2}(-m_i) \to I_{Y_2/X_2}$. Since a global section of $I_{Y_2/X_2}(\nu)$ maps to zero in $I_{Y/X}$ only if it is a multiple of f, by Nakayama's lemma, this lift is surjective at the level of global sections in different twists, and hence on the level of sheaves. Hence there is a commuting diagram of exact sequences:

$$
\begin{array}{ccccccc}
0 & & 0 & & 0 & & \\
\uparrow & & \uparrow & & \uparrow & & \\
I_{Y_2/X_2}(-d) & \to & I_{Y_2/X_2} & \to & I_{Y/X} & \to & 0 \\
\uparrow & & \uparrow & & \uparrow & & \\
\oplus_{i=1}^{k} \mathcal{O}_{X_2}(-m_i - d) & \to & \oplus_{i=1}^{k} \mathcal{O}_{X_2}(-m_i) & \to & \oplus_{i=1}^{k} \mathcal{O}_X(-m_i) & \to & 0 \\
\uparrow & & \uparrow & & \uparrow & & \\
\mathcal{G}(-d) & \to & \mathcal{G} & \to & G & \to & 0 \\
\uparrow & & \uparrow & & \uparrow & & \\
0 & & 0 & & 0 & &
\end{array}
$$

where the sheaf $\mathcal{G}$ is defined as the kernel of the lift, and the map from the left column to the middle column is multiplication by f. It is easy to verify that the lowest row induces an exact sequence

$$0 \to G(-d) \to \mathcal{G} \to G \to 0.$$

By Nakayama's lemma, $\mathcal{G}$ is a vector bundle on X_2. □

Proposition 2.4 *Let E be an ACM bundle on X. If E extends to a bundle $\mathcal{E}$ on X_2, then E is a sum of line bundles.*

Proof There is an exact sequence $0 \to E(-d) \to \mathcal{E} \to E \to 0$, where the left hand map is induced by multiplication by f on $\mathcal{E}$. Let $F_0 = \oplus \mathcal{O}_{\mathbb{P}^n}(a_i) \twoheadrightarrow E$ be a surjection induced by the minimal generators of E. Since E is ACM, this lifts to a map $F_0 \twoheadrightarrow \mathcal{E}$. This lift is surjective on global sections by Nakayama's lemma (since the sections of $\mathcal{E}$ which are sent to 0 in E are multiples of f). Thus we have

a diagram

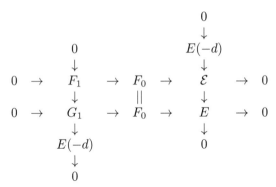

G_1 and F_1 are sums of line bundles on $\mathbb{P}^n$ by Horrocks' Theorem. Furthermore, $G_1 \cong F_0(-d)$. Thus $0 \to F_0(-d) \xrightarrow{\Phi} F_0 \to E \to 0$ is a minimal resolution for E on $\mathbb{P}^n$. As a consquence of this, one checks that $\det \Phi = f^{\operatorname{rank} E}$. On the other hand, the degree of $\det \Phi = d \operatorname{rank} F_0$ and so we have $\operatorname{rank} F_0 = \operatorname{rank} E$. Restricting, this resolution to X, we get a surjection $F_0 \otimes \mathcal{O}_X \twoheadrightarrow E$. The ranks of both vector bundles being the same, this implies that this is an isomorphism. $\qquad \square$

Corollary 2.5 *Let $Y \subset X$ be a codimension 2 ACM subvariety. If the conormal sheaf sequence (2.1) splits, then*
- *the ACM bundle G associated to Y is a sum of line bundles,*
- *there is a codimension 2 subvariety S in $\mathbb{P}^n$ such that $Y = X \cap S$.*

Proof The first statement follows from Lemma 2.3 and Proposition 2.4. For the second statement, since the bundle G associated to Y is a sum of line bundles $\oplus_{i=1}^{k-1} \mathcal{O}_X(-l_i)$ on X, the map $G \to \oplus_{i=1}^{k} \mathcal{O}_X(-m_i)$ can be lifted to a map $\oplus_{i=1}^{k-1} \mathcal{O}_{\mathbb{P}}(-l_i) \to \oplus_{i=1}^{k} \mathcal{O}_{\mathbb{P}}(-m_i)$. The determinantal variety S of codimension 2 in $\mathbb{P}^n$ determined by this map has the property that $Y = X \cap S$. $\qquad \square$

In conclusion, we obtain the following collection of counterexamples:

Corollary 2.6 *If G is an ACM bundle on X which is not a sum of line bundles, and if Y is a subvariety of codimension 2 in X constructed from G as in Proposition 2.1, then Y does not satisfy the conclusion of either Question 2 or Question 3.*

Buchweitz-Greuel-Schreyer have shown [BGS] that any hypersurface of degree at least 2 supports (usually many) non-split ACM bundles. We will give another construction in the next section.

3 Remarks

3.1 The infinitesimal Question 3 was treated by studying the extension of the bundle to the thickened hypersurface X_2. This method goes back to Ellingsrud, Gruson, Peskine and Strømme [EGPS]. If we are not interested in the infinitesimal Question 3, but just in the more geometric Question 2, a geometric argument gives an even easier proof of the existence of codimension 2 ACM subvarieties $Y \subset X$ which are not of the form $Y = X \cap Z$ for some codimension 2 subvariety $Z \subset \mathbb{P}^n$.

Proposition 3.1 *Let E be an ACM bundle on a hypersurface X in $\mathbb{P}^n$ which extends to a sheaf $\mathcal{E}$ on $\mathbb{P}^n$; i.e. there is an exact sequence*

$$0 \to \mathcal{E}(-d) \xrightarrow{f} \mathcal{E} \to E \to 0. \tag{3.1}$$

Then E is a sum of line bundles.

Proof At each point p on X, over the local ring $\mathcal{O}_{\mathbb{P},p}$ the sheaf $\mathcal{E}$ is free, of the same rank as E. Hence $\mathcal{E}$ is locally free except at finitely many points. Let $\mathbb{H}$ be a general hyperplane not passing through these points. Let $X' = X \cap \mathbb{H}$, and $\mathcal{E}', E'$ be the restrictions of $\mathcal{E}, E$ to $\mathbb{H}, X'$.

It is enough to show that E' is a sum of line bundles on X'. This is because any isomorphism $\oplus \mathcal{O}_{X'}(a_i) \to E'$ can be lifted to an isomorphism $\oplus \mathcal{O}_X(a_i) \to E$, as $H^1(E(\nu)) = 0, \forall \nu \in \mathbb{Z}$. The bundle E' on X' is ACM and from the sequence

$$0 \to \mathcal{E}'(-d) \to \mathcal{E}' \to E' \to 0,$$

it is easy to check that $H^i(\mathcal{E}'(\nu)) = 0, \forall \nu \in \mathbb{Z}$, for $2 \le i \le n-2$. Since $\mathcal{E}'$ is a vector bundle on $\mathbb{H}$, we can dualize the sequence to get

$$0 \to \mathcal{E}'^{\vee}(-d) \to \mathcal{E}'^{\vee} \to E'^{\vee} \to 0.$$

$E'^{\vee}$ is still an ACM bundle, hence $H^i(\mathcal{E}'^{\vee}(\nu)) = 0, \forall \nu \in \mathbb{Z}$, and $2 \le i \le n-2$.

By Serre duality, we conclude that $\mathcal{E}'$ is an ACM bundle on $\mathbb{H}$, and by Horrocks' theorem, $\mathcal{E}'$ is a sum of line bundles. Hence, its restriction E' is also a sum of line bundles on X'. □

Proposition 3.2 *Let Y be an ACM subvariety of codimension 2 in the hypersurface X such that the associated ACM bundle G is not a sum of line bundles. Then there is no pure subvariety Z of codimension 2 in $\mathbb{P}^n$ such that $Z \cap X = Y$.*

Proof Suppose there is such a Z. Then there is an exact sequence $0 \to I_{Z/\mathbb{P}}(-d) \to I_{Z/\mathbb{P}} \to I_{Y/X} \to 0$, where the inclusion is multiplication by f, the polynomial defining X. Since Z has no embedded points, $H^1(I_{Z/\mathbb{P}}(\nu)) = 0$ for $\nu << 0$. Combining this with $H^1(I_{Y/X}(\nu)) = 0, \forall \nu \in \mathbb{Z}$, and using the long exact sequence of cohomology, we get $H^1(I_{Z/\mathbb{P}}(\nu)) = 0, \forall \nu \in \mathbb{Z}$.

Now suppose Y has the resolution $0 \to G \to \oplus \mathcal{O}_X(-m_i) \to I_{Y/X} \to 0$. From the vanishing just proved, the right hand map can be lifted to a map $\oplus \mathcal{O}_{\mathbb{P}}(-m_i) \to I_{Z/\mathbb{P}}$, which is easily checked to be surjective (at the level of global sections). It follows that if $\mathcal{G}$ is the kernel of this lift, $\mathcal{G}$ is an extension of G to $\mathbb{P}^n$. By the previous proposition, G is a sum of line bundles. This is a contradiction. □

3.2 Voisin's original example was as follows. Let P_1 and P_2 be two planes meeting at a point p in $\mathbb{P}^4$. The union Σ is a surface which is not locally Cohen-Macaulay at p. Let X be a smooth hypersurface of degree $d > 1$ which passes through p. $X \cap \Sigma$ is a curve Z in X with an embedded point at p. The reduced subscheme Y has the form $Y = C_1 \cup C_2$, where C_1 and C_2 are plane curves. Voisin argues that Y itself does not have the form $X \cap S$ for any surface S in $\mathbb{P}^4$.

We can treat this example from the point of view of ACM bundles. $I_{Z/X}$ has a resolution on X which is just the restriction of the resolution of the ideal of the union $P_1 \cup P_2$ in $\mathbb{P}^4$, *viz.*

$$0 \to \mathcal{O}_X(-4) \to 4\mathcal{O}_X(-3) \to 4\mathcal{O}_X(-2) \to I_{Z/X} \to 0.$$

From the sequence $0 \to I_{Z/X} \to I_{Y/X} \to k_p \to 0$, it is easy to see that Y is ACM, with a resolution

$$0 \to G \to 4\mathcal{O}_X(-2) \oplus \mathcal{O}_X(-d) \to I_{Y/X} \to 0.$$

G is an ACM bundle. If it were a sum of line bundles, comparing the two resolutions, we find that $h^0(G(2)) = 0$ and $h^0(G(3)) = 4$, hence $G = 4\mathcal{O}_X(-3)$. But then

$G \to 4\mathcal{O}_X(-2) \oplus \mathcal{O}_X(-d)$ cannot be an inclusion. Thus G is an ACM bundle which is not a sum of line bundles.

Voisin's subsequent smooth examples were obtained by placing Y on a smooth surface T contained in X and choosing divisors Y' in the linear series $|Y + mH|$ on T. When m is large, Y' can be chosen smooth. In fact, such curves Y' are doubly linked to the original curve Y in X, hence they have a similar resolution $G' \to L \to I_{D'/X} \to 0$, where L is a sum of line bundles and where G' equals G up to a twist and a sum of line bundles.

The fact that G above is not a sum of line bundles is related (via the mapping cone of the map of resolutions) to the fact that k_p itself cannot have a finite resolution by sums of line bundles on X. This follows from the following proposition which provides another argument for the existence of ACM bundles on arbitrary smooth hypersurfaces of degree ≥ 2.

Proposition 3.3 *Let X be a smooth hypersurface in $\mathbb{P}^n$ of degree ≥ 2 with homogeneous coordinated ring S_X. Let L be a linear space (possibly a point or even empty) inside X of codimension r, with homogeneous ideal $I(L)$ in S_X. A free presentation of $I(L)$ of length $r - 2$ will have a kernel whose sheafification is an ACM bundle on X which is not a sum of line bundles.*

Proof It should first be understood that the homogeneous ideal $I(L)$ of the empty linear space will be taken as the irrelevant ideal $(X_0, X_1, \ldots, X_n)$. Let the free presentation of $I(L)$ together with the kernel be

$$0 \to M \to F_{r-2} \to \cdots \to F_0 \to I(L) \to 0,$$

where F_i are free graded S_X modules. Its sheafification looks like

$$0 \to \tilde{M} \to \tilde{F}_{r-2} \to \cdots \to \tilde{F}_0 \to I_{L/X} \to 0.$$

Since L is locally Cohen-Macaulay, $\tilde{M}$ is a vector bundle on X, and since L is ACM, so is $\tilde{M}$. M equals $\oplus_{\nu \in \mathbb{Z}} H^0(\tilde{M}(\nu))$. Hence, $\tilde{M}$ is a sum of line bundles only if M is a free S_X module.

If $\mathbb{H}$ is a general hyperplane in $\mathbb{P}^n$ which meets X and L transversally along $X_{\mathbb{H}}$ and $L_{\mathbb{H}}$ respectively, the above sequences of modules and sheaves can be restricted to give similar sequences in $\mathbb{H}$. The restriction $\tilde{M}_{\mathbb{H}}$ is an ACM bundle on $X_{\mathbb{H}}$.

Repeat this successively to find a maximal and general linear space $\mathbb{P}$ in $\mathbb{P}^n$ which does not meet L. If $X' = X \cap \mathbb{P}$, the restriction of the sequence of S_X modules to X' gives a resolution

$$0 \to M' \to F'_{r-2} \to \cdots \to F'_0 \to S_{X'} \to k \to 0.$$

Localize this sequence of graded $S_{X'}$ modules at the irrelevant ideal $I(L) \cdot S_{X'}$, to look at its behaviour at the vertex of the affine cone over X'. k is the residue field of this local ring. Since X and hence X' has degree ≥ 2, the cone is not smooth at the vertex. By Serre's theorem ([Se], IV-C-3-Cor 2), k cannot have finite projective dimension over this local ring. Hence M' is not a free module. Therefore neither is M. $\square$

3.3   We make a few concluding remarks about Question 1, the Degree Conjecture of Griffiths and Harris. A vector bundle G on a smooth hypersurface X in $\mathbb{P}^4$ has a second Chern class $c_2(G) \in A^2(X)$, the Chow group of codimension 2 cycles. If $h \in A^1(X)$ is the class of the hyperplane section of X, the degree of any element $c \in A^2(X)$ will be defined to be the degree of the zero cycle $c \cdot h \in A^3(X)$. (Note that by the Lefschetz theorem, all classes in $A^1(X)$ are multiples of h.)

With this notation, if E is any bundle on X and Y is a curve obtained from E with the sequence (*vide* Proposition 2.1)

$$0 \to E \to \oplus_{i=1}^{k} \mathcal{O}_X(m_i) \to I_{Y/X}(m) \to 0,$$

a calculation tells us that the degree d of X divides the degree of Y if and only if d divides the degree of $c_2(E)$.

More generally: let Y be any curve in X and resolve $I_{Y/X}$ to get

$$0 \to E \to \oplus_{i=1}^{l} \mathcal{O}_X(b_i) \to \oplus_{i=1}^{k} \mathcal{O}_X(a_i) \to I_{Y/X} \to 0,$$

where E is an ACM bundle on X. Then a similar calculation tells us that the degree d of X divides the degree of Y if and only if d divides the degree of $c_2(E)$.

Hence we may ask the following question which is equivalent to the Degree Conjecture:

ACM Degree Conjecture If X is a general hypersurface in $\mathbb{P}^4$ of degree $d \geq 6$, then for any indecomposable ACM vector bundle E on X, d divides the degree of $c_2(E)$.

The examples created above in Proposition 3.3 satisfy this, when L has codimension > 2 in X. In [MRR], this conjecture is settled for ACM bundles of rank 2 on X.

References

[BCC]   Ballico, E.; Catanese, F.; Ciliberto, C. (Editors) *Classification of irregular varieties. Minimal models and abelian varieties*, Proceedings of the conference held in Trento, December 17–21, 1990. Lecture Notes in Mathematics, 1515. Springer-Verlag, Berlin, 1992. vi+149 pp.

[Ban]   Bǎnicǎ, C., *Smooth reflexive sheaves*, Rev. Roumaine Math. Pures Appl. 36 (1991), no. 9-10, 571–593.

[BGS]   Buchweitz, R.-O.; Greuel, G.-M.; Schreyer, F.-O., *Cohen-Macaulay modules on hypersurface singularities. II*, Invent. Math. 88 (1987), no. 1, 165–182.

[EGPS] Ellingsrud, G.; Gruson, L.; Peskine, C.; Strømme, S. A., *On the normal bundle of curves on smooth projective surfaces*, Invent. Math. 80 (1985), no. 1, 181–184.

[GH]   Griffiths, Phillip; Harris, Joseph., *On the Noether-Lefschetz theorem and some remarks on codimension-two cycles*, Math. Ann. 271 (1985), no. 1, 31–51.

[Kl]   Kleiman, Steven L. *Geometry on Grassmannians and applications to splitting bundles and smoothing cycles*, Inst. Hautes Études Sci. Publ. Math. No. 36 1969 281–297.

[Mat]   Matsumura, H., *Commutative Ring Theory*, Cambridge Univ. Press, 1986.

[MRR]   Mohan Kumar, N.; Rao, A. P.; Ravindra, G. V., *Arithmetically Cohen-Macaulay bundles on three dimensional hypersurfaces*, Int. Math. Res. Not. IMRN 2007, no. 8, Art. ID rnm025, 11 pp.

[Se]   Serre, J.-P., *Algèbre locale. Multiplicités*, Lecture Notes in Mathematics, 11 Springer-Verlag, Berlin-New York 1965 vii+188 pp.

[Vo]   Voisin, Claire, *Sur une conjecture de Griffiths et Harris*, Algebraic curves and projective geometry (Trento, 1988), 270–275, Lecture Notes in Math., 1389, Springer, Berlin, 1989.

Fields Institute Communications
Volume **56**, 2009

Smooth Motives

Marc Levine

Department of Mathematics
Northeastern University
Boston, MA 02115 USA
`marc@neu.edu`

Abstract. Following ideas of Bondarko [4], we construct a DG category whose homotopy category is equivalent to the full subcategory of motives over a base-scheme S generated by the motives of smooth projective S-schemes, assuming that S is itself smooth over a perfect field.

Contents

Introduction

Recently, Bondarko [4] has given a construction of a DG category of motives over a field k, built out of "higher finite correspondences" between smooth projective varieties over k. Assuming resolution of singularities, the homotopy category of this DG category is equivalent to Voevodsky's category of effective geometric motives. The main goal in this paper is to extend this construction to the case of motives over a base-scheme S. For simplicity, we restrict to the case of a regular S, essentially

2000 *Mathematics Subject Classification.* Primary 14C25, 19E15; Secondary 19E08 14F42, 55P42.

Key words and phrases. Motives, motivic cohomology, correspondences, DG category.

The author gratefully acknowledges the support of the Humboldt Foundation, and the support of the NSF via the grant DMS-0457195.

of finite type over a field, although many aspects of the construction should be possible in a more general setting.

If S has positive Krull dimension, one would not expect that the motive of an arbitrary smooth S-scheme be expressible in terms of motives of smooth projective S-schemes. Thus, the category of motives we construct represents a special type of motive over S. Since the Betti realization of our motives will land in the derived category of local systems on S^{an} rather than in the derived category of constructible sheaves, we call our category $SmMot_{gm}^{\mathrm{eff}}(S)$ the category of *smooth effective geometric motives over S*. We have as well a version with the Tate motive inverted, $SmMot_{gm}(S)$. Both $SmMot_{gm}^{\mathrm{eff}}(S)$ and $SmMot_{gm}(S)$ are constructed by taking the homotopy category of suitable DG categories, and then taking an idempotent completion.

We do not construct a tensor structure on $SmMot_{gm}^{\mathrm{eff}}(S)$ or $SmMot_{gm}(S)$. However, after passing to $\mathbb{Q}$-coefficients, we replace our cubical construction with alternating cubes, which makes possible a tensor structure on DG categories whose homotopy categories are equivalent to $SmMot_{gm}^{\mathrm{eff}}(S)_{\mathbb{Q}}$ and $SmMot_{gm}(S)_{\mathbb{Q}}$ (up to idempotent completion).

Our main comparison result involves the categories of motives $DM^{\mathrm{eff}}(S)$ and $DM(S)$ constructed by Cisinski-Déglise. We construct exact functors

$$\rho_S^{\mathrm{eff}} : SmMot_{gm}^{\mathrm{eff}}(S) \to DM^{\mathrm{eff}}(S); \ \rho_S : SmMot_{gm}(S) \to DM(S)$$

and we show

Theorem 1 *Let k be a field. Suppose that S is a smooth k-scheme, essentially of finite type over k. Then ρ_S is a fully faithful embedding.*

This of course implies that ρ_S^{eff} is a faithful embedding, but due to the lack of a cancellation theorem in $DM^{\mathrm{eff}}(S)$, we do not know if ρ_S^{eff} is full.

Our main technical tool is an extension of the Friedlander-Lawson-Voevodsky moving lemmas to the case of cycles on a smooth projective scheme over a regular semi-local base scheme B, with $\mathcal{O}_B$ containing a field. This enables us to extend the fundamental duality theorem for equi-dimensional cycles on smooth projective varieties to smooth projective schemes over a regular semi-local base (over a field). We pass from the semi-local case to an arbitrary regular base (over a field) by making a Zariski sheafification; we were not able to extend the available techniques beyond the semi-local case, so the Zariski sheafification is forced upon us. We do not know if the duality theorem over a general base holds before making the Zariski sheafification.

As hinted above, our interest in these constructions arose from our desire to construct a refined realization functor on the subcategory of $DM(S)$ generated by smooth projective S-schemes. One example, given above, is that we should have a realization functor

$$\Re_B : SmMot_{gm}(S) \to D^b(\mathrm{Loc}/S^{\mathrm{an}}),$$

where $\mathrm{Loc}/S^{\mathrm{an}}$ is the abelian category of local systems of abelian groups on S^{an}, refining the usual Betti realization of $DM_{gm}(S)$ into the derived category of constructible sheaves. Similarly, one should have realizations of $SmMot_{gm}(S)$ to the derived categories of smooth l-adic étale sheaves on $S^{\mathrm{ét}}$ or variations of mixed Hodge structures on S^{an}. By our main theorem, we can view the triangulated category $DTM(S)$ of mixed Tate motives over S as the full subcategory of $SmMot_{gm}(S)$ generated by the Tate twists of the motive of S. Our construction of $SmMot_{gm}(S)$

as the homotopy of a DG category (after taking an idempotent completion) gives a similar DG description of $DTM(S)$. Thus, we can hope to refine the realization functors for $SmMot_{gm}(S)$ even further if we restrict to $DTM(S)$. This should give us a Betti realization functor on $DTM(S)$ to the derived category of uni-potent local systems on S^{an}, an étale realization functor to relatively uni-potent étale sheaves on $S^{\text{ét}}$ and a Hodge realization to uni-potent variations of mixed Hodge structures on S^{an}. The paper of Deligne-Goncharov [6] and our work with Esnault [9], giving constructions of the mixed Tate fundamental group for some types of schemes S, gave us the motivation for the construction of categories of smooth motives and refined realization functors. As this paper is long enough already, we will postpone the construction of these realization functors to a future work.

The paper is organized as follows. We begin with a resumé of the cubical category and cubical constructions. This is a more convenient setting for constructing commutative DG structures than the simplicial one; we took the opportunity here of collecting a number of useful results on cubical constructions that are scattered throughout the literature. We also discuss a variant of cubical structures involving the extended cubical category. This variation on the cubical theme adds the cubical analog of the simplicial degeneracy maps; many of the most useful results on cubical objects that arise in nature actually use the extended cubical structure, so we thought it would be useful to give an abstract discussion.

The next section recalls from [8, 15, 17, 22] various versions of complexes over a DG category and the associated triangulated derived categories, as well as some useful foundational results.

In §4 we apply this machinery to the category of correspondences, endowed with the algebraic n-cubes as a cubical object. This leads to our construction of the DG category of higher correspondences, $dgCor_S$, the full DG subcategory $dgPrCor_S$ of correspondences on smooth projective S-scheme, the Zariski sheafified version $R\Gamma(S, \underline{dgPrCor}_S)$, the DG category of motivic complexes $dgSmMot_S^{\mathrm{eff}} :=$ $\mathcal{C}^b(R\Gamma(S, \underline{dgPrCor}_S))$, and finally the triangulated category of smooth effective motives over S, $SmMot_{gm}^{\mathrm{eff}}(S)_S$, defined by taking the idempotent completion of the homotopy category $\mathcal{K}^b(R\Gamma(S, \underline{dgPrCor}_S))$. We also define the $\mathbb{Q}$-version with alternating cubes, $dgSmMot_{S\mathbb{Q}}^{\mathrm{eff}}$, which is a DG tensor category, leading to the tensor triangulated category $SmMot_{gm}^{\mathrm{eff}}(S)_\mathbb{Q}$. Finally, we consider versions of these categories, $dgSmMot_S$, $dgSmMot_{S\mathbb{Q}}$, $SmMot_{gm}(S)$ and $SmMot_{gm}(S)_\mathbb{Q}$, formed by inverting the Lefschetz motive.

In §5 we state our main duality theorem for equi-dimensional cycles over a semi-local base (theorem 5.4), as well as the the projective bundle formula (theorem 5.5). We derive the consequences of these results for duality in the categories $SmMot_{gm}^{\mathrm{eff}}(S)$ and $SmMot_{gm}(S)$. In §6, we briefly recall some aspects of the definition of the Cisinski-Déglise categories of motives over a base, $DM^{\mathrm{eff}}(S)$ and $DM(S)$, define exact functors

$$\rho_S^{\mathrm{eff}} : SmMot_{gm}^{\mathrm{eff}}(S) \to DM^{\mathrm{eff}}(S),$$
$$\rho_S : SmMot_{gm}(S) \to DM(S),$$

and prove our main result (corollary 6.14). Finally, in §7 we prove our extension of the Friedlander-Lawson-Voevodsky moving lemmas and give the proofs of theorems 5.4 and 5.5.

As I have already mentioned, a major motivation for this paper arose out of the joint work [9] with Hélène Esnault, whom I would like to thank for her encouragement and suggestions. Also, among a number of helpful comments of the referee, the suggestion that I should use the work of Keller and Toën on DG categories for the discussion in §2 gave me a chance to broaden my DG horizons, for which I am very grateful. Finally, I want to express my heartfelt gratitude to Spencer Bloch. As so much of my work has been inspired by Spencer's remarkably original ideas, his steady encouragement over the years has been more important to me than I can possibly put into words.

1 Cubical objects and DG categories

1.1 Cubical objects. We recall some notions discussed in e.g. [18]. We introduce the "cubical category" **Cube**. This is the subcategory of **Sets** with objects $\underline{n} := \{0, 1\}^n$, $n = 0, 1, 2, \ldots$, and morphisms generated by

1. Inclusions: $\eta_{n,i,\epsilon} : \underline{n} \to \underline{n+1}$, $\epsilon = 0, 1$, $i = 1, \ldots, n+1$,

$$\eta_{n,i,\epsilon}(y_1, \ldots, y_{n-1}) = (y_1, \ldots, y_{i-1}, \epsilon, y_i, \ldots, y_{n-1}).$$

2. Projections: $p_{n,i} : \underline{n} \to \underline{n-1}$, $i = 1, \ldots, n$,

$$p_{n,i}(y_1, \ldots, y_n) = (y_1, \ldots, y_{i-1}, y_{i+1}, \ldots, y_n).$$

3. Permutations of factors: $(\epsilon_1, \ldots, \epsilon_n) \mapsto (\epsilon_{\sigma(1)}, \ldots, \epsilon_{\sigma(n)})$ for $\sigma \in S_n$.
4. Involutions: $\tau_{n,i}$ exchanging 0 and 1 in the ith factor of $\underline{n}$.

Clearly all the Hom-sets in **Cube** are finite. For a category $\mathcal{A}$, we call a functor $F : \mathbf{Cube}^{\mathrm{op}} \to \mathcal{A}$ a *cubical object* of $\mathcal{A}$ and a functor $F : \mathbf{Cube} \to \mathcal{A}$ a *co-cubical object* of $\mathcal{A}$.

Remark 1.1 The permutations and involutions in (3) and (4) give rise to a subgroup of $\mathrm{Aut}_{\mathbf{Sets}}(\underline{n})$ isomorphic to the semi-direct product $F_n := (\mathbb{Z}/2)^n \rtimes S_n$, where S_n acts on $(\mathbb{Z}/2)^n$ by permuting the factors.

We extend the standard sign representation of Σ_n to the sign representation

$$\mathrm{sgn} : F_n \to \{\pm 1\}$$

by

$$\mathrm{sgn}(\epsilon_1, \ldots, \epsilon_n, \sigma) := (-1)^{\sum_j \epsilon_j} \mathrm{sgn}(\sigma).$$

Example 1.2 Let S be a scheme, set $\mathbb{A}_S^1 := \mathrm{Spec}\,_S \mathcal{O}_S[y]$. We set $\square_S^n := (\mathbb{A}_S^1)^n$. S_n acts on $\square_S^n$ by permuting the factors. We let $\mathbb{Z}/2$ act on $\mathbb{A}_S^1$ by $x \mapsto 1 - x$. This gives us an action of F_n on $\square_S^n$.

Let $p_{n,i} : (\mathbb{A}_S^1)^n \to (\mathbb{A}_S^1)^{n-1}$ be the projection omitting the ith factor. Letting $y_{n,i} : (\mathbb{A}_S^1)^n \to \mathbb{A}_S^1$ be the projection on the ith factor, we use the coordinate system $(y_1, \cdots, y_n)$ on $\square_S^n$, with $y_i := y \circ y_{n,i}$.

Let $\eta_{n,i,\epsilon} : \square_S^{n-1} \to \square_S^n$ be the inclusion

$$\eta_{n,i,\epsilon}(y_1, \ldots, y_{n-1}) = (y_1, \ldots, y_{i-1}, \epsilon, y_i, \ldots, y_{n-1})$$

This gives us the co-cubical object $n \mapsto \square_S^n$ in $\mathbf{Sm}/S$.

A *face* of $\square_S^n$ is a subscheme F defined by equations of the form

$$y_{i_1} = \epsilon_1, \ldots, y_{i_s} = \epsilon_s; \; \epsilon_j \in \{0, 1\}.$$

1.2 Cubical objects in a pseudo-abelian category. Let $\mathcal{A}$ be a pseudo-abelian category, $\underline{A} : \mathbf{Cube}^{\mathrm{op}} \to \mathcal{A}$ a cubical object. For $\epsilon \in \{0,1\}$, let $\pi_{n,i}^\epsilon : \underline{A}(\underline{n}) \to \underline{A}(\underline{n})$ be the endomorphism $p_{n,i}^* \circ \eta_{n-1,i,\epsilon}^*$, and set

$$\pi_n := (\mathrm{id} - \pi_{n,n}^1) \circ \ldots \circ (\mathrm{id} - \pi_{n,1}^1).$$

Note that the $\pi_{n,i}^\epsilon$ are commuting idempotents, and that the subobject $(\mathrm{id} - \pi_{n,i}^\epsilon)^*(\underline{A}(\underline{n})) \subset \underline{A}(\underline{n})$ is a kernel for $\eta_{n,i,\epsilon}$. Since $\mathcal{A}$ is pseudo-abelian, the objects

$$\underline{A}(\underline{n})^0 := \cap_{i=1}^n \ker \eta_{n-1,i,1}^* \subset \underline{A}(\underline{n})$$

and

$$\underline{A}(\underline{n})^{\mathrm{degn}} := \sum_{i=1}^n p_{n,i}^*(\underline{A}(\underline{n-1})) \subset \underline{A}(\underline{n})$$

are well-defined.

Let $(\underline{A}_*, d)$ be the complex with $\underline{A}_n := \underline{A}(\underline{n})$ and with

$$d_n := \sum_{i=1}^n (-1)^i (\eta_{n,i,1}^* - \eta_{n,i,0}^*) : \underline{A}_{n+1} \to \underline{A}_n.$$

Write $\underline{A}_n^0, \underline{A}_n^{\mathrm{degn}}$ for $\underline{A}^0(\underline{n}), \underline{A}^{\mathrm{degn}}(\underline{n})$, respectively.

The following result is the basis of all "cubical" constructions; the proof is elementary and is left to the reader.

Lemma 1.3 *Let $\underline{A} : \mathbf{Cube}^{\mathrm{op}} \to \mathcal{A}$ be a cubical object in a pseudo-abelian category $\mathcal{A}$. Then*

1. For each n, π_n maps $\underline{A}_n$ to $\underline{A}_n^0$ and defines a splitting

$$\underline{A}_n = \underline{A}_n^{\mathrm{degn}} \oplus \underline{A}_n^0.$$

2. $d_n(\underline{A}_n^{\mathrm{degn}}) = 0$, $d_n(\underline{A}_n^0) \subset \underline{A}_{n-1}^0$

Definition 1.4 Let $\underline{A} : \mathbf{Cube}^{\mathrm{op}} \to \mathcal{A}$ be a cubical object in a pseudo-abelian category $\mathcal{A}$. Define the complex (A_*, d) to be the quotient complex

$$\underline{A}_*/\underline{A}_*^{\mathrm{degn}}$$

of $\underline{A}_*$.

Lemma 1.3 shows that A_* is well-defined and is isomorphic to the subcomplex $\underline{A}_*^0$ of $\underline{A}_*$. We often use cohomological notation, with $A^n := A_{-n}$, etc.

1.3 Products. Suppose we have two cubical objects

$$\underline{A}, \underline{B} : \mathbf{Cube}^{\mathrm{op}} \to \mathcal{A}$$

in a tensor category $(\mathcal{A}, \otimes)$. Form the diagonal cubical object $\underline{A} \otimes \underline{B}$ by

$$\underline{A} \otimes \underline{B}(\underline{n}) := \underline{A}(\underline{n}) \otimes \underline{B}(\underline{n})$$

and on morphisms by

$$\underline{A} \otimes \underline{B}(f) := \underline{A}(f) \otimes \underline{B}(f).$$

Let $p_{n,m}^1 : \underline{n+m} \to \underline{n}$, $p_{n,m}^2 : \underline{n+m} \to \underline{m}$ be the projections on the first n and last m factors, respectively. Let

$$\cup_{A,B}^{n,m} : \underline{A}(\underline{n}) \otimes \underline{B}(\underline{m}) \to \underline{A}(\underline{n+m}) \otimes \underline{B}(\underline{n+m})$$

be the map $\underline{A}(p_{n,m}^1) \otimes \underline{B}(p_{n,m}^2)$. One easily checks that the direct sum of the maps $\cup_{A,B}^{n,m}$ defines a map of complexes

$$\cup_{A,B} : \underline{A}^* \otimes \underline{B}^* \to \underline{A \otimes B}^*. \tag{1.1}$$

It is easy to see that we have an associativity property

$$\cup_{A \otimes B, C} \circ (\cup_{A,B} \otimes \mathrm{id}_{\underline{C}^*}) = \cup_{A, B \otimes C} \circ (\mathrm{id}_{\underline{A}^*} \otimes \cup_{B,C}) \tag{1.2}$$

but not in general a commutativity property.

1.4 Alternating cubes. Recall the semi-direct product $F_n := (\mathbb{Z}/2)^n \ltimes S_n$ and the sign representation $\mathrm{Sgn} : F_n \to \{\pm 1\}$. If $\mathcal{A}$ is a pseudo-abelian category and M an F_n-module in $\mathcal{A}$ (i.e., we are given a homomorphism $F_n \to \mathrm{Aut}_{\mathcal{A}}(M)$), we let M^{Sgn} be the largest subobject of M on which F_n acts by the sign representation:

$$M^{\mathrm{Sgn}} := \cap_{g \in F_n} \ker((g - \mathrm{Sgn}(g)\mathrm{id}_M).$$

Similarly, if S_n acts on M, we let M^{sgn} be the subobject of M

$$M^{\mathrm{sgn}} := \cap_{g \in \Sigma_n} \ker((g - \mathrm{sgn}(g)\mathrm{id}_M).$$

Let $\underline{A} : \mathbf{Cube}^{\mathrm{op}} \to \mathcal{A}$ be a cubical object in a pseudo-abelian category $\mathcal{A}$. For each $n = 0, 1, 2, \ldots$, define the subobject $\underline{A}^{\mathrm{Alt}}(\underline{n})$ of $\underline{A}(\underline{n})$ by

$$\underline{A}^{\mathrm{Alt}}(\underline{n}) := \underline{A}(\underline{n})^{\mathrm{Sgn}}.$$

Similarly, let $A^{\mathrm{alt}}(n) := A(\underline{n})^{\mathrm{sgn}}$.

Lemma 1.5 *1. $n \mapsto \underline{A}^{\mathrm{Alt}}(\underline{n})$ defines a sub-cubical object of $\underline{A}$.*

2. $n \mapsto A^{\mathrm{alt}}(\underline{n})$ defines a sub-complex of A_*.

3. Suppose $2 \times$ id is invertible on all the objects $\underline{A}(\underline{n})$. Then the map $\underline{A}_* \to A_*$ induces an isomorphism of complexes $\underline{A}_*^{\mathrm{Alt}} \to A_*^{\mathrm{alt}}$.

Proof This is straightforward, noting that the degenerate subcomplex is killed by the idempotent

$$\frac{1}{2^n} \prod_{i=1}^{n} (1 - \tau_i)$$

where τ_i is the involution in the ith factor of $\underline{n}$. $\square$

1.5 Extended cubes. We note that product of sets makes **Sets** a symmetric monoidal category, and that **Cube** is a symmetric monoidal subcategory. Let **ECube** be the smallest symmetric monoidal subcategory of **Sets** having the same objects as **Cube**, containing **Cube** and containing the morphism

$$\mu : \underline{2} \to \underline{1}$$

defined by the multiplication of integers:

$$\mu((1,1)) = 1; \quad \mu(a,b) = 0 \text{ for } (a,b) \neq (1,1).$$

An *extended cubical object* in a category $\mathcal{C}$ is a functor $F : \mathbf{ECube}^{\mathrm{op}} \to \mathcal{C}$.

Let $\underline{F} : \mathbf{Cube}^{\mathrm{op}} \to \mathcal{A}$ be a cubical object in a pseudo-abelian category. Let $NF(\underline{n}) \subset \underline{F}(\underline{n})$ be the subobject

$$NF(\underline{n}) := \cap_{i=2}^n \ker(\eta_{n,i,0}^*) \cap \cap_{i=1}^n \ker(\eta_{n,i,1}^*).$$

This defines the *normalized subcomplex* NF^* of $\underline{F}^*$. Note that NF^* is a subcomplex of $\underline{F}(*)_0$.

Lemma 1.6 *Let* $\underline{F} : \mathbf{ECube}^{\mathrm{op}} \to \mathcal{A}$ *be an extended cubical object in a pseudo-abelian category* $\mathcal{A}$. *Then the inclusion* $i : NF^* \to \underline{F}(*)_0$ *is a homotopy equivalence.*

Proof Let

$$N^M F_n := \begin{cases} \cap_{i=n-M}^{n} \ker(\eta_{n,i,0}^*) \cap \cap_{i=1}^{n} \ker(\eta_{n,i,1}^*) & \text{for } n - M > 2, \\ NF^n & \text{for } n - M \leq 2. \end{cases}$$

For each M, the subobjects $N^M F_n \subset \underline{F}_n^0$ form a subcomplex $N^M F_*$ of $\underline{F}_*^0$ which contains NF_* and agrees with NF_* in degrees $n \leq M + 2$. Since $N^{-1}F_* = \underline{F}_*^0$, it thus suffices to show that the inclusion

$$i^M : N^M F_* \to N^{M-1} F_*$$

is a homotopy equivalence, such that the chosen homotopy inverse p^M and chosen homotopy h^M between $i^M \circ p^M$ and id satisfy:

1. $p^M \circ i^M = \mathrm{id}$
2. $h_n^M : N^{M-1} F_{n-1} \to N^{M-1} F_n$ is the zero map for $n \leq M$

Indeed, in this case, the infinite composition

$$p := \ldots p^M \circ p^{M-1} \circ \ldots \circ p^0$$

makes sense, as does the infinite sum

$$h := \sum_M i^0 \circ \ldots i^{M-1} \circ h^M \circ p^{M-1} \circ \ldots p^0.$$

The map p gives a homotopy inverse to i, with $pi = \mathrm{id}$ and h defines a homotopy between ip and the identity. We proceed to define the maps p^M and h^M.

Define the map $q : \underline{2} \to \underline{1}$ by

$$q(x, y) = 1 - \mu(1 - x, 1 - y) = 1 - (1 - x)(1 - y).$$

For $1 \leq i \leq n - 1$, let $q_{n,i} : \underline{n} \to \underline{n-1}$ be the map

$$q_{n,i}(x_1, \ldots, x_n) := (x_1, x_2, \ldots, x_{i-1}, q(x_i, x_{i+1}), x_{i+2}, \ldots, x_n).$$

Then

$$q_{n,i} \circ \eta_{n,i,0} = q_{n,i} \circ \eta_{n,i+1,0} = \mathrm{id}_{\underline{n-1}}, \tag{1.3}$$

$$q_{n,i} \circ \eta_{n,i,1} = q_{n,i} \circ \eta_{n,i+1,1} = \eta_{n,i,1} \circ p_{n-1,i},$$

$$q_{n,i} \circ \eta_{n,j,\epsilon} = \begin{cases} \eta_{n-1,j,\epsilon} \circ q_{n-1,i-1} & \text{for } 1 \leq j < i, \\ \eta_{n-1,j-1,\epsilon} \circ q_{n-1,i} & \text{for } i + 1 < j \leq n. \end{cases}$$

Defining $q_{n,j}^* = 0$ for $j \leq 0$ or $j \geq n$, this implies that, for $i \geq 1$, the maps

$$p_n^M := \mathrm{id} - q_{n,n-M-1}^* \circ \eta_{n-1,n-M,0}^*$$

define a map of complexes $p^M : \underline{F}_*^0 \to \underline{F}_*^0$ which restricts to the inclusion $N^M F_* \to \underline{F}_*^0$ on $N^M F_*$, and maps $N^{M-1} F_*$ to $N^M F_*$. We let

$$p^M : N^{M-1} F_* \to N^M F_*$$

be the restriction.

Let $h_n^M : \underline{F}_{n-1}^0 \to \underline{F}_n^0$ be the map $(-1)^{n-M} q_{n,n-M-1}^*$. The relations (1.3) imply that h_n^M restricts to a map

$$h_n^M : N^{M-1} F_{n-1} \to N^{M-1} F_n$$

and, on $\underline{F}_n^0$, we have

$$d_n h_M^{n+1} + h_M^n d_{n-1}$$

$$= (-1)^{n-M+1} \left[\sum_{j=1}^{n+1} (-1)^j \eta_{n,j,0}^* \circ q_{n+1,n-M}^* - \sum_{l=1}^{n} (-1)^l q_{n,n-M-1}^* \circ \eta_{n-1,l,0}^* \right]$$

$$= (-1)^{n-M+1} \sum_{j=1}^{n-M-1} (-1)^j q_{n,n-M-1}^* \circ \eta_{n-1,j,0}^*$$

$$+ (-1)^{n-M+1} \sum_{j=n-M+2}^{n+1} (-1)^j q_{n,n-M}^* \circ \eta_{n-1,j-1,0}^*$$

$$- (-1)^{n-M+1} \sum_{l=1}^{n} (-1)^l q_{n,n-M-1}^* \circ \eta_{n-1,l,0}^*$$

$$= q_{n,n-M-1}^* \eta_{n-1,n-M,0}^*$$

$$+ (-1)^{n-M} \sum_{j=n-M+1}^{n} (-1)^j (q_{n,n-M}^* + q_{n,n-M-1}^*) \circ \eta_{n-1,j,0}^*.$$

Since $\eta_{n-1,j,0}^* = 0$ on $N^{M-1} F_n$ for $j \geq n-M+1$, the h_n^M give the desired homotopy.
□

Let $\underline{F} : \mathbf{ECube}^{\mathrm{op}} \to \mathcal{A}$ be an extend cubical object in an abelian category $\mathcal{A}$. Then we have the following description of $H_n(NF_*)$:

$$H_n(NF_*) = \frac{\cap_{i=1}^n \ker \eta_{n-1,i,0}^* \cap \cap_{i=1}^n \ker \eta_{n-1,i,1}^*}{\eta_{n,1,0}^* [\cap_{i=2}^{n+1} \ker \eta_{n,i,0}^* \cap \cap_{i=1}^{n+1} \ker \eta_{n,i,1}^*]}.$$

From this description, we see that the symmetric group S_n acts on $H_n(NF_*)$ through the permutation action on $\underline{n}$ and the action $\sigma \mapsto \mathrm{id} \times F(\sigma)$ on $\underline{F}(\underline{n+1})$. Via lemma 1.6, this gives us an S_n-action on $H_n(F_*)$.

Proposition 1.7 $\underline{F} : \mathbf{ECube}^{\mathrm{op}} \to \mathcal{A}$ *be an extended cubical object in an abelian category* $\mathcal{A}$. *Suppose that the Hom-groups in* $\mathcal{A}$ *are* $\mathbb{Q}$-*vector spaces. Then the inclusion*

$$F_*^{\mathrm{alt}} \to F_*$$

is a quasi-isomorphism.

Proof Since the Hom-groups in $\mathcal{A}$ are $\mathbb{Q}$-vector spaces, the natural map

$$H_n(F_*^{\mathrm{alt}}) \to H_n(F_*)^{\mathrm{alt}}$$

is an isomorphism for all n. Thus, we need only show that the symmetric group S_n acts on $H_n(F_*)$ by the sign representation.

Fix an element in $Z_n(NF_*)$ representing a class $[z] \in H_n(NF_*)$, i.e.

$$\eta_{n-1,i,\epsilon}^*(z) = 0$$

for all i and for $\epsilon = 0, 1$. Let $\tau : \underline{n} \to \underline{n}$ be the permutation exchanging the first two factors. Let $h_n : \underline{n+1} \to \underline{n}$ be the map

$$h_n(x_1, x_2, x_3, \dots, x_{n+1}) := (x_2, q(x_1, x_3), x_4, \dots, x_{n+1}),$$

and let $b := h_n^*(z)$. Then

$$h_n \circ \eta_{n,1,0} = \mathrm{id}$$
$$h_n \circ \eta_{n,2,0}(x_1, \dots, x_n) = (0, q(x_1, x_2), x_3, \dots, x_n)$$
$$h_n \circ \eta_{n,3,0} = \tau$$
$$h_n \circ \eta_{n,j,0} = \eta_{n-1,j-1,0} \circ h_{n-1} \text{ for } j \geq 4.$$

Similarly, $h_n \circ \eta_{n,j,1} = \eta_{n-1,j',1} \circ f_{n,j}$ for some j' and some map $f_{n,j} : \underline{n} \to \underline{n-1}$. Thus

$$db = z + \tau^*(z)$$

proving the result. $\square$

1.6 Cubical enrichments and DG categories. The category of cubical abelian groups $\mathbf{Ab}^{\mathbf{Cube}^{\mathrm{op}}}$ inherits the structure of a symmetric monoidal category from $\mathbf{Ab}$. Explicitly, if we have two cubical abelian groups $n \mapsto A(n)$, $n \mapsto B(n)$, the tensor product $A \otimes B$ is the cubical abelian group $n \mapsto A(n) \otimes B(n)$, with morphisms acting by

$$g(a \otimes b) = g(a) \otimes g(b).$$

A *cubical category* is a category $\mathcal{C}$ enriched in cubical abelian groups, that is, for each pair of objects X, Y of $\mathcal{C}$, we have a cubical abelian group

$$\mathcal{H}om(X, Y, -) : \mathbf{Cube}^{\mathrm{op}} \to \mathbf{Ab},$$
$$n \mapsto \mathcal{H}om_{\mathcal{C}}(X, Y, n)$$

for each object X of $\mathcal{C}$, an identity element $\mathrm{id}_X \in \mathcal{H}om_{\mathcal{C}}(X, X, 0)$ and for objects X, Y, Z an associative composition law

$$\circ_{X,Y,Z} : \mathcal{H}om_{\mathcal{C}}(Y, Z, -) \otimes \mathcal{H}om_{\mathcal{C}}(X, Y, -) \to \mathcal{H}om_{\mathcal{C}}(X, Z, -),$$

with $f \circ_{X,Y,Z} \mathrm{id}_Z = f$, $\mathrm{id}_X \circ_{X,Y,Z} g = g$.

We have the functor $\mathcal{C} \mapsto \mathcal{C}_0$ from cubical categories to pre-additive categories, where $\mathcal{C}_0$ has the same objects as $\mathcal{C}$ and

$$\mathrm{Hom}_{\mathcal{C}_0}(X, Y) := \mathcal{H}om(X, Y, 0).$$

A *cubical enrichment* of a pre-additive category $\mathcal{C}$ is a cubical category $\tilde{\mathcal{C}}$ together with an isomorphism $\mathcal{C} \cong \tilde{\mathcal{C}}_0$.

Next, we recall some basic facts about DG categories. For a complex $C \in C(\mathbf{Ab})$, we have the group of cycles in degree n, $Z^n C$ and the cohomology $H^n C$. For complexes C, C', we have the Hom-complex $\mathcal{H}om_{C(\mathbf{Ab})}(C, C')^*$, with

$$\mathcal{H}om_{C(\mathbf{Ab})}(C, C')^n := \prod_p \mathrm{Hom}_{\mathbf{Ab}}(C^p, C'^{n+p},$$

and with differential

$$d_{C',C} f := d_{C'} \circ f - (-1)^{\deg f} f \circ d_C.$$

We have as well as the group of maps of complexes

$$\mathrm{Hom}_{C(\mathbf{Ab})}(C, C') := Z^0 \mathcal{H}om_{C(\mathbf{Ab})}(C, C')^*.$$

This forms the additive category $C(\mathbf{Ab})$.

$C(\mathbf{Ab})$ has the *shift* functor $C \mapsto C[1]$ with $C[1]^n := C^{n+1}$ and differential $d^n_{C[1]} := -d^{n+1}_C$; for a morphism $f = \{f^n : C^n \to D^n\}$, $f[1] : C[1] \to D[1]$ is the collection $f[1]^n := f^{n+1}$. For a morphism $f : A \to B$, we have the *cone* complex $\mathrm{Cone}(f)$ and the *cone sequence*

$$A \xrightarrow{f} B \xrightarrow{i} \mathrm{Cone}(f) \xrightarrow{p} A[1] \tag{1.4}$$

with $\mathrm{Cone}(f)^n := B^n \oplus A^{n+1}$ and differential $d(b, a) := (db + f(a), -da)$; i and p are the evident inclusion and projection. Clearly this sequence is natural with respect to morphisms $(u, v) : f \to g$.

Tensor product of complex $A^* \otimes B^*$ is defined as usual by $(A^* \otimes B^*)^n := \oplus_{i+j=n} A^i \otimes B^j$, with differential given by the Leibniz rule

$$d(a \otimes b) = da \otimes b + (-1)^{\deg a} a \otimes db.$$

The commutativity constraint $\tau_{A,B} : A^* \otimes B^* \to B^* \otimes A^*$ is

$$\tau_{A,B}(a \otimes b) = (-1)^{\deg a \cdot \deg b} b \otimes a.$$

This makes $C(\mathbf{Ab})$ into a symmetric monoidal category (in fact, a tensor category).

More generally, for a commutative ring k, we have the k-linear tensor category of complexes $C(k)$, and complexes of k-modules $\mathcal{H}om_{C(k)}(A, B)^*$.

A *DG category* is a category enriched in complexes of abelian groups. Concretely, for objects X, Y in a DG category $\mathcal{C}$, one has the *Hom complex*

$$\mathcal{H}om_{\mathcal{C}}(X, Y)^* \in C(\mathbf{Ab}),$$

and for X, Y, Z in $\mathcal{C}$, a composition law

$$\circ_{X,Y,Z} : \mathcal{H}om_{\mathcal{C}}(Y, Z)^* \otimes \mathcal{H}om_{\mathcal{C}}(X, Y)^* \to \mathcal{H}om_{\mathcal{C}}(X, Z).$$

The map $\circ_{X,Y,Z}$ is a map of complexes; equivalently, we have the Leibniz rule:

$$d(f \circ g) = df \circ g + (-1)^{\deg f} f \circ dg.$$

One has associativity of composition and an identity morphism $\mathrm{id}_X \in \mathcal{H}om_{\mathcal{C}}(X, X)^0$ with $d\,\mathrm{id}_X = 0$.

Each DG category $\mathcal{C}$ has an underlying pre-additive category $\mathrm{frg}\,\mathcal{C}$ with the same objects and with morphisms the underlying group of the complex $\mathcal{H}om_{\mathcal{C}}(X, Y)^*$, i.e., forget the grading and differential. A functor of DG categories $F : \mathcal{A} \to \mathcal{B}$ is a functor of the underlying pre-additive categories such that $F : \mathcal{H}om_{\mathcal{A}}(X, Y)^* \to \mathcal{H}om_{\mathcal{B}}(F(X), F(Y))^*$ is a map of complexes for each X, Y in $\mathcal{A}$. This defines the category of (small) DG categories, with forgetful functor frg to pre-additive categories.

For a DG category $\mathcal{C}$, one has the additive category $Z^0\mathcal{C}$, with the same objects as $\mathcal{C}$ and with

$$\mathrm{Hom}_{Z^0\mathcal{C}}(X, Y) := Z^0\mathcal{H}om_{\mathcal{C}}(X, Y).$$

We have the category $H^0\mathcal{C}$, with the same objects as $\mathcal{C}$ and with

$$\mathrm{Hom}_{H^0\mathcal{C}}(X, Y) := H^0\mathcal{H}om_{\mathcal{C}}(X, Y),$$

and the graded version $H^*\mathcal{C}$ with graded group of morphisms

$$\mathrm{Hom}_{H^*\mathcal{C}}(X, Y)^* := H^*\mathcal{H}om_{\mathcal{C}}(X, Y).$$

Clearly each DG functor $F : \mathcal{C} \to \mathcal{C}'$ induces functors of additive categories $H^0F : H^0\mathcal{C} \to H^0\mathcal{C}'$ and $Z^0F : H^0\mathcal{C} \to Z^0\mathcal{C}'$, making H^0 and Z^0 functors from

DG categories to pre-additive categories; similarly, $\mathcal{C} \mapsto H^*\mathcal{C}$ defines a functor from DG categories to graded pre-additive categories;

More generally, for a commutative ring k, we have the notion of a k-DG category, this being a category enriched in complexes $C(k)$, with functors Z^0 and H^0 to k-additive categories.

Example 1.8 With the Hom-complexes described above and the evident composition law, we have the structure of a DG category on $C(\mathbf{Ab})$, which we denote by $C_{dg}(\mathbf{Ab})$; we recover $C(\mathbf{Ab})$ as $Z^0 C_{dg}(\mathbf{Ab})$. More generally, for a commutative ring k, we have the k-tensor category of complexes $C(k)$ and the k-DG category $C_{dg}(k)$.

We now show how to associate a DG category to a cubical category. Let $\mathcal{C}$ be a cubical category. For objects X, Y in $\mathcal{C}$, let $\mathcal{H}om_{dg\mathcal{C}}(X,Y)^*$ be the complex $\underline{\mathcal{H}om}_{\mathcal{C}}(X,Y)^* / \underline{\mathcal{H}om}_{\mathcal{C}}(X,Y)^*_{\text{degn}}$ associated to the cubical abelian group

$$n \mapsto \underline{\mathcal{H}om}_{\mathcal{C}}(X,Y,n)$$

following definition 1.4. For objects X, Y, Z in $\mathcal{C}$, let

$$\circ_{X,Y,Z} : \mathcal{H}om_{dg\mathcal{C}}(Y,Z)^* \otimes \mathcal{H}om_{dg\mathcal{C}}(X,Y)^* \to \mathcal{H}om_{dg\mathcal{C}}(X,Z)^*$$

be the map of complexes induced by the composition $\underline{\circ}_{X,Y,Z}$ and the product (1.1).

Lemma 1.9 *The complexes $\mathcal{H}om_{dg\mathcal{C}}(X,Y)^*$ and composition law $\circ_{X,Y,Z}$ define a DG category $dg\mathcal{C}$.*

The proof is easy and is left to the reader. We sometimes write $\mathcal{C}$ for the DG category $dg\mathcal{C}$, if the context makes the meaning clear.

Here is a method for constructing a cubical category from a tensor category which we will use to construct our DG category of motives.

Definition 1.10 Let $n \mapsto \square^n$ be a co-cubical object (denoted $\square^*$) in a tensor category $(\mathcal{C}, \otimes)$ such that $\square^0$ is the unit object with respect to $\otimes$.

A *co-multiplication* δ^* on $\square^*$ is a morphism of co-cubical objects

$$\delta^* : \square^* \to \square^* \otimes \square^*,$$

where $\square^* \otimes \square^*$ is the diagonal co-cubical object $n \mapsto \square^n \otimes \square^n$, which is

1. *co-associative*: $(\delta^* \otimes \mathrm{id}_{\square^*}) \circ \delta^* = (\mathrm{id}_{\square^*} \otimes \delta^*) \circ \delta^*$.

2. *co-unital*: $\square^0 \xrightarrow{\delta(0)} \square^0 \otimes \square^0 \xrightarrow{\mu} \square^0 = \mathrm{id}_{\square^0}$ where μ is the unit isomorphism for $\otimes$.

3. *symmetric*: Let t be the commutativity constraint in $(\mathcal{C}, \otimes)$. Then $t_{\square^*,\square^*} \circ \delta^* = \delta^*$.

Given a co-multiplication on a co-cubical object $\square^*$, define

$$\underline{\mathcal{H}om}(X,Y,n) := \mathcal{H}om_{\mathcal{C}}(X \times \square^n, Y),$$

giving us the cubical object $n \mapsto \underline{\mathcal{H}om}(X,Y,n)$ of $\mathbf{Ab}$; we denote the associated complex by $\underline{\mathcal{H}om}(X,Y)^*$. The co-multiplication gives us the map of cubical objects

$$\hat{\circ}_{X,Y,Z} : \underline{\mathcal{H}om}(Y,Z,*) \otimes \underline{\mathcal{H}om}(X,Y,*) \to \underline{\mathcal{H}om}(X,Z,*)$$

sending $f : X \otimes \square^n \to Y$ and $g : Y \otimes \square^n \to Z$ to the composition

$$X \otimes \square^n \xrightarrow{\delta(n)} X \otimes \square^n \otimes \square^n \xrightarrow{f \otimes \mathrm{id}} Y \otimes \square^n \xrightarrow{g} Z.$$

The following proposition is proved by a straightforward computation.

Proposition 1.11 *Let* $\square^* : \mathbf{Cube} \to \mathcal{C}$ *be a co-cubical object in a tensor category* $\mathcal{C}$, *with a co-multiplication* δ. *Then* $(X, Y, n) \mapsto \mathcal{H}om_{\mathcal{C}}(X \times \square^n, Y)$, *with the composition law* $\underline{\circ}_{X,Y}$ *defined above, defines a cubical enrichment of* $\mathcal{C}$.

We denote the DG category formed by the cubical enrichment described above by $(\mathcal{C}, \otimes, \square^*, \delta)$, or just $(\mathcal{C}, \square^*)$ when the context makes the meaning clear.

Suppose now that, in addition to the assumptions used above, the Hom-groups in $\mathcal{C}$ are $\mathbb{Q}$-vector spaces, i.e., $\mathcal{C}$ is $\mathbb{Q}$-linear. We may then define the *alternating projection*

$$\mathrm{Alt} : \mathcal{H}om_{\mathcal{C}}(X \times \square^n, Y) \to \mathcal{H}om_{\mathcal{C}}(X \times \square^n, Y)^{\mathrm{alt}}$$

by applying the idempotent Alt in the rational group ring $\mathbb{Q}[F_n]$ corresponding to the sign representation.

Definition 1.12 Suppose that $\mathcal{C}$ is $\mathbb{Q}$-linear. Define the sub-DG category $(\mathcal{C}, \otimes, \square^*, \delta)^{\mathrm{alt}}$ of $(\mathcal{C}, \otimes, \square^*, \delta)$, with the same objects as $(\mathcal{C}, \otimes, \square^*, \delta)$, and with complex of morphisms given by the subcomplex

$$\mathcal{H}om_{\mathcal{C}}(X, Y)^{\mathrm{alt}\,n} := \mathcal{H}om_{\mathcal{C}}(X \times \square^n, Y)^{\mathrm{Alt}} \subset \mathcal{H}om_{\mathcal{C}}(X, Y, n).$$

The composition law is defined by the composition

$$\mathcal{H}om_{\mathcal{C}}(Y, Z)^{\mathrm{alt}*} \otimes \mathcal{H}om_{\mathcal{C}}(X, Y)^{\mathrm{alt}*} \xrightarrow{\circ} \mathcal{H}om_{\mathcal{C}}(X, Z)^* \xrightarrow{\mathrm{alt}} \mathcal{H}om_{\mathcal{C}}(X, Z)^{\mathrm{alt}*}.$$

Proposition 1.13 *Let* $\square^* : \mathbf{Cube} \to \mathcal{C}$ *be a co-cubical object in a tensor category* $\mathcal{C}$, *with a co-multiplication* δ. *Suppose that* $\square^*$ *extends to a functor*

$$\square^* : \mathbf{ECube} \to \mathcal{C}$$

and that $\mathcal{C}$ *is* $\mathbb{Q}$-*linear. Then the natural inclusion functor*

$$(\mathcal{C}, \square^*)^{\mathrm{alt}} \to (\mathcal{C}, \square^*)$$

is a quasi-equivalence[1] of $\mathbb{Q}$-*DG categories.*

Proof This follows immediately from proposition 1.7. $\qquad\qquad\qquad\square$

1.7 Tensor structure. There is a natural tensor structure on the $\mathbb{Q}$-DG category $(\mathcal{C}, \otimes, \square^*, \delta)^{\mathrm{alt}}$, which we now describe.

Given $f : X \times \square^n \to Y$, $f' : X' \times \square^{n'} \to Y'$, define

$$f \tilde{\otimes} g : X \otimes X' \otimes \square^{n+n'} \to Y \otimes Y'$$

as the composition

$$X \otimes X' \otimes \square^{n+n'} \xrightarrow{\mathrm{id} \otimes \delta_{n,n'}} X \otimes X' \otimes \square^n \otimes \square^{n'}$$

$$\xrightarrow{\tau_{X', \square^n}} X \otimes \square^n \otimes X' \otimes \square^{n'} \xrightarrow{f \otimes f'} Y \otimes Y'.$$

Assuming that $\mathcal{C}$ is $\mathbb{Q}$-linear, we define $f \otimes_{\square} g$ by applying the alternating projection:

$$f \otimes_{\square} g := (f \tilde{\otimes} g) \circ (\mathrm{id}_{X \otimes X'} \otimes \mathrm{alt}_{n+n'}).$$

Proposition 1.14 *Let* $\mathcal{C}$ *be a* $\mathbb{Q}$-*tensor category,* $\square^*$ *a co-cubical object of* $\mathcal{C}$ *and* $\delta : \square \to \square \otimes \square$ *a co-multiplication. Then* $((\mathcal{C}, \otimes, \square^*, \delta)^{\mathrm{alt}}, \otimes_{\square})$ *is a* $\mathbb{Q}$-*DG tensor category, with commutativity constraints induced by the commutativity constraints in* $\mathcal{C}$.

---

[1]for the definition of the term "quasi-equivalence", see definition 2.13

Proof One checks easily that the integral operation $\tilde{\otimes}$ satisfies the Leibniz rule:

$$\partial(f\tilde{\otimes}g) = \partial f\tilde{\otimes}g + (-1)^{\deg f}f \otimes \partial g.$$

Let $\delta_{n,m} : \square^{n+m} \to \square^n \otimes \square^m$ denote the composition $(p_1^{n,m} \otimes p_2^{n,m}) \circ \delta^{n+m}$. Let $\pi_{\mathrm{Alt}}^N : \square^N \to \square^N$ be the map induced by the alternating idempotent in $\mathbb{Q}[F_N]$.

It follows from the properties of co-associativity and symmetry of δ, together with the fact that δ is a map of co-cubical objects, that

$$(\mathrm{id}_{\square^n} \otimes t_{m,n'} \otimes \mathrm{id}_{\square^{m'}})(\delta_{n,m} \otimes \delta_{n',m'}) \circ \delta_{n+m,n'+m'}$$
$$= (\delta_{n,n'} \otimes \delta_{m,m'}) \circ \delta_{n+n',m+m'} \circ \square(\mathrm{id}_{\underline{n}} \times \tau_{m,n'} \times \mathrm{id}_{\underline{m'}}),$$

where $\tau_{m,n'} : \underline{m} \times \underline{n'} \to \underline{n'} \times \underline{m}$ is the symmetry in **Cube**, and t_{**} is the symmetry in $\mathcal{C}$. Composing on the right with $\pi_{\mathrm{Alt}}^{n+m+n'+m'}$ yields the identity

$$(\mathrm{id}_{\square^n} \otimes t_{m,n'} \otimes \mathrm{id}_{\square^{m'}})(\delta_{n,m} \otimes \delta_{n',m'}) \circ \delta_{n+m,n'+m'} \circ \pi_{\mathrm{Alt}}^{n+m+n'+m'}$$
$$= (-1)^{mn'}(\delta_{n,n'} \otimes \delta_{m,m'}) \circ \delta_{n+n',m+m'} \circ \pi_{\mathrm{Alt}}^{n+m+n'+m'}.$$

The identity

$$(f \otimes_\square g) \circ (f' \otimes_\square g') = (-1)^{\deg g \deg f'} ff' \otimes_\square gg'$$

follows directly from this.

One shows by a similar argument that, for $f \in \mathrm{Hom}_\mathcal{C}(X \otimes \square^p, Y)^{\mathrm{Alt}}$, $g \in \mathrm{Hom}_\mathcal{C}(X' \otimes \square^q, Y')^{\mathrm{Alt}}$, we have

$$t_{Y,Y'} \circ (f \otimes_\square g) = (-1)^{pq}(g \otimes_\square f) \circ (t_{X,X'} \otimes \mathrm{id}_{\square^{p+q}}),$$

completing the proof. $\qquad\qquad\qquad\qquad\qquad\qquad\qquad\qquad\qquad\qquad\qquad\square$

Remark 1.15 One could hope that the operations $\tilde{\otimes}$ define at least a monoidal structure on $(\mathcal{C}, \otimes, \square^*, \delta)$, but this is in general not the case. In fact, as we noted in the proof of proposition 1.14, sending f, g to $f\tilde{\otimes}g$ does satisfy the Leibniz rule, but we do not have the identity

$$(f\tilde{\otimes}g) \circ (f'\tilde{\otimes}g') = (-1)^{\deg f' \deg g} ff'\tilde{\otimes}gg'$$

in general: the cubes on the two sides of this equation are in a different order.

In spite of this, the operation $\otimes$ *does* extend to an action of $\mathcal{C}$ on $(\mathcal{C}, \otimes, \square^*, \delta)$. Consider $\mathcal{C}$ as a DG category with all morphisms of degree zero (and zero differential). Define

$$\otimes : \mathcal{C} \otimes (\mathcal{C}, \otimes, \square^*, \delta) \to (\mathcal{C}, \otimes, \square^*, \delta)$$

to be the same as $\otimes$ on objects, and on morphisms by

$$\otimes : \mathrm{Hom}_\mathcal{C}(A, B) \otimes \mathrm{Hom}_\mathcal{C}(X \otimes \square^n, Y) \to \mathrm{Hom}_\mathcal{C}(A \otimes X \otimes \square^n, B \otimes Y)$$

2 Complexes over a DG category

We recall various constructions of categories of complexes over a DG category, the triangulated categories arising from these DG categories, and their main properties.

2.1 The category of $\mathcal{A}^{\mathrm{op}}$-modules and the derived category. We begin with an abstract approach, following [16, 17]. Let $\mathcal{C}$ be a DG category. The opposite category $\mathcal{C}^{\mathrm{op}}$ is the DG category with Hom complexes

$$\mathcal{H}om_{\mathcal{C}^{\mathrm{op}}}(X, Y)^* := \mathcal{H}om_{\mathcal{C}}(Y, X)^*$$

and with composition law given by

$$\mathcal{H}om_{\mathcal{C}^{\mathrm{op}}}(Y, Z)^* \otimes \mathcal{H}om_{\mathcal{C}^{\mathrm{op}}}(X, Y)^* = \mathcal{H}om_{\mathcal{C}^{\mathrm{op}}}(Z, Y)^* \otimes \mathcal{H}om_{\mathcal{C}^{\mathrm{op}}}(Y, X)^*$$
$$\xrightarrow{\tau} \mathcal{H}om_{\mathcal{C}}(Y, X)^* \otimes \mathcal{H}om_{\mathcal{C}}(Z, Y)^*$$
$$\xrightarrow{\circ_{Z,Y,X}} \mathcal{H}om_{\mathcal{C}}(Z, X)^* = \mathcal{H}om_{\mathcal{C}^{\mathrm{op}}}(X, Z)^*.$$

A *DG module* M over a (small) DG category $\mathcal{A}$ is a DG functor

$$M : \mathcal{A} \to C_{dg}(\mathbf{Ab}).$$

For DG modules M, N, we have the Hom-complex $\mathrm{Hom}(M, N)^*$. $\mathrm{Hom}(M, N)^n$ is the degree n natural transformations $f : M \to N$, and with differential induced by the differential in the complexes $\mathrm{Hom}_{C(\mathbf{Ab})}(M(a), N(a))^*$, $a \in \mathcal{A}$.

We let $\mathcal{C}_{dg}(\mathcal{A})$ denote the DG category of $\mathcal{A}^{\mathrm{op}}$-modules. We have as well additive category $\mathcal{C}(\mathcal{A}) := Z^0 \mathcal{C}_{dg}(\mathcal{A})$ and the homotopy category $\mathcal{K}(\mathcal{A}) := H^0(\mathcal{C}_{dg}(\mathcal{A}))$.

$\mathcal{C}(\mathcal{A})$ inherits a shift functor $M \mapsto M[1]$ from $C(\mathbf{Ab})$; similarly for each morphism $f : M \to N$, we have the cone $\mathrm{Cone}(f)$ and the sequence of morphisms in $\mathcal{C}(\mathcal{A})$

$$M \xrightarrow{f} N \xrightarrow{i} \mathrm{Cone}(f) \xrightarrow{p} M[1]$$

induced from the sequence (1.4).

The shift operator defines a translation functor $M \mapsto M[1]$ on $\mathcal{K}(\mathcal{A})$. One makes $\mathcal{K}(\mathcal{A})$ into a triangulated category by taking the distinguished triangles to be those isomorphic to a cone sequence (see e.g. [16, §2.2]). Note that $\mathcal{C}(\mathcal{A})$ and $\mathcal{K}(\mathcal{A})$ admits arbitrary direct sums.

2.2 Idempotent completion. Recall that an additive category $\mathcal{A}$ is *pseudo-abelian* if each idempotent endomorphism admits a kernel and cokernel. If $\mathcal{A}$ is an additive category, one has the *idempotent completion* $\mathcal{A} \to \mathcal{A}^\natural$ of $\mathcal{A}$, which is the universal functor of $\mathcal{A}$ to a pseudo-abelian category; $\mathcal{A}^\natural$ has objects (M, p), where M is an object of $\mathcal{A}$ and $p : M \to M$ is an idempotent endomorphism,

$$\mathrm{Hom}_{\mathcal{A}^\natural}((M, p), (N, q)) := p^* q_* \mathrm{Hom}_{\mathcal{A}}(M, N) = q_* p^* \mathrm{Hom}_{\mathcal{A}}(M, N) \subset \mathrm{Hom}_{\mathcal{A}}(M, N),$$

and the composition law in $\mathcal{A}^\natural$ is induced by that of $\mathcal{A}$. If $\mathcal{A}$ is a tensor category, $\mathcal{A}^\natural$ inherits the tensor structure, making $\mathcal{A} \to \mathcal{A}^\natural$ is a tensor functor.

One extends the definition to DG categories and DG tensor categories in the evident manner: if $\mathcal{C}$ is a DG category, then $\mathcal{C}^\natural$ has objects (M, p) with $p : M \to M$ an idempotent endomorphism in $Z^0 \mathcal{C}$ The Hom-complex is given by

$$\mathcal{H}om_{\mathcal{C}^\natural}((M, p), (N, q))^* := p^* q_* \mathcal{H}om_{\mathcal{C}}(M, N) = q_* p^* \mathcal{H}om_{\mathcal{C}}(M, N).$$

A theorem of Balmer-Schlichting [1] tells us that, for $\mathcal{A}$ a triangulated category, $\mathcal{A}^\natural$ has a canonical structure of a triangulated category for which $\mathcal{A} \to \mathcal{A}^\natural$ is exact; the same holds in the setting of tensor triangulated categories.

2.3 The derived category. A morphism $f : M \to N$ in $\mathcal{K}(\mathcal{A})$ is a *quasi-isomorphism* if $H^*(f(a)) : H^*(M(a)) \to H^*(N(a))$ is an isomorphism for each $a \in \mathcal{A}$. Call M *acyclic* if the canonical map $0 \to M$ is a quasi-isomorphism; the full subcategory of acyclic complexes is a thick subcategory of $\mathcal{K}(\mathcal{A})$, closed under direct sums.

Definition 2.1 ([16, §4.1]) The *derived category* $\mathcal{D}(\mathcal{A})$ is the localization of $\mathcal{K}(\mathcal{A})$ with respect to the full subcategory of acyclic complexes.

Remark 2.2 As $\mathcal{K}(\mathcal{A})$ and $\mathcal{D}(\mathcal{A})$ admit arbitrary direct sums, these triangulated categories are idempotently complete.

In $\mathcal{C}(\mathcal{A})$, we have the representable $\mathcal{A}^{\mathrm{op}}$-modules: for $A \in \mathcal{A}$, let $A^\vee \in \mathcal{C}(\mathcal{A})$ be the $\mathcal{A}^{\mathrm{op}}$-module

$$A^\vee(X) := \mathcal{H}om_{\mathcal{A}}(X, A).$$

We let $\mathcal{C}^b_{dg}(\mathcal{A})$ denote the smallest full DG subcategory of $\mathcal{C}_{dg}(\mathcal{A})$ containing the modules $A^\vee$, $A \in \mathcal{A}$, and closed under shift and taking cones. Let $\mathcal{K}^b(\mathcal{A}) := H^0 \mathcal{C}^b_{dg}(\mathcal{A})$.

Let $\mathcal{C}^\infty_{dg}(\mathcal{A}) \subset \mathcal{C}_{dg}(\mathcal{A})$ be the full DG subcategory of *semi-free* $\mathcal{A}^{\mathrm{op}}$-modules, that is, DG functors $M : \mathcal{A}^{\mathrm{op}} \to C_{dg}(\mathbf{Ab})$ such that,

1. After composing M with the forgetful functor from $C_{dg}(\mathbf{Ab})$ to graded abelian groups, $M = \oplus_{i \in I} A_i^\vee[n_i]$.
2. The index set I can be written as a nested union $I = \cup_{j \geq 0} I_j$, $I_0 \subset I_1 \subset \ldots$, such that, letting $M_a := \oplus_{i \in I_a} A_i^\vee[n_i]$, we have $d_M = 0$ on M_0 and $d_M(M_{a+1}) \subset M_a$.

It is easy to see that $\mathcal{C}^\infty_{dg}(\mathcal{A}) \supset \mathcal{C}^b_{dg}(\mathcal{A})$. Set $\mathcal{K}^\infty(\mathcal{A}) := H^0 \mathcal{C}^\infty_{dg}(\mathcal{A})$.

Let $\mathrm{per}(\mathcal{A}) \subset \mathcal{D}(\mathcal{A})$ be the full subcategory of compact objects of $\mathcal{D}(\mathcal{A})$ (recall that an object X in an additive category is compact if $\mathrm{Hom}(X, \oplus_\alpha A_\alpha) = \oplus_\alpha \mathrm{Hom}(X, A_\alpha)$). It is easy to see that $\mathrm{per}(\mathcal{A})$ is a triangulated subcategory of $\mathcal{D}(\mathcal{A})$.

Theorem 2.3 *1. The inclusion $\mathcal{K}^\infty(\mathcal{A}) \to \mathcal{K}(\mathcal{A})$ followed by the quotient map $\mathcal{K}(\mathcal{A}) \to \mathcal{D}(\mathcal{A})$ induces an equivalence $\varphi : \mathcal{K}^\infty(\mathcal{A}) \to \mathcal{D}(\mathcal{A})$.*

2. $\varphi(\mathcal{K}^b(\mathcal{A})) \subset \mathrm{per}(\mathcal{A})$ and φ identifies $\mathrm{per}(\mathcal{A})$ with the pseudo-abelian hull of $\mathcal{K}^b(\mathcal{A})$.

For a proof, we refer the reader to [17, theorem 3.2, corollary 3.7].

Remark 2.4 If $\mathcal{A}$ is an additive category, we consider $\mathcal{A}$ as a DG category with all Hom-complexes supported in degree zero. Then $\mathcal{C}^b_{dg}(\mathcal{A})$ is equivalent to the usual category of bounded complexes $C^b_{dg}(\mathcal{A})$ and $\mathcal{K}^b(\mathcal{A})$ is equivalent to $K^b(\mathcal{A}) := H^0(C^b_{dg}(\mathcal{A}))$.

2.4 Tensor structure. A *DG tensor category* is a DG category $\mathcal{A}$ together with the structure of a tensor category on the underlying pre-additive category frg $\mathcal{A}$ which is compatible with the DG structure, meaning:

1. Tensor product of morphisms defines a map of complexes

$$\otimes : \mathcal{H}om_{\mathcal{A}}(X, Y) \otimes \mathrm{Hom}_{\mathcal{A}}(X', Y') \to \mathcal{H}om_{\mathcal{A}}(X \otimes X', Y \otimes Y').$$

2. The associativity, commutativity and unit constraints are morphisms in $Z^0 \mathcal{A}$.

Given a commutative ring k, we have the analogous notion of a k-DG tensor category; functors of DG tensor categories are the evident notion.

Let $\mathcal{A}$ be a DG tensor category. We explain how to extend the tensor product construction from $\mathcal{A}$ to $\mathcal{C}_{dg}(\mathcal{A})$. Start by defining the tensor product of representable functors by $A^\vee \otimes B^\vee := (A \otimes B)^\vee$. Extend to translations as follows: set $A^\vee[n] \otimes B^\vee := (A \otimes B)^\vee[n]$. For an $\mathcal{A}^{\mathrm{op}}$-module M, let $-M$ be the $\mathcal{A}$ with the same underlying functor to graded groups, and with differential $d_{-M} = -d_M$. Set $A^\vee \otimes (B^\vee[n]) := (-1)^n (A \otimes B)^\vee[n]$. Extend the tensor product to arbitrary $\mathcal{A}^{\mathrm{op}}$-modules by taking the Kan extension, that is: For an arbitrary $\mathcal{A}^{\mathrm{op}}$-module M, let I_M be the category with objects $(A, n, \varphi : A^\vee[n] \to M)$, where the morphisms are $f : A \to B$ that yield a commutative diagram

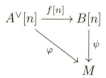

Define $M \otimes N$ as the colimit over $I_M \times I_N$ of $A^\vee[n] \otimes B^\vee[m]$. The tensor product of morphisms is defined in the evident manner.

One easily shows that this does indeed define a tensor structure on $\mathcal{C}_{dg}(\mathcal{A})$, making $\mathcal{C}_{dg}(\mathcal{A})$ a DG tensor category, making $\mathcal{C}^b_{dg}(\mathcal{A})$ and $\mathcal{C}^\infty_{dg}(\mathcal{A})$ DG tensor subcategories. Similarly, one shows that the tensor product commutes with shift and is compatible with the formation of cone sequences (both up to canonical isomorphism).

Passing to the homotopy categories, we have the tensor triangulated category $\mathcal{K}(\mathcal{A})$ with tensor triangulated subcategories $\mathcal{K}^b(\mathcal{A})$ and $\mathcal{K}^\infty(\mathcal{A})$; via theorem 2.3, this makes $\mathcal{D}(\mathcal{A})$ a tensor triangulated category with tensor triangulated subcategory $\mathrm{per}(\mathcal{A})$.

Remark 2.5 In the setting of DG tensor categories, the equivalences of proposition 2.16 and remark 2.17 are equivalences of tensor triangulated categories.

2.5 The category $\mathcal{A}^{\mathrm{pretr}}$. For the reader who prefers a less abstract approach, we give the concrete version. We review a version of Kapranov's explicit construction [15] of complexes over a DG category.

Definition 2.6 Let $\mathcal{A}$ be a DG category. The DG category Pre-Tr$(\mathcal{A})$ has objects $\mathcal{E}$ consisting of the following data:

1. A finite collection of objects of $\mathcal{A}$, $\{E_i, N \le i \le M\}$ (N and M depending on $\mathcal{E}$).
2. Morphisms $e_{ij} : E_j \to E_i$ in $\mathcal{A}$ of degree $j-i+1$, for $N \le j, i \le M$, satisfying

$$(-1)^i d e_{ij} + \sum_k e_{ik} e_{kj} = 0.$$

For $\mathcal{E} := \{E_i, e_{ij}\}$, $\mathcal{F} := \{F_i, f_{ij}\}$, a morphism $\varphi : \mathcal{E} \to \mathcal{F}$ of degree n is a collection of morphisms $\varphi_{ij} : E_j \to F_i$ in $\mathcal{A}$, such that φ_{ij} has degree $n+j-i$. The composition of morphisms $\varphi : \mathcal{E} \to \mathcal{F}$, $\psi : \mathcal{F} \to \mathcal{G}$ is defined by

$$(\psi \circ \varphi)_{ij} := \sum_k \psi_{ik} \circ \varphi_{kj}.$$

Given a morphism $\varphi : \mathcal{E} \to \mathcal{F}$ of degree n, define

$$\partial_{\mathcal{F},\mathcal{E}}(\varphi) \in \mathrm{Hom}_{\mathrm{Pre\text{-}Tr}(\mathcal{A})}(\mathcal{E},\mathcal{F})^{n+1}$$

to be the collection $\partial_{\mathcal{F},\mathcal{E}}(\varphi)_{ij} : E_j \to F_i$ with

$$\partial_{\mathcal{F},\mathcal{E}}(\varphi)_{ij} := (-1)^i d(\varphi_{ij}) + \sum_k f_{ik}\varphi_{kj} - (-1)^n \sum_k \varphi_{ik}e_{kj}.$$

For $\mathcal{E} = \{E_i, e_{ij} : E_j \to E_i\}$ in Pre-Tr($\mathcal{A}$), and $n \in \mathbb{Z}$, define $\mathcal{E}[n]$ by

$$(\mathcal{E}[n])_i := E_{i+n}, \ e[n]_{ij} := (-1)^n e_{i+n,j+n} : (\mathcal{E}[n])_j \to (\mathcal{E}[n])_i.$$

For $\varphi \in \mathrm{Hom}(\mathcal{E},\mathcal{F})^p$, define $\varphi[n] \in \mathrm{Hom}(\mathcal{E}[n],\mathcal{F}[n])^p$ by setting

$$\varphi[n]_{ij} : (\mathcal{E}[n])_j \to (\mathcal{F}[n])_i$$

equal to $(-1)^{np}\varphi_{i+n,j+n}$. It follows directly from the definitions that $(\mathcal{E},\varphi) \mapsto (\mathcal{E}[n],\varphi[n])$ defines a DG auto-isomorphism with inverse the shift by $-n$.

Let $\varphi : \mathcal{E} \to \mathcal{F}$ be a degree 0 morphism in Pre-Tr($\mathcal{A}$) with $\partial\varphi = 0$. We define the *cone* of φ, Cone(φ), as the object in Pre-Tr($\mathcal{A}$) with

$$\mathrm{Cone}(\varphi)_i := F_i \oplus E_{i+1}$$

and with morphisms $\mathrm{Cone}(\varphi)_{ij} : \mathrm{Cone}(\varphi)_j \to \mathrm{Cone}(\varphi)_i$ given by the usual matrix

$$\begin{pmatrix} f_{ij} & 0 \\ \varphi_{ij} & -e_{ij} \end{pmatrix}.$$

The inclusions $F_i \to F_i \oplus E_{i+1}$ define the morphism $i_\varphi : \mathcal{F} \to \mathrm{Cone}(\varphi)$ and the projections $F_i \oplus E_{i+1} \to E_{i+1}$ define the morphism $p_\varphi : \mathrm{Cone}(\varphi) \to \mathcal{E}[1]$. Note that $\partial(i_\varphi) = 0 = \partial(p_\varphi)$. This gives us the *cone sequence* in Z^0Pre-Tr($\mathcal{A}$):

$$\mathcal{E} \xrightarrow{\varphi} \mathcal{F} \xrightarrow{i_\varphi} \mathrm{Cone}(\varphi) \xrightarrow{p_\varphi} \mathcal{E}[1].$$

For a detailed verification that the above definitions satisfy the necessary properties, see [19, §2.1, 2.2].

We have the fully faithful embedding of DG categories

$$\tilde{i} : \mathcal{A} \to \mathrm{Pre\text{-}Tr}(\mathcal{A})$$

sending an object E of $\mathcal{A}$ to $\tilde{i}(E) := \{E_0 = E\}$; noting that

$$\mathrm{Hom}_{\mathcal{A}^{\mathrm{pretr}}}(i(E), i(F))^* = \mathrm{Hom}_{\mathcal{A}}(i(E), i(F))^*$$

defines i on morphisms and shows that $\tilde{i}$ is a fully faithful embedding.

Definition 2.7 Let $\mathcal{A}$ be a DG category. $\mathcal{A}^{\mathrm{pretr}}$ is the smallest full *DG* subcategory of Pre-Tr($\mathcal{A}$) containing $\tilde{i}(\mathcal{A})$ and closed under translation and the formation of cones.

We let

$$i : \mathcal{A} \to \mathcal{A}^{\mathrm{pretr}}$$

be the DG functor induced by $\tilde{i}$.

Remark 2.8 $\mathcal{A}^{\mathrm{pretr}}$ is denoted Pre-Tr$^+(\mathcal{A})$ in [15]. Also, one can explicitly describe the objects in $\mathcal{A}^{\mathrm{pretr}}$ as follows: $\mathcal{E} = \{E_i, e_{ij}\}$ is in $\mathcal{A}^{\mathrm{pretr}}$ if there is a finite ordered set K such that $E_i \cong \oplus_{j \in J} E_i^k$ and the k, k' component $e_{ij}^{kk'} : E_j^{k'} \to E_i^k$ is zero unless $k' < k$. See also [8, §2.4] for a slightly different description of $\mathcal{A}^{\mathrm{pretr}}$.

As remarked in [8, Remark 2.6], the inclusion $\mathcal{A}^{\mathrm{pretr}} \to \mathrm{Pre\text{-}Tr}(\mathcal{A})$ is in general *not* an equivalence of DG categories. However, we do have

Definition 2.9 (Bondarko [4, §2.7.2]) Let $\mathcal{A}$ be a DG category. Call $\mathcal{A}$ *negative* if $\mathrm{Hom}_{\mathcal{A}}(X,Y)^n = 0$ for $n > 0$.

Remark 2.10 If $\mathcal{A}$ is negative, then the inclusion $\mathcal{A}^{\mathrm{pretr}} \to \mathrm{Pre\text{-}Tr}(\mathcal{A})$ is an isomorphism. Indeed, given $\{E_i, e_{ij}\}$ in $\mathrm{Pre\text{-}Tr}(\mathcal{A})$, $e_{ij} : E_j \to E_i$ has degree $j - i + 1$, hence is zero if $j \geq i$.

For $\mathcal{A}$ negative, $\mathcal{A}^{\mathrm{pretr}}$ has "stupid truncations" $\sigma_{\leq n}$, $\sigma_{\geq n}$, where

$$\sigma_{\leq n}(\{E_i, e_{ij}\}) = \{E_i, e_{ij}, i, j \leq n\}; \quad \sigma_{\geq n}(\{E_i, e_{ij}\}) = \{E_i, e_{ij}, i, j \geq n\},$$

generalizing the stupid truncation functors for complexes over an additive category. Bondarko uses these to construct a "weight filtration" on his category of motives, see [4, §7].

We are only interested in the subcategory $\mathcal{A}^{\mathrm{pretr}}$, so for the remainder of this section, we will limit ourselves to constructions involving $\mathcal{A}^{\mathrm{pretr}}$, although all the constructions extend directly to $\mathrm{Pre\text{-}Tr}(\mathcal{A})$.

Now suppose that $(\mathcal{A}, \otimes)$ is a DG tensor category. We give $\mathcal{A}^{\mathrm{pretr}}$ a tensor structure as follows. On objects $\mathcal{E} = \{E_i, e_{ij}\}$, $\mathcal{F} = \{F_i, f_{ij}\}$, let $\mathcal{E} \otimes \mathcal{F}$ be the object with terms

$$(\mathcal{E} \otimes \mathcal{F})_i := \oplus_k E_k \otimes F_{i-k}$$

and maps $g_{ij} : (\mathcal{E} \otimes \mathcal{F})_j \to (\mathcal{E} \otimes \mathcal{F})_i$ given by the sum of the maps

$$(-1)^{(l-k+1)n} e_{kl} \otimes \mathrm{id}_{F_n} \quad \text{or} \quad (-1)^k \mathrm{id}_{E_k} \otimes f_{mn}.$$

We also need to define a cup product map

$$\cup : \mathrm{Hom}(\mathcal{E}, \mathcal{F}) \otimes \mathrm{Hom}(\mathcal{E}', \mathcal{F}') \to \mathrm{Hom}(\mathcal{E} \otimes \mathcal{E}', \mathcal{F} \otimes \mathcal{F}')$$

For this, suppose $\mathcal{E} = \{E_i, e_{ij}\}$, $\mathcal{F} = \{F_i, f_{ij}\}$, $\mathcal{E} = \{E_i', e_{ij}'\}$, $\mathcal{F}' = \{F_i', f_{ij}'\}$. Given $\varphi_{ij} : E_j \to E_i'$ of degree $j - i + p$, $\psi_{kl} : F_l \to F_k'$ of degree $l - k + q$ (the degree taken in $\mathcal{C}$), define

$$\varphi_{ij} \cup \psi_{kl} := (-1)^{(j-i+p)k+qj} \varphi_{ij} \otimes \psi_{kl} : E_j \otimes F_l \to E_i' \otimes F_k'.$$

Taking the sum over all components defines the graded map

$$\cup : \mathrm{Hom}_{\mathcal{A}^{\mathrm{pretr}}}(\mathcal{E}, \mathcal{F}) \otimes \mathrm{Hom}_{\mathcal{A}^{\mathrm{pretr}})}(\mathcal{E}', \mathcal{F}') \to \mathrm{Hom}_{\mathcal{A}^{\mathrm{pretr}}}(\mathcal{E} \otimes \mathcal{E}', \mathcal{F} \otimes \mathcal{F}').$$

Finally, we need a commutativity constraint. Let $t_{E,F} : E \otimes F \to F \otimes E$ be the commutativity constraint in $\mathcal{A}$. Define

$$\tau_{\mathcal{E},\mathcal{F}} : \mathcal{E} \otimes \mathcal{F} \to \mathcal{F} \otimes \mathcal{E}$$

as the collection of maps

$$(-1)^{jl} t_{E_i, F_l} : E_i \otimes F_l \to F_l \otimes E_i.$$

A number of direct computations (see [19, §2.3] for details) show

Proposition 2.11 Let $(\mathcal{A}, \otimes)$ be a DG tensor category. Then the operations $(\mathcal{E}, \mathcal{F}) \mapsto \mathcal{E} \otimes \mathcal{F}$, $(\varphi, \psi) \mapsto \varphi \cup \psi$ and commutativity constraints τ_{**} makes $\mathcal{A}^{\mathrm{pretr}}$ into a DG tensor category.

Via the embedding $i : \mathcal{A} \to \mathcal{A}^{\mathrm{pretr}}$, we have the functor

$$i^{\mathrm{pretr}} : \mathcal{A}^{\mathrm{pretr}} \to \mathcal{C}_{dg}(\mathcal{A})$$

sending $\mathcal{E}$ in $\mathcal{A}^{\mathrm{pretr}}$ to the functor

$$i^{\mathrm{pretr}}(\mathcal{E})(X) := \mathrm{Hom}_{\mathcal{A}^{\mathrm{pretr}}}(i(X), \mathcal{E}),$$

and similarly for morphisms. One easily checks that i^{pretr} is compatible with translation and cone sequences, and that $i^{\mathrm{pretr}} \circ i$ is the functor $E \mapsto E^{\vee}$.

Proposition 2.12 *The functor i^{pretr} is a fully faithful embedding, inducing an isomorphism of DG categories*

$$\varphi : \mathcal{A}^{\mathrm{pretr}} \to \mathcal{C}^{b}_{dg}(\mathcal{A}).$$

If $\mathcal{A}$ is a DG tensor category, then i^{pretr} is a DG tensor functor and φ is a DG tensor equivalence.

Proof For a morphism $f : X \to Y$ in $\mathcal{C}(\mathcal{A})$, and for Z in $\mathcal{C}(\mathcal{A})$, we have the natural isomorphism of graded groups

$$\mathrm{Hom}_{\mathcal{C}(\mathcal{A})}(Z, \mathrm{cone}(f))^{*} = \mathrm{Hom}_{\mathcal{C}(\mathcal{A})}(Z, Y)^{*} \otimes \mathrm{Hom}_{\mathcal{C}(\mathcal{A})}(Z, X)^{*}[1].$$

As the Yoneda functor $E \mapsto E^{\vee}$ is a fully faithful embedding $\mathcal{A} \to \mathcal{C}_{dg}(\mathcal{A})$, and i^{pretr} is compatible with translations and cone sequences it follows that i^{pretr} is fully faithful, and φ is an isomorphism. If $\mathcal{A}$ is a DG tensor category, then the Yoneda embedding $\mathcal{A} \to \mathcal{C}_{dg}(\mathcal{A})$ is compatible with tensor products, by definition; as this compatibility extends to translations, direct sums and cone sequences, it follows that i^{pretr} is a DG tensor functor and hence φ is a DG tensor equivalence. $\square$

2.6 Quasi-equivalence. Equivalence of DG categories is too rigid a notion, instead, we have the relation of quasi-equivalence.

Definition 2.13 A DG functor $f : \mathcal{A} \to \mathcal{B}$ of small DG categories is a *quasi-equivalence* if

1. f induces a surjection on isomorphism classes.
2. $H^{*}f : H^{*}\mathcal{A} \to H^{*}\mathcal{B}$ is fully faithful, in other words, f induces a quasi-isomorphism

$$\mathrm{Hom}_{\mathcal{A}}(X, Y)^{*} \to \mathrm{Hom}_{\mathcal{B}}(f(X), f(Y))^{*}$$

for all X, Y in $\mathcal{A}$.

The main result on quasi-equivalence is the following:

Theorem 2.14 *If $f : \mathcal{A} \to \mathcal{B}$ is a quasi-equivalence, then the induced functors $\mathcal{D}(f) : \mathcal{D}(\mathcal{A}) \to \mathcal{D}(\mathcal{B})$, $\mathcal{K}(f) : \mathrm{per}(\mathcal{A}) \to \mathrm{per}(\mathcal{B})$, $\mathcal{K}^{b}(f) : \mathcal{K}^{b}(\mathcal{A}) \to \mathcal{K}^{b}(\mathcal{B})$ are equivalences of triangulated categories. If f is in addition a DG tensor functor, then all these equivalences are equivalences of tensor triangulated categories.*

The result for $\mathcal{D}(f) : \mathcal{D}(\mathcal{A}) \to \mathcal{D}(\mathcal{B})$ can be found in [22, proposition 3.2]. Passing to the subcategory of compact objects implies the result for the restriction to $\mathrm{per}(\mathcal{A}) \to \mathrm{per}(\mathcal{B})$. For $\mathcal{K}^{b}(f) : \mathcal{K}^{b}(\mathcal{A}) \to \mathcal{K}^{b}(\mathcal{B})$, we can use the fact that

$$\mathrm{Hom}_{\mathcal{K}(\mathcal{A})}(X^{\vee}, Y^{\vee}[n]) \cong H^{n}(\mathrm{Hom}_{\mathcal{A}}(X, Y)^{*})$$

and that the subcategory of $\mathcal{K}(\mathcal{A})^{2}$ of pairs of objects (X, Y) for which f induces an isomorphism

$$\mathrm{Hom}_{\mathcal{K}(\mathcal{A})}(X, Y^{\vee}[n]) \to \mathrm{Hom}_{\mathcal{K}(\mathcal{B})}(f(X), f(Y)[n])$$

is triangulated. One can also derive the result for $\mathrm{per}(\mathcal{A}) \to \mathrm{per}(\mathcal{B})$ from this, using theorem 2.3 to identify $\mathrm{per}(\mathcal{A})$ and $\mathrm{per}(\mathcal{B})$ with the pseudo-abelian hulls of $\mathcal{K}^{b}(\mathcal{A})$ and $\mathcal{K}^{b}(\mathcal{B})$, respectively.

2.7 Morita theory. A weaker equivalence than quasi-equivalence of DG categories, *Morita equivalence*, induces a equivalence on the derived category. This equivalence relation is induced from the relation given by the existence of a "tilting module" with some good properties. We first recall the general setting from [17, §3.5].

Let $\mathcal{A}$ and $\mathcal{B}$ be small DG categories. The DG category $\mathcal{A} \otimes \mathcal{B}$ has objects (A, B), with A in $\mathcal{A}$, B in $\mathcal{B}$, morphisms

$$\mathrm{Hom}((A, B), (A', B'))^* := \mathrm{Hom}_{\mathcal{A}}(A, A')^* \otimes \mathrm{Hom}_{\mathcal{B}}(B, B')^*$$

with composition law

$$(f \otimes g) \circ (f' \otimes g') = (-1)^{\deg g \cdot \deg f'} f f' \otimes g g'.$$

Let X an $\mathcal{B}$-$\mathcal{A}$-bimodule, that is, an $\mathcal{A} \otimes \mathcal{B}^{\mathrm{op}}$-module. For each $\mathcal{B}^{\mathrm{op}}$-module M, we have the $\mathcal{A}^{\mathrm{op}}$-module

$$GM := \mathrm{Hom}(X, M) : A \mapsto \mathrm{Hom}_{\mathcal{C}_{dg}(\mathcal{B})}(X(A, -), M)^*,$$

giving the DG functor $G : \mathcal{C}_{dg}(\mathcal{B}) \to \mathcal{C}_{dg}(\mathcal{A})$. G has the left adjoint F, $F(L) := L \otimes_{\mathcal{A}} X$,

$$(L \otimes_{\mathcal{A}^{\mathrm{op}}} X)(M) := \oplus_N L(N) \otimes X(N, M)/\sim$$

where $\sim$ is the quotient identifying

$$\oplus_{f:N' \to N} L(N') \otimes X(N, M) \xrightarrow{\Sigma_f f_* \otimes \mathrm{id}} L(N) \otimes X(N, M)$$

$$\Sigma_f \mathrm{id} \otimes f^* \downarrow$$

$$L(N') \otimes X(N', M)$$

Taking derived functors, one has the pair of adjoint functors $LF : \mathcal{D}(\mathcal{A}) \to \mathcal{D}(\mathcal{B})$, $RG : \mathcal{D}(\mathcal{B}) \to \mathcal{D}(\mathcal{B})$. We recall the following result of Keller:

Lemma 2.15 (Keller [17, Lemma 3.10]) *LF is an equivalences if the following conditions hold:*

1. *$X(A, -) \in \mathcal{D}(\mathcal{B})$ is perfect for all $A \in \mathcal{A}$.*
2. *The morphism*

   $$\mathrm{Hom}_{\mathcal{A}}(A, A')^* \to \mathrm{Hom}_{\mathcal{C}(\mathcal{B})}(X(A, -), X(A', -))^*$$

   is a quasi-isomorphism for all A, A' in $\mathcal{A}$
3. *The DG $\mathcal{B}^{\mathrm{op}}$-modules $X(A, -)$, $A \in \mathcal{A}$, form a set of compact generators for $\mathcal{D}(\mathcal{B})$.*

We will use this general theory to prove

Proposition 2.16 *Let $\mathcal{A}_0$ be a small DG category, and let $\mathcal{A}$ be either $\mathcal{C}_{dg}^b(\mathcal{A}_0)$ or $\mathcal{C}_{dg}(\mathcal{A}_0)$. If $\mathcal{A}_0$ is an additive category, we also allow $\mathcal{A} = C_{dg}^?(\mathcal{A}_0)$, where $? = +, -, \emptyset$. Then there are a natural equivalences*

$$\mathcal{D}(\mathcal{A}) \sim \mathcal{D}(\mathcal{A}_0); \ \mathrm{per}(\mathcal{A}) \sim \mathrm{per}(\mathcal{A}_0).$$

Proof As an exact equivalence $\mathcal{D}(\mathcal{A}) \sim \mathcal{D}(\mathcal{A}_0)$ induces an exact equivalence on the subcategories of perfect objects, we need only verify the hypotheses of lemma 2.15.

In case $\mathcal{A} = \mathcal{C}_{dg}^b(\mathcal{A}_0)$ or $\mathcal{C}_{dg}(\mathcal{A}_0)$, let $i_0 : \mathcal{A}_0 \to \mathcal{A}$ be the Yoneda embedding; in case $\mathcal{A}_0$ is an additive category and $\mathcal{A} = C_{dg}^?(\mathcal{A}_0)$, let $i_0 : \mathcal{A}_0 \to \mathcal{A}$ be the embedding sending $A \in \mathcal{A}_0$ to A considered as a complex supported in degree zero. In all cases, let X be the $\mathcal{A}$-$\mathcal{A}_0$-bimodule $(A, B) \mapsto \mathrm{Hom}_\mathcal{A}(B, i_0(A))^*$. Then $X(A, -) = i_0(A)^\vee$; property (1) follows from theorem 2.3 and (2) follows from the Yoneda lemma and the fact that i_0 is fully faithful. It remains to show that the $X(A, -)$ form a set of generators for $\mathcal{D}(\mathcal{A})$. Using theorem 2.3, it suffices to show that the $X(A, -)$ form a set of generators for $\mathcal{K}^\infty(\mathcal{A})$.

For this, we first consider the case $\mathcal{A} = \mathcal{C}_{dg}(\mathcal{A}_0)$. It follows easily from the definition of semi-free modules that the smallest triangulated subcategory of $\mathcal{K}^\infty(\mathcal{A})$ closed under direct sums and containing the image of the Yoneda embedding of $\mathcal{A}$ is all of $\mathcal{K}^\infty(\mathcal{A})$. Replacing $\mathcal{A}$ with $\mathcal{A}_0$, we see that the same holds for the image of $i_0(\mathcal{A}_0)$ in $\mathcal{K}^\infty(\mathcal{A}_0)$. Combining these two facts, we see that smallest triangulated subcategory of $\mathcal{K}^\infty(\mathcal{A})$ closed under direct sums and containing the objects $X(A, -)$, A in $\mathcal{A}_0$, is all of $\mathcal{K}^\infty(\mathcal{A})$, and hence the $X(A, -)$ form a set of generators for $\mathcal{K}^\infty(\mathcal{A})$. The proof for $\mathcal{A} = C_{dg}^?(\mathcal{A}_0)$ is similar. $\qquad\square$

Remark 2.17 Let $\mathcal{A}_0$ be the additive category of injective objects in a Grothendieck abelian category $\mathcal{C}$. Then the homotopy category $H^0(C_{dg}^+(\mathcal{A}_0))$ of $C_{dg}^+(\mathcal{A}_0)$ is equivalent to the derived category $D^+(\mathcal{C})$, and $H^0(C_{dg}^+(\mathcal{A}_0))$ is in turn equivalent to the full triangulated subcategory of $\mathcal{K}^\infty(\mathcal{A}_0) \sim \mathcal{D}(\mathcal{A}_0)$ of objects with bounded below cohomology. Using proposition 2.16, we have an equivalence

$$\mathcal{D}(C_{dg}^+(\mathcal{A}_0)) \to \mathcal{D}(\mathcal{A}_0)$$

inducing an equivalence of $\mathcal{D}^b(C_{dg}^+(\mathcal{A}_0))$ with $D^+(\mathcal{C})$.

Remark 2.18 There is a *total complex functor*

$$\mathrm{Tot} : (\mathcal{A}^{\mathrm{pretr}})^{\mathrm{pretr}} \to \mathcal{A}^{\mathrm{pretr}}$$

defined as follows: if $\mathcal{E} = \{\mathcal{E}^l, \varphi^{kl} : \mathcal{E}^l \to \mathcal{E}^k\}$, where $\mathcal{E}^l = \{E_i^l, e_{ij}^l : E_j^l \to E_i^l\}$ and $\varphi^{kl} = \{\varphi_{ij}^{kl} : E_j^l \to E_i^k\}$, then $\mathrm{Tot}(\mathcal{E})$ is given by the collection of objects

$$\mathrm{Tot}(\mathcal{E})_n := \oplus_{j+l=n} E_j^l,$$

together with the morphisms $\mathrm{Tot}(\mathcal{E})_{mn} : \mathrm{Tot}(\mathcal{E})_n \to \mathrm{Tot}(\mathcal{E})_m$ defined as the sum of terms $(-1)^l e_{ij}^l$ and φ_{ij}^{kl}.

For an additive category $\mathcal{A}_0$, exactly the same formulas define a total complex functor

$$\mathrm{Tot} : C_{dg}^?(\mathcal{A}_0)^{\mathrm{pretr}} \to C_{dg}^?(\mathcal{A}_0); \quad ? = +, -, \infty$$

which is an equivalence of DG categories with inverse the evident extension of the inclusion functor i_0.

One easily checks that Tot respects translation and cone sequences, and hence induces exact equivalences of triangulated categories

$$\mathrm{Tot} : \mathcal{K}^b(C_{dg}^?(\mathcal{A}_0)) \to K^?(\mathcal{A}_0); \quad ? = +, -, \infty.$$

If $\mathcal{A}$ is a DG tensor category, or $\mathcal{A}_0$ is a tensor category, the above equivalences are equivalences of DG tensor categories, resp. tensor triangulated categories. Thus, we have in this case a "concrete" version of Morita equivalence.

Corollary 2.19 *Let $\mathcal{C}$ be a DG category, $\mathcal{A}$ an additive category, and*

$$F : \mathcal{C} \to C^?(\mathcal{A})$$

a DG functor, $? = b, +, -, \infty$. Then F defines a canonical exact functor

$$K^b(F) : \mathcal{K}^b(\mathcal{C}) \to K^?(\mathcal{A});$$

if F is a DG tensor functor, then $K^?(F)$ is an exact tensor functor for $? = b, +, -$.

3 Sheafification of DG categories

We show how to construct a DG category $R\Gamma\mathcal{C}$ out of a presheaf of DG categories $U \mapsto \mathcal{C}(U)$ so that the Hom complexes in $R\Gamma\mathcal{C}$ compute the hypercohomology of the presheaf of Hom complexes for $\mathcal{C}$.

3.1 Sheafifying. Fix a Grothendieck topology τ on some full subcategory Opn_S^τ of $\mathbf{Sch}_S$; we assume that Opn_S^τ is closed under fiber product over S, that τ is subcanonical, and that τ has a conservative set of points $Pt_\tau(S)$. Let $Sh_\tau(S)$ be the category of sheaves of abelian groups on Opn_S for the topology τ, and $PSh_\tau(S)$ the category of presheaves on Opn_S. We let

$$i : \prod_{x \in Pt_\tau(S)} \mathbf{Ab} \to Sh_\tau(S)$$

be the canonical map of topoi, i.e, we have the functor $i^* : Sh_\tau(S) \to \prod_{x \in Pt_\tau(S)}$ (the factor i_x^* sends f to the stalk f_x at x) and the right adjoint of i^*, $i_* : \prod_{x \in Pt_\tau(S)} \to Sh_\tau(S)$. Let $\mathcal{G}^0 := i_* \circ i^*$, $\mathcal{G}^n := (\mathcal{G}^0)^n$. Combining the unit $\epsilon : \mathrm{id} \to i_* i^*$ and the co-unit $\eta : i^* i_* \to \mathrm{id}$ of the adjunction gives us the cosimplicial object in the category of endofunctors of $Sh_\tau(S)$:

$$
\mathcal{G}^0 \leftarrow \mathcal{G}^1 \rightrightarrows \dots
$$

with augmentation $\epsilon : \mathrm{id} \to \mathcal{G}^0$. For a sheaf $\mathcal{F}$, we let

$$\mathcal{F} \xrightarrow{\epsilon} \mathcal{G}^*\mathcal{F}$$

be the augmented complex associated to the augmented cosimplicial sheaf $n \mapsto \mathcal{G}^n(\mathcal{F})$; we extend this construction to complexes by taking the total complex of the associated double complex.

Lemma 3.1 *For each complex of sheaves $\mathcal{F}$, the augmentation $\epsilon : \mathcal{F} \to \mathcal{G}^*(\mathcal{F})$ is a quasi-isomorphism, and $\mathcal{G}^n(\mathcal{F})$ is acyclic for the functor $\Gamma(S, -)$.*

Thus, we have a functorial $\Gamma(S, -)$-acyclic resolution for complexes of sheaves of abelian groups on S. In addition, i^* is a tensor functor, and the isomorphism

$$i^*(i_*(\mathcal{A}) \otimes i_*(\mathcal{B})) \to i^* i_*(\mathcal{A}) \otimes i^* i_*(\mathcal{B})$$

followed by the co-unit

$$i^* i_*(\mathcal{A}) \otimes i^* i_*(\mathcal{B}) \to \mathcal{A} \otimes \mathcal{B}$$

induces a natural transformation $i_*(\mathcal{A}) \otimes i_*(\mathcal{B}) \to i_*(\mathcal{A} \otimes \mathcal{B})$. This gives us a natural map

$$\mathcal{G}^0(\mathcal{F}) \otimes \mathcal{G}^0(\mathcal{F}') \to \mathcal{G}^0(\mathcal{F} \otimes \mathcal{F}')$$

and thus by iteration a map of cosimplicial sheaves

$$\mu : [n \mapsto \mathcal{G}^n(\mathcal{F}) \otimes \mathcal{G}^n(\mathcal{F}')] \to [n \mapsto \mathcal{G}^n(\mathcal{F} \otimes \mathcal{F}')].$$

We have the Alexander-Whitney map

$$AW : \mathcal{G}^*(\mathcal{F}) \otimes \mathcal{G}^*(\mathcal{F}') \to [n \mapsto \mathcal{G}^n(\mathcal{F}) \otimes \mathcal{G}^n(\mathcal{F}')]^*$$

sending $\mathcal{G}^p(\mathcal{F}) \otimes \mathcal{G}^q(\mathcal{F}') \to \mathcal{G}^{p+q}(\mathcal{F}) \otimes \mathcal{G}^{p+q}(\mathcal{F}')$ by $\mathcal{G}(i_1^{p,q}) \otimes \mathcal{G}(i_2^{p,q})$, where

$$i_1^{p,q} : \{0,\dots,p\} \to \{0,\dots,p+q\}; \quad i_2^{p,q} : \{0,\dots,q\} \to \{0,\dots,p+q\}$$

are the maps

$$i_1^{p,q}(j) = j, \ i_2^{p,q}(j) = j + p.$$

The composition

$$\mu_{\mathcal{F},\mathcal{F}'}^* := \mu \circ AW : \mathcal{G}^*(\mathcal{F}) \otimes \mathcal{G}^*(\mathcal{F}') \to \mathcal{G}^*(\mathcal{F} \otimes \mathcal{F}')$$

makes $\mathcal{F} \mapsto \mathcal{G}^*(\mathcal{F})$ a weakly monoidal functor, i.e., we have the associativity

$$\mu_{\mathcal{F} \otimes \mathcal{F}',\mathcal{F}''}^* \circ (\mu_{\mathcal{F},\mathcal{F}'}^* \otimes \mathrm{id}_{\mathcal{G}^*(\mathcal{F}'')}) = \mu_{\mathcal{F},\mathcal{F}' \otimes \mathcal{F}''}^* \circ (\mathrm{id}_{\mathcal{G}^*(\mathcal{F})} \otimes \mu_{\mathcal{F}',\mathcal{F}''}^*).$$

These facts enable the following construction: Let $U \mapsto \mathcal{C}(U)$ be a presheaf of DG categories on Opn_S^τ. For objects X and Y in $\mathcal{C}(S)$, we have the presheaf

$$[f : U \to S] \mapsto \mathcal{H}om_{\mathcal{C}(U)}(f^*(X), f^*(Y));$$

we denote the associated sheaf by $\underline{\mathcal{H}om}_{\mathcal{C}}^\tau(X, Y)$.

Let $R\Gamma(S, \mathcal{C})$ be the DG category with the same objects as $\mathcal{C}(S)$, and with Hom-complex

$$\mathcal{H}om_{R\Gamma(S,\mathcal{C})}(X, Y)^* := \mathcal{G}^*(\underline{\mathcal{H}om}_{\mathcal{C}}^\tau(X, Y))(S).$$

Our remarks on the Godement resolution show that the composition law for the sheaf of DG categories $\mathcal{C}$ defines canonically an associative composition law

$$\circ : \mathcal{H}om_{R\Gamma(S,\mathcal{C})}(Y, Z)^* \otimes \mathcal{H}om_{R\Gamma(S,\mathcal{C})}(X, Y)^* \to \mathcal{H}om_{R\Gamma(S,\mathcal{C})}(X, Z)^*$$

for $R\Gamma(S, \mathcal{C})$, making $R\Gamma(S, \mathcal{C})$ a DG category. In addition we have

Proposition 3.2 *Let $\mathcal{C}$ be a presheaf of DG categories on Opn_S^τ. Then for each pair of objects X, Y of $\mathcal{C}(S)$, we have a canonical isomorphism*

$$\mathbb{H}^n(S_\tau, \underline{\mathcal{H}om}_{\mathcal{C}}^\tau(X, Y)) \cong H^n(\mathcal{H}om_{R\Gamma(S,\mathcal{C})}(X, Y)).$$

Indeed, since $\mathcal{F} \to \mathcal{G}^*(\mathcal{F})$ is a $\Gamma(S, -)$-acyclic resolution of $\mathcal{F}$ for any complex of sheaves $\mathcal{F}$, we have a canonical isomorphism $\mathbb{H}^n(S_\tau, \mathcal{F}) \cong H^n(\mathcal{G}^*(\mathcal{F})(S))$ for any complex of presheaves $\mathcal{F}$ on Opn_S^τ.

3.2 Thom-Sullivan co-chains. Suppose now that we are given a presheaf of DG tensor categories $U \mapsto \mathcal{C}(U)$ on Opn_S^τ. The Alexander-Whitney maps gives us a natural associative pairing of Hom-complexes

$$\otimes : \mathcal{H}om_{R\Gamma(S,\mathcal{C})}(X, Y)^* \otimes \mathcal{H}om_{R\Gamma(S,\mathcal{C})}(X', Y')^* \to \mathcal{H}om_{R\Gamma(S,\mathcal{C})}(X \otimes X', Y \otimes Y')^*$$

satisfying the Leibniz rule and compatible with the tensor product operation on $\mathcal{C}(S)$, however, this map is not commutative. For this, we need assume that $U \mapsto \mathcal{C}(U)$ is a sheaf of $\mathbb{Q}$-DG category; we can then modify the construction of $R\Gamma(S, \mathcal{C})$, giving a quasi-equivalent DG category $R\Gamma(S, \mathcal{C})^\otimes$ with a well-defined DG tensor structure. The construction uses the *Thom-Sullivan cochains*; we have taken this material from [13].

Let $|\Delta_n|$ be the real n-simplex

$$|\Delta_n| := \{(t_0,\dots,t_n) \in \mathbb{R}^{n+1} \mid \sum_{i=0}^n t_i = 1, 0 \le t_i\},$$

and let Δ_n be the simplicial set $\mathrm{Hom}_\Delta(-, [n]) : \Delta^{\mathrm{op}} \to \mathbf{Sets}$. For $f : [m] \to [n]$ in Δ, let $|f| : |\Delta_m| \to |\Delta_n|$ denote the corresponding affine-linear map.

For a cosimplicial abelian group G, we have the associated complex G^*, and the *normalized* subcomplex, NG^*, with NG^p the subgroup of those $g \in G([p])$ such $G(f)(g) = 0$ for all $f : [p] \to [q]$ in Δ which are not injective. The inclusion of $NG^* \hookrightarrow G^*$ is a homotopy equivalence. In particular, we have the complex of (simplicial) cochains of Δ_n, and the subcomplex of normalized cochains $Z^*(\Delta_n)$.

Let $\Omega^*(|\Delta_n|)$ denote the complex of $\mathbb{Q}$-polynomial differential forms on $|\Delta_n|$:

$$\Omega^*(|\Delta_n|) := \Omega^*_{\mathbb{Q}[t_0,\dots,t_n]/\sum_{i=0}^n t_i - 1}.$$

Sending $[n]$ to $Z^*(\Delta_n)$, $\Omega^*(|\Delta_n|)$ determines functors

$$Z^* : \Delta^{\mathrm{op}} \to C^{\geq 0}(\mathbf{Ab}), \quad \Omega^* : \Delta^{\mathrm{op}} \to C^{\geq 0}(\mathrm{Mod}_{\mathbb{Q}}),$$

where $C^{\geq 0}(\mathrm{Mod}_{\mathbb{Q}})$ is the category of complexes of $\mathbb{Q}$-vector spaces concentrated in degrees ≥ 0, and $C^{\geq 0}(\mathbf{Ab})$ is the integral version. There is a natural homotopy equivalence

$$\int : \Omega^* \to Z^* \otimes \mathbb{Q}$$

defined by

$$\int(\omega)(\sigma) = \int_{|\sigma|} \omega$$

for $\omega \in \Omega^m(|\Delta_n|)$ and σ an m-simplex of Δ_n.

We have the category $\mathrm{Mor}(\Delta)$, with objects the morphisms in Δ, where a morphism $f \to g$ is a commutative diagram

For functors $F : \Delta^{\mathrm{op}} \to C^{\geq 0}(\mathbf{Ab})$ and $G : \Delta \to \mathbf{Ab}$, we have the functor

$$F(\mathrm{domain}) \otimes G(\mathrm{range}) : \mathrm{Mor}(\Delta) \to C^{\geq 0}$$
$$f \mapsto F(\mathrm{domain}(f)) \otimes G(\mathrm{range}(f));$$

let $F \otimes_{\leftarrow} G$ be the projective limit

$$F \otimes_{\leftarrow} G := \varprojlim_{\mathrm{Mor}(\Delta)} F(\mathrm{domain}) \otimes G(\mathrm{range}).$$

Explicitly, an element ϵ of $(F \otimes_{\leftarrow} G)^m$ is given by a collection $p \mapsto \epsilon_p \in F^m([p]) \otimes G([p])$ such that

$$F(f) \otimes G(\mathrm{id})(\epsilon_p) = F(\mathrm{id}) \otimes G(f)(\epsilon_q)$$

for each $f : [q] \to [p]$ in Δ. The operation $\otimes_{\leftarrow}$ is functorial and respects homotopy equivalence.

For a cosimplicial abelian group G, we have the well-defined map $e : Z^* \otimes_{\leftarrow} G \to NG^*$ defined by sending $\epsilon := (\dots \epsilon_p \dots) \in (Z^* \otimes_{\leftarrow} G)^q$ to $\epsilon_q(\mathrm{id}_{[q]}) \in G([q])$. In [13, Lemma 3.1] it is shown that this map is well-defined, lands in NG^*, and gives an isomorphism of complexes. Thus, we have the natural homotopy equivalences

$$\int \otimes_{\leftarrow} \mathrm{id} : \Omega^* \otimes_{\leftarrow} G \to NG^* \otimes \mathbb{Q} \to G^* \otimes \mathbb{Q}.$$

The operation of wedge product makes Ω^* into a simplicial commutative differential graded algebra. Suppose we have two cosimplicial abelian groups G, G' : $\Delta \to \mathbf{Ab}$, giving us the diagonal cosimplicial abelian group $G \otimes G' : \Delta \to \mathbf{Ab}$. We have the map

$$\cup : (\Omega^* \underset{\leftarrow}{\otimes} G) \otimes (\Omega^* \underset{\leftarrow}{\otimes} G') \to \Omega^* \underset{\leftarrow}{\otimes} (G \otimes G')$$

induced by the map

$$(\omega \otimes g) \otimes (\omega' \otimes g') \mapsto (\omega \wedge \omega') \otimes (g \otimes g'),$$

for $\omega \otimes g \in \Omega^q(|\Delta_p|) \otimes G([p])$, $\omega' \otimes g' \in \Omega^{q'}(|\Delta_p|) \otimes G([p])$. It is easy to check that this gives a well-defined functorial product of cochain complexes and that $\cup$ is associative and commutative, in the evident sense.

We have as well the product map

$$\cup_{AW} : G^* \otimes G'^* \to (G \otimes G')^*$$

induced by the Alexander-Whitney maps $i_1^p : [p] \to [p+q]$, $i_2^q : [q] \to [p+q]$

$$(g \in G^p) \otimes (g' \in G'^q) \mapsto G(i_1^p)(g) \otimes G'(i_2^q)(g') \in G^{p+q} \otimes G'^{p+q}.$$

It follows easily from the change of variables formula for integration that the diagram

$$
\begin{array}{ccc}
(\Omega^* \underset{\leftarrow}{\otimes} G) \otimes (\Omega^* \underset{\leftarrow}{\otimes} G') & \xrightarrow{\ \cup\ } & \Omega^* \underset{\leftarrow}{\otimes} (G \otimes G') \\
{\scriptstyle \int \otimes \int} \downarrow & & \downarrow {\scriptstyle \int} \\
(G^* \otimes \mathbb{Q}) \otimes (G'^* \otimes \mathbb{Q}) & \xrightarrow[\cup_{AW}]{} & (G \otimes G')^* \otimes \mathbb{Q}
\end{array}
\tag{3.1}
$$

commutes.

We extend the operation $\Omega^* \underset{\leftarrow}{\otimes} (-)$ to functors $G^* : \Delta^{\mathrm{op}} \to C^+(\mathbf{Ab})$ by taking the total complex of the double complex

$$\ldots \to \Omega^* \underset{\leftarrow}{\otimes} G^0 \to \Omega^* \underset{\leftarrow}{\otimes} G^1 \to \ldots .$$

All the properties of $\Omega^* \underset{\leftarrow}{\otimes} (-)$ described above extend to the case of complexes.

3.3 Sheafifying DG tensor categories. Let $U \mapsto \mathcal{C}(U)$ be a presheaf of $\mathbb{Q}$-DG tensor categories on Opn_S^τ.

Define

$$\mathcal{H}om_{R\Gamma(S,\mathcal{C})^\otimes}(X,Y)^* := \Omega^* \underset{\leftarrow}{\otimes} \mathcal{G}^*(\mathcal{H}om_{\mathcal{C}}^\tau(X,Y))(S)$$

Using the functoriality of the Godement resolution, the composition law for $U \mapsto \mathcal{C}(U)$ together with the product $\cup$ defined in §3.2 defines a composition law

$$\circ : \mathcal{H}om_{R\Gamma(S,\mathcal{C})^\otimes}(Y,Z)^* \otimes \mathcal{H}om_{R\Gamma(S,\mathcal{C})^\otimes}(X,Y)^* \to \mathcal{H}om_{R\Gamma(S,\mathcal{C})^\otimes}(X,Z)^*$$

giving us the DG category $R\Gamma(S,\mathcal{C})^\otimes$. By the commutativity of (3.1) the maps

$$\int \underset{\leftarrow}{\otimes} \mathrm{id} : \mathcal{H}om_{R\Gamma(S,\mathcal{C})^\otimes}(X,Y)^* \to \mathcal{H}om_{R\Gamma(S,\mathcal{C})}(X,Y)^*$$

define a DG functor

$$\int : R\Gamma(S,\mathcal{C})^\otimes \to R\Gamma(S,\mathcal{C})$$

which is a quasi-equivalence.

Finally, the tensor product operation on $\mathcal{C}$ gives rise to maps of cosimplicial objects

$$\otimes : \mathcal{G}^*(\mathcal{H}om_{\mathcal{C}}^{\tau}(X,Y))(S) \otimes \mathcal{G}^*(\mathcal{H}om_{\mathcal{C}}^{\tau}(X',Y'))(S) \to \mathcal{G}^*(\mathcal{H}om_{\mathcal{C}}^{\tau}(X \otimes X', Y \otimes Y'))(S).$$

Applying the Thom-Sullivan cochain construction gives the map

$$\tilde{\otimes} : \mathcal{H}om_{R\Gamma(S,\mathcal{C})^{\otimes}}(X,Y)^* \otimes \mathcal{H}om_{R\Gamma(S,\mathcal{C})^{\otimes}}(X',Y')^* \to \mathcal{H}om_{R\Gamma(S,\mathcal{C})^{\otimes}}(X \otimes X', Y \otimes Y')^*$$

that makes $R\Gamma(S,\mathcal{C})^{\otimes}$ a DG tensor category, with commutativity constraint induced from $\mathcal{C}(S)$.

Remark 3.3 Let $U \mapsto \mathcal{C}(U)$ be a presheaf of $\mathbb{Q}$-DG tensor categories on Opn_S^{τ}. The quasi-equivalence $\int : R\Gamma(S,\mathcal{C})^{\otimes} \to R\Gamma(S,\mathcal{C})$ induces an equivalence of triangulated categories

$$\mathcal{K}^b\left(\int\right) : \mathcal{K}^b(R\Gamma(S,\mathcal{C})^{\otimes}) \to \mathcal{K}^b(R\Gamma(S,\mathcal{C}))$$

(theorem 2.14). In addition, the DG tensor structure we have defined on $R\Gamma(S,\mathcal{C})^{\otimes}$ endows $\mathcal{C}^b(R\Gamma(S,\mathcal{C})^{\otimes})$ with the structure of a DG tensor category, and makes $\mathcal{K}^b(R\Gamma(S,\mathcal{C})^{\otimes})$ a tensor triangulated category. Similarly, for the idempotent completions, we have equivalences of (tensor) triangulated categories

$$\mathcal{D}\left(\int\right) : \mathrm{per}(R\Gamma(S,\mathcal{C})^{\otimes}) \to \mathrm{per}(R\Gamma(S,\mathcal{C})).$$

4 DG categories of motives

Let S be a fixed base-scheme; we assume that S is a regular scheme of finite Krull dimension. Let $\mathbf{Sm}/S$ denote the category of smooth S-schemes of finite type, $\mathbf{Proj}/S \subset \mathbf{Sm}/S$ be full subcategory of $\mathbf{Sm}/S$ consisting of the smooth projective S-schemes. Our construction of the DG category of motives goes as follows: We form the DG category of correspondences $dgCor_S$ from a cubical enhancement of the category of finite correspondences Cor_S, using the algebraic n-cubes as the cubical object; restricting to smooth projective S-schemes gives us our basic DG category $dgPrCor_S$. We sheafify the construction over S, then take the homotopy category of complexes over the sheafified DG category to form the triangulated category $\mathcal{K}^b(R\Gamma(S,\underline{dgPrCor}_S))$. Taking the idempotent completion gives us our category of smooth effective motives

$$SmMot_{gm}^{\mathrm{eff}}(S) := \mathrm{per}(R\Gamma(S,\underline{dgPrCor}_S));$$

inverting tensor product with the Lefschetz motive gives us the category of smooth motives $SmMot_{gm}(S)$.

We also have a parallel version with $\mathbb{Q}$-coefficients; using alternating cubes endows everything with a tensor structure. Now for the details.

4.1 DG categories of correspondences. We begin by recalling the definition of the category of finite correspondences.

Definition 4.1 For $X, Y \in \mathbf{Sm}/S$, $Cor_S(X,Y)$ is the free abelian group on the integral closed subschemes $W \subset X \times_S Y$ such that the projection $W \to X$ is finite and surjective onto an irreducible component of X.

Now let X, Y and Z be in $\mathbf{Sm}/S$. Take generators $W \in Cor_S(X,Y)$, $W' \in Cor_S(Y,Z)$. As in [24, Chap. V], each component T of the intersection $W \times_S Z \cap X \times_S W' \subset X \times_S Y \times_S Z$ is finite over $X \times_S Z$ and over X (via the projections) and the map

$$T \to X$$

is surjective over some irreducible component of X. Thus, letting p_{XY}, p_{YZ}, etc., denote the projections from the triple product $X \times_S Y \times_S Z$, and $\cdot_{XYZ}$ the intersection product of cycles on $X \times_S Y \times_S Z$, the expression

$$W \circ W' := p_{XZ*}(p_{XY}^*(W) \cdot_{XYZ} p_{YZ}^*(W'))$$

gives a well-defined and associative composition law

$$\circ : Cor_S(X,Y) \otimes Cor_S(Y,Z) \to Cor_S(X,Z),$$

defining the pre-additive category Cor_S. Cor_S is an additive category, with disjoint union being the direct sum, and product over S makes Cor_S into a tensor category.

Sending $X \in \mathbf{Sm}/S$ to $X \in Cor_S$ and sending $f : X \to Y$ to the graph of f, $\Gamma_f \subset X \times_S Y$, defines a faithful embedding $i_S : \mathbf{Sm}/S \to Cor_S$. Making $\mathbf{Sm}/S$ a symmetric monoidal category using the product over S makes i_S a symmetric monoidal functor.

From example 1.2, we have the co-cubical object $\square_S^*$,

$$n \mapsto \square_S^n \cong \mathbb{A}_S^n.$$

In fact, $\square_S^*$ extends to a functor

$$\square_S^* : \mathbf{ECube} \to \mathbf{Sm}/S$$

by sending the multiplication map $\mu : \underline{2} \to \underline{1}$ to the usual multiplication

$$\mu_S : \square_S^2 \to \square_S^1; \; \mu_S(x,y) = xy.$$

Define the co-multiplication $\delta : \square_S^* \to \square_S^* \otimes \square_S^*$ by taking the collection of diagonal maps $\delta^n := \delta_{\square^n} : \square_S^n \to \square_S^n \times_S \square_S^n$. One easily verifies the properties of definition 1.10.

Definition 4.2 Let $dgCor_S$ denote the DG category $(Cor_S, \otimes, \square_S^*, \delta)$. We denote the DG tensor category $((Cor_{S\mathbb{Q}}, \otimes, \square_S^*, \delta)^{\mathrm{alt}}, \otimes_\square)$ by $dgCor_S^{\mathrm{alt}}$.

Proposition 4.3 *1. The DG categories $dgCor_S$ and $dgCor_S^{\mathrm{alt}}$ are negative (see definition 2.9).*

2. The functor $dgCor_S^{\mathrm{alt}} \to dgCor_{S\mathbb{Q}}$ is a quasi-equivalence.

Proof Indeed, (1) is obvious, and (2) follows from proposition 1.13. $\square$

Definition 4.4 Let $dgPrCor_S$ be the full DG subcategory of $dgCor_S$ with objects $X \in \mathbf{Proj}/S$. $dgPrCor_S^{\mathrm{alt}}$ is similarly defined as the full DG subcategory of $dgCor_S^{\mathrm{alt}}$ with objects $X \in \mathbf{Proj}/S$.

Note that $dgPrCor_S^{\mathrm{alt}}$ is a DG tensor subcategory of $dgCor_S^{\mathrm{alt}}$, and the functor $dgPrCor_S^{\mathrm{alt}} \to dgPrCor_{S\mathbb{Q}}$ induced by $dgCor_S^{\mathrm{alt}} \to dgCor_{S\mathbb{Q}}$ is a quasi-equivalence.

4.2 Functoriality. Let $f : S' \to S$ be a k-morphism of regular schemes. We have the well-defined pull-back functor $f^* : Cor_S \to Cor_{S'}$, with $f^*(X) = X \times_S S'$ for $X \in \mathbf{Sm}/S$ and using Serre's intersection multiplicity formula to define the pull-back of cycles

$$f^* : Cor_S(X, Y) \to Cor_{S'}(f^*(X), f^*(Y)).$$

Note that the cycle pull-back is always well-defined for finite correspondences, using the isomorphism

$$(X \times_S S') \times_{S'} (Y \times_S S') \cong (X \times_S S') \times_S Y.$$

Since $f^*(X) \times_{S'} \square_{S'}^n \cong f^*(X \times_S \square_S^n)$, the pull-back extends to the map of complexes

$$f^* : \mathcal{H}om_{dgCor_S}(X, Y) \to \mathcal{H}om_{dgCor_{S'}}(f^*(X), f^*(Y)),$$

defining the functor $S \mapsto dgCor_S$ from regular schemes to DG categories. We have as well the sub-functor $S \mapsto dgPrCor_S$.

A similar construction defines the functor $S \mapsto dgCor_S^{\mathrm{alt}}$, from regular schemes to DG tensor categories, and the subfunctor $S \mapsto dgPrCor_S^{\mathrm{alt}}$.

4.3 Complexes and smooth motives.

Definition 4.5 Let S be a regular scheme. Let $\underline{dgPrCor}_S$ denote the Zariski presheaf

$$U \mapsto \underline{dgPrCor}_S(U) := dgPrCor_U.$$

The DG category of *smooth effective geometric motives over* S, $dgSmMot_S^{\mathrm{eff}}$, is defined as

$$dgSmMot_S^{\mathrm{eff}} := \mathcal{C}_{dg}^b(R\Gamma(S, \underline{dgPrCor}_S)).$$

The triangulated category, $SmMot_{gm}^{\mathrm{eff}}(S)$, of smooth effective motives over S is defined as the idempotent completion of the homotopy category

$$\mathcal{K}^b(R\Gamma(S, \underline{dgPrCor}_S^{\mathrm{eff}})) = H^0 dgSmMot_S^{\mathrm{eff}},$$

in other words,

$$SmMot_{gm}^{\mathrm{eff}}(S) = \mathrm{per}(R\Gamma(S, \underline{dgPrCor}_S^{\mathrm{eff}})) \subset \mathcal{D}(R\Gamma(S, \underline{dgPrCor}_S^{\mathrm{eff}})).$$

The Zariski presheaf, $\underline{dgPrCor}_S^{\mathrm{alt}}$, is defined by

$$U \mapsto \underline{dgPrCor}_S^{\mathrm{alt}}(U) := dgPrCor_U^{\mathrm{alt}}.$$

The DG tensor category of smooth effective geometric motives over S with $\mathbb{Q}$-coefficients, $dgSmMot_{S\mathbb{Q}}^{\mathrm{eff}}$, is defined as

$$dgSmMot_{S\mathbb{Q}}^{\mathrm{eff}} := \mathcal{C}_{dg}^b(R\Gamma(S, \underline{dgPrCor}_S^{\mathrm{alt}})^{\otimes}).$$

The tensor triangulated category, $SmMot_{gm}^{\mathrm{eff}}(S)_{\mathbb{Q}}$, of smooth effective motives over S with $\mathbb{Q}$-coefficients is defined as the idempotent completion of the homotopy category $\mathcal{K}^b(R\Gamma(S, \underline{dgSmMot}_S^{\mathrm{alt}})^{\otimes}) = H^0 dgSmMot_{S\mathbb{Q}}^{\mathrm{eff}}$,

$$SmMot_{gm}^{\mathrm{eff}}(S)_{\mathbb{Q}} := \mathrm{per}(R\Gamma(S, \underline{dgSmMot}_S^{\mathrm{alt}})^{\otimes}) \subset \mathcal{D}(R\Gamma(S, \underline{dgSmMot}_S^{\mathrm{alt}})^{\otimes}).$$

Remark 4.6 We can also define unbounded versions: the triangulated category of effective motives over S

$$SmMot^{\text{eff}}(S) := \mathcal{D}(R\Gamma(S, \underline{dgPrCor}_S^{\text{eff}})),$$

and the tensor triangulated category of effective motives over S with $\mathbb{Q}$-coefficients

$$SmMot^{\text{eff}}(S)_{\mathbb{Q}} := \mathcal{D}(R\Gamma(S, \underline{dgSmMot}_S^{\text{alt}})^{\otimes}).$$

We will mainly restrict our discussion to the geometric categories.

Proposition 4.7 *The functor* $\underline{dgPrCor}_S^{\text{alt}} \to \underline{dgPrCor}_S \otimes \mathbb{Q}$ *defined by the natural functors*

$$dgPrCor_U^{\text{alt}} \to dgPrCor_U \otimes \mathbb{Q}$$

gives rise to a functor of DG categories

$$R\Gamma(S, \underline{dgPrCor}_S^{\text{alt}})^{\otimes} \to R\Gamma(S, \underline{dgPrCor}_S) \otimes \mathbb{Q},$$

which in turn induces an equivalence of $SmMot_{gm}^{\text{eff}}(S)_{\mathbb{Q}}$ *with the idempotent completion of* $SmMot_{gm}^{\text{eff}}(S) \otimes \mathbb{Q}$, *as triangulated categories.*

Proof This follows from proposition 4.3 and remark 3.3. $\qquad\qquad\square$

Remark 4.8 The DG categories $R\Gamma(S, \underline{dgPrCor}_S)$ and $dgSmMot_S^{\text{eff}}$ are functorial in the regular scheme S, as is the triangulated category $SmMot_{gm}^{\text{eff}}(S)$. The same holds for the DG tensor categories $R\Gamma(S, \underline{dgPrCor}_S^{\text{alt}})^{\otimes}$, $dgSmMot_{S\mathbb{Q}}^{\text{eff}}$ and the tensor triangulated category $SmMot_{gm}^{\text{eff}}(S)_{\mathbb{Q}}$.

4.4 Lefschetz motives. In **Proj**/S, we have the idempotent endomorphism α of $\mathbb{P}_S^1$ defined as the composition

$$\mathbb{P}_S^1 \xrightarrow{p} S \xrightarrow{i_\infty} \mathbb{P}_S^1.$$

Since Cor_S is a tensor category, we have the object $\mathbb{L} := (\mathbb{P}_S^1, 1 - \alpha)$ in $Cor_S^{\natural}$, as well as the nth tensor power $\mathbb{L}^n$ of $\mathbb{L}$ and, for each $X \in \mathbf{Sm}/S$, the object $X \otimes \mathbb{L}^n$. We thus have these objects in the DG categories of correspondences $dgPrCor_S^{\natural}$ and $R\Gamma(S, \underline{dgPrCor}_S)^{\natural}$.

Definition 4.9 The DG category of Lefschetz motives over S, $dgLCor_S^{\text{eff}}$, is defined to be the full DG subcategory of $R\Gamma(S, \underline{dgPrCor}_S)^{\natural}$ with objects $\mathbb{L}^d$, $d \geq 0$. The DG category of effective mixed Tate motives is

$$dgTMot_S^{\text{eff}} := \mathcal{C}_{dg}^b(dgLCor_S^{\text{eff}}).$$

The triangulated category of effective mixed Tate motives over S, $DTMot^{\text{eff}}(S)$, is the homotopy category $\mathcal{K}^b(dgLCor_S^{\text{eff}}) = H^0 dgTMot_S^{\text{eff}}$.

We have parallel definitions of DG tensor categories and a tensor triangulated category with $\mathbb{Q}$-coefficients. The DG tensor category of Lefschetz motives over S with $\mathbb{Q}$-coefficients, $dgLCor_{S\mathbb{Q}}^{\text{eff}}$, is the full DG subcategory of $R\Gamma(S, \underline{dgSmMot}_S^{\text{alt}})^{\otimes\natural}$ with objects finite direct sums of the $\mathbb{L}^d$, $d \geq 0$, The DG tensor category of effective mixed Tate motive with $\mathbb{Q}$-coefficients is

$$dgTMot_{S\mathbb{Q}}^{\text{eff}} := \mathcal{C}_{dg}^b(dgLCor_{S\mathbb{Q}}^{\text{eff}}),$$

and the tensor triangulated category of effective mixed Tate motives over S with , $DTMot^{\text{eff}}(S)_{\mathbb{Q}}$, is the homotopy category $\mathcal{K}^b(dgLCor_{S\mathbb{Q}}^{\text{eff}}) = H^0 dgTMot_{S\mathbb{Q}}^{\text{eff}}$.

5 Duality

In section 7, we will prove an extension of the moving lemmas of Friedlander-Lawson to smooth projective schemes over a regular, semi-local base. In this section, we use these moving lemmas to define a twisted duality for Hom-complexes in $R\Gamma(S, \underline{dgSmMot}_S)$, and extend this to a duality in various categories of motives.

In this section S will be a regular scheme over a fixed base-field k.

5.1 Equi-dimensional cycles.

Definition 5.1 Let X and Y be smooth over S, $r \geq 0$ an integer. The group $z^S_{equi}(Y, r)(X)$ is the free abelian group on the integral subschemes $W \subset X \times_S Y$ such that the projection $W \to X$ dominates an irreducible component X' of X, and such that, for each $x \in X$, the fiber W_x over x has pure dimension r over $k(x)$, or is empty.

We let $z^S_{equi}(Y, r)^{\text{eff}}(X) \subset z^S_{equi}(Y, r)(X)$ be the submonoid of effective cycles, that is, the free monoid on the generators W for $z^S_{equi}(Y, r)(X)$ described above.

For an S-morphism $f : X' \to X$, and for $W \in z^S_{equi}(Y, r)(X)$, the pull-back cycle $(f \times \mathrm{id}_Y)^*(W)$ is well-defined and in $z^S_{equi}(Y, r)(X')$. Thus $z^S_{equi}(Y, r)$ is a presheaf on $\mathbf{Sm}/S$. For $x \in X \in \mathbf{Sm}/S$, and for $W \in z^S_{equi}(Y, r)(X)$, we denote the pull-back $i_x^*(W)$ by the inclusion $i_x : x \to X$ by W_x.

Suppose Y is projective over S, with a fixed embedding $Y \hookrightarrow \mathbb{P}^N_S$. Then we have a well-defined degree homomorphism

$$\deg : z^S_{equi}(Y, r)(X) \to H^0(X_{\text{Zar}}, \mathbb{Z})$$

which sends a cycle $W \in z^S_{equi}(Y, r)(X)$ to the locally constant function on X

$$x \mapsto \deg(W_x),$$

where $\deg(W_x)$ is the usual degree of the cycle W_x in $\mathbb{P}^N$.

For each integer $e \geq 1$, we let $z^S_{equi}(Y, r)^{\text{eff}}_{\leq e}(X)$ be the subset of $z^S_{equi}(Y, r)^{\text{eff}}(X)$ consisting of those W with $\deg(W) \leq e$ on X. We let

$$z^S_{equi}(Y, r)_{\leq e}(X) \subset z^S_{equi}(Y, r)(X)$$

be the subgroup generated by the set $z^S_{equi}(Y, r)^{\text{eff}}_{\leq e}(X)$. Thus we have the presheaf of sets $z^S_{equi}(Y, r)^{\text{eff}}_{\leq e}(X)$, and the presheaf of abelian groups $z^S_{equi}(Y, r)_{\leq e}(X)$.

Definition 5.2 For X, Y in $\mathbf{Sm}/S$, the *cubical Suslin complex* $C^S(Y, r)^*(X)$ is the complex associated to the cubical object

$$n \mapsto z^S_{equi}(Y, r)(X \times_S \square^n_S),$$

i.e.

$$C^S(Y, r)^n(X) := z^S_{equi}(Y, r)(X \times_S \square^{-n}_S)/\text{degn}.$$

For $f : X' \to X$, define

$$f^* : C^S(Y, r)(X) \to C^S(Y, r)(X')$$

via the pull-back maps

$$(f \times \mathrm{id}_{\square^n})^* : z^S_{equi}(Y, r)(X \times \square^n) \to z^S_{equi}(Y, r)(X' \times \square^n);$$

this defines the presheaf of complexes $C^S(Y, r)$ on $\mathbf{Sm}/S$.

Suppose that Y is in $\mathbf{Proj}/S$ and we are given an embedding $Y \hookrightarrow \mathbb{P}^N_S$ over S. Let $C^S(Y,r)_{\leq e}(X) \subset C^S(Y,r)(X)$ be the subcomplex corresponding to the cubical abelian group

$$n \mapsto z^S_{equi}(Y,r)_{\leq e}(X \times_S \square^n_S) \subset z^S_{equi}(Y,r)(X \times_S \square^n_S).$$

Remarks 5.3 1. Suppose that Y is in $\mathbf{Proj}/S$. Then

$$C^S(Y,0)^*(X) = \mathcal{H}om_{dgCor_S}(X,Y)^*,$$

and $C^S(Y,0)$ is the presheaf $\mathcal{H}om_{dgCor_S}(-,Y)^*$ on $\mathbf{Sm}/S$.

2. Let $f : Y \to Y'$ be a proper morphism. Push-forward by the projection $\mathrm{id} \times f : X \times \square^n \times Y \to X \times \square^n \times Y'$ defines the map of complexes

$$f_*(X) : C^S(Y,r)(X) \to C^S(Y',r)(X)$$

giving us the map of presheaves $f_* : C^S(Y,r) \to C^S(Y',r)$. Thus $Y \mapsto C^S(Y,r)$ defines a functor from $\mathbf{Proj}/S$ to complexes of sheaves on $\mathbf{Sm}/S$.

3. Correspondences act on $z^S_{equi}(Y,r)(X)$, both in Y (covariantly) and in X (contravariantly). Thus sending (Y,X) to $z^S_{equi}(Y,r)(X)$ extends to a functor

$$z^S_{equi}(-,r)(-) : Cor_S \times Cor_S^{\mathrm{op}} \to \mathbf{Ab}.$$

We have a similar extension of the complexes $C^S(Y,r)(X)$ to

$$C^S(-,r)(-) : Cor_S \times Cor_S^{\mathrm{op}} \to C^-(\mathbf{Ab}).$$

As $\mathbf{Ab}$ is abelian, the bi-functors $z^S_{equi}(-,r)(-) :$ and $C^S(-,r)(-)$ extend to the idempotent completion of Cor_S; in particular, the presheaves $z^S_{equi}(Y \otimes \mathbb{L}^n, r)$ and $C^S(Y \otimes \mathbb{L}^n, r)$ are defined.

Now take $Y, X \in \mathbf{Sm}/S$ with $X \to S$ equi-dimensional of dimension p over S. Each integral $W \in z^S_{equi}(Y,r)(U \times_S X)$ gives us an integral subscheme W of $U \times_S X \times_S Y$ which is equi-dimensional of relative dimension $r + p$ over some component of U. This defines the map of complexes

$$\int_X : C^S(Y,r)(U \times_S X) \to C^S(X \times_S Y, r + p)(U).$$

We can now state our main result, to be proven in §7.

Theorem 5.4 *Let S be a regular semi-local k-scheme, essentially of finite type over k. Then for all $X, Y \in \mathbf{Proj}/S$, $U \in \mathbf{Sm}/S$, with $X \to S$ of relative dimension p, the map*

$$\int_X : C^S(Y,r)(U \times_S X) \to C^S(X \times_S Y, r + p)(U)$$

is a quasi-isomorphism.

In addition, we will need a computation of $C_*(Y \times \mathbb{P}^n, r)$ for $r \geq n$. Fix linear inclusions $\iota_j : \mathbb{P}^j \to \mathbb{P}^n$, $j = 0, \ldots, n$. For each i, $0 \leq i \leq n \leq r$, define the map

$$\alpha_j : C^S(Y, r-j)(X) \to C^S(Y \times \mathbb{P}^n, r)(X)$$

by sending $W \subset Y \times_S X \times \square^n$ to $(\iota_j \times \mathrm{id})_*(\mathbb{P}^j \times W)$.

Theorem 5.5 *Let S be a regular semi-local k-scheme, essentially of finite type over k. Then for all $X, Y \in \mathbf{Proj}/S$, and for $r \geq n$, the map*

$$\sum_{j=0}^{n} \alpha_j : \oplus_{j=0}^{n} C^S(Y, r-j)(X) \to C^S(Y \times \mathbb{P}^n, r)(X)$$

is a quasi-isomorphism.

Theorem 5.5 will also be proven in §7. As consequence, we have

Corollary 5.6 *Let S be a regular semi-local k-scheme, essentially of finite type over k.*

1. For $Y \in \mathbf{Proj}/S$, $X \in \mathbf{Sm}/S$ and $n \geq 0$, let

$$\varphi : C_*^S(Y, r)(X) \to C^S(Y \otimes \mathbb{L}^n, r+n)(X)$$

be the map sending $W \subset X \times \square^n \times_S Y$ to $W \times (\mathbb{P}^1)^n \subset X \times (\mathbb{P}^1)^n \times \square^n \times_S Y$. Then φ is an isomorphism in $D^-(\mathbf{Ab})$.

2. For $Y, Z \in \mathbf{Proj}/S$, $X \in \mathbf{Sm}/S$, with Y of dimension d over S, there are natural isomorphisms in $D^-(\mathbf{Ab})$

$$C^S(Y \times_S Z, r)(X) \cong C^S(Z \otimes \mathbb{L}^d, r)(X \times_S Y).$$

For $Z = S \times_k Z_0$, with $Z_0 \in \mathbf{Proj}/k$, we have an isomorphism

$$C^S(Z \otimes \mathbb{L}^d, r)(X \times_S Y) \cong C^k(Z_0 \otimes \mathbb{L}^d, r)(p_*(X \times_S Y)).$$

3. For $X \in \mathbf{Sm}/S$, $Y \in \mathbf{Proj}/S$ and $r \geq n \geq 0$ integers, there are natural isomorphisms in $D^-(\mathbf{Ab})$

$$C^S(Y, r)(X \otimes \mathbb{L}^n) \cong C^S(Y, r+n)(X).$$

Proof For (1), the isomorphism

$$C^S(Y \otimes \mathbb{L}^n, r+n)(X) \cong C^S(Y, r)(X),$$

follows by induction and the projective bundle formula (theorem 5.5). Indeed, it suffices to define a natural isomorphism

$$C^S(Y \otimes \mathbb{L}^n, r+1)(X) \to C^S(Y \otimes \mathbb{L}^{n-1}, r)(X). \tag{5.1}$$

Since the projection $\mathbb{P}^1 \to \operatorname{Spec} k$ is split by the inclusion $i_\infty : \operatorname{Spec} k \to \mathbb{P}^1$, we get a direct sum decomposition

$$C^S(Y \times \mathbb{P}^1, r) \cong C^S(Y, r) \oplus C^S(Y \otimes \mathbb{L}, r).$$

By induction, we have a similar direct sum decomposition of $C^S(Y \times (\mathbb{P}^1)^n, r)$ for all n; this reduces the proof of (5.1) to showing that

$$C^S(Y \times (\mathbb{P}^1)^{n-1} \otimes \mathbb{L}, r+1)(X) \cong C^S(Y \times (\mathbb{P}^1)^{n-1}, r)(X),$$

which reduces us to the case $n = 1$. Comparing with the isomorphism

$$\alpha_0 + \alpha_1 : C^S(Y, r+1) \oplus C^S(Y, r) \to C^S(Y \times \mathbb{P}^1, r+1)$$

and noting that $p_* \circ \alpha_1 = 0$, $p_* \circ \alpha_0 = \operatorname{id}$, we see that α_1 gives an isomorphism

$$\alpha_1 : C^S(Y, r) \to C^S(Y \otimes \mathbb{L}, r+1),$$

completing the proof of the first assertion.

The first isomorphism in (2) follows from the first assertion and theorem 5.4:

$$C^S(Z \otimes \mathbb{L}^d, r)(X \times_S Y) \cong C^S(Y \times_S Z \otimes \mathbb{L}^d, d+r)(X) \cong C^S(Y \times_S Z, r)(X).$$

For the second isomorphism, $C^S(Z \otimes \mathbb{L}^d, r)(X \times_S Y) = C^k(Z_0 \otimes \mathbb{L}^d, r)(X \times_S Y)$, since we have the isomorphism of schemes (over $\operatorname{Spec} k$)

$$T \times_S (S \times_k Z_0) \times_S \mathbb{P}^1_S \cong p_* T \times_k Z_0 \times_k \mathbb{P}^1_k$$

for all S-schemes T.

For (3), it suffices to prove the case $n = 1$. By theorem 5.4, we have the quasi-isomorphism

$$\int_{\mathbb{P}^1} : C^S(Y, r)(X \times \mathbb{P}^1) \to C^S(Y \times \mathbb{P}^1, r+1)(X).$$

Since $Y \times \mathbb{P}^1 \cong Y \oplus Y \otimes \mathbb{L}$ in $Cor^\natural_S$, and similarly for $X \times \mathbb{P}^1$, we have the quasi-isomorphism

$$\int_{\mathbb{P}^1} : C^S(Y, r)(X) \oplus C^S(Y, r)(X \otimes \mathbb{L}) \to C^S(Y \times \mathbb{P}^1, r+1)(X).$$

Since $\mathbb{L} = (\mathbb{P}^1, 1 - i_\infty p)$, the summand $C^S(Y, r)(X \otimes \mathbb{L})$ of $C^S(Y, r)(X \times \mathbb{P}^1)$ is the image of $\operatorname{id} - p^* i_\infty^*$, which is the same as the kernel of i_∞^*. Similarly, the map $C^S(Y, r)(X) \to C^S(Y, r)(X \times \mathbb{P}^1)$ is defined by p^*. Using the flag $0 \subset \mathbb{P}^1$ to define the quasi-isomorphism of theorem 5.5,

$$\alpha_0 + \alpha_1 : C^S(Y, r+1)(X) \oplus C^S(Y, r)(X) \to C^S(Y \times \mathbb{P}^1, r+1)(X),$$

the inverse isomorphism (in $D^-(\mathbf{Ab})$) is given by the map of complexes

$$C^S(Y \times \mathbb{P}^1, r+1)_{Y \times \infty}(X) \xrightarrow{(p_*, i_\infty^*)} C^S(Y, r+1)(X) \oplus C^S(Y, r)(X).$$

composed with the inverse of the quasi-isomorphism

$$C^S(Y \times \mathbb{P}^1, r+1)_{Y \times \infty}(X) \hookrightarrow C^S(Y \times \mathbb{P}^1, r+1)(X).$$

We note that the image of $\int_{\mathbb{P}^1}$ lands in $C^S(Y \times \mathbb{P}^1, r+1)_{Y \times \infty}(X)$, and sends $C^S(Y, r)(X \otimes \mathbb{L})$ to $\ker i_\infty^*$ and $C^S(Y, r)(X)$ to $\ker p_*$. Thus $p_* \circ \int_{\mathbb{P}^1}$ defines a quasi-isomorphism

$$p_* \circ \int_{\mathbb{P}^1} : C^S(Y, r)(X \otimes \mathbb{L}) \to C^S(Y, r+1)(X),$$

as desired. $\qquad\qquad\qquad\qquad\qquad\qquad\qquad\qquad\qquad\qquad\qquad\qquad\qquad\qquad\quad\square$

Remark 5.7 For later use, we extract from the proof a description of the isomorphism in corollary 5.6(2,3). In (2), the isomorphism is the composition of

$$\int_Y : C^S(Z \otimes \mathbb{L}^d, r)(X \times_S Y) \to C^S(Y \times_S Z \otimes \mathbb{L}^d, r+d)(X)$$

with the inverse of

$$- \times (\mathbb{P}^1)^n : C^S(Y \times_S Z, r)(X) \to C^S(Y \times_S Z \otimes \mathbb{L}^d, r+d)(X).$$

To describe the map in (3), let $p : Y \times \mathbb{P}^1 \to Y$ be the projection. We have the composition

$$C^S(Y, r)(X \times \mathbb{P}^1) \xrightarrow{\int_{\mathbb{P}^1}} C^S(Y \times \mathbb{P}^1, r+1)(X) \xrightarrow{p_*} C^S(Y, r+1)(X).$$

Then the restriction of $p_* \circ \int_{\mathbb{P}^1}$ to the summand $C^S(Y, r)(X \otimes \mathbb{L})$ of $C^S(Y, r)(X \times \mathbb{P}^1)$,

$$p_* \circ \int_{\mathbb{P}^1} : C^S(Y, r)(X \otimes \mathbb{L}) \to C^S(Y, r+1)(X),$$

is the isomorphism in (3). Iterating, we have the map

$$[p_* \circ \int_{\mathbb{P}^1}]^n : C^S(Y, r)(X \otimes \mathbb{L}^n) \to C^S(Y, r+n)(X),$$

giving the isomorphism in (3).

For $Y, Z \in \mathbf{Proj}/S$, $X \in \mathbf{Sm}/S$, we have the natural map

$$\times_S Z : z^S_{equi}(Y, r)(X) \to z^S_{equi}(Y \times_S Z, r)(X \times_S Z)$$

defined by sending a cycle W on $X \times_S Y$ to the image of W in $X \times_S Z \times_S Y \times_S Z$ via the "diagonal" embedding $X \times_S Y \to X \times_S Z \times_S Y \times_S Z$. This gives us the map

$$\times_S Z : C^S(Y, r)(X) \to C^S(Y \times_S Z, r)(X \times_S Z).$$

Taking $Z = \mathbb{P}^1$ and extending to the pseudo-abelian hull gives the map

$$\otimes \mathbb{L} : C^S(Y, r)(X) \to C^S(Y \otimes \mathbb{L}, r)(X \otimes \mathbb{L}),$$

sending $W \subset X \times \square^n \times_S Y$ to $W \times \Delta_{\mathbb{P}^1} \subset X \times \mathbb{P}^1 \times \square^n \times_S Y \times \mathbb{P}^1$.

Corollary 5.8 *Let S be a regular semi-local k-scheme, essentially of finite type over k. For $X, Y \in \mathbf{Proj}/S$, the map $\otimes \mathbb{L} : C^S(Y, r)(X) \to C^S(Y \otimes \mathbb{L}, r)(X \otimes \mathbb{L})$ is a quasi-isomorphism.*

Proof By the projective bundle formula (theorem 5.5), the map $W \mapsto W \times \mathbb{P}^1$ gives a quasi-isomorphism

$$- \times \mathbb{P}^1 : C^S(Y, r)(X) \to C^S(Y \otimes \mathbb{L}, r+1)(X).$$

By corollary 5.6(3) and remark 5.7, the map

$$p_* \circ \int_{\mathbb{P}^1} : C^S(Y \otimes \mathbb{L}, r)(X \otimes \mathbb{L}) \to C^S(Y \otimes \mathbb{L}, r+1)(X)$$

is a quasi-isomorphism. Now, for $W \in C^S(Y, r)(X)$, we have

$$p_* \circ \int_{\mathbb{P}^1} (W \otimes \mathbb{L}) = p_* \circ \int_{\mathbb{P}^1} (W \times \Delta_{\mathbb{P}^1}) = W \times \mathbb{P}^1,$$

hence

$$\otimes \mathbb{L} : C^S(Y, r)(X) \to C^S(Y \otimes \mathbb{L}, r)(X \otimes \mathbb{L})$$

is a quasi-isomorphism. $\qquad\qquad\qquad\qquad\qquad\qquad\qquad\qquad\qquad\qquad\qquad\square$

5.2 Duality for smooth motives. The duality results of the previous section extend to the category $SmMot_{gm}(S)$ and defines a duality on the tensor triangulated category $SmMot_{gm}(S)_\mathbb{Q}$.

Proposition 5.9 *Let S be a regular scheme, essentially of finite type over a field k. For $X \in \mathbf{Proj}/S$, the natural map*

$$\mathcal{H}om_{dgPrCor^\natural_S}(X, \mathbb{L}^d)^* \to \mathcal{H}om_{R\Gamma(S, \underline{dgSmMot}_S)^\natural}(X, \mathbb{L}^d)^*$$

is a quasi-isomorphism.

Proof Let $p_* X \in \mathbf{Sm}/k$ denote the scheme X, considered as a k-scheme via the structure morphism $p : S \to \operatorname{Spec} k$. We have a canonical isomorphism

$$\mathcal{H}om_{dgPrCor^{\natural}_S}(X, \mathbb{L}^d)^* \cong \mathcal{H}om_{dgPrCor^{\natural}_k}(p_* X, \mathbb{L}^d)^*,$$

induced by the isomorphisms

$$(\mathbb{P}^1)^d_S \times_S \square^n_S \times_S X \cong (\mathbb{P}^1_k)^d \times_k \square^n_k \times_k p_* X.$$

On the other hand, the complex $\mathcal{H}om_{dgPrCor^{\natural}_k}(p_* X, \mathbb{L}^d)^*$ is just a cubical version of the weight d Friedlander-Suslin complex of the k-scheme $p_* X$, hence we have isomorphisms [23]

$$H^n(\mathcal{H}om_{dgPrCor^{\natural}_k}(p_* X, \mathbb{L}^d)^*) \cong H^{2d+n}(X, \mathbb{Z}(d)) \cong \operatorname{CH}^d(X, -n).$$

Since the higher Chow groups of smooth k-schemes satisfy the Mayer-Vietoris property for the Zariski topology [3], the presheaf of complexes on S

$$U \mapsto \mathcal{H}om_{dgPrCor^{\natural}_U}(X \times_S U, \mathbb{L}^d)^*$$

satisfies the Mayer-Vietoris property on S_{Zar}. Thus, by the Brown-Gersten theorem [2] the natural map

$$\mathcal{H}om_{dgPrCor^{\natural}_S}(X, \mathbb{L}^d)^* \to R\Gamma(S, [U \mapsto \mathcal{H}om_{dgPrCor^{\natural}_U}(X \times_S U, \mathbb{L}^d)^*])$$

is a quasi-isomorphism. Since $R\Gamma(S, [U \mapsto \mathcal{H}om_{dgPrCor^{\natural}_U}(X \times_S U, \mathbb{L}^d)^*])$ is by definition equal to $\mathcal{H}om_{R\Gamma(S, dgSmMot_S)^{\natural}}(X, \mathbb{L}^d)^*$, the result is proven. $\square$

Following remark 1.15, the tensor structure on Cor_S extends to an action of Cor_S on the presheaf of DG categories $\underline{dgSmMot}_S$, giving us an action of $Cor^{\natural}_S$ on the DG categories $R\Gamma(S, \underline{dgSmMot}_S)^{\natural}$ and $dgSmMot^{\text{eff}\natural}_S$. Thus, we have an action of $Cor^{\natural}_S$ on the triangulated category $SmMot^{\text{eff}}_{gm}(S)$:

$$\otimes : Cor^{\natural}_S \otimes SmMot^{\text{eff}}_{gm}(S) \to SmMot^{\text{eff}}_{gm}(S).$$

In particular, each object $A \in Cor^{\natural}_S$ gives an exact functor

$$(-) \otimes A : SmMot^{\text{eff}}_{gm}(S) \to SmMot^{\text{eff}}_{gm}(S)$$

sending a morphism $f : X \to Y$ in $SmMot^{\text{eff}}_{gm}(S)$ to $f \otimes \operatorname{id} : X \otimes A \to Y \otimes A$.

Theorem 5.10 *Let S be a regular scheme, essentially of finite type over a field k.*

1. Let X, Y and Z be in $\mathbf{Proj}/S$, $d = \dim_S Y$. Then there is a natural quasi-isomorphism

$$\mathcal{H}om_{R\Gamma(S, \underline{dgSmMot}_S)}(X, Y \times_S Z)^* \cong \mathcal{H}om_{R\Gamma(S, \underline{dgSmMot}_S)^{\natural}}(X \times_S Y, Z \otimes \mathbb{L}^d)^*.$$

2. Let X and Y be in $\mathbf{Proj}/S$, $d = \dim_S Y$, and take Z in $\mathbf{Proj}/k$. Then there is a natural quasi-isomorphism

$$\mathcal{H}om_{R\Gamma(S, \underline{dgSmMot}_S)}(X, Y \times_k Z)^* \cong \mathcal{H}om_{dgSmMot^{\natural}_k}(p_*(X \times_S Y), Z \otimes \mathbb{L}^d)^*.$$

3. Let X and Y be in $\mathbf{Proj}/S$. Then the map

$$\otimes \mathbb{L} : \mathcal{H}om_{R\Gamma(S, \underline{dgSmMot}_S)}(X, Y)^* \to \mathcal{H}om_{R\Gamma(S, \underline{dgSmMot}_S)^{\natural}}(X \otimes \mathbb{L}, Y \otimes \mathbb{L})^*$$

is a quasi-isomorphism.

Proof For $X, Y \in \mathbf{Proj}/S$, let $\mathcal{H}om_{\underline{dgSmMot}_S}(X, Y)$ denote the Zariski sheaf on S associated to the presheaf

$$U \mapsto \mathcal{H}om_{dgSmMot_U}(X_U, Y_U)^*;$$

extend the notation to define the sheaf $\mathcal{H}om_{\underline{dgSmMot}_S^\natural}(X, Y \otimes \mathbb{L}^d)$, etc.

From corollary 5.6 and remark 5.7, we have isomorphisms in $D^-(Sh_S^{\mathrm{Zar}})$

$$\mathcal{H}om_{\underline{dgSmMot}_S}(X, Y \times_S Z)^* \cong \mathcal{H}om_{\underline{dgSmMot}_S^\natural}(X \times_S Y, Z \otimes \mathbb{L}^d)^*$$

and

$$\otimes\mathbb{L} : \mathcal{H}om_{\underline{dgSmMot}_S}(X, Y)^* \to \mathcal{H}om_{\underline{dgSmMot}_S^\natural}(X \otimes \mathbb{L}, Y \otimes \mathbb{L})^*$$

These induce isomorphisms (in $D(\mathbf{Ab})$) after applying $\Gamma(S, G^*(-)) \cong R\Gamma(S, -)$, proving (1) and (3). (2) follows from (1), noting that we have the isomorphism of $\mathcal{H}om_{\underline{dgSmMot}_S}(X \times_S Y, S \times_k Z \otimes \mathbb{L}^d)^*$ with the constant sheaf (on S_{Zar}) with value $\mathcal{H}om_{\underline{dgSmMot}_k^\natural}(p_*(X \times_S Y), Z \otimes \mathbb{L}^d)^*$, and that for a constant sheaf of complexes $\mathcal{F}$ on S_{Zar}, the natural map

$$\Gamma(S, \mathcal{F}) \to R\Gamma(S, \mathcal{F})$$

is an isomorphism in $D(\mathbf{Ab})$.   □

Definition 5.11 The DG category of motives over S, $dgSmMot_S$, is the DG category formed by inverting $\otimes\mathbb{L}$ on $dgSmMot_S^{\mathrm{eff}\natural}$. The triangulated category of motives over S, $SmMot_{gm}(S)$ is the triangulated category formed by inverting $\otimes\mathbb{L}$ on $SmMot_{gm}^{\mathrm{eff}}(S)$ (we will see in corollary 5.14 below that $SmMot_{gm}(S)$ has a natural triangulated structure).

Explicitly, $dgSmMot_S$ has objects $X \otimes \mathbb{L}^n$, $n \in \mathbb{Z}$, X in $dgSmMot_S^{\mathrm{eff}\natural}$, with

$$\mathcal{H}om_{dgSmMot_S}(X \otimes \mathbb{L}^n, Y \otimes \mathbb{L}^m)^* := \varinjlim_N \mathcal{H}om_{dgSmMot_S^{\mathrm{eff}\natural}}(X \otimes \mathbb{L}^{N+n}, Y \otimes \mathbb{L}^{N+m})^*.$$

Similarly, $SmMot_{gm}(S)$ has objects $X \otimes \mathbb{L}^n$, $n \in \mathbb{Z}$, X in $SmMot_{gm}^{\mathrm{eff}}(S)$, with

$$\mathcal{H}om_{SmMot_{gm}(S)}(X \otimes \mathbb{L}^n, Y \otimes \mathbb{L}^m)^* := \varinjlim_N \mathcal{H}om_{SmMot_{gm}^{\mathrm{eff}}(S)}(X \otimes \mathbb{L}^{N+n}, Y \otimes \mathbb{L}^{N+m})^*.$$

Remarks 5.12 We can also invert $\otimes\mathbb{L}$ on $dgPrCor_U^\natural$ for all open $U \subset S$, giving us the presheaf $\underline{dgPrCor}_S^\natural[\otimes\mathbb{L}^{-1}]$ on S_{Zar}, and the associated DG category $R\Gamma(S, \underline{dgPrCor}_S^\natural[\otimes\mathbb{L}^{-1}])$. We have an isomorphism of DG categories

$$R\Gamma(S, \underline{dgPrCor}_S^\natural[\otimes\mathbb{L}^{-1}]) \cong R\Gamma(S, \underline{dgPrCor}_S)^\natural[\otimes\mathbb{L}^{-1}],$$

giving us the isomorphism of DG categories

$$dgSmMot_S \cong \mathcal{C}_{dg}^b(R\Gamma(S, \underline{dgPrCor}_S^\natural[\otimes\mathbb{L}^{-1}])).$$

2. Inverting $\otimes\mathbb{L}$ on the various Lefschetz/Tate categories of definition 4.9 gives us full sub-DG categories $dgLCor_S$, $dgTMot(S)$ of $R\Gamma(S, \underline{dgPrCor}_S^\natural[\otimes\mathbb{L}^{-1}])$ and $dgSmMot_S$, resp., with

$$dgTMot(S) \cong \mathcal{C}_{dg}^b(dgLCor_S),$$

and the full triangulated subcategory $DTMot(S)$ of $SmMot_{gm}(S)$.

With $\mathbb{Q}$-coefficients we have the analogous DG tensor sub-categories $dgLCor_{S\mathbb{Q}}^{\mathrm{alt}}$, $dgTMot_{S\mathbb{Q}}$ of $R\Gamma(S, \underline{dgSmMot}_S^{\mathrm{alt}})^{\otimes\natural}[\otimes\mathbb{L}^{-1}])$, $dgSmMot_S$, resp., with

$$dgTMot_{S\mathbb{Q}} \cong \mathcal{C}_{dg}^b(dgLCor_{S\mathbb{Q}}^{\mathrm{alt}}),$$

and the full tensor triangulated subcategory $DTMot(S)_{\mathbb{Q}}$ of $SmMot_{gm}(S)_{\mathbb{Q}}$.

Theorem 5.13 *For $X, Y \in dgSmMot_S^{\mathrm{eff}\natural}$, the natural map*

$$\mathcal{H}om_{dgSmMot_S^{\mathrm{eff}\natural}}(X, Y)^* \to \mathcal{H}om_{dgSmMot_S}(X, Y)^*$$

is a quasi-isomorphism, and the natural map

$$\mathrm{Hom}_{SmMot_{gm}^{\mathrm{eff}}(S)}(X, Y[n]) \to \mathrm{Hom}_{SmMot_{gm}(S)}(X, Y[n])$$

is an isomorphism for all n. Furthermore, the isomorphism

$$H^n\mathcal{H}om_{dgSmMot_S^{\mathrm{eff}\natural}}(X, Y)^* \cong \mathrm{Hom}_{SmMot_{gm}^{\mathrm{eff}}(S)^\natural}(X, Y[n])$$

induces an isomorphism

$$H^n\mathcal{H}om_{dgSmMot_S}(X \otimes \mathbb{L}^p, Y \otimes \mathbb{L}^m)^* \cong \mathrm{Hom}_{SmMot_{gm}(S)}(X \otimes \mathbb{L}^p, Y \otimes \mathbb{L}^m[n]).$$

for each $n, m, p \in \mathbb{Z}$.

Proof For $X, Y \in \mathbf{Proj}/S$, the first assertion is theorem 5.10(2). Since $dgSmMot_S^{\mathrm{eff}}$ is generated by $\mathbf{Proj}/S$ by taking translation, cone sequences and isomorphisms, the first assertion for $X, Y \in dgSmMot_S^{\mathrm{eff}}$ follows; this immediately implies the result for $dgSmMot_S^{\mathrm{eff}\natural}$.

Since cohomology commutes with filtered inductive limits, we have the isomorphism

$$H^n\mathcal{H}om_{dgSmMot_S}(X, Y)^* \cong \mathrm{Hom}_{SmMot_{gm}(S)}(X, Y[n]).$$

as claimed. Thus the first assertion implies the second. □

As immediate consequence we have

Corollary 5.14 *1. $SmMot_{gm}(S)$ is equivalent to the idempotent completion of $H^0dgSmMot_S$,*

$$SmMot_{gm}(S) \sim (H^0dgSmMot_S)^\natural \sim \mathrm{per}(R\Gamma(S, \underline{dgPrCor}_S^\natural[\otimes\mathbb{L}^{-1}])).$$

In particular, $SmMot_{gm}(S)$ has the natural structure of a triangulated category.

2. The canonical functor $SmMot_{gm}^{\mathrm{eff}}(S) \to SmMot_{gm}(S)$ is an exact fully faithful embedding.

Remark 5.15 We can also invert $\otimes\mathbb{L}$ on the DG tensor categories $dgCor_S^{\mathrm{alt}\natural}$, $R\Gamma(S, \underline{dgPrCor}_S^{\mathrm{alt}})^{\otimes\natural}$ and $dgSmMot_{S\mathbb{Q}}^{\mathrm{eff}\natural}$, and on the tensor triangulated category $SmMot_{gm}^{\mathrm{eff}}(S)_{\mathbb{Q}}$. Setting

$$dgSmMot_{S\mathbb{Q}} := dgSmMot_{S\mathbb{Q}}^{\mathrm{eff}\natural}[\otimes\mathbb{L}^{-1}], \quad SmMot_{gm}(S)_{\mathbb{Q}} := SmMot_{gm}^{\mathrm{eff}}(S)_{\mathbb{Q}}[\otimes\mathbb{L}^{-1}],$$

this gives us the DG tensor functor $dgSmMot_{S\mathbb{Q}}^{\mathrm{eff}\natural} \to dgSmMot_{S\mathbb{Q}}$, and the exact tensor functor $SmMot_{gm}^{\mathrm{eff}}(S)_{\mathbb{Q}} \to SmMot_{gm}(S)_{\mathbb{Q}}$. The analogs of theorem 5.13 and corollary 5.14 hold in this setting. Similarly, the analog of theorem 5.13 and corollary 5.14 hold for the respective subcategories of Tate motives.

5.3 Chow motives. We recall the definition of the category of Chow motives over S.

Definition 5.16 Let S be a regular scheme. Let $Ch\tilde{M}ot^{\text{eff}}(S)$ be the category with the same objects as **Proj**/S, and with morphisms (for $X \to S$ of pure dimension d_X over S)

$$\text{Hom}_{Ch\tilde{M}ot^{\text{eff}}(S)}(X, Y) := \text{CH}_{d_X}(X \times_S Y).$$

The composition law is the usual one of composition of correspondence classes: for $W \in \text{Hom}_{Ch\tilde{M}ot^{\text{eff}}(S)}(X, Y)$, $W' \in \text{Hom}_{Ch\tilde{M}ot^{\text{eff}}(S)}(Y, Z)$, define

$$W' \circ W := p_{13*}(p_{12}^*(W) \cdot_{XYZ} p_{23}^*(W')),$$

where p_{ij} is the projection of $X \times_S Y \times_S Z$ on the ij factors. The operation of product over S makes $Ch\tilde{M}ot^{\text{eff}}(S)$ a tensor category. Sending $f : X \to Y$ to the graph of f defines a functor

$$\tilde{m}_S : \mathbf{Proj}/S \to Ch\tilde{M}ot^{\text{eff}}(S).$$

The category $ChMot^{\text{eff}}(S)$ of *effective Chow motives over S* is the idempotent completion of $Ch\tilde{M}ot^{\text{eff}}(S)$. We let $\mathbb{L} = (\mathbb{P}^1, 1 - i_\infty \circ p)$, and define the category of Chow motives over S as

$$ChMot^{\text{eff}}(S) := Ch\tilde{M}ot^{\text{eff}}(S)[\otimes \mathbb{L}^{-1}].$$

Take $X, Y \in \mathbf{Proj}/S$. By theorem 5.10 and [23], we have the isomorphism

$$\text{Hom}_{SmMot_{gm}^{\text{eff}}(S)}(X, Y) = H^0 \mathcal{H}om_{R\Gamma(S, \underline{dgSmMot}_S)}(X, Y)^*$$
$$\cong H^0 \mathcal{H}om_{dgSmMot_k}(p_*(X \times_S Y), \mathbb{L}^{d_Y})^* = H^{2d_Y}(X \times_S Y, \mathbb{Z}(d_Y))$$
$$\cong \text{CH}_{d_X}(X \times_S Y) = \text{Hom}_{ChMot^{\text{eff}}(S)}(X, Y),$$

which we denote as

$$\varphi_{X,Y} : \text{Hom}_{SmMot_{gm}^{\text{eff}}(S)}(X, Y) \to Hom_{ChMot^{\text{eff}}(S)}(X, Y).$$

Lemma 5.17 φ_{**} *respects the composition: for* $X, Y, Z \in \mathbf{Proj}/S$,

$$f \in \text{Hom}_{SmMot_{gm}^{\text{eff}}(S)}(X, Y), \ g \in \text{Hom}_{SmMot_{gm}^{\text{eff}}(S)}(Y, Z),$$

we have

$$\varphi_{X,Z}(g \circ f) = \varphi_{Y,Z}(g) \circ \varphi_{X,Y}(f).$$

Proof If f is in the image of $z_{equi}^S(Y, 0)(X) \to \text{Hom}_{SmMot_{gm}^{\text{eff}}(S)}(X, Y)$ and g is in the image of $z_{equi}^S(Z, 0)(Y) \to \text{Hom}_{SmMot_{gm}^{\text{eff}}(S)}(Y, Z)$, the result is obvious, so it suffices to show that

$$z_{equi}^S(Y, 0)(X) \to \text{Hom}_{SmMot_{gm}^{\text{eff}}(S)}(X, Y) = \text{CH}_{d_X}(X \times_S Y)$$

is surjective for all $X, Y \in \mathbf{Proj}/S$; this is lemma 5.18 below. $\square$

Lemma 5.18 *Take* $X \in \mathbf{Sm}/S$ *of dimension* d_X *over* S. *For* $Y \in \mathbf{Proj}/S$, , *the map*

$$z_{equi}^S(Y, r)(X) \to \text{CH}_{d_X+r}(X \times_S Y),$$

sending a cycle W *on* $X \times_S Y$ *to its cycle class, is surjective for all* $r \geq 0$.

Proof Take a locally closed immersion $X \times_S Y$ in a projective space $\mathbb{P}_k^N$ and let $T \subset \mathbb{P}_k^N$ be the closure of $X \times_S Y$. Let γ be a dimension $d_X + r$ cycle on $X \times_S Y$ and let $\bar{\gamma}$ be the closure of γ on T. By [10, theorem 1.7], $\bar{\gamma}$ is rationally equivalent to a cycle $\bar{\gamma}'$ such that each component of $\bar{\gamma}'$ intersects the cycle $x \times Y$ properly on T, for each point $x \in X$ (note that $x \times Y$ is closed on $T_{k(x)}$ and is contained in the smooth locus of $T_{k(x)}$). Thus, the restriction γ' of $\bar{\gamma}'$ to $X \times_S Y$ is rationally equivalent to γ and is in the image of $z_{equi}^S(Y,r)(X) \to \mathrm{CH}_{d_X+r}(X \times_S Y)$, completing the proof. $\qquad\square$

Thus, we have shown

Proposition 5.19 *1. There is a fully faithful embedding* $\psi^{\mathrm{eff}} : ChMot^{\mathrm{eff}}(S) \to SmMot_{gm}^{\mathrm{eff}}(S)$, *such that the diagram*

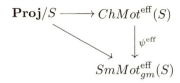

commutes, and such that

$$\psi^{\mathrm{eff}}(X,Y) : \mathrm{Hom}_{ChMot^{\mathrm{eff}}(S)}(X,Y) \to \mathrm{Hom}_{SmMot_{gm}^{\mathrm{eff}}(S)}(X,Y)$$

is the inverse of the isomorphism $\varphi_{X,Y}$.

2. ψ^{eff} *extends to a fully faithful embedding* $\psi : ChMot(S) \to SmMot_{gm}(S)$.

3. The maps ψ^{eff} *and* ψ *are compatible with the* $\otimes$ *action of* Cor_S. *Also, the maps* $\psi^{\mathrm{eff}} : ChMot^{\mathrm{eff}}(S) \to SmMot_{gm}^{\mathrm{eff}}(S)_{\mathbb{Q}}$, $\psi^{\mathrm{eff}} : ChMot(S) \to SmMot_{gm}(S)_{\mathbb{Q}}$ *induced by* ψ^{eff} *and* ψ *are tensor functors.*

Lemma 5.20 *For* $X \in \mathbf{Proj}/S$, *there are maps*

$$\tilde{\delta}_X : \mathbb{L}^d \to X \otimes X; \quad \tilde{\epsilon}_X : X \otimes X \otimes \mathbb{L}^d$$

in $SmMot_{gm}^{\mathrm{eff}}(S)$ *such that the composition*

$$X \otimes \mathbb{L}^d \xrightarrow{\mathrm{id}\otimes\delta_X} X \otimes X \otimes X \xrightarrow{\epsilon_X \otimes \mathrm{id}} \mathbb{L}^d \otimes X \xrightarrow{\tau} X \otimes \mathbb{L}^d$$

is the identity.

Proof By proposition 5.19, it suffices to construct $\tilde{\delta}_X$ and $\tilde{\epsilon}_X$ in $ChMot(S)$. Identifying $\mathrm{Hom}_{ChMot(S)}(\mathbb{L}^d, X \otimes X)$ with the appropriate summand of $\mathrm{CH}_d((\mathbb{P}^1)^d \otimes X \times_S X)$, $\tilde{\delta}_X$ is represented by the cycle $0 \times \Delta_X$. Similarly

$$\tilde{\epsilon}_X \in \mathrm{Hom}_{ChMot(S)}(X \otimes X, \mathbb{L}^d) \subset \mathrm{CH}_{2d}(X \times_S X \times (\mathbb{P}^1)^d)$$

is represented by the cycle $\Delta_X \times (\mathbb{P}^1)^d$; the desired identity is a straightforward computation. $\qquad\square$

Finally, we have the duality theorem:

Theorem 5.21 *Sending* $X \in \mathbf{Proj}/S$ *of dimension d over S to* $X^D := X \otimes \mathbb{L}^{-d}$ *extends to an exact duality*

$$D : SmMot_{gm}(S)_{\mathbb{Q}}^{\mathrm{op}} \to SmMot_{gm}(S)_{\mathbb{Q}}.$$

Proof For $X \in \mathbf{Proj}/S$, the maps $\tilde{\delta}_X$ and $\tilde{\epsilon}_X$ define maps

$$\delta_X : S \to X \otimes (X \otimes \mathbb{L}^{-d_X}); \ \epsilon_X : (X \otimes \mathbb{L}^{-d}) \otimes X \to S.$$

The identity of lemma 5.20 tells us that $(X \otimes \mathbb{L}^{-d}, \delta_X, \epsilon_X)$ defines a dual of X in $SmMot_{gm}(S)_{\mathbb{Q}}$. Since $SmMot_{gm}(S)_{\mathbb{Q}}$ is generated by the Tate twists of objects in $\mathbf{Proj}/S$ (as an idempotently complete triangulated category), this implies that every object of $SmMot_{gm}(S)_{\mathbb{Q}}$ admits a dual (see e.g. [20, Part I, Chap. IV, theorem 1.2.5]). Since a dual of an object in a tensor category is unique up to unique isomorphism, this suffices to define the exact tensor duality involution D. □

6 Smooth motives and motives over a base

Cisinski-Déglise have defined a tensor triangulated category of effective motives over a base-scheme S, $DM^{\mathrm{eff}}(S)$, and a tensor triangulated category of motives over S, $DM(S)$, with an exact tensor functor $DM^{\mathrm{eff}}(S) \to DM(S)$ that inverts $\otimes \mathbb{L}$. In this section, we show how to define exact functor

$$\rho_S^{\mathrm{eff}} : SmMot_{gm}^{\mathrm{eff}}(S) \to DM^{\mathrm{eff}}(S), \ \rho_S : SmMot_{gm}(S) \to DM(S)$$

which give equivalences of $SmMot_{gm}^{\mathrm{eff}}(S)$, $SmMot_{gm}(S)$ with the full triangulated subcategories of $DM^{\mathrm{eff}}(S)$ and $DM(S)$ generated by the motives of smooth projective S schemes, resp. the Tate twists of smooth projective S-schemes. Working with $\mathbb{Q}$ coefficients, we have the same picture, with ρ_S^{eff}, ρ_S replaced by exact tensor functors

$$\rho_{S\mathbb{Q}}^{\mathrm{eff}} : SmMot_{gm}^{\mathrm{eff}}(S)_{\mathbb{Q}} \to DM^{\mathrm{eff}}(S)_{\mathbb{Q}} \ \rho_{S\mathbb{Q}} : SmMot_{gm}(S)_{\mathbb{Q}} \to DM(S)_{\mathbb{Q}},$$

giving analogous equivalences.

6.1 Cisinski-Déglise categories of motives. We summarize the construction of the category $DM^{\mathrm{eff}}(S)$ of effective motives over S, and the category $DM(S)$ of motives over S, from [5]. Although S is allowed to be a quite general scheme in [5], we restrict ourselves to the case of a base-scheme S that is separated, smooth and essentially of finite type over a field.

Define the abelian category of *presheaves with transfer* on $\mathbf{Sm}/S$, $PST(S)$, as the category of additive presheaves of abelian groups on Cor_S, "additive" meaning taking disjoint union to direct sum. We have the representable presheaves $\mathbb{Z}_S^{tr}(Z)$ for $Z \in \mathbf{Sm}/S$ by $\mathbb{Z}_S^{tr}(Z)(X) := Cor_S(X, Z)$ and pull-back maps given by the composition of correspondences.

One gives the category of complexes $C(PST(S))$ the *Nisnevich local* model structure (which we won't need to specify). The homotopy category is equivalent to the (unbounded) derived category $D(Sh_{\mathrm{Nis}}^{tr}(S))$, where $Sh_{\mathrm{Nis}}^{tr}(S)$ is the full subcategory of $PST(S)$ consisting of the presheaves with transfer which restrict to Nisnevich sheaves on $\mathbf{Sm}/S$.

The operation

$$\mathbb{Z}_S^{tr}(X) \otimes_S^{tr} \mathbb{Z}_S^{tr}(X') := \mathbb{Z}_S^{tr}(X \times_S X')$$

extends to a tensor structure $\otimes_S^{tr}$ making $PST(S)$ a tensor category: one forms the *canonical left resolution* $\mathcal{L}(\mathcal{F})$ of a presheaf $\mathcal{F}$ by taking the canonical surjection

$$\mathcal{L}_0(\mathcal{F}) := \bigoplus_{X \in \mathbf{Sm}/S, s \in \mathcal{F}(X)} \mathbb{Z}_S^{tr}(X) \xrightarrow{\varphi_0} \mathcal{F}$$

setting $\mathcal{F}_1 := \ker \varphi_0$ and iterating. One then defines

$$\mathcal{F} \otimes_S^{tr} \mathcal{G} := H_0(\mathcal{L}(\mathcal{F}) \otimes_S^{tr} \mathcal{L}(\mathcal{G}))$$

noting that $\mathcal{L}(\mathcal{F}) \otimes_S^{tr} \mathcal{L}(\mathcal{G})$ is defined since both complexes are degreewise direct sums of representable presheaves.

The restriction of $\otimes_S^{tr}$ to the subcategory of cofibrant objects in $C(Sh_{\mathrm{Nis}}^{tr}(S))$ induces a tensor operation $\otimes_S^L$ on $D(Sh_{\mathrm{Nis}}^{tr}(S))$ which makes $D(Sh_{\mathrm{Nis}}^{tr}(S))$ a tensor triangulated category.

Definition 6.1 ([5, definition 10.1]) $DM^{\mathrm{eff}}(S)$ is the localization of the triangulated category $D(Sh_{\mathrm{Nis}}^{tr}(S))$ with respect to the localizing category generated by the complexes $\mathbb{Z}_S^{tr}(X \times \mathbb{A}^1) \to \mathbb{Z}_S^{tr}(X)$. Denote by $m_S(X)$ the image of $\mathbb{Z}_S^{tr}(X)$ in $DM^{\mathrm{eff}}(S)$.

Remark 6.2 1. $DM^{\mathrm{eff}}(S)$ is a tensor triangulated category with tensor product $\otimes_S$ induced from the tensor product $\otimes_S^L$ via the localization map

$$Q_S : D(Sh_{\mathrm{Nis}}^{tr}(S)) \to DM^{\mathrm{eff}}(S),$$

and satisfying $m_S(X) \otimes_S m_S(Y) = m_S(X \times_S Y)$.

2. There is a model category $C(PST_{\mathbb{A}^1}(S))$ having $C(PST(S))$ as underlying category, defined as the left Bousfield localization of $C(PST(S))$ with respect to the following complexes

1. For each *elementary Nisnevich square* with $X \in \mathbf{Sm}/S$:

$$
\begin{array}{ccc}
W & \hookrightarrow & X' \\
\| & & \downarrow f \\
W & \hookrightarrow & X
\end{array}
$$

one has the complex

$$\mathbb{Z}_S^{tr}(X' \setminus W) \to \mathbb{Z}_S^{tr}(X \setminus W) \oplus \mathbb{Z}_S^{tr}(X') \to \mathbb{Z}_S^{tr}(X)$$

Recall that the square above is an elementary Nisnevich square if f is étale, the horizontal arrows are closed immersions of reduced schemes and the square is cartesian.

2. For $X \in \mathbf{Sm}/S$, one has the complex $\mathbb{Z}_S^{tr}(X \times \mathbb{A}^1) \to \mathbb{Z}_S^{tr}(X)$.

The homotopy category of $C(PST_{\mathbb{A}^1}(S))$ is equivalent to $DM^{\mathrm{eff}}(S)$.

Definition 6.3 Let T^{tr} be the presheaf with transfers

$$T^{tr} := \mathrm{coker}(\mathbb{Z}_S^{tr}(S) \xrightarrow{i_{\infty *}} \mathbb{Z}_S^{tr}(\mathbb{P}_S^1))$$

and let $\mathbb{Z}_S(1)$ be the image in $DM^{\mathrm{eff}}(S)$ of $T^{tr}[-2]$. Let

$$\otimes T^{tr} : C(PST(S)) \to C(PST(S))$$

be the functor $C \mapsto C \otimes_S^{tr} T^{tr}$.

Let $\mathbf{Spt}_{T^{tr}}(S)$ be the model category of $\otimes T^{tr}$ spectra in $C(PST_{\mathbb{A}^1}(S))$, i.e., an object is a sequence $E := (E_0, E_1, \ldots)$, $E_n \in C(PST(S))$, with bonding maps

$$\epsilon_n : E_n \otimes_S^{tr} T^{tr} \to E_{n+1}.$$

Morphisms are given by sequences of maps in $C(PST(S))$ which strictly commute with the respective bonding maps.

The model structure on the category of T^{tr}-spectra is the *stable model structure*, defined by following the construction of Hovey [14]. One first defines the projective model structure on $\mathbf{Spt}_{T^{tr}}(S)$, with weak equivalences maps $E \to F$ for which $E_n \to F_n$ is a weak equivalence for all n; let $\mathcal{H}_{proj}\mathbf{Spt}_{T^{tr}}(S)$ be the associated homotopy category. Next, for each $E \in \mathbf{Spt}_{T^{tr}}(S)$ there is a canonical fibrant model $E \to E^f$, where $E^f := (E_0^f, E_1^f, \ldots)$ with each E_n^f fibrant in $C(PST_{\mathbb{A}^1}(S))$ and the map

$$E_n^f \to \mathcal{H}om(T^{tr}, E_{n+1}^f)$$

adjoint to the bonding map $E_n^f \otimes_S^{tr} T^{tr} \to E_{n+1}^f$ is a weak equivalence in the model category $C(PST_{\mathbb{A}^1}(S))$. A *stable weak equivalence* in $\mathbf{Spt}_{T^{tr}}(S)$ is a map $f : A \to B$ such that

$$f^* : \mathrm{Hom}_{\mathcal{H}_{proj}\mathbf{Spt}_{T^{tr}}(S)}(B, E^f) \to \mathrm{Hom}_{\mathcal{H}_{proj}\mathbf{Spt}_{T^{tr}}(S)}(A, E^f)$$

is an isomorphism for all E. Hovey's stable model structure on $\mathbf{Spt}_{T^{tr}}(S)$ has weak equivalences the stable weak equivalences; we refer the interested reader to [5] or [14] for a description of the cofibrations and fibrations.

Definition 6.4 The category of triangulated motives over S, $DM(S)$, is the homotopy category of $\mathbf{Spt}_{T^{tr}}(S)$ for the stable model structure.

We will use the following results from [5].

Theorem 6.5 ([5, section 10.3, corollary 6.12]) *Suppose that S is in $\mathbf{Sm}/k$ for a field k, take X in $\mathbf{Sm}/S$, and let $m_k(X)$, $m_S(X)$ denote the motives of X in $DM(k)$, $DM(S)$, respectively. Then there is a natural isomorphism*

$$\mathrm{Hom}_{DM(S)}(m_S(X), \mathbb{Z}(n)[m]) \cong \mathrm{Hom}_{DM(k)}(m_k(X), \mathbb{Z}(n)[m]).$$

In addition, the natural map

$$\varinjlim_N \mathrm{Hom}_{DM^{\mathrm{eff}}(S)}(m_S(X) \otimes \mathbb{Z}(N), \mathbb{Z}(n+N)[m]) \to \mathrm{Hom}_{DM(S)}(m_S(X), \mathbb{Z}(n)[m])$$

is an isomorphism. Finally, the cancellation theorem holds in this setting: the natural map

$$\mathrm{Hom}_{DM^{\mathrm{eff}}(S)}(m_S(X), \mathbb{Z}(n)[m])] \to \mathrm{Hom}_{DM^{\mathrm{eff}}(S)}(m_S(X) \otimes \mathbb{Z}(1), \mathbb{Z}(n+1)[m])$$

is an isomorphism.

Remarks 6.6 1. By [23] $\mathrm{Hom}_{DM(k)}(m_k(X), \mathbb{Z}(n)[m])$ is motivic cohomology in the sense of Voevodsky [24, chapter V], that is

$$\mathrm{Hom}_{DM(k)}(m_k(X), \mathbb{Z}(n)[m]) = H^m(X, \mathbb{Z}(n)) \cong \mathrm{CH}^n(X, 2n - m).$$

In fact, this follows immediately from [23] in the case of a perfect field; the general case follows by using the usual trick of viewing a field k as a limit of finitely generated extensions of the prime field k_0, and the fact that, for a projective systems of regular noetherian schemes S_α with affine transition maps, the functor

$$S_\alpha \mapsto \mathrm{Hom}_{DM(S)}(m_{S_\alpha}(X_{S_\alpha}), \mathbb{Z}(n)[m])$$

transforms the projective limit $\varprojlim S_\alpha$ to the inductive limit (see [7, prop 4.2.19]).

2. The isomorphisms in theorem 6.5 are proven as follows: Let $p : S \to \mathrm{Spec}\, k$

be the (smooth) structure morphism. The limit argument mentioned above reduces us to the case of k a perfect field. The restriction of base functor induces an exact functor

$$p_\sharp : DM^{\text{eff}}(S) \to DM^{\text{eff}}(k)$$

and the pull-back $X \mapsto X \times_k S$ induces an exact functor

$$p^* : DM^{\text{eff}}(k) \to DM^{\text{eff}}(S),$$

right adjoint to $p_\sharp$. In addition, one has $p^*(\mathbb{Z}(n)) \cong \mathbb{Z}(n)$, and $p_\sharp(m_S(X) \otimes \mathbb{Z}(N)) \cong m_k(X) \otimes \mathbb{Z}(N)$ for $X \in \mathbf{Sm}/S$, $N \in \mathbb{Z}$. We have similar functors on $DM(S)$, $DM(k)$. Thus, we have

$$\operatorname{Hom}_{DM^{\text{eff}}(S)}(m_S(X) \otimes \mathbb{Z}(N), \mathbb{Z}(n+N)[m])$$
$$\cong \operatorname{Hom}_{DM^{\text{eff}}(S)}(m_S(X) \otimes \mathbb{Z}(N), p^*\mathbb{Z}(n+N)[m])$$
$$\cong \operatorname{Hom}_{DM^{\text{eff}}(k)}(p_\sharp(m_S(X) \otimes \mathbb{Z}(N)), \mathbb{Z}(n+N)[m])$$
$$\cong \operatorname{Hom}_{DM^{\text{eff}}(k)}(m_k(X) \otimes \mathbb{Z}(N), \mathbb{Z}(n+N)[m]),$$

and similarly with DM^{eff} replaced by DM. This already proves the first isomorphism. For the next two, the above isomorphisms reduce us to the case of $S = \operatorname{Spec} k$. Also, $DM_-^{\text{eff}}(k)$ is a full subcategory of $DM^{\text{eff}}(k)$. Finally, Voevodsky's identification of motivic cohomology with the higher Chow groups [23], together with the projective bundle formula, gives the limited version of the cancellation theorem we need to finish the proof.

6.2 Tensor structure. The tensor structure on $C(PST(S))$ induces a "tensor operation" on the spectrum category by the usual device of choosing a cofinal subset $\mathbb{N} \subset \mathbb{N} \times \mathbb{N}$, $i \mapsto (n_i, m_i)$, with $n_{i+1} + m_{i+1} = n_i + m_i + 1$ for each i: each pair of T^{tr} spectra $E := (E_0, E_1, \ldots)$ and $F := (F_0, F_1, \ldots)$ gives rise to a T^{tr} bispectrum

$$E \boxtimes_S^{tr} F := \begin{pmatrix} & \vdots & \\ \cdots & E_i \otimes_S^{tr} F_j & \cdots \\ & \vdots & \end{pmatrix}$$

with vertical and horizontal bonding maps induced by the bonding maps for E and F, respectively. The vertical bonding maps use in addition the symmetry isomorphism in $C(PST_{\mathbb{A}^1}(S))$. Finally, the choice of the cofinal $\mathbb{N} \subset \mathbb{N} \times \mathbb{N}$ converts a bispectrum to a spectrum.

Of course, this is not even associative, so one does not achieve a tensor operation on $\mathbf{Spt}_{T^{tr}}(S)$, but $\boxtimes_S^{tr}$ (on cofibrant objects) does pass to the localization $DM(S)$, and gives rise there to a tensor structure, making $DM(S)$ a tensor triangulated category. We write this tensor operation as $\otimes_S$, as before.

Remark 6.7 One can also define a "Spanier-Whitehead" category $DM^{\text{S-W}}(S)$ by inverting the functor $- \otimes T^{tr} = - \otimes \mathbb{Z}_S(1)[2]$ on $DM^{\text{eff}}(S)$; this is clearly equivalent to inverting $- \otimes \mathbb{Z}_S(1)$. Concretely, $DM^{\text{S-W}}(S)$ has objects $X(n)$, $n \in \mathbb{Z}$, with morphisms

$$\operatorname{Hom}_{DM^{\text{S-W}}(S)}(X(n), Y(m)) := \varinjlim_N \operatorname{Hom}_{DM^{\text{eff}}(S)}(X \otimes \mathbb{Z}_S(N+n), Y \otimes \mathbb{Z}_S(N+m)).$$

Sending X to $X(n)$ clearly defines an auto-equivalence of $DM^{\text{S-W}}(S)$.

$DM^{\text{S-W}}(S)$ inherits the structure of a triangulated category from $DM^{\text{eff}}(S)$, by declaring a triangle $\mathcal{T}$ in $DM^{\text{S-W}}(S)$ to be distinguished if $\mathcal{T}(N)$ is the image of a distinguished triangle in $DM^{\mathrm{eff}}(S)$ for some $N >> 0$. One shows that the symmetry isomorphism $\mathbb{Z}_S(1) \otimes \mathbb{Z}_S(1) \to \mathbb{Z}_S(1) \otimes \mathbb{Z}_S(1)$ is the identity, which implies that $DM^{\mathrm{S\text{-}W}}(S)$ inherits a tensor structure from $DM^{\mathrm{eff}}(S)$.

Since $- \otimes T^{tr}$ is isomorphic to the shift operator in $DM(S)$, this functor is invertible on $DM(S)$, hence $DM^{\mathrm{eff}}(S) \to DM(S)$ factors through a canonical exact functor $\varphi_S : DM^{\mathrm{S\text{-}W}}(S) \to DM(S)$. Giving $DM(S)$ the tensor structure described above, it is easy to see that φ_S is a tensor functor.

6.3 Motives and smooth motives. For $X \in \mathbf{Sm}/S$, we have the presheaf $U \mapsto C^S(X,0)(U)$ (definition 5.2), giving us the object $C^S(X)$ in $C^-(Sh^{tr}_{\mathrm{Nis}}(S))$. Suppose X is in $\mathbf{Proj}/S$. Then by remark 5.3(1), $C^S(X)^0 = \mathbb{Z}^{tr}_S(X)$, giving the natural map
$$\iota_X : \mathbb{Z}^{tr}_S(X) \to C^S(X).$$

Lemma 6.8 *For $X \in \mathbf{Proj}/S$, ι_X is an isomorphism in $DM^{\mathrm{eff}}(S)$.*

Proof Let $\mathbb{Z}^{tr}_S(X)^*$ be the complex which is $\mathbb{Z}^{tr}_S(X)$ in each degree $n \leq 0$, and with differential $d^n : \mathbb{Z}^{tr}_S(X)^n \to \mathbb{Z}^{tr}_S(X)^{n+1}$ the identity if n is even and 0 if n is odd. The projection $U \times \square^n \to U$ gives a map of complexes
$$\pi_U : \mathbb{Z}^{tr}(X)^*(U) \to C^S(X)(U)$$
functorially in U, hence a map of complexes of presheaves
$$\pi : \mathbb{Z}^{tr}(X)^* \to C^S(X).$$
On the other hand, we have
$$C^S(X)^n(U) \cong \mathrm{Hom}_{C(PST(S))}(\mathbb{Z}^{tr}(U \times \square^n), \mathbb{Z}^{tr}(X))$$
and, in degree n π is the map induced on $\mathrm{Hom}_{C(PST(S))}(-, \mathbb{Z}^{tr}(X))$ by
$$p_* : \mathbb{Z}^{tr}(U \times \square^n) \to \mathbb{Z}^{tr}(U).$$
As $\mathbb{Z}^{tr}(U)$ is cofibrant for all $U \in \mathbf{Sm}/S$ and p_* is a weak equivalence (both in the model category $C(PST_{\mathbb{A}^1}(S))$) it follows that $\pi : \mathbb{Z}^{tr}(X)^* \to C^S(X)$ is an isomorphism in the homotopy category $DM^{\mathrm{eff}}(S)$. Since ι factors through π via the homotopy equivalence
$$\iota_0 : \mathbb{Z}^{tr}(X) \to \mathbb{Z}^{tr}(X)^*$$
ι is an isomorphism as well. $\square$

Let $G^*_S : C^-(Sh^{tr}_{\mathrm{Nis}}(S)) \to C(Sh^{tr}_{\mathrm{Nis}}(S))$ be the Godement resolution functor, with respect to S_{Zar}. Concretely, for $\mathcal{F} \in Sh^{tr}_{\mathrm{Nis}}(S)$, $G^0_S(\mathcal{F})$ is the sheaf
$$G^0_S(\mathcal{F})(U) := \prod_{s \in S} \mathcal{F}(U \otimes_S \mathcal{O}_{S,s}).$$
where we define $\mathcal{F}(U \otimes_S \mathcal{O}_{S,s})$ as the limit of $\mathcal{F}(U \otimes_S V)$, where V runs over the Zariski open neighborhoods of s in S. We have the natural transformation $\mathrm{id} \to G_S$ induced by the inclusion $\mathcal{F} \to G^0_S\mathcal{F}$. Since $\mathcal{F} \to G^*_S\mathcal{F}$ is a Zariski local weak equivalence, this map induces an isomorphism in $D(Sh^{tr}_{\mathrm{Nis}})$. Thus, lemma 6.8 gives

Lemma 6.9 *For each $X \in \mathbf{Proj}/S$, the composition*
$$\mathbb{Z}^{tr}_S(X) \xrightarrow{\iota_X} C^S(X) \to G^*_S C^S(X)$$
define an isomorphism $m_S(X) \cong G^*_S C^S(X)$ in $DM^{\mathrm{eff}}(S)$.

Remark 6.10 Since S has finite Krull dimension, say d, Grothendieck's theorem [12] tells us that S has Zariski cohomological dimension $\leq d$. Thus, for $\mathcal{F} \in C^-(Sh_{\mathrm{Nis}}^{tr}(S))$, $G_S(\mathcal{F})$ is cohomologically bounded above. Therefore, the composition of G_S with the quotient functor $C(Sh_{\mathrm{Nis}}^{tr}(S)) \to D(Sh_{\mathrm{Nis}}^{tr}(S))$ factors (up to natural isomorphism) through a modified Godement resolution

$$G_S^- : C^-(Sh_{\mathrm{Nis}}^{tr}(S)) \to D^-(Sh_{\mathrm{Nis}}^{tr}(S)).$$

Remark 6.11 One can also define categories of motives with coefficients; for simplicity, we restrict to the case of a coefficient ring A being a localization of $\mathbb{Z}$. One simply replaces the category $PST(S)$ with the category $PST(S)_A$ of additive presheaves of A-modules on Cor_S, and then follows the same procedure as above, forming the tensor triangulated category of effective S-motives with A coefficients, $DM^{\mathrm{eff}}(S)_A$, and tensor triangulated category of S-motives with A coefficients, $DM(S)_A$. All the results described above for $DM^{\mathrm{eff}}(S)$ and $DM(S)$ hold for $DM^{\mathrm{eff}}(S)_A$ and $DM(S)_A$, suitably modified to reflect the A-module structure. We note that $DM^{\mathrm{eff}}(S)_A$ and $DM(S)_A$ are idempotently complete.

We now proceed to construct an exact functor

$$\rho_S : SmMot_{gm}^{\mathrm{eff}}(S) \to DM^{\mathrm{eff}}(S).$$

For this, consider the presheaf of DG categories $\underline{dgCor}_S$ on S_{Zar}

$$U \mapsto dgCor_U.$$

Recall that $dgCor_U$ has objects $X \in \mathbf{Sm}/U$, and the full subcategory with objects $X \in \mathbf{Proj}/U$ is our category $dgPrCor_U$. We have the "Godement extension" $R\Gamma(S, \underline{dgCor}_S)$ with objects $X \in \mathbf{Sm}/S$, containing $R\Gamma(S, \underline{dgPrCor}_S)$ as the full DG subcategory with objects $X \in \mathbf{Proj}/S$.

Take $X \in \mathbf{Sm}/S$. By construction, the presheaf on $\mathbf{Sm}/S$ of Hom-complexes

$$U \mapsto \mathcal{H}om_{R\Gamma(S,\underline{dgCor}_S)}(U, X)^*$$

is just $G_S^* C^S(X)$. Thus, for $X, Y \in \mathbf{Sm}/S$, the composition law in $R\Gamma(S, \underline{dgCor}_S)$ defines a natural map

$$\tilde{\rho}_{X,Y} : \mathcal{H}om_{R\Gamma(S,\underline{dgCor}_S)}(X, Y)^* \to \mathcal{H}om_{C(Sh_{\mathrm{Nis}}^{tr}(S))}(G_S^* C^S(X), G_S^* C^S(Y)).$$

This defines for us the functor of DG categories

$$\tilde{\rho} : R\Gamma(S, \underline{dgCor}_S) \to C_{dg}(Sh_{\mathrm{Nis}}^{tr}(S)).$$

Applying the functor $\mathcal{K}^b$ and composing with the total complex functor

$$\mathrm{Tot} : \mathcal{K}^b(C_{dg}(Sh_{\mathrm{Nis}}^{tr}(S))) \to K(Sh_{\mathrm{Nis}}^{tr}(S))$$

and the quotient functor

$$K(Sh_{\mathrm{Nis}}^{tr}(S)) \to DM^{\mathrm{eff}}(S)$$

gives us the exact functor

$$\mathcal{K}^b(\tilde{\rho}) : \mathcal{K}^b(R\Gamma(S, \underline{dgCor}_S)) \to DM^{\mathrm{eff}}(S).$$

If we restrict to the full subcategory $\mathcal{K}^b(R\Gamma(S, \underline{dgPrCor}_S))$ of $\mathcal{K}^b(R\Gamma(S, \underline{dgCor}_S))$ and extend canonically to the idempotent completion, we have defined an exact functor

$$\rho_S^{\mathrm{eff}} : SmMot_{gm}^{\mathrm{eff}}(S) \to DM^{\mathrm{eff}}(S).$$

with $\rho_S^{\mathrm{eff}}(X) = G_S^* C^S(X)$ for all $X \in \mathbf{Proj}/S$. The natural isomorphism $m_S(X) \to G_S^* C^S(X)$ in $DM^{\mathrm{eff}}(S)$ constructed in lemma 6.9 shows that the diagram

$$\mathbf{Proj}/S \longrightarrow SmMot_{gm}^{\mathrm{eff}}(S)$$

$$m_S \searrow \qquad \downarrow \rho_S^{\mathrm{eff}}$$

$$DM^{\mathrm{eff}}(S)$$

commutes up to natural isomorphism.

We note that $\rho_S^{\mathrm{eff}}(\mathbb{L}) = T^{tr}$, hence the composition

$$SmMot_{gm}^{\mathrm{eff}}(S) \xrightarrow{\rho_S} DM^{\mathrm{eff}}(S) \to DM^{\mathrm{S\text{-}W}}(S)$$

sends $- \otimes \mathbb{L}$ to the invertible endomorphism $- \otimes T^{tr}$ on $DM^{\mathrm{S\text{-}W}}(S)$, hence this composition factors through a canonical extension

$$\rho_S^{\mathrm{S\text{-}W}} : SmMot_{gm}(S) \to DM^{\mathrm{S\text{-}W}}(S).$$

Let

$$\rho_S : SmMot_{gm}(S) \to DM(S).$$

be the composition of $\rho^{\mathrm{S\text{-}W}}$ with the canonical functor $DM^{\mathrm{S\text{-}W}}(S) \to DM(S)$.

Remark 6.12 It is easy to check that the restriction of ρ_S^{eff} to $H^0 dgCor_S$ is a tensor functor. By the surjectivity of the cycle class map (lemma 5.18) this shows that the composition

$$\rho_S^{\mathrm{eff}} \circ \psi^{\mathrm{eff}} : ChMot^{\mathrm{eff}}(S) \to DM^{\mathrm{eff}}(S)$$

is a tensor functor. Similarly,

$$\rho_S \circ \psi : ChMot(S) \to DM(S)$$

is a tensor functor.

Theorem 6.13 *Let Y be in $\mathbf{Proj}/S$, $d = dim_S Y$, and take A, B in $DM(S)$. Then for every $n \in \mathbb{Z}$, there is a natural isomorphism*

$$\mathrm{Hom}_{DM(S)}(A, m_S(Y) \otimes B[n]) \cong \mathrm{Hom}_{DM(S)}(A \otimes m_S(Y), B \otimes \mathbb{Z}(d)[2d + n])$$

Proof $Y \in ChMot(S)$ has the dual $(Y \otimes \mathbb{L}^{-d}, \delta_Y, \epsilon_Y)$ and $m_S(Y) \cong \rho_S \psi_S(Y)$, hence $m_S(Y)$ has the dual $(m_S(Y)(-d)[-2d], \rho_S \psi_S(\delta_Y), \rho_S \psi_S(\epsilon_Y))$ in $DM(S)$. This gives us the desired isomorphism. $\square$

Corollary 6.14 *The functors*

$$\rho_S^{\mathrm{eff}} : SmMot_{gm}^{\mathrm{eff}}(S) \to DM^{\mathrm{eff}}(S),$$

$$\rho_S : SmMot_{gm}(S) \to DM(S).$$

are faithful embeddings and ρ_S is fully faithful.

Proof Since

$$SmMot_{gm}^{\mathrm{eff}}(S) \to SmMot_{gm}(S)$$

is fully faithful (corollary 5.14), it suffices to show that ρ_S is fully faithful.

Take $X, Y \in \mathbf{Proj}/S$, let $d = dim_S Y$. By theorem 6.13, we have the duality isomorphism

$$\mathrm{Hom}_{DM(S)}(m_S(X), m_S(Y)) \cong \mathrm{Hom}_{DM(S)}(m_S(X) \otimes m_S(Y), \mathbb{Z}^{tr}(d)[2d + n]).$$

As the construction of the duality isomorphism arises from the duality in $ChMot(S)$, the diagram

$$\begin{array}{ccc} \mathrm{Hom}_{SmMot_{gm}(S)}(X, Y[n]) & \xrightarrow{\ \sim\ } & \mathrm{Hom}_{SmMot_{gm}(S)}(X \times_S Y, \mathbb{L}^d[n]) \\ \rho_S \downarrow & & \downarrow \rho_S \\ \mathrm{Hom}_{DM(S)}(m_S(X), m_S(Y)[n]) & \xrightarrow{\ \sim\ } & \mathrm{Hom}_{DM(S)}(m_S(X \times_S Y), \mathbb{Z}^{tr}(d)[2d+n]) \end{array}$$

commutes. But by theorem 6.5 and remark 6.6, we have

$$\mathrm{Hom}_{DM(S)}(m_S(X \times_S Y), \mathbb{Z}^{tr}(d)[2d+n])$$
$$\cong \mathrm{Hom}_{DM(k)}(m_k(p_* X \times_S Y), \mathbb{Z}^{tr}(d)[2d+n])$$
$$\cong H^{2d+n}(p_* X \times_S Y, \mathbb{Z}(d)).$$

It is easy to check that the natural map

$$H^{2d+n}(p_* X \times_S Y, \mathbb{Z}(d)) \cong H^n(C^k(\mathbb{L}^d)(p_*(X \times_S Y))$$
$$= H^n(C^S(\mathbb{L}^d)(X \times_S Y)) \to \mathrm{Hom}_{DM(S)}(m_S(X \times_S Y), \mathbb{Z}^{tr}(d)[2d+n])$$

inverts this isomorphism, from which it follows that the composition

$$H^{2d+n}(X \times_S Y, \mathbb{Z}(d)) \cong H^n(C^k(\mathbb{L}^d))(p_*(X \times_S Y)) = H^n(C^S(\mathbb{L}^d)(X \times_S Y))$$
$$\to \mathrm{Hom}_{SmMot_{gm}(S)}(X \times_S Y, \mathbb{L}^d[n]) \xrightarrow{\rho_S} \mathrm{Hom}_{DM(S)}(m_S(X \times_S Y), \mathbb{Z}^{tr}(d)[2d+n])$$
$$\cong \mathrm{Hom}_{DM(k)}(m_k(p_* X \times_S Y), \mathbb{Z}^{tr}(d)[2d+n]) \cong H^{2d+n}(X \times_S Y, \mathbb{Z}(d))$$

is the identity. Therefore,

$$\rho_S : \mathrm{Hom}_{SmMot_{gm}(S)}(X \times_S Y, \mathbb{L}^d[n]) \to \mathrm{Hom}_{DM(S)}(m_S(X \times_S Y), \mathbb{Z}^{tr}(d)[2d+n])$$

is an isomorphism. $\qquad\square$

Definition 6.15 Let $DTM(S)$ be the full triangulated subcategory of $DM(S)$ generated by the Tate objects $\mathbb{Z}_S^{tr}(n)$, $n \in \mathbb{Z}$, and let $DTM^{\mathrm{eff}}(S)$ be the full triangulated subcategory of $DM^{\mathrm{eff}}(S)$ generated by the Tate objects $\mathbb{Z}_S^{tr}(n)$, $n \geq 0$

Restricting to the Tate categories gives

Corollary 6.16 *The restriction of ρ_S to*

$$\rho_S^{Tate} : DTMot(S) \to DM(S)$$

is a fully faithful embedding, giving an equivalence of $DTMot(S)$ with $DTM(S)$. The version with $\mathbb{Q}$-coefficients

$$\rho_S^{Tate} : DTMot(S)_{\mathbb{Q}} \to DM(S)_{\mathbb{Q}}$$

defines an equivalence of $DTMot(S)_{\mathbb{Q}}$ with $DTM(S)_{\mathbb{Q}}$ as tensor triangulated categories.

Similarly, we have the equivalences

$$\rho_S^{\mathrm{eff}\,Tate} : DTMot^{\mathrm{eff}}(S) \to DTM^{\mathrm{eff}}(S)$$

and

$$\rho_{S\mathbb{Q}}^{\mathrm{eff}\,Tate} : DTMot^{\mathrm{eff}}(S)_{\mathbb{Q}} \to DTM^{\mathrm{eff}}(S)_{\mathbb{Q}}.$$

Proof This is immediate consequence of corollary 6.14, except for the assertion that ρ_S^{effTate} and $\rho_{S\mathbb{Q}}^{\text{effTate}}$ are full. But it follows from the weak cancellation theorem 6.5 that $DTM^{\text{eff}}(S) \to DTM(S)$ is fully faithful, whence the result. $\square$

7 Moving lemmas

We proceed to extend the Friedlander-Lawson-Voevodsky moving lemmas to the case of a semi-local regular base scheme, and use these results to prove theorem 5.4 and theorem 5.5. We refer the reader to §5.1 for the notation involving the various presheaves of equi-dimensional cycles.

7.1 Friedlander-Lawson-Voevodsky moving lemmas. The Friedlander-Lawson moving lemmas [10] are applied in [11] to prove the main moving lemmas and duality statements for the cycle complexes of equi-dimensional cycles. This involves a two-step process: one moves cycles on $\mathbb{P}^n$ by one process (which we will not need to recall explicitly) and then one uses a variation of the projecting cone argument to extend the moving process to a smooth projective variety. The main result following from the first step is

Theorem 7.1 ([11, theorem 6.1]) *Let k be a field, and n, r, d, e integers. Then there are maps of presheaves of abelian monoids on $\mathbf{Sm}/k$*

$$H_U^{\pm} : z_{equi}^k(\mathbb{P}^n, r)^{\text{eff}}(U) \to z_{equi}^k(\mathbb{P}^n, r)^{\text{eff}}(U \times \mathbb{A}^1)$$

such that

1. *There is an integer $m \geq 0$ such that*

$$i_0^* \circ H_U^+ = (m+1) \cdot \text{id}_{z_{equi}^k(\mathbb{P}^n, r)^{\text{eff}}(U)}$$

$$i_0^* \circ H_U^- = m \cdot \text{id}_{z_{equi}^k(\mathbb{P}^n, r)^{\text{eff}}(U)}$$

   where $i_0 : U \to U \times \mathbb{A}^1$ is the zero-section.
2. *Let $i_x : \text{Spec} \, k' \to U \times \mathbb{A}^1 \setminus \{0\}$ be a k'-point of $U \times \mathbb{A}^1 \setminus \{0\}$, where $k' \supset k$ is some extension field of k. Then for each $W \in z_{equi}^k(\mathbb{P}^n, s)_{\leq d}^{\text{eff}}(k')$, $Z \in z_{equi}^k(\mathbb{P}^n, r)_{\leq e}^{\text{eff}}(U)$, with $0 \leq r, s \leq n$, $r + s \geq n$, the cycles $x^*(H_U^{\pm}(Z))$ and W intersect properly on $\mathbb{P}_{k'}^n$.*

The second step involves the projecting cone construction, which enables one to go from moving cycles on $\mathbb{P}^n$ to moving cycles on a smooth projective $X \subset \mathbb{P}^N$ of dimension n. We first recall the situation over a base-field k.

Let $D_0, \ldots, D_n \subset \mathbb{P}^N$ be degree f hypersurfaces, and let $F_0, \ldots, F_n$ be the corresponding defining equations. If $X \cap D_0 \cap \ldots \cap D_n = \emptyset$, then

$$F := (F_0 : \ldots : F_n) : \mathbb{P}^N \setminus \cap_{i=0}^n D_i \to \mathbb{P}^n$$

restricts to X to define a finite surjective morphism

$$F_X : X \to \mathbb{P}^n$$

For Z an effective cycle of dimension r on X, we thus have the effective cycle $F_X^*(F_{X*}(Z))$, which can be written as

$$F_X^*(F_{X*}(Z)) = Z + R_F(Z)$$

for a uniquely determined effective cycle $R_F(Z)$ of dimension r. Sending Z to $R_F(Z)$ thus gives a map

$$R_F : z_{equi}^k(X, r)^{\text{eff}}(k) \to z_{equi}^k(X, r)^{\text{eff}}(k).$$

Let $\mathcal{U}_X(d)$ be the open subscheme of $H^0(\mathbb{P}^N, \mathcal{O}(d))^n$ consisting of $D_0, \ldots, D_n$ such that $\cap_{i=0}^n D_i \cap X = \emptyset$. It is easy to see that the complement of $\mathcal{U}_X(d)$ in $H^0(\mathbb{P}^N, \mathcal{O}(d))^n$ is a hypersurface (in fact, the defining equation of this complement is exactly the classical Chow form of X, whose coefficients give the Chow point of X in the Chow variety of dimension n, degree $\deg X$ effective cycles on $\mathbb{P}^N$).

For $F \in \mathcal{U}_X(f)$, let $\operatorname{ram}_X(F) \subset X$ be the ramification locus of F_X. Choose integers $f_0, \ldots, f_n \geq 1$ and let $\mathcal{R}_X(d_*) \subset \prod_{i=0}^n \mathcal{U}_X(d_i)$ be the open subscheme consisting of those tuples $(F^0, \ldots, F^n)$, $F^i := (F_0^i : \ldots : F_n^i) \in \mathcal{U}_X(d_i)$ for which

$$\cap_{i=0}^n \operatorname{ram}_X(F^i) = \emptyset.$$

Remark 7.2 It follows from Bertini's theorem that, if $d_i \geq 2$ for all i, then $\mathcal{R}_X(d_*)$ is a dense open subscheme of $\prod_{i=0}^n \mathcal{U}_X(d_i)$ (even in positive characteristic).

The next moving result is a translation of [10, proposition 1.3 and theorem 1.7].

Proposition 7.3 *Let $X \subset \mathbb{P}_k^N$ be a smooth closed subscheme of dimension n. Fix dimensions $0 \leq r, s \leq n$ with $r + s \geq n$ and a degree $e \geq 1$. Then there is an increasing function*

$$G := G_{r,s,N,e,X} : \mathbb{N} \to \mathbb{N},$$

depending on X and the integers r, s, N and e, and, for each $d_* := (d_0, \ldots, d_n)$, there is a closed subset $\mathcal{B}_X(d_*, e, r, s)$ of $\mathcal{R}_X(d_*)$, such that the following holds: Let k' be a field extension of k, and take $F^0, \ldots, F^n \in \mathcal{R}_X(d_*)(k') \setminus \mathcal{B}_X(d_*, e, r, s)(k')$. Suppose that $d_0 \geq G(e)$ and $d_i \geq G(d_{i-1}^n \cdot e)$ for $i = 1, \ldots, n$. Then

1. *For each pair of effective cycles Z, W on $X_{k'}$ with Z of dimension r, W of dimension s and both having degree $\leq e$, the cycles W and $R_X(F^n) \circ \ldots \circ R_X(F^0)(Z)$ intersect properly on $X_{k'}$.*

2. *Let Z, W be a pair of effective cycles on $X_{k'}$ with Z of dimension r, W of dimension s and both having degree $\leq e$. Suppose that Z and W intersect properly on $X_{k'}$. Then for every $j = 0, \ldots, n$, the cycles W and $(F_X^{j*} F_{X*}^j) \circ \ldots \circ (F_X^{0*} F_{X*}^0)(Z)$ intersect properly on $X_{k'}$.*

Remark 7.4 If we have two closed subsets $\mathcal{B}_X^1(d_*, e, r, s), \mathcal{B}_X^2(d_*, e, r, s)$ satisfying the conditions of proposition 7.3, then the intersection $\mathcal{B}_X^1(d_*, e, r, s) \cap \mathcal{B}_X^2(d_*, e, r, s)$ also satisfies proposition 7.3. Thus, without loss of generality, we may assume that $\mathcal{B}_X(d_*, e, r)$ is the *minimal* closed subset of $\mathcal{R}_X(d_*)$ satisfying the conditions of proposition 7.3.

Assuming this to be the case, we have the following description of $\mathcal{B}_X(d_*, e, r, s)$:

$$\mathcal{B}_X(d_*, e, r, s) = \{F^* := (F^0, \ldots, F^n) \in \mathcal{R}_X(d_*) \mid \exists \text{ an extension field } k' \supset k(F^*),$$

$Z \in z_{equi}(X, r)_{\leq e}^{\mathrm{eff}}(k'), W \in z_{equi}(X, s)_{\leq e}^{\mathrm{eff}}(k')$ such that either

1. W and $R_X(F^n) \circ \ldots \circ R_X(F^0)(Z)$ do not intersect properly on $X_{k'}$

   or

2. W and Z intersect properly on $X_{k'}$, but this is not the case for

   W and $(F_X^{j*} F_{X*}^j) \circ \ldots \circ (F_X^{0*} F_{X*}^0)(Z)$ for some $j, 0 \leq j \leq n\}$.

To see this, it suffices to see that the set described is closed; for this it suffices to see that this set is closed under specialization. This follows from the fact the the Chow varieties parametrizing effective cycles of fixed degree and dimension on X are proper over k.

7.2 Extending the moving lemma. We now consider the two moving lemmas in the setting of a smooth projective scheme over a regular base-scheme B. The extension of the first moving lemma is a direct corollary of theorem 7.1.

Corollary 7.5 *Let k be a field, and n, r, d, e integers, B a regular k-scheme, essentially of finite type over k. Then there are maps of presheaves of abelian monoids on $\mathbf{Sm}/B$*

$$H_U^\pm : z_{equi}^B(\mathbb{P}_B^n, r)^{\text{eff}}(U) \to z_{equi}^B(\mathbb{P}_B^n, r)^{\text{eff}}(U \times \mathbb{A}^1)$$

such that

1. *Let $i_0 : U \to U \times \mathbb{A}^1$ be the zero-section. There is an integer $m \geq 0$ such that*

$$i_0^* \circ H_U^+ = (m+1) \cdot \text{id}_{z_{equi}^B(\mathbb{P}^n, r)^{\text{eff}}(U)}$$
$$i_0^* \circ H_U^- = m \cdot \text{id}_{z_{equi}^B(\mathbb{P}^n, r)^{\text{eff}}(U)}.$$

2. *Let $i : T \to U \times \mathbb{A}^1 \setminus \{0\}$ be a morphism of B-schemes. Then for each $W \in z_{equi}^B(\mathbb{P}_B^n, s)_{\leq d}^{\text{eff}}(T)$, $Z \in z_{equi}^B(\mathbb{P}_B^n, r)_{\leq e}^{\text{eff}}(U)$, with $0 \leq r, s \leq n$, $r + s \geq n$, the cycles $i^*(H_U^\pm(Z))$ and W intersect properly on $\mathbb{P}_T^n$.*

Proof First of all, it suffices to prove the result if B is of finite type over k. Let $p : B \to \text{Spec}\, k$ be the structure morphism and $p_* : \mathbf{Sm}/B \to \mathbf{Sm}/k$ the base-restriction functor, $p^* : \mathbf{Sm}/k \to \mathbf{Sm}/B$ the pull-back functor $p^*(Y) := Y \times_k B$. Then we have the evident identifications

$$z_{equi}^B(\mathbb{P}_B^n, r)(U) = z_{equi}^k(\mathbb{P}_k^n, r)(p_*U),$$

induced by the canonical isomorphism $U \times_B \mathbb{P}_B^n \cong p_*U \times_k \mathbb{P}_k^n$. Thus the maps $H_{p_*U}^\pm$ given by theorem 7.1 give rise to maps

$$H_U^\pm : z_{equi}^B(\mathbb{P}_B^n, r)(U) \to z_{equi}^B(\mathbb{P}_B^n, r)(U \times \mathbb{A}^1)$$

which clearly satisfy our condition (1).

For (2), if we have two cycles $A \in z_{equi}^B(\mathbb{P}_B^n, r)^{\text{eff}}(T)$ and $A' \in z_{equi}^B(\mathbb{P}_B^n, s)^{\text{eff}}(T)$ such that, for each point $t \in T$, the fibers $t^*(A), t^*(A')$ intersect properly on $\mathbb{P}_t^n$, then A and A' intersect properly on $T \times_B \mathbb{P}_B^n$ and $A \cdot A'$ is in $z_{equi}^B(\mathbb{P}_B^n, r+s-n)^{\text{eff}}(T)$. This, together with theorem 7.1(2), proves (2). $\qquad\square$

We now show how to extend the second step of the moving lemma. Take $X \in \mathbf{Proj}/B$ with a fixed embedding $X \hookrightarrow \mathbb{P}_B^N$ over B. For each $b \in B$, let $\mathcal{C}_{X_b}(d) = \mathbb{P}(H^0(\mathbb{P}_s^N, \mathcal{O}(d)))^{n+1} \setminus \mathcal{U}_{X_b}(d)$. Let $V \subset \text{Proj}_B(p_*\mathcal{O}(d))^{n+1} \times \mathbb{P}^N$ be the incidence correspondence with points $\{((f_0 : \ldots : f_n), x) \mid x \in \cap_{i=0}^n (f_i = 0)\}$. Let $\mathcal{C}_X(d) = p_1(V \cap p_2^{-1}(X))$, and let $\mathcal{U}_X(d) = \mathbb{P}(p_*\mathcal{O}(d)))^{n+1} \setminus \mathcal{C}_X(d)$. Then clearly $\mathcal{C}_X(d) \subset \mathbb{P}(H^0(\mathbb{P}_s^N, \mathcal{O}(d)))^{n+1}$ is a closed subset with fiber $\mathcal{C}_{X_b}(d)$ over $b \in B$, hence $\mathcal{U}_X(d)$ has fiber $\mathcal{U}_{X_b}(d)$ over $b \in B$.

Let $F_d : \mathcal{U}_X(d) \times_B X \to \mathcal{U}_X(d) \times_B \mathbb{P}^n$ be the $\mathcal{U}_X(d)$-morphism parametrized by $\mathcal{U}_X(d)$, i.e., $F_d((f_0 : \ldots : f_n), x) := ((f_0 : \ldots : f_n), (f_0(x) : \ldots : f_n(x)))$. We have the ramification locus $\text{ram}_X(F)_\mathcal{U} \subset \mathcal{U}_X(d) \times_B X$.

For a sequence $d_* := (d_0, \ldots, d_n)$, let

$$\mathcal{U}_X(d_*) := \mathcal{U}_X(d_0) \times_B \ldots \times_B \mathcal{U}_X(d_n)$$

let $p_i : \mathcal{U}_X(d_*) \to \mathcal{U}_X(d_i)$ be the projection, and let $\mathcal{F}_X(d_*) \subset \mathcal{U}_X(d_*) \times_B X$ be the intersection $\cap_{i=0}^n (p_i \times \text{id}_X)^{-1}(\text{ram}_X(F_{d_i})_\mathcal{U})$. Finally, let $q : \mathcal{U}_X(d_*) \times_B X \to \mathcal{U}_X(d_*)$

be the projection and set

$$\mathcal{R}_X(d_*) := \mathcal{U}_X(d_*) \setminus q(\mathcal{F}_X(d_*)).$$

Clearly $\mathcal{R}_X(d_*)$ is an open subscheme of $\mathcal{U}_X(d_*)$ with fiber $\mathcal{R}_{X_b}(d_*)$ over $b \in B$.

Each $F \in \mathcal{U}_X(d_*)(B)$ thus determines a finite B-morphism

$$F_X : X \to \mathbb{P}^n_B$$

and gives rise to the map

$$R_F : z^B(X, r)^{\text{eff}} \to z^B(X, r)^{\text{eff}}$$

with

$$F_X^*(F_{X*})(Z) = Z + R_F(Z)$$

for each $Z \in z^B(X, r)^{\text{eff}}$.

Lemma 7.6 *Fix integers e, r, s and a sequence of integers $d_0, \ldots, d_n$, with $d_i \geq 2$, $0 \leq r, s \leq n$, $r + s \geq n$ and $e \geq 1$. For each $b \in B$, we have the closed subset $\mathcal{B}_{X_b}(d_*, e, r, s)$ of $\mathcal{R}_{X_b}(d_*) \subset \mathcal{R}_X(d_*)$ given by proposition 7.3; we assume following remark 7.4 that $\mathcal{B}_{X_b}(d_*, e, r, s)$ is minimal for each b. Let*

$$\mathcal{B}_X(d_*, e, r, s) := \cup_{b \in B} \mathcal{B}_{X_b}(d_*, e, r, s).$$

Then $\mathcal{B}_X(d_*, e, r, s)$ is closed in $\mathcal{R}_X(d_*)$.

Proof Let $b \to b'$ be a specialization of points of B, x a point of $\mathcal{B}_{X_b}(d_*, e, r, s)$, and $x \to x'$ an extension of $b \to b'$ to a specialization of x to a point of $\mathcal{R}_{X_{b'}}(d_*)$. Using the properness of Chow varieties, it follows from the description of $\mathcal{B}_X(d_*, e, r, s)$ in remark 7.4 that x' is in $\mathcal{B}_{X_{b'}}(d_*, e, r, s)$, proving the result. $\square$

We specialize to the case of a semi-local B, still regular and essentially of finite type over k. Recall the function $G_{r,s,N,e,X} : \mathbb{N} \to \mathbb{N}$ from proposition 7.3, defined for $X \in \mathbf{Proj}/k$, with a given embedding $X \hookrightarrow \mathbb{P}^N_k$, and integers r, s, e. Given $X \in \mathbf{Proj}/B$ with a given embedding over B, $X \hookrightarrow \mathbb{P}^N_B$, let $b_1, \ldots, b_m$ be the closed points of B and define

$$G_{r,s,N,e,X}(m) := \max_i \{G_{r,s,N,e,X_{b_i}}(m)\}.$$

Proposition 7.7 *Let B be a semi-local regular k-scheme, with k an infinite field. Take X in $\mathbf{Proj}/B$ of relative dimension n over B, with a fixed closed immersion $X \hookrightarrow \mathbb{P}^N_B$ over B. Fix dimensions $0 \leq r, s \leq n$ and a degree $e \geq 1$ with $r + s \geq n$ and let $G := G_{r,s,N,e,X}$. Then for all tuple of integers $d_0, \ldots, d_n$ with $d_0 \geq G(e)$, $d_i \geq G(d_{i-1}^n \cdot e)$ for $i = 1, \ldots, n$, there is a point $(F^0, \ldots, F^n) \in \mathcal{R}_X(d_*)(B) \setminus \mathcal{B}_X(d_*, e, r, s)(B)$.*

Proof We note that the fiber of $\mathcal{R}_X(d_*)$ over $b \in B$ is $\mathcal{R}_{X_b}(d_*)$ and similarly for $\mathcal{B}_X(d_*, e, r, s)$. Then $\mathcal{R}_X(d_*) \setminus \mathcal{B}_X(d_*, e, r, s) \to B$ is an open subscheme of the product of projective spaces $\prod_{i=0}^n \mathrm{Proj}_B(p_* \mathcal{O}_{\mathbb{P}^N}(d_i))$ over B; by proposition 7.3, the fiber $\mathcal{R}_X(d_*) \setminus \mathcal{B}_X(d_*, e, r, s)$ is non-empty for each closed point b of B. As k is assumed infinite, B has infinite residue fields over each of its closed points. Thus, for each closed point b of B, there is a $k(b)$-point F_b^* in $\mathcal{R}_X(d_*)_b \setminus \mathcal{B}_X(d_*, e, r, s)_b$. Since B is semi-local, there is a B-point F^* of $\prod_{i=0}^n \mathrm{Proj}_B(p_* \mathcal{O}_{\mathbb{P}^N}(d_i))$ with $F^*(b) = F_b^*$ for each closed point b; F^* is automatically a B point of $\mathcal{R}_X(d_*) \setminus \mathcal{B}_X(d_*, e, r, s)$, using again the fact that B is semi-local. $\square$

Remark 7.8 Take $(F^0, \ldots, F^n) \in \mathcal{R}_X(d_*)(B) \setminus \mathcal{B}_X(d_*, e, r, s)(B)$ as given by proposition 7.7. Then for each B-scheme $T \to B$, and each pair of cycles

$$Z \in z^B(X, r)^{\text{eff}}_{\le e}(T), \ W \in z^B(X, s)^{\text{eff}}_{\le e}(T),$$

the cycles W and $R_X(F^n) \circ \ldots \circ R_X(F^0)(Z)$ intersect properly on X_T and the intersection $W \cdot_{X_T} R_X(F^n) \circ \ldots \circ R_X(F^0)(Z)$ is in $z^B(X, r+s-n)^{\text{eff}}(T)$.

Indeed, this follows from the fact that the operation $R_X(F)$ is compatible with taking fibers, hence, for all $t \in T$, the cycles W_t and $R_{X_t}(F_t^n) \circ \ldots \circ R_{X_t}(F_t^0)(Z_t)$ intersect properly on X_t. Thus W and $R_X(F^n) \circ \ldots \circ R_X(F^0)(Z)$ intersect properly on X_T and the intersection $W \cdot_{X_T} R_X(F^n) \circ \ldots \circ R_X(F^0)(Z)$ is equi-dimensional (of dimension $r+s-n$) over T.

Similarly, if W and Z intersect properly on X_T and $W \cdot_{X_T} Z$ is in $z^B(X, r+s-n)^{\text{eff}}(T)$, then for each $j = 0, \ldots, n$, W and $Z_j := (F_X^{j*} F_{X*}^j) \circ (F_X^{0*} F_{X*}^0)(Z)$ intersect properly on X_T and $W \cdot Z_j$ is in $z^B(X, r+s-n)^{\text{eff}}(T)$.

We can now prove our extension of [11, theorem 6.3].

Theorem 7.9 *Let k be a field, B a regular k-scheme, essentially of finite type over k, $X \in \mathbf{Proj}/B$ of dimension n over B with a given embedding $X \hookrightarrow \mathbb{P}^N_B$. Let r, s, e, d be given integers with $e \ge 1$, $0 \le r, s \le n$, $r + s \ge n$. Then is a map of presheaves*

$$H_{X,U} : z^B_{equi}(X, r)(U) \to z^B_{equi}(X, r)(U \times \mathbb{A}^1); \ U \in \mathbf{Sm}/B,$$

such that

1. *Let $i_0 : U \to U \times \mathbb{A}^1$ be the zero-section. Then $i_0^* \circ H_{X,U} = \text{id}_{z^B_{equi}(X,r)(U)}$.*

2. *Let $i : T \to U \times \mathbb{A}^1 \setminus \{0\}$ be a morphism of B-schemes. Then for each $W \in z^B_{equi}(X, s)^{\text{eff}}_{\le e}(T)$, $Z \in z^B_{equi}(X, r)^{\text{eff}}_{\le e}(U)$, with $0 \le r, s \le n$, $r + s \ge n$, the cycles $i^*(H_{X,U}(Z))$ and W intersect properly on X_T and $i^*(H_{X,U}(Z)) \cdot_{X_T} W$ is in $z^B_{equi}(X, r+s-n)^{\text{eff}}(T)$.*

3. *Let $i : T \to U \times \mathbb{A}^1$ be a morphism of B-schemes and take*

$$W \in z^B_{equi}(X, s)^{\text{eff}}_{\le e}(T),$$
$$Z \in z^B_{equi}(X, r)^{\text{eff}}_{\le e}(U),$$

with $0 \le r, s \le n$, $r + s \ge n$. Suppose that $(p_U \circ i)^*(Z)$ and W intersect properly on X_T and $(p_U \circ i)^*(Z) \cdot_{X_T} W$ is in $z^B_{equi}(X, r+s-n)^{\text{eff}}(T)$. Then the cycles $i^*(H_{X,U}(Z))$ and W intersect properly on X_T and $i^*(H_{X,U}(Z)) \cdot_{X_T} W$ is in $z^B_{equi}(X, r+s-n)^{\text{eff}}(T)$.

Proof The proof follows that in [11] and [10]. If k is a finite field, we use the fact that k admits a infinite pro-l extension k_l for every prime l prime to the characteristic, plus the usual norm argument, to reduce to the case of an infinite field. We may assume that B is irreducible. Thus for each $b \in B$, the embedded subscheme $X_b \subset \mathbb{P}^N_b$ has degree d_X independent of b.

We denote the maps

$$H_U^{\pm} : z^B_{equi}(\mathbb{P}^n_B, r)^{\text{eff}}(U) \to z^B_{equi}(\mathbb{P}^n_B, r)^{\text{eff}}(U \times \mathbb{A}^1)$$

given by corollary 7.5 by $H_U^{\pm}(d)$, noting explicitly the dependence on the degree d. Similarly, we write $m(d)$ for the integer m that appears in corollary 7.5(1).

Choose some $F^* \in \mathcal{R}_X(d_*)(B) \setminus \mathcal{B}_X(d_*)(B)$, as given by proposition 7.7. Let $p_U : U \times \mathbb{A}^1 \to U$ be the projection. We have the maps of presheaves

$$F_{X*}^i : z_{equi}^B(X, r)_d^{\mathrm{eff}} \to z_{equi}^B(\mathbb{P}^n, r)_d^{\mathrm{eff}},$$

$$F_X^{i*} : z_{equi}^B(\mathbb{P}^n, r)_d^{\mathrm{eff}} \to z_{equi}^B(X, r)_{d_i^n ed}^{\mathrm{eff}}$$

and similarly without the degree restrictions. Also, for $W \in z_{equi}^B(X, s)^{\mathrm{eff}}(U)$, $Z \in z_{equi}^B(\mathbb{P}_B^n, r)^{\mathrm{eff}}(U)$, $W \cdot_{U \times_B X} F_X^{i*}(Z)$ is defined and is in $z_{equi}^B(X, s+r-n)^{\mathrm{eff}}(U)$ if and only if $F_{X*}^i(W) \cdot_{U \times_B \mathbb{P}_B^n} Z$ is defined and is in $z_{equi}^B(\mathbb{P}_B^n, s+r-n)^{\mathrm{eff}}(U)$. Finally, we note that, for $Z \in z_{equi}^B(X, r)_d^{\mathrm{eff}}(U)$, $R_{F^i}(Z)$ is in $z_{equi}^B(X, r)_{d(d_i^n e-1)}^{\mathrm{eff}}(U)$.

We now define a sequence of maps of presheaves on $\mathbf{Sm}/B$,

$$H_{X,U,j}^{\pm} : z_{equi}^B(X, r)^{\mathrm{eff}}(U) \to z_{equi}^B(X, r)^{\mathrm{eff}}(U \times \mathbb{A}^1); \; U \in \mathbf{Sm}/B.$$

Define the integers δ_j inductively by $\delta_0 = e$, and

$$\delta_j = (m(\delta_{j-1}) + 1) d_i^n e \delta_{j-1}$$

for $j \geq 1$. These numbers have the property that, if Z is in $z_{equi}^B(X, r)_{\leq \delta_j}^{\mathrm{eff}}(U)$, then $(m(\delta_j) + 1) \cdot R_{F^j}(Z)$ is in $z_{equi}^B(X, r)_{\leq \delta_{j+1}}^{\mathrm{eff}}(U)$. For $Z \in z_{equi}^B(X, r)^{\mathrm{eff}}(U)$, define

$$H_{X,U,0}^{\pm}(Z) := F_X^{0*} \circ H_U^{\pm}(\delta_0) \circ F_{X*}^0(Z).$$

For $j = 1, \ldots, n$, let

$$H_{X,U,j}^{\pm}(Z) := F_X^{j*} \circ H_U^{\pm}(\delta_j) \circ F_{X*}^j(R_{F^{j-1}} \circ \ldots \circ R_{F^0}(Z)).$$

Finally, we let

$$H_{X,U,n+1}^+(Z) = p_U^*(R_{F^n} \circ \ldots \circ R_{F^0}(Z))$$

and $H_{X,U,n+1}^-(Z) = 0$. Set

$$H_{X,U}(Z) := \sum_{j=0}^{n+1} (-1)^j (H_{X,U,j}^+(Z) - H_{X,U,j}^-(Z)).$$

Thus, $U \mapsto H_{X,U}$ defines a map of presheaves on $\mathbf{Sm}/B$

$$H_X : z_{equi}^B(X, r) \to z_{equi}^B(X, r)((-) \times \mathbb{A}^1).$$

We have

$$i_0^*(H_{X,U,j}^+(Z) - H_{X,U,j}^-(Z)) = \begin{cases} F_X^{*0} \circ F_{X*}^0(Z) & \text{for } j = 0 \\ F_X^{j*} \circ F_{X*}^j(R_{F^{j-1}} \circ \ldots \circ R_{F^0}(Z)) & \text{for } j = 1, \ldots, n \\ R_{F^n} \circ \ldots \circ R_{F^0}(Z) & \text{for } j = n + 1. \end{cases}$$

Since $F_X^{j*} \circ F_{X*}^j = \mathrm{id} + R_{F^j}$, and $i_0^* \circ (H_U^+ - H_U^-) = \mathrm{id}$, this implies that

$$i_0^* \circ H_{X,U}(Z) = Z$$

for all $Z \in z_{equi}^B(X, r)(U)$. This verifies the property (1).

For (2), let $i : T \to U \times \mathbb{A}^1 \setminus \{0\}$ be a morphism of B-schemes, take $W \in z_{\leq e}^B(X, s)_{\leq e}^{\mathrm{eff}}(T)$ and $Z \in z_{equi}^k(X, r)_{\leq e}^{\mathrm{eff}}(U)$, with $0 \leq r, s \leq n$, $r + s \geq n$. Take $j = 0, \ldots, n$. By corollary 7.5 and our remarks above, the cycles $i^*(H_{X,U,j}^{\pm}(Z))$ and W intersect properly on X_T and $i^*(H_{X,U,j}^{\pm}(Z)) \cdot_{X_T} W$ is in $z_{equi}^B(X, r+s-n)^{\mathrm{eff}}(T)$. By remark 7.8, the same holds for $j = n + 1$. This proves (2).

For (3), under the given assumptions for Z and W, it follows from remark 7.8 that W and

$$Z_j := i^* \circ p_U^* (F_X^{j*} F_{X*}^j) \circ \ldots \circ (F_X^{0*} F_{X*}^0)(Z)$$

intersect properly on X_T. Since $H_{X,U,j}^{\pm}(Z)$ is effective and $i_0^* \circ H_{X,U,j}^{\pm}(Z)$ plus some effective cycle is a multiple of $(F_X^{j*} F_{X*}^j) \circ \ldots \circ (F_X^{0*} F_{X*}^0)(Z)$, (3) follows from (2). $\square$

Fix a field extension k' of k, and let $Y \subset \mathbb{P}_{k'}^N$ be a closed subset. Let $\mathcal{C}_Y(s,e)$ be the Chow scheme parametrizing effective cycles of dimension s and degree $\leq e$ on Y. For $L \supset k'$ an extension field and W an effective dimension s cycle on $\mathbb{P}_L^N$ of degree $\leq e$, and with support in Y_L, we let $\mathrm{chow}(W) \in \mathcal{C}(s,e)(\bar{L})$ denote the Chow point of W.

For $Y \subset \mathbb{P}_B^N$, we have the relative Chow scheme $\mathcal{C}_{Y/B}(s,e) \to B$, with (reduced) fiber over $b \in B$ the Chow scheme of $Y_b \subset \mathbb{P}_b^N$.

Definition 7.10 For a fixed regular noetherian base-scheme B, take $Y \in$ **Proj**$/B$ and fix an embedding $Y \hookrightarrow \mathbb{P}_B^N$ over B. Let $\mathcal{C} \subset \mathcal{C}_{Y/B}(s,e)$ be any collection of locally closed subsets of $\mathcal{C}_{Y/B}(s,e)$.

Let

$$z_{equi}^B(Y,r)_{\mathcal{C}}(X) \subset z_{equi}^B(Y,r)(X)$$

be the subgroup generated by integral closed subschemes $Z \subset X \times_B Y$ such that,

1. Z is in $z_{equi}^B(Y,r)(X)$.
2. Take $W \in z_{equi}^B(Y,s)_{\leq e}^{eff}(X)$ and suppose that, for each $x \in X$, the cycle $i_x^*(W)$ on $Y_{k(x)}$ has Chow point $\mathrm{chow}(W)$ in $\mathcal{C}(k(\bar{x}))$. Then the intersection $W \cdot_{X \times_B Y} Z$ is defined and is in $z_{equi}^B(Y, r+s-n)(X)$.

This defines the subpresheaf $z_{equi}^B(Y,r)_{\mathcal{C}}$ of $z_{equi}^B(Y,r)$. The subpresheaves

$$z_{equi}^B(Y,r)_{\mathcal{C},\leq e}^{eff} \subset z_{equi}^B(Y,r)_{\leq e}^{eff}; \ z_{equi}^B(Y,r)_{\mathcal{C},\leq e} \subset z_{equi}^B(Y,r)_{\leq e},$$

etc., are defined similarly.

We let $C^B(Y,r)_{\mathcal{C}}(X) \subset C^B(Y,r)(X)$ be the subcomplex associated to the cubical object

$$n \mapsto z_{equi}^B(Y,r)_{\mathcal{C}}(X \times \square^n).$$

This gives us the presheaf of subcomplexes $C^B(Y,r)_{\mathcal{C}} \subset C^B(Y,r)$. The subcomplex $C^B(Y,r)_{\mathcal{C},\leq e} \subset C^B(Y,r)_{\leq e}$ (see definition 5.2) is defined similarly.

Theorem 7.11 *Let B be a semi-local regular scheme, essentially of finite type over some field k. Take $X \in$ **Proj**$/B$ of relative dimension n over B, and integers r,s,e with $e \geq 1$, $0 \leq r, s \leq n$, $r+s \geq n$. Fix an embedding $X \hookrightarrow \mathbb{P}_B^N$. Let $\mathcal{C} \subset \mathcal{C}_X(s,e)$ be a collection of locally closed subsets. Then the inclusion $C^B(X,r)_{\mathcal{C}} \subset C^B(X,r)$ is a quasi-isomorphism.*

Proof Since

$$C^B(X,r)_{\mathcal{C}} = \cup_{e \geq 1} C^B(X,r)_{\mathcal{C},\leq e}; \ C^B(X,r) = \cup_{e \geq 1} C^B(X,r)_{\leq e},$$

it suffices to show that the map

$$\iota : \frac{C^B(X,r)_{\leq e}}{C^B(X,r)_{\mathcal{C},\leq e}} \to \frac{C^B(X,r)}{C^B(X,r)_{\mathcal{C}}}$$

induced by the inclusions $C^B(X,r)_{\mathcal{C},\leq e} \subset C^B(X,r)_{\mathcal{C}}$ and $C^B(X,r)_{\leq e} \subset C^B(X,r)$ gives the zero-map on homology.

Let

$$H_X : z^B_{equi}(X,r) \to z^B_{equi}(X,r)((-) \times \mathbb{A}^1)$$

be the map given by theorem 7.9 for the given values of r, s, e. By theorem 7.9(3), H_X restricts to a map

$$H_X : z^B_{equi}(X,r)_{\mathcal{C}} \to z^B_{equi}(X,r)((-) \times \mathbb{A}^1)_{\mathcal{C}}.$$

By theorem 7.9(2), $i_1^* \circ H_X$ defines a map

$$G_X : z^B_{equi}(X,r) \to z^B_{equi}(X,r)_{\mathcal{C}}$$

and by theorem 7.9(1), $i_0^* \circ H_X = \mathrm{id}$.

Identifying $U \times \mathbb{A}^1 \times_B X \times \square^n$ with $U \times_B X \times \square^{n+1}$ via the exchange of factors

$$U \times \mathbb{A}^1 \times_B X \times \square^n \to U \times_B X \times \square^n \times \mathbb{A}^1 = U \times_B X \times \square^{n+1},$$

$(-1)^n H_X$ gives us maps

$$h_X^{-n} : C^B(X,r)^{-n}_{\le e} \to C^B(X,r)^{-n-1}, \quad h_{X\mathcal{C}}^{-n} : C^B(X,r)^{-n}_{\le e,\mathcal{C}} \to C^B(X,r)^{-n-1}_{\mathcal{C}}$$

and thus induces a degree -1 map

$$\bar{h}_X : \frac{C^B(X,r)_{\le e}}{C^B(X,r)_{\le e,\mathcal{C}}} \to \frac{C^B(X,r)}{C^B(X,r)_{\mathcal{C}}}$$

which gives a homotopy between the map ι and the zero-map, completing the proof. $\qquad\square$

We use theorem 7.11 to prove theorem 5.4:

Corollary 7.12 *Let B be a semi-local regular scheme, essentially of finite type over some field k. Take $X, Y \in \mathbf{Proj}/B$, $U \in \mathbf{Sm}/B$ and let $p = dim_B X$. Then for $0 \le r \le dim_B Y$, the map*

$$\int_X : C^B(Y,r)(U \times_B X) \to C^B(X \times_B Y, r+p)(U)$$

is a quasi-isomorphism.

Proof Let $s = dim_B Y$. Fix embeddings of Y, X into some $\mathbb{P}^N_B$, with Y of degree e. The Segre embedding $\mathbb{P}^N \times \mathbb{P}^N \to \mathbb{P}^M$ gives us an embedding of $X \times_B Y$ in $\mathbb{P}^M_B$ such that $x \times Y$ has degree e for each $x \in X$. Take $\mathcal{C} \subset \mathcal{C}_{X \times_B Y/B}(s, e)$ to be the family of cycles $x \times Y$, $x \in X$. Note that the map $\int_X$ identifies $z_{equi}(Y,r)(U \times_B X)$ with $z_{equi}(X \times_B Y, r+p)_{\mathcal{C}}(U)$, and thus gives an isomorphism

$$\int_X : C^B(Y,r)(U \times_B X) \to C^B(X \times_B Y, r+p)_{\mathcal{C}}(U).$$

Since $r + p + s \ge p + s = dim_B X \times_B Y$, we may apply theorem 7.11 to conclude that the inclusion

$$C^B(X \times_B Y, r+p)_{\mathcal{C}}(U) \subset C^B(X \times_B Y, r+p)(U)$$

is a quasi-isomorphism, completing the proof. $\qquad\square$

To complete this section, we prove the projective bundle formula.

Proof of theorem 5.5 We proceed by induction on n. Let $\pi : Y \times \mathbb{P}^n \to Y$ be the projection. By theorem 7.11, the inclusion

$$C^B(Y \times \mathbb{P}^n, r)_{\{Y \times \mathbb{P}^{n-1}\}} \subset C^B(Y \times \mathbb{P}^n, r)$$

is a quasi-isomorphism for $r \geq 1$ Here $\mathbb{P}^{n-1} \subset \mathbb{P}^n$ is the hyperplane $X_n = 0$. We have the intersection map

$$i^*_{\mathbb{P}^{n-1}} : z_{equi}(Y \times \mathbb{P}^n, r)_{\{Y \times \mathbb{P}^{n-1}\}} \to z_{equi}(Y \times \mathbb{P}^{n-1}, r - 1)$$

and the cone-map

$$C_{p_0}(-) : z_{equi}(Y \times \mathbb{P}^{n-1}, r - 1) \to z_{equi}(Y \times \mathbb{P}^n, r)_{\{Y \times \mathbb{P}^{n-1}\}}$$

where $p_0 := (0 : \ldots, 0 : 1)$. Let $\iota_0 : \operatorname{Spec} k \to \mathbb{P}^n$ be the inclusion of the point p_0 and let $\pi : \mathbb{P}^n \to \operatorname{Spec} k$ be the projection. Let

$$\mu : \mathbb{P}^n \times (\mathbb{A}^1 \setminus \{0\}) \to \mathbb{P}^n$$

be the multiplication map

$$\mu((x_0 :, \ldots : x_n), t) := (x_0 : \ldots : x_{n-1} : t x_n).$$

We have as well the natural transformation

$$H_U : z_{equi}(Y \times \mathbb{P}^n, r)_{\{Y \times \mathbb{P}^{n-1}\}}(U) \to z_{equi}(Y \times \mathbb{P}^n, r)_{\{Y \times \mathbb{P}^{n-1}\}}(U \times \mathbb{A}^1)$$

which sends a cycle $Z \in z_{equi}(Y \times \mathbb{P}^n, r)_{\{Y \times \mathbb{P}^{n-1}\}}(U)$ to the closure of $\mu^*(Z)$. One checks that this is well-defined and satisfies

$$i^*_1 \circ H_U = \operatorname{id}_{z_{equi}(Y \times \mathbb{P}^n, r)_{\{Y \times \mathbb{P}^{n-1}\}}(U)}; \ i^*_0 \circ H_U = C_{p_0} \circ i^*_{\mathbb{P}^{n-1}} + \alpha_0 \circ \pi_*,$$

where $\alpha_0 := \iota_{0*}$. As in the proof of theorem 7.11, the maps

$$h_U := (-1)^n H_U : C^B_n(Y \times \mathbb{P}^n, r)_{\{Y \times \mathbb{P}^{n-1}\}}(U) \to C^B_{n+1}(Y \times \mathbb{P}^n, r)_{\{Y \times \mathbb{P}^{n-1}\}}(U)$$

define a homotopy between the identity and $C_{p_0} \circ i^*_{\mathbb{P}^{n-1}} + \alpha_0 \circ \pi_*$.

On the other hand, since $p_0 \cap \mathbb{P}^{n-1} = \emptyset$, $i^*_{\mathbb{P}^{n-1}}$ is the zero map on the image of α_{0*}. Also, since $\pi(C_{p_0}(Z))$ has dimension $< \dim_B Z$ for any equi-dimensional closed subset Z of $Y \times_B \mathbb{P}^{n-1}$, π_* is zero on the image of C_{p_0}. Finally,

$$\pi_* \circ \alpha_0 = \operatorname{id}; \ i^*_{\mathbb{P}^{n-1}} \circ C_{p_0} = \operatorname{id}$$

hence, for $r \geq n \geq 1$,

$$\alpha_0 + C_{p_0} : C^B(Y, r) \oplus C^B(Y \times \mathbb{P}^{n-1}, r - 1) \to C^B(Y \times \mathbb{P}^n, r)_{\{Y \times \mathbb{P}^{n-1}\}}$$

is a homotopy equivalence. By induction

$$\sum_{j=0}^{n-1} \alpha_j : C^B(Y, r - j - 1) \to C^B(Y \times \mathbb{P}^{n-1}, r - 1)$$

is a quasi-isomorphism; since $C_{p_0} \circ \alpha_j = \alpha_{j+1}$ (where we use p_0 and the subspaces $C_{p_0}(\mathbb{P}^j \subset \mathbb{P}^{n-1})$ for the flag of linear subspaces of $\mathbb{P}^n_k$ needed to define the maps α_j), the induction goes through.   □

References

[1] Balmer, P. and Schlichting, M., *Idempotent completion of triangulated categories.* J. Algebra **236** (2001), no. 2, 819–834.

[2] Brown, K. S. and Gersten, S. M., *Algebraic K-theory as generalized sheaf cohomology.* Algebraic K-theory, I: Higher K-theories (Proc. Conf., Battelle Memorial Inst., Seattle, Wash., 1972), pp. 266–292. Lecture Notes in Math., **341**, Springer, Berlin, 1973.

[3] Bloch, S., *The moving lemma for higher Chow groups.* J. Algebraic Geom. **3** (1994), no. 3, 537–568.

[4] Bondarko, M.V., *Differential graded motives: weight complex, weight filtrations and spectral sequences for realizations; Voevodsky versus Hanamura.* J. Inst. Math. Jussieu **8** (2009), no. 1, 39–97.

[5] Cisinski, D.-C. and Déglise, F., *Triangulated categories of motives*, preprint 2007.

[6] Deligne, P. and Goncharov, A. *Groupes fondamentaux motiviques de Tate mixtes.* Ann. Sci. Éc. Norm. Sup. (4) **38** (2005), no 1, 1–56.

[7] Déglise, F., *Finite correspondences and transfers over a regular base.* Nagel, Jan (ed.) et al., **Algebraic cycles and motives.** Volume 1. Selected papers of the EAGER conference, Leiden, Netherlands, August 30–September 3, 2004 on the occasion of the 75th birthday of Professor J. P. Murre. Cambridge: Cambridge University Press. London Mathematical Society Lecture Note Series **343**, 138-205 (2007).

[8] Drinfeld, V., *DG quotients of DG categories.* Journal of Algebra **272** (2004) 643–691.

[9] Esnault, H. and Levine, M., *Tate motives and the fundamental group*, preprint (2007) http://arxiv.org/abs/0708.4034

[10] Friedlander, E. M. and Lawson, H. B., *Moving algebraic cycles of bounded degree.* Invent. Math. **132** (1998), no. 1, 91–119.

[11] Friedlander, E. M. and Voevodsky, V., *Bivariant cycle cohomology.* In: **Cycles, transfers, and motivic homology theories**, 138–187, Ann. of Math. Stud., **143**, Princeton Univ. Press, Princeton, NJ, 2000.

[12] Grothendieck, A., *Sur quelques points d'algèbre homologique.* Tôhoku Math. J. (2) **9** 1957 119–221.

[13] Hinich, V. A. and Schechtman, V. V., *On homotopy limit of homotopy algebras.* K-theory, arithmetic and geometry (Moscow, 1984–1986), 240–264, Lecture Notes in Math., Vol. 1289, Springer, Berlin-New York, 1987.

[14] Hovey, M., *Spectra and symmetric spectra in general model categories.* J. Pure Appl. Algebra **165** (2001), no. 1, 63–127.

[15] Kapranov, M. M., *On the derived categories of coherent sheaves on some homogeneous spaces.* Invent. Math. **92** (1988), no. 3, 479–508.

[16] Keller, B., *Deriving DG categories.* Ann.Sci. Éc. Norm. Sup., 4e série, **27** (1994) 63–102.

[17] Keller, B., *On differential graded categories.* International Congress of Mathematicians. Vol. II, 151–190, Eur. Math. Soc., Zrich, 2006.

[18] Krishna, A. and Levine, M., *Additive Chow groups of schemes*, to appear, J. f.d Reine u. Ang. Math., http://arxiv.org/abs/0702138

[19] Levine, M., *Smooth motives*, preprint April 2008. arXiv:0807.2265

[20] Levine, M., **Mixed Motives**. Math. Surveys and Monographs **57**, AMS, Prov. 1998.

[21] Thomason, R. W., *Algebraic K-theory and étale cohomology.* Ann. Sci. École Norm. Sup. (4) **18** (1985), no. 3, 437–552.

[22] Toën, B., *The homotopy theory of dg-categories and derived Morita theory.* Invent. Math. **167** (2007), 615–667.

[23] Voevodsky, V., *Motivic cohomology groups are isomorphic to higher Chow groups in any characteristic.* Int. Math. Res. Not. 2002, no. 7, 351–355.

[24] Voevodsky, V., Suslin, A. and Friedlander, E. M., **Cycles, transfers, and motivic homology theories**. Annals of Mathematics Studies, **143**. Princeton University Press, Princeton, NJ, 2000.

Fields Institute Communications
Volume **56**, 2009

Cycles on Varieties over Subfields of $\mathbb{C}$ and Cubic Equivalence

James D. Lewis

Department of Mathematical and Statistical Sciences
University of Alberta
Edmonton, Alberta T6G 2G1, Canada
lewisjd@ualberta.ca

To Spencer, on the occasion of his 63rd birthday.

Abstract. Let S be a smooth projective surface defined over a number field k, with positive (geometric) genus. Generalizing the work of Schoen ([Sch2]), in [G-G-P] it is shown that there exists a subfield $K \subset \mathbb{C}$ with $K \supset k$ of transcendence degree one over k, and a nontorsion class $\xi \in F^2\mathrm{CH}^2(S_K) := \ker\big(\mathrm{CH}^2_{\hom}(S_K) \to \mathrm{Alb}(S_{\mathbb{C}})\big)$, where $\mathrm{Alb}(S_{\mathbb{C}})$ is the Albanese variety of $S_{\mathbb{C}}$. The proof of Griffiths-Green-Paranjape (op. cit.) relies partly on the work of Terosoma ([T]). In this paper we take up a certain case, where we use Hodge theory instead of [T] to arrive at related infinite rank results.

1 Introduction

Let $X/\mathbb{C}$ be a projective algebraic manifold. In [L1] (and others, e.g., Griffiths/Green, J. P. Murre, M. Saito/M. Asakura, S. Saito, W. Raskind, Uwe Jannsen, et al.), there is constructed a candidate Bloch-Beilinson filtration $F^\nu\mathrm{CH}^r(X;\mathbb{Q})$, $\nu = 0, ..., r$, on the Chow group $\mathrm{CH}^r(X;\mathbb{Q}) := \mathrm{CH}^r(X) \otimes \mathbb{Q}$, of codimension r algebraic cycles, modulo rational equivalence. The defining features of such a filtration are explained in §2(v), as well as a cursory construction, mentioned in §3. If $K \subset \mathbb{C}$ is a subfield, and if $X := X_{\mathbb{C}} = X_K \times \mathbb{C}$, then from the inclusion $\mathrm{CH}^r(X_K;\mathbb{Q}) \subset \mathrm{CH}^r(X_{\mathbb{C}};\mathbb{Q})$, one has an induced filtration on $\mathrm{CH}^r(X_K;\mathbb{Q})$. In [Sch2], there is constructed an example of a surface X defined over a field K of transcendence degree one over $\mathbb{Q}$, for which $F^2\mathrm{CH}^2(X_K;\mathbb{Q})$ has infinite rank. In light of the recent developments in [G-G-P] and [MSa], as well as recent works of

2000 *Mathematics Subject Classification.* Primary 14C25; Secondary 14C30.

Key words and phrases. Abel-Jacobi map, Chow group, mixed Hodge structure, Bloch-Beilinson filtration.

Partially supported by a grant from the Natural Sciences and Engineering Research Council of Canada.

[As], [R-S] and [K], the motivation for this paper is an expansion of a theorem by the author, mentioned in the final paragraphs in §7 of [K], and refined in the form of Theorem 1.2 below. Here the method of proof really only involves a fine tuning of the arguments presented in §7 of [L-S], where the emphasis there was placed on constructing an uncountable number of cycles with trivial Mumford-Griffiths invariant for a certain class of smooth projective varieties over $\mathbb{C}$. Inspired by some of the ideas of H. Saito in [HSa] (also see [L3]), we arrive at a "cubic equivalence" analog of [MSa] (Theorem 0.3) in Corollary 1.4 below, that in this case requires only Hodge theory.

From this point on, let us assume given smooth projective varieties $S/\overline{\mathbb{Q}}$, $C/\overline{\mathbb{Q}}$, $X/\overline{\mathbb{Q}}$, with $d = \dim X$, $d_S = \dim S$ and $\dim C = 1$. Consider $w \in \mathrm{CH}^r\big((C \times S \times X)/\overline{\mathbb{Q}}\big)$. One has a composite map

$$P(C,S,w) : H^1(C,\mathbb{Q}) \otimes H^{2d_S - \nu + 1}(S,\mathbb{Q}) \xrightarrow{[w]_*} H^{2r-\nu}(X,\mathbb{Q}) \to H^{r-\nu,r}(X),$$

where singular cohomology is defined on the associated complex space (see §2 (iv)). We will further assume that $[w] \in H^1(C,\mathbb{Q}) \otimes H^{\nu-1}(S,\mathbb{Q}) \otimes H^{2r-\nu}(X,\mathbb{Q})$. For example, if we assume the standard conjecture regarding the algebraicity of Künneth components of a given diagonal class, then such a situation can be arranged. More specifically, if for smooth projective W of dimension d_W, we denote by $[\Delta_W(2d_W - i, i)] \in H^{2d_W - i}(W,\mathbb{Q}) \otimes H^i(W,\mathbb{Q})$, the appropriate Künneth component of the diagonal class, and assume $\Delta_W(2d_W - i, i)$ is algebraic, then for any algebraic cycle $[w] \in H^{r,r}(C \times S \times X, \mathbb{Q})$, $[\delta \circ w] \in H^1(C,\mathbb{Q}) \otimes H^{\nu-1}(S,\mathbb{Q}) \otimes H^{2r-\nu}(X,\mathbb{Q})$, where

$$\delta = \mathrm{Pr}_{14}^*(\Delta_C(1,1)) \bigcap \mathrm{Pr}_{25}^*(\Delta_S(2d_S - \nu + 1, \nu - 1)) \bigcap \mathrm{Pr}_{36}^*(\Delta_X(2d - 2r + \nu, 2r - \nu))$$

$$\in \mathrm{CH}^{1+d_S+d}\big((C \times S \times X) \times (C \times S \times X); \mathbb{Q}\big).$$

Further, $P(C,S,w) = P(C,S,\delta \circ w)$.

Definition 1.1 Let $H_{\overline{\mathbb{Q}}}^{r-\nu,r}(X/\overline{\mathbb{Q}}) = \mathbb{C}$-vector space generated by the image of $P(C,S,w)$ over all such C, S, w.

Note that $H_{\overline{\mathbb{Q}}}^{r,r}(X/\overline{\mathbb{Q}}) = 0$. Hence we can assume that $\nu \geq 1$.

Theorem 1.2 If $H_{\overline{\mathbb{Q}}}^{r-\nu,r}(X/\overline{\mathbb{Q}}) \neq 0$, for some $\nu \geq 1$, then there exists a subfield $K \subset \mathbb{C}$ of transcendence degree $\nu - 1$ over $\mathbb{Q}$ such that $Gr_F^\nu \mathrm{CH}^r(X_K; \mathbb{Q})$ has infinite rank.

In §5 we construct a class of varieties satisfying the hypothesis of the above theorem. As a consequence, we arrive at:

Corollary 1.3 Assume given smooth projective curves $C_1, ..., C_d$ over $\overline{\mathbb{Q}}$ with $X/\overline{\mathbb{Q}}$ given above of dimension d, and a dominating rational map (over $\overline{\mathbb{Q}}$) $Z : C_1 \times \cdots \times C_d \to X$. Let $r \geq 2$, $I := \{i_1 < \cdots < i_r\} \subset \{1, ..., d\}$, and let $\mathrm{Pr}_I : C_1 \times \cdots \times C_d \to C_{i_1} \times \cdots \times C_{i_r}$ be the projection. Fix an embedding $\overline{\mathbb{Q}}(C_{i_2} \times \cdots \times C_{i_r}) \overset{\sim}{\hookrightarrow} K \subset \mathbb{C}$, and let $(\xi_2, ..., \xi_r) \in C_{i_2}(\mathbb{C}) \times \cdots \times C_{i_r}(\mathbb{C})$ be the point corresponding to this embedding. Fix $e_j \in C_{i_j}(\overline{\mathbb{Q}})$, $j = 1, ..., r$. Finally assume that $H^{r,0}(X) \neq 0$. Then for some $I \subset \{1, ..., d\}$ with $|I| = r$,

$$\big\{ \big(Z \circ {}^\mathrm{T}(\mathrm{graph}(\mathrm{Pr}_I))\big)_* \big((p - e_1) \times (\xi_2 - e_2) \times \cdots \times (\xi_r - e_r)\big) \mid p \in C_{i_1}(\overline{\mathbb{Q}}) \big\},$$

generates an infinite rank subspace of $Gr_F^r \mathrm{CH}^r(X_K; \mathbb{Q})$.

It is important to mention here that similar results for zero-cycles have been obtained by [As], [K] and [R-S]. For example, the reader should compare Corollary 1.3 to similar (and possibly more specific) results on zero-cycles, such as Theorem 3.5 and §6 in [K], together with the references cited there.

Next, in §7, we apply the basic ideas behind the proofs of Theorem 1.2 and Corollary 1.3 to the situation involving a descending filtration $F_c^\nu \mathrm{CH}^r(X)$ derived from cubic equivalence (see [HSa]), to a certain subspace $\underline{Gr}_{F_c}^\nu \mathrm{CH}^r(X;\mathbb{Q}) \subset Gr_F^\nu \mathrm{CH}^r(X;\mathbb{Q})$, and prove:

Corollary 1.4 *Let us assume the general Hodge conjecture (GHC). If*

$$\underline{Gr}_{F_c}^\nu \mathrm{CH}^r(X/\mathbb{C};\mathbb{Q}) \neq 0$$

for some $\nu \geq 1$, then there exists a subfield $K \subset \mathbb{C}$ of transcendence degree $\nu - 1$ over $\mathbb{Q}$ such that $\underline{Gr}_{F_c}^\nu \mathrm{CH}^r(X_K;\mathbb{Q})$ has infinite rank.

We also give an application of [G-G-P]/[MSa] to a certain class of hypersurfaces, viz.,

Proposition 1.5 *Let $X \subset \mathbb{P}^{d+1}/\overline{\mathbb{Q}}$ be a smooth hypersurface of degree $m \geq 3$. Put*

$$k = \left[\frac{d+1}{m}\right], \quad r = d - k, \quad \nu = 2r - d = d - 2k,$$

and assume $\nu \geq 1$ and the numerical condition:

$$k(d + 2 - k) + 1 - \binom{m+k}{k} \geq 0.$$

Then $\exists$ a field K of transcendence degree $\nu - 1$ over $\mathbb{Q}$ for which $Gr_F^\nu \mathrm{CH}^r(X_K;\mathbb{Q}) \neq 0$.

Remarks 1.6

(i) In Corollary 1.4, we do not need the GHC assumption in the case that X is a surface.

(ii) Suppose that $W/\overline{\mathbb{Q}}$ is a variety, and $w \in \mathrm{CH}^r(W_{\mathbb{C}})$, where $W_{\mathbb{C}} = W \times_{\overline{\mathbb{Q}}} \mathbb{C}$. Then one can spread w to a cycle $\tilde{w} \in \mathrm{CH}^r((V \times W)/\overline{\mathbb{Q}})$, where $V/\overline{\mathbb{Q}}$ is a variety. More precisely, if η is the generic point of V, then for a suitable embedding $\overline{\mathbb{Q}}(\eta) \subset \mathbb{C}$, we have $w = \tilde{w}_\eta \in \mathrm{CH}^r(W_\eta \times \mathbb{C}) = \mathrm{CH}^r(W_{\mathbb{C}})$. Let $t \in V(\mathbb{C})$ correspond to the embedding $\overline{\mathbb{Q}}(\eta) \subset \mathbb{C}$, and $q \in V(\overline{\mathbb{Q}})$ a closed point. (Note that q exists by the Nullstellensatz theorem.) Then we have $\tilde{w}[q] \sim_{\mathrm{alg}} w[t] =: w$ on $W_{\mathbb{C}}$. The upshot of all of this is that in the definition of $H_{\overline{\mathbb{Q}}}^{r-\nu,r}(X/\overline{\mathbb{Q}})$, we could allow w to range through $\mathrm{CH}^r((C \times S \times X)_{\mathbb{C}};\mathbb{Q})$ (with C, S, X still defined over $\overline{\mathbb{Q}}$), and still arrive at the same invariant $H_{\overline{\mathbb{Q}}}^{r-\nu,r}(X/\overline{\mathbb{Q}})$.

(iii) The main result in [G-G-P] states that for a smooth projective surface X of positive geometric genus, defined over a number field k, then for some overfield K of transcendence degree one over k, the kernel of the Albanese map $\mathrm{CH}^2_{\mathrm{hom}}(X_K) \to \mathrm{Alb}(X)$ contains a nontorsion element. This statement itself is more general than our theorem in the case when $(r,d) = (2,2)$, provided we replace "infinite rank" by "nonzero". One wonders if "infinite rank" requires the nonvanishing $H_{\overline{\mathbb{Q}}}^{r-\nu,r}(X/\overline{\mathbb{Q}}) \neq 0$.

(iv) In light of [Sch1], it is clear that our invariant $H^{r-\nu,r}_{\overline{\mathbb{Q}}}(X/\overline{\mathbb{Q}})$ can only be expected to be nonzero for a special class of varieties.

(v) In our case the infinite rank of $Gr^\nu_F \mathrm{CH}^r(X_K;\mathbb{Q})$ for $\nu \geq 1$ involves cycles with "trivial" Mumford-Griffiths invariant, as defined for example in [As], or in [L-S].

This paper began when I was visiting Jacob Murre in Leiden in October of 2006, where the discussion of an 'interesting algebraic cycle" took place. I want to thank Jacob for his encouragement and warm hospitality. I have also benefited from conversations with Matt Kerr and Andreas Rosenschon. I am grateful to the referee for the suggested improvements to this paper, and in particular the footnote pertaining to Lemma 4.1.

2 Notation

(i) The term variety will mean a integral separated scheme of finite type over a field k.

(ii) As pointed out earlier, we assume given smooth projective varieties $S/\overline{\mathbb{Q}}$, $C/\overline{\mathbb{Q}}$, $X/\overline{\mathbb{Q}}$, with $d = \dim X$, $d_S = \dim S$ and $\dim C = 1$.

(iii) For a variety W, $\mathrm{CH}^r(W)$ is the Chow group of cycles of codimension r on W, modulo rational equivalence. The subgroup in $\mathrm{CH}^r(W)$ of cycles algebraically equivalent to zero is denoted by $\mathrm{CH}^r_{\mathrm{alg}}(W)$. We put $\mathrm{CH}^r(W;\mathbb{Q}) := \mathrm{CH}^r(W) \otimes_{\mathbb{Z}} \mathbb{Q}$. Further, we write $\mathrm{CH}^r(W_{\mathbb{C}}) = \mathrm{CH}^r(W/\mathbb{C})$ to mean the same group.

(iv) If W is a variety defined over a subfield $k \subset \mathbb{C}$, and if $H^\bullet(-)$ is singular cohomology, we write $H^\bullet(W)$ to mean $H^\bullet(W(\mathbb{C}))$, where $W(\mathbb{C})$ is the associated complex space.

(v) In this paper, we work with the candidate Bloch-Beilinson filtration as defined in [L1]. It has the following properties for any given smooth projective variety X of dimension d. For all r, there is a filtration

$$\mathrm{CH}^r(X;\mathbb{Q}) \ = \ F^0 \supset F^1 \supset \cdots \supset F^\nu \supset F^{\nu+1} \supset \cdots$$
$$\supset \ F^r \supset F^{r+1} = F^{r+2} = \cdots,$$

which satisfies the following

(a) $F^1 = \mathrm{CH}^r_{\mathrm{hom}}(X;\mathbb{Q})$

(b) $F^2\mathrm{CH}^r(X;\mathbb{Q}) \subset \ker\big(AJ : \mathrm{CH}^r_{\mathrm{hom}}(X;\mathbb{Q}) \to J^r(X) \otimes \mathbb{Q}\big)$, where AJ is the Abel-Jacobi map, and $J^r(X)$ is the Griffiths jacobian.

(c) $F^\ell \bullet F^\nu \subset F^{\ell+\nu}$, where $\bullet$ is the intersection product.

(d) F^ν is preserved under push-forwards f_* and pull-backs f^*, where $f : X \to Y$ is a morphism of smooth projective varieties. [In short, F^ν is preserved under the action of correspondences between smooth projective varieties.]

(e) $Gr^\nu_F := F^\nu/F^{\nu+1}$ factors through the Grothendieck motive. More specifically, let us assume that the Künneth components of the diagonal class $[\Delta] = \bigoplus_{p+q=2d}[\Delta(p,q)] \in H^{2d}(X \times X, \mathbb{Q})$ are algebraic. Then

$$\Delta(2d - 2r + \ell, 2r - \ell)_* \Big|_{Gr^\nu_F \mathrm{CH}^r(X;\mathbb{Q})} = \delta_{\ell,\nu}.$$

(f) Let $D^r(X) := \bigcap_\nu F^\nu$. If a variant of the Bloch-Beilinson conjecture holds on the injectivity of the Abel-Jacobi map, for smooth quasiprojective varieties over a number field, then $D^r(X) = 0$.

(vi) If V is a $\mathbb{Q}$-vector space, we put $V(r) := V \otimes_{\mathbb{Q}} \mathbb{Q}(r)$, where $\mathbb{Q}(r) := \mathbb{Q} \cdot (2\pi\sqrt{-1})^r$ is the Tate twist.

(vii) If $H_{\mathbb{Q}}$ is a $\mathbb{Q}$-Hodge structure of weight $2r - 1$ and $H_{\mathbb{C}} = H_{\mathbb{Q}} \otimes \mathbb{C}$, we put

$$J^r(H_{\mathbb{Q}}) = \frac{H_{\mathbb{C}}}{F^r H_{\mathbb{C}} + H_{\mathbb{Q}}}.$$

3 A filtration and arithmetic normal functions

Consider a $\overline{\mathbb{Q}}$-spread $\rho : \mathcal{X} \to \mathcal{S}$, where ρ is smooth and proper, with $\mathcal{S}$ affine. Let η be the generic point of $\mathcal{S}$, and put $K := \overline{\mathbb{Q}}(\eta)$. Write $X_K := \mathcal{X}_\eta$. From [L1] we introduced a decreasing filtration $\mathcal{F}^\nu \mathrm{CH}^r(\mathcal{X}; \mathbb{Q})$, with the property that $Gr_{\mathcal{F}}^\nu \mathrm{CH}^r(\mathcal{X}; \mathbb{Q}) \hookrightarrow E_\infty^{\nu, 2r-\nu}(\rho)$, where $E_\infty^{\nu, 2r-\nu}(\rho)$ is the ν-th graded piece of the Leray filtration on the lowest weight part $\underline{H}_{\mathcal{H}}^{2r}(\mathcal{X}, \mathbb{Q}(r))$ of Beilinson's absolute Hodge cohomology $H_{\mathcal{H}}^{2r}(\mathcal{X}, \mathbb{Q}(r))$ associated to ρ. That lowest weight part $\underline{H}_{\mathcal{H}}^{2r}(\mathcal{X}, \mathbb{Q}(r)) \subset H_{\mathcal{H}}^{2r}(\mathcal{X}, \mathbb{Q}(r))$ is given by the image $H_{\mathcal{H}}^{2r}(\overline{\mathcal{X}}, \mathbb{Q}(r)) \to H_{\mathcal{H}}^{2r}(\mathcal{X}, \mathbb{Q}(r))$, where $\overline{\mathcal{X}}$ is a smooth compactification of $\mathcal{X}$. There is a cycle class map $\mathrm{CH}^r(\mathcal{X}; \mathbb{Q}) := \mathrm{CH}^r(\mathcal{X}/\overline{\mathbb{Q}}; \mathbb{Q}) \to \underline{H}_{\mathcal{H}}^{2r}(\mathcal{X}, \mathbb{Q}(r))$, which is conjecturally injective under the Bloch-Beilinson conjecture assumption in §2(v)(f) above. (Injectivity would imply $D^r(X) = 0$.) Regardless of whether or not injectivity holds, the filtration $\mathcal{F}^\nu \mathrm{CH}^r(\mathcal{X}; \mathbb{Q})$ is given by the pullback of the Leray filtration on $\underline{H}_{\mathcal{H}}^{2r}(\mathcal{X}, \mathbb{Q}(r))$ to $\mathrm{CH}^r(\mathcal{X}; \mathbb{Q})$. It is proven in [L1] that the term $E_\infty^{\nu, 2r-\nu}(\rho)$ fits in a short exact sequence:

$$0 \to \underline{E}_\infty^{\nu, 2r-\nu}(\rho) \to E_\infty^{\nu, 2r-\nu}(\rho) \to \underline{\underline{E}}_\infty^{\nu, 2r-\nu}(\rho) \to 0,$$

where

$$\underline{E}_\infty^{\nu, 2r-\nu}(\rho) = \mathrm{hom}_{\mathrm{MHS}}(\mathbb{Q}(0), H^\nu(\mathcal{S}, R^{2r-\nu}\rho_* \mathbb{Q}(r))),$$

$$\underline{\underline{E}}_\infty^{\nu, 2r-\nu}(\rho) = \frac{\mathrm{Ext}_{\mathrm{MHS}}^1(\mathbb{Q}(0), W_{-1}H^{\nu-1}(\mathcal{S}, R^{2r-\nu}\rho_* \mathbb{Q}(r)))}{\mathrm{hom}_{\mathrm{MHS}}(\mathbb{Q}(0), Gr_W^0 H^{\nu-1}(\mathcal{S}, R^{2r-\nu}\rho_* \mathbb{Q}(r)))}$$

$$\subset \mathrm{Ext}_{\mathrm{MHS}}^1(\mathbb{Q}(0), H^{\nu-1}(\mathcal{S}, R^{2r-\nu}\rho_* \mathbb{Q}(r))).$$

[Here the latter inclusion is a result of the short exact sequence:

$$W_{-1}H^{\nu-1}(\mathcal{S}, R^{2r-\nu}\rho_* \mathbb{Q}(r)) \hookrightarrow W_0 H^{\nu-1}(\mathcal{S}, R^{2r-\nu}\rho_* \mathbb{Q}(r)),$$
$$\to Gr_W^0 H^{\nu-1}(\mathcal{S}, R^{2r-\nu}\rho_* \mathbb{Q}(r)).]$$

One then has (by definition)

$$F^\nu \mathrm{CH}^r(X_K; \mathbb{Q}) = \varinjlim_{U \subset S/\overline{\mathbb{Q}}} \mathcal{F}^\nu \mathrm{CH}^r(\mathcal{X}_U; \mathbb{Q}), \quad \mathcal{X}_U := \rho^{-1}(U),$$

$$F^\nu \mathrm{CH}^r(X_{\mathbb{C}}; \mathbb{Q}) = \varinjlim_{K \subset \mathbb{C}} F^\nu \mathrm{CH}^r(X_K; \mathbb{Q}).$$

Further, since direct limits preserve exactness,

$$Gr_F^\nu \mathrm{CH}^r(X_K; \mathbb{Q}) = \varinjlim_{U \subset S/\overline{\mathbb{Q}}} Gr_{\mathcal{F}}^\nu \mathrm{CH}^r(\mathcal{X}_U; \mathbb{Q}),$$

$$Gr_F^\nu \mathrm{CH}^r(X_{\mathbb{C}}; \mathbb{Q}) = \varinjlim_{K \subset \mathbb{C}} Gr_F^\nu \mathrm{CH}^r(X_K; \mathbb{Q}).$$

Let $V \subset \mathcal{S}(\mathbb{C})$ be smooth, irreducible, closed subvariety of dimension $\nu - 1$ (note that $\mathcal{S}$ affine $\Rightarrow V$ affine). One has a commutative square

$$
\begin{array}{ccc}
\mathcal{X}_V & \hookrightarrow & \mathcal{X}(\mathbb{C}) \\
\rho_V \downarrow & & \downarrow \rho \\
V & \hookrightarrow & \mathcal{S}(\mathbb{C}),
\end{array}
$$

and a commutative diagram

$$\xi \in \ Gr_{\mathcal{F}}^{\nu}CH^r(\mathcal{X};\mathbb{Q}) \ \mapsto \ Gr_F^{\nu}CH^r(X_K;\mathbb{Q})$$

$$\downarrow$$

$$
\begin{array}{ccccccccc}
0 & \to & \underline{E}_{\infty}^{\nu,2r-\nu}(\rho) & \to & E_{\infty}^{\nu,2r-\nu}(\rho) & \to & \underline{\underline{E}}_{\infty}^{\nu,2r-\nu}(\rho) & \to & 0 \\
 & & \downarrow & & \downarrow & & \downarrow & & \\
0 & \to & \underline{E}_{\infty}^{\nu,2r-\nu}(\rho_V) & \to & E_{\infty}^{\nu,2r-\nu}(\rho_V) & \to & \underline{\underline{E}}_{\infty}^{\nu,2r-\nu}(\rho_V) & \to & 0 \\
 & & & & & & \| & & \\
 & & & & & & 0 & &
\end{array}
$$

where $\underline{\underline{E}}_{\infty}^{\nu,2r-\nu}(\rho_V) = 0$ follows from the weak Lefschetz theorem for locally constant systems over affine varieties (see for example [Ar], and the references cited there). Thus for any $\xi \in Gr_{\mathcal{F}}^{\nu}CH^r(\mathcal{X};\mathbb{Q})$, we have a "normal function" η_{ξ} with the property that for any such smooth irreducible closed $V \subset S(\mathbb{C})$ of dimension $\nu - 1$, we have a value $\eta_{\xi}(V) \in \underline{E}_{\infty}^{\nu,2r-\nu}(\rho_V)$. Here we think of V as a point on a suitable open subset of the Chow variety of dimension $\nu - 1$ subvarieties of $\mathcal{S}(\mathbb{C})$ and η_{ξ} defined on that subset. This idea of normal functions, which is *only* introduced here in this section for philosophical reasons, is discussed in much greater detail in [K-L] and [K]. Note that it is rather clear from this that

$$F^2 CH^r(X;\mathbb{Q}) \subset \ker \left(AJ : CH^r_{\mathrm{hom}}(X;\mathbb{Q}) \to J^r(X) \otimes \mathbb{Q}\right).$$

Now suppose that $\mathcal{S}$ is affine of dimension $\nu - 1$. Then in this case $V = \mathcal{S}$, and $\xi \in Gr_{\mathcal{F}}^{\nu}CH^r(\mathcal{X};\mathbb{Q})$ induces a "single point" normal function

$$\eta_{\xi}(V) = \eta_{\xi}(\mathcal{S}) \in \mathrm{Ext}^1_{\mathrm{MHS}}(\mathbb{Q}(0), H^{\nu-1}(\mathcal{S}, R^{2r-\nu}\rho_*\mathbb{Q}(r))).$$

4 Some Hodge theory and proof of the theorem

Consider the setting of §1. If $\dim S < \nu - 1$, then one easily checks that for Hodge type reasons $P(C, S, w) = 0$. If $\dim S \geq \nu - 1$, then by the weak Lefschetz theorem one can find a smooth subvariety $S_0 \subset S/\overline{\mathbb{Q}}$ of dimension $\nu - 1$ such that if w_0 is the restriction of w to $C \times S_0 \times X$, then $P(C, S, w)$ and $P(C, S_0, w_0)$ have the same image. Thus we may assume that $\dim S = \nu - 1$ with $P(C, S, w) \neq 0$. Further, by Remark 1.6 (ii), we can assume that w is defined over $\overline{\mathbb{Q}}$. Note that there is a commutative diagram:

$$J^1(C(\mathbb{C})) \otimes \mathbb{Q} \xrightarrow{[w]_*} \operatorname{Ext}^1_{\text{MHS}}(\mathbb{Q}(0), H^{\nu-1}(S, \mathbb{Q}) \otimes H^{2r-\nu}(X, \mathbb{Q})(r)))$$

$$\uparrow \qquad \alpha \nearrow$$

$$J^1(C/\overline{\mathbb{Q}}) \otimes \mathbb{Q}$$

Now even though $C(\overline{\mathbb{Q}})$ is dense in $C(\mathbb{C})$ (a fortiori $J^1(C/\overline{\mathbb{Q}})$ dense in $J^1(C(\mathbb{C}))$), we still have to show that $J^1(C/\overline{\mathbb{Q}})$ doesn't map into the torsion subgroup of $\operatorname{Ext}^1_{\text{MHS}}(\mathbb{Z}(0), H^{\nu-1}(S, \mathbb{Z}) \otimes H^{2r-\nu}(X, \mathbb{Z})(r)))$, which would imply that α is zero. However, it is rather easy to show that:

Lemma 4.1 *There exists an Abelian subvariety $A/\overline{\mathbb{Q}} \subset J^1(C/\overline{\mathbb{Q}})$ of positive dimension such that $\alpha|_{A(\overline{\mathbb{Q}}) \otimes \mathbb{Q}}$ is injective.*

Proof This is a consequence of [L-S] (Lemma 7.5); however the idea is straight forward. By a spread argument similar to that in Remark 1.6 (i), one argues that $A/\overline{\mathbb{Q}}$ is simple $\Leftrightarrow A \times_{\overline{\mathbb{Q}}} \mathbb{C}$ is simple. The assertion of the lemma then follows from Poincaré's complete reducibility theorem.[1]   □

Now use the fact that $A(\overline{\mathbb{Q}})$ has infinite rank ([F-G]) to deduce that α is nonzero, (and more to the point, that the image of α likewise has infinite rank). In particular, there exists $\mu \in J^1(C/\overline{\mathbb{Q}}) \otimes \mathbb{Q}$ such that if we put $\Gamma_\mu := w[\mu] \in \operatorname{CH}^r((S \times X)/\overline{\mathbb{Q}}; \mathbb{Q})$, then $[\Gamma_\mu] \neq 0$ in $\operatorname{Ext}^1_{\text{MHS}}(\mathbb{Q}(0), H^{\nu-1}(S, \mathbb{Q}) \otimes H^{2r-\nu}(X, \mathbb{Q})(r)))$. It suffices to show that after possibly replacing A by a smaller Abelian subvariety over $\overline{\mathbb{Q}}$ of positive dimension (which we will still denote by A), the composite

$$A(\overline{\mathbb{Q}}) \otimes \mathbb{Q} \xrightarrow{\alpha} \operatorname{Ext}^1_{\text{MHS}}(\mathbb{Q}(0), H^{\nu-1}(S, \mathbb{Q}) \otimes H^{2r-\nu}(X, \mathbb{Q})(r)))$$

$$\longrightarrow \operatorname{Ext}^1_{\text{MHS}}(\mathbb{Q}(0), H^{\nu-1}(U, \mathbb{Q}) \otimes H^{2r-\nu}(X, \mathbb{Q})(r))),$$

remains injective for all nonempty Zariski open $U \subset S/\overline{\mathbb{Q}}$. For in this case, we will have exhibited an injection $A(\overline{\mathbb{Q}}) \otimes \mathbb{Q} \hookrightarrow Gr_F^\nu \operatorname{CH}^r(X_K; \mathbb{Q})$, where $K := \overline{\mathbb{Q}}(S)$, given by $\mu \mapsto \Gamma_{\mu,\eta}$, η the generic point of S. Towards this goal, we need the following:

Lemma 4.2 ([L-S]) *Consider the map*

$$[w]_* : H^1(C, \mathbb{Q}) \to H^{\nu-1}(S, \mathbb{Q}) \otimes H^{2r-\nu}(X, \mathbb{Q}),$$

induced by w. Then its image is not contained in

$$N_H^1 H^{\nu-1}(S, \mathbb{Q}) \otimes H^{2r-\nu}(X, \mathbb{Q}) + H^{\nu-1}(S, \mathbb{Q}) \otimes N_H^{r-\nu+1} H^{2r-\nu}(X, \mathbb{Q}),$$

where $N_H^p H^\bullet(X, \mathbb{Q})$ is the largest subHodge structure lying in

$$F^p H^\bullet(X, \mathbb{C}) \cap H^\bullet(X, \mathbb{Q}).$$

Proof This is immediate from our assumption that $P(C, S, w) \neq 0$.   □

---

[1] Alternatively, and as the referee pointed out, the fact that any simple factor A of $J^1(C/\mathbb{C})$ is actually defined over $\overline{\mathbb{Q}}$ is due to the fact that the endomorphism algebra of J^1 and its decomposition as a product of simple algebras is the same over $\overline{\mathbb{Q}}$ and over $\mathbb{C}$. If there were an endomorphism defined over $\mathbb{C}$ that did not come from $\overline{\mathbb{Q}}$, it would have to fix the points of prime power order, since these points are the same over $\overline{\mathbb{Q}}$ as over $\mathbb{C}$. Thus it would have to be the identity.

Now put

$$W_{U,0} := W_0\big(H^{\nu-1}(U,\mathbb{Q}) \otimes H^{2r-\nu}(X,\mathbb{Q})(r)\big)$$
$$W_{U,-1} := W_{-1}\big(H^{\nu-1}(U,\mathbb{Q}) \otimes H^{2r-\nu}(X,\mathbb{Q})(r)\big)$$
$$= \mathrm{Im}\big(H^{\nu-1}(S,\mathbb{Q}) \otimes H^{2r-\nu}(X,\mathbb{Q})(r)$$
$$\to H^{\nu-1}(U,\mathbb{Q}) \otimes H^{2r-\nu}(X,\mathbb{Q})(r)\big).$$

From the short exact sequence

$$0 \to W_{U,-1} \to W_{U,0} \to Gr^0_{U,W} \to 0,$$

we have the exact sequence:

$$\cdots \to \quad \mathrm{hom}_{\mathrm{MHS}}(\mathbb{Q}(0), Gr^0_{U,W}) \xrightarrow{h} \mathrm{Ext}^1_{\mathrm{MHS}}(\mathbb{Q}(0), W_{U,-1})$$
$$\to \mathrm{Ext}^1_{\mathrm{MHS}}(\mathbb{Q}(0), W_{U,0}) \to \cdots . \tag{4.1}$$

Next, put

$$W^H_{U,0} := W_0\big(H^{\nu-1}(U,\mathbb{Q}) \otimes N_H^{r-\nu+1} H^{2r-\nu}(X,\mathbb{Q})(r)\big)$$

$$W^H_{U,-1} := W_{-1}\big(H^{\nu-1}(U,\mathbb{Q}) \otimes N_H^{r-\nu+1} H^{2r-\nu}(X,\mathbb{Q})(r)\big).$$

An elementary Hodge theory argument shows that:

Lemma 4.3 ([L-S]) $\mathrm{hom}_{\mathrm{MHS}}(\mathbb{Q}(0), Gr^0_{U,W^H}) = \mathrm{hom}_{\mathrm{MHS}}(\mathbb{Q}(0), Gr^0_{U,W}).$

Thus from the commutative diagram:

$$0 \to W_{U,-1} \to W_{U,0} \to Gr^0_{U,W} \to 0$$

$$\uparrow \qquad\qquad \uparrow \qquad\qquad \uparrow$$

$$0 \to W^H_{U,-1} \to W^H_{U,0} \to Gr^0_{U,W^H} \to 0,$$

we deduce that the image of h in (4.1) above lies in the image of:

$$h_{U,J} : \mathrm{Ext}^1_{\mathrm{MHS}}(\mathbb{Q}(0), W^H_{U,-1}) \to \mathrm{Ext}^1_{\mathrm{MHS}}(\mathbb{Q}(0), W_{U,-1}). \tag{4.2}$$

Next, after taking the limit over open subsets $U \subset S/\overline{\mathbb{Q}}$, the cokernel of the map in (4.2), viz.,

$$\mathrm{coker}\, h_J := \varinjlim_U \mathrm{coker}\, h_{U,J},$$

is a compact complex torus (tensored with $\mathbb{Q}$). This is due to the observation that

$$\varinjlim_U W^{(H)}_{U,-1} = \left(\frac{H^{\nu-1}(S,\mathbb{Q})}{N_H^1 H^{\nu-1}(S,\mathbb{Q})}\right) \otimes (N_H^{r-\nu+1}) H^{2r-\nu}(X,\mathbb{Q})(r),$$

is a finite dimensional Hodge structure. (Even though the same cannot be said for $W^{(H)}_{U,0}$.) By our assumptions on w, viz., $P(C,S,w) \neq 0$, the map

$$[w]_* : J^1(C/\mathbb{C}) \otimes \mathbb{Q} \to \mathrm{coker}\, h_J,$$

is nontrivial. Moreover by the same reasoning as in Lemma 4.1, there is an Abelian variety $A/\overline{\mathbb{Q}} \subset J^1(C/\overline{\mathbb{Q}})$ of positive dimension, such that the composite

$$A(\overline{\mathbb{Q}}) \otimes \mathbb{Q} \to J^1(C/\mathbb{C}) \otimes \mathbb{Q} \xrightarrow{[w]_*} \mathrm{coker}\, h_J,$$

is injective. So in summary, if we put $K := \overline{\mathbb{Q}}(S)$, then we have a commutative diagram:

$$A(\overline{\mathbb{Q}}) \otimes \mathbb{Q} \qquad \hookrightarrow \qquad \operatorname{coker} h_J$$

$$\searrow \qquad\qquad\qquad \nearrow$$

$$Gr_F^\nu \mathrm{CH}^r(X_K; \mathbb{Q}),$$

hence the theorem, using again [F-G].

5 Application to a special class of varieties

Recall that $X/\overline{\mathbb{Q}}$ is smooth projective of dimension d. Let us assume that for some $\nu \geq 2$, $H^{0,\nu}(X) \neq 0$. Further, we are going to assume that there exists smooth projective curves $\{C_1, ..., C_d\}$ over $\overline{\mathbb{Q}}$, and a rational dominating map $Z : C_1 \times \cdots \times C_d \to X$. Here we identify $Z \subset (C_1 \times \cdots \times C_d \times X)/\overline{\mathbb{Q}}$ as a correspondence of codimension d. Consider $I = \{i_1, ..., i_\nu\} \subset \{1, ..., d\}$, a given subset of ν distinct elements satisfying the property that the composite:

$$\bigotimes_{j=1}^{\nu} H^{0,1}(C_{i_j}) \xrightarrow{\mathrm{Pr}_I^*} H^{0,\nu}(C_1 \times \cdots \times C_d) \xrightarrow{[Z]_*} H^{0,\nu}(X),$$

is nonzero, where $\mathrm{Pr}_I : C_1 \times \cdots \times C_d \to C_I := C_{i_1} \times \cdots \times C_{i_\nu}$ is the projection. Note that:

$$Z \circ {}^{\mathrm{T}}\big(\mathrm{graph}(\mathrm{Pr}_I)\big) \in \mathrm{CH}^\nu(C_I \times X).$$

For all $j \in I$, fix $e_j \in C_j(\overline{\mathbb{Q}})$ and put

$$\Delta_{C_j}(1,1) = \Delta_{C_j} - e_j \times C_j - C_j \times e_j,$$

where $\Delta_{C_j} \in \mathrm{CH}^1(C_j \times C_j)$ is the diagonal class of C_j. Note that $\Delta_{C_j}(1,1)$ corresponds to the Künneth component of $[\Delta_{C_j}]$ in $H^1(C_j, \mathbb{Q}) \otimes H^1(C_j, \mathbb{Q})$. Now put

$$\Xi_I \quad := \quad \bigcap_{j=1}^{\nu} \mathrm{Pr}_{i,\nu+i}^* \big(\Delta_{C_{i_j}}(1,1)\big) \in \mathrm{CH}^\nu(C_I \times C_I),$$

$$w \quad := \quad \big\{ Z \circ {}^{\mathrm{T}}\big(\mathrm{graph}(\mathrm{Pr}_I)\big) \circ \Xi_I \big\} \in \mathrm{CH}^\nu(C_I \times X).$$

Then w has the property that:

$$[w] \in H^1(C_{i_1}, \mathbb{Q}) \otimes \cdots \otimes H^1(C_{i_\nu}, \mathbb{Q}) \otimes H^\nu(X, \mathbb{Q}) \subset H^{2\nu}(C_I \times X, \mathbb{Q}).$$

Now put $C := C_{i_1}$, $S = C_{i_2} \times \cdots \times C_{i_\nu}$. Thus $w \in \mathrm{CH}^r(C \times S \times X)$, where $r = \nu$, and by construction

$$[w] \in H^1(C, \mathbb{Q}) \otimes H^{2d_S - \nu + 1}(S, \mathbb{Q}) \otimes H^{2r-\nu}(X, \mathbb{Q}),$$

and

$$P(C, S, w) : H^1(C, \mathbb{Q}) \otimes H^{2d_S - \nu + 1}(S, \mathbb{Q}) \xrightarrow{[w]_*} H^{2r-\nu}(X, \mathbb{Q}) \to H^{r-\nu,r}(X),$$

is nontrivial, where $d_S = \nu - 1$, and $r = \nu$. According to our Theorem 1.2, $Gr_F^r \mathrm{CH}^r(X_K; \mathbb{Q})$ has infinite rank, where $K = \overline{\mathbb{Q}}(S)$. However we can be more explicit. With regard to a choice of embedding $\overline{\mathbb{Q}}(S) \subset \mathbb{C}$, the generic point η of S

corresponds to a point $\xi = (\xi_2, ..., \xi_\nu) \in S(\mathbb{C}) = C_{i_2}(\mathbb{C}) \times \cdots \times C_{i_\nu}(\mathbb{C})$. Then for $p \in C_{i_1}(\overline{\mathbb{Q}})$,

$$\Xi_{I,*}(p, \xi_2, ..., \xi_\nu) = (p - e_1) \times (\xi_2 - e_2) \times \cdots \times (\xi_\nu - e_\nu) \in \mathrm{CH}^\nu(C_I).$$

Thus with $\nu = r$, we easily arrive at Corollary 1.3 above.

6 A comparison of two invariants on X

As is clearly evident in [G-G-P] and pointed out in [MSa], the proof of the main result in [G-G-P] uses that more than a Hodge theory argument. The purpose of this section is to explain our result in this context. We first introduce:

Proposition 6.1 *Let $X/\overline{\mathbb{Q}}$ be smooth and projective. For any smooth projective variety W, let $d_W := \dim W$. The following spaces are equal:*

(i) $\{[w]_* H^{2d_W - \nu}(W, \mathbb{Q}) \subset H^{2r - \nu}(X, \mathbb{Q}) \mid W/\overline{\mathbb{Q}},\ w \in \mathrm{CH}^r((W \times X)/\overline{\mathbb{Q}}),$
     $d_W = \nu \}$,

(ii) $\{[w]_* H^{2d_W - \nu}(W, \mathbb{Q}) \subset H^{2r - \nu}(X, \mathbb{Q}) \mid W/\overline{\mathbb{Q}},\ w \in \mathrm{CH}^r((W \times X)/\overline{\mathbb{Q}})\}$,

(iii) $\{[w]_* H^{2d_W - \nu}(W, \mathbb{Q}) \subset H^{2r - \nu}(X, \mathbb{Q}) \mid W/\mathbb{C},\ w \in \mathrm{CH}^r((W \times X)/\mathbb{C})\}$.

Proof Omitted. (For example (ii) $\Leftrightarrow$ (iii) follows from Remark 1.6 (ii).) $\square$

Definition 6.2 Let $H_W^{(r,\nu)}$ be the subspace introduced in (i) above. We define $H_{\mathrm{alg}}^{r-\nu,r}(X)$ to be the complex subspace generated by the composite image of:

$$H_W^{(r,\nu)} \to H^{2r-\nu}(X, \mathbb{Q}) \xrightarrow{\text{Hodge projection}} H^{r-\nu,r}(X),$$

over all smooth projective W.

Remarks 6.3 (i) One has an inclusion $H_{\overline{\mathbb{Q}}}^{r-\nu,r}(X/\overline{\mathbb{Q}}) \subset H_{\mathrm{alg}}^{r-\nu,r}(X)$, which in general is not an equality by [Sch1].

(ii) Note that for example if $X/\overline{\mathbb{Q}}$ is a surface (i.e. $d = 2$), then by choosing $W = X$ and $w = \Delta_X$, it follows that $H_{\mathrm{alg}}^{0,2}(X) = H^{0,2}(X)$.

We prove the following:

Proposition 6.4 (i) $H_{\overline{\mathbb{Q}}}^{r-\nu,r}(X/\overline{\mathbb{Q}}) \neq 0 \Rightarrow \exists\ V/\overline{\mathbb{Q}}, \dim V = \nu - 1$, and $\xi \in \mathrm{CH}_{\mathrm{alg}}^r(V \times X)$ *such that*

$$0 \neq AJ(\xi) \in J^r \left(\frac{H^{\nu-1}(V, \mathbb{Q})}{N_H^1 H^{\nu-1}(V, \mathbb{Q})} \otimes \frac{H^{2r-\nu}(X, \mathbb{Q})}{N_H^{r-\nu+1} H^{2r-\nu}(X, \mathbb{Q})} \right) \otimes \mathbb{Q}.$$

(ii) $H_{\mathrm{alg}}^{r-\nu,r}(X) \neq 0 \Rightarrow \exists\ V/\overline{\mathbb{Q}}, \dim V = \nu - 1$, and $\xi \in \mathrm{CH}_{\mathrm{hom}}^r(V \times X)$ *such that*

$$0 \neq AJ(\xi) \in J^r \left(\frac{H^{\nu-1}(V, \mathbb{Q})}{N_H^1 H^{\nu-1}(V, \mathbb{Q})} \otimes \frac{H^{2r-\nu}(X, \mathbb{Q})}{N_H^{r-\nu+1} H^{2r-\nu}(X, \mathbb{Q})} \right) \otimes \mathbb{Q}.$$

Proof Part (i) is obvious, and the nonvanishing of $AJ(\xi)$ in

$$J^r \left(\frac{H^{\nu-1}(V, \mathbb{Q})}{N_H^1 H^{\nu-1}(V, \mathbb{Q})} \otimes \frac{H^{2r-\nu}(X, \mathbb{Q})}{N_H^{r-\nu+1} H^{2r-\nu}(X, \mathbb{Q})} \right) \otimes \mathbb{Q},$$

can be used to replace the hypothesis in our Theorem 1.2. Part (ii) is a little more subtle, and can be deduced from the discussion in §7 of [K]. The basic idea is this. $H_{\mathrm{alg}}^{r-\nu,r}(X) \neq 0$ implies that there is a nonzero class in

$$\varinjlim_{U \subset S/\overline{\mathbb{Q}}} \underline{E}_{\infty}^{r,2r-\nu},$$

which gives rise to a normal function with nonzero Abel-Jacobi invariant in (6.1) below over "transcendental" $V \subset S(\mathbb{C})$. But by the work of Terosoma, this is also the case for some $V \subset S/\overline{\mathbb{Q}}$. We observe that the assumption of the nonvanishing of $AJ(\xi)$ in

$$J^r\left(\frac{H^{\nu-1}(V,\mathbb{Q})}{N_H^1 H^{\nu-1}(V,\mathbb{Q})} \otimes \frac{H^{2r-\nu}(X,\mathbb{Q})}{N_H^{r-\nu+1} H^{2r-\nu}(X,\mathbb{Q})}\right) \otimes \mathbb{Q}, \qquad (6.1)$$

which in a special case, is a key assumption in [G-G-P], and plays a role in the generalization in [MSa]. ☐

Corollary 6.5 *Let* $X \subset \mathbb{P}^{d+1}/\overline{\mathbb{Q}}$ *be a smooth hypersurface of degree* $m \geq 3$. *Put*

$$k = \left[\frac{d+1}{m}\right], \ r = d - k, \ \nu = 2r - d = d - 2k,$$

and assume $\nu \geq 1$ *and the numerical condition:*

$$k(d+2-k) + 1 - \binom{m+k}{k} \geq 0.$$

Then there exists a field K *of transcendence degree* $\nu - 1$ *over* $\mathbb{Q}$ *for which*

$$Gr_F^\nu \mathrm{CH}^r(X_K;\mathbb{Q}) \neq 0.$$

Proof The essential idea is this. First of all, by the Lefschetz theorem, one has a Chow-Künneth decomposition of the diagonal class in the sense of Murre [M], and hence one can make use of Chow-Künneth projectors. Secondly, for X given in Corollary 6.5, one can choose $W/\overline{\mathbb{Q}}$ to be a suitable smooth projective subvariety of the Fano variety of $\mathbb{P}^k$'s in X, with $\dim W = d - 2k$. $w \in \mathrm{CH}^r(W \times X)$ is given by $w := \{(c, \mathbb{P}_c^k) \subset W \times X \mid c \in W\}$. The rest follows from an application of the results of [L2] to Proposition 6.4 (ii). ☐

7 Cubic equivalence

Much of this section is inspired by the work of H. Saito ([HSa]).

Definition 7.1 (cf. [HSa] or [L3].) Let $W/\mathbb{C}$ be a projective variety, ℓ be a positive integer, and ξ_1, ξ_2 cycles of codimension r in W. We say that ξ_1, ξ_2 are ℓ-*cubic equivalent*, denoted by $\xi_1 \sim_{(\ell)} \xi_2$, if there exists ℓ smooth projective curves $C_1, ..., C_\ell$ and a cycle z on $C_1 \times \cdots \times C_\ell \times W$ of codimension r, points $a_i^{(0)}$, $a_i^{(1)}$ on C_i ($i = 1, ..., \ell$) such that $z(a_1^{(i_1)}, ..., a_\ell^{(i_\ell)})$ are defined for all $i_1, .., i_\ell = 0$, 1, and that

$$\xi_1 - \xi_2 = \sum_{i_1,...,i_\ell = 0, \ 1} (-1)^{i_1 + \cdots + i_\ell} z(a_1^{(i_1)}, ..., a_\ell^{(i_\ell)}).$$

Let $X/\overline{\mathbb{Q}}$ be given as above, and consider the cubic equivalence filtration

$$\mathrm{CH}^r(X/\mathbb{C}) := F_c^0 \supset F_c^1 \supset F_c^2 \supset \cdots \supset F_c^\nu \supset F_c^{\nu+1} \supset \cdots,$$

where

$$F_c^\ell \mathrm{CH}^r(X/\mathbb{C}) = \{\xi \in \mathrm{CH}^r(X/\mathbb{C}) \mid \xi \sim_{(\ell)} 0\}.$$

We observe that

$$F_c^1 = \mathrm{CH}_{\mathrm{alg}}^r(X/\mathbb{C}) \subset \mathrm{CH}_{\mathrm{hom}}^r(X/\mathbb{C}; \mathbb{Q}) =: F^1 \mathrm{CH}^r(X/\mathbb{C}; \mathbb{Q}).$$

Note that

$$\sum_{i_1,\ldots,i_\ell = 0,\, 1} (-1)^{i_1 + \cdots + i_\ell} z_*(a_1^{(i_1)}, \ldots, a_\ell^{(i_\ell)})$$

$$= z_*\left(\bigcap_{j=1}^\ell \mathrm{Pr}_j^*(a_j^{(0)} - a_j^{(1)})\right) \subset F^\ell \mathrm{CH}^r(X/\mathbb{C}; \mathbb{Q}),$$

using the mutiplicative properties of the filtration under the intersection product. Here $\mathrm{Pr}_j : C_1 \times \cdots \times C_\ell \to C_j$ is the canonical projection. We deduce that for all $\nu \geq 0$,

$$F_c^\nu \mathrm{CH}^r(X/\mathbb{C}; \mathbb{Q}) \subset F^\nu \mathrm{CH}^r(X/\mathbb{C}; \mathbb{Q}),$$

although equality is not the case in general ([L3]). We introduce another invariant:

Definition 7.2 Let $H_{\overline{\mathbb{Q}},c}^{r-\nu,r}(X/\overline{\mathbb{Q}}) = \mathbb{C}$-vector space generated by the images

$$[w]_* : H^\nu(C_1 \times \cdots \times C_\nu, \mathbb{Q}) \to H^{2r-\nu}(X, \mathbb{Q}) \to H^{r-\nu,r}(X),$$

over all smooth projective curves $C_1, \ldots, C_\nu$ over $\overline{\mathbb{Q}}$ and all $w \in \mathrm{CH}^r(C_1 \times \cdots \times C_\nu \times X)$.

Remarks 7.3

(i) As before, we can allow $C_1, \ldots, C_\nu$ over $\mathbb{C}$ without any any change in $H_{\overline{\mathbb{Q}},c}^{r-\nu,r}(X/\overline{\mathbb{Q}})$, and furthermore the composite image above is the same as the restriction of the composite map to $H^1(C_1, \mathbb{Q}) \otimes \cdots \otimes H^1(C_\nu, \mathbb{Q})$ (for Hodge type reasons).

(ii) By the same construction in §5, we can modify w without affecting the aforementioned composite image, so that

$$[w] \in H^1(C_1, \mathbb{Q}) \otimes \cdots \otimes H^1(C_\nu, \mathbb{Q}) \otimes H^{2r-\nu}(X, \mathbb{Q}).$$

(iii) It is obvious that

$$H_{\overline{\mathbb{Q}},c}^{r-\nu,r}(X/\overline{\mathbb{Q}}) \subset H_{\overline{\mathbb{Q}}}^{r-\nu,r}(X/\overline{\mathbb{Q}}),$$

where again equality does not hold in general.

(iv) A simple consequence of the results in [HSa] is that

$$H_{\overline{\mathbb{Q}},c}^{r-\nu,r}(X/\overline{\mathbb{Q}}) \neq 0 \Rightarrow Gr_{F_c}^\nu \mathrm{CH}^r(X/\mathbb{C}; \mathbb{Q}) \neq 0.$$

Let us introduce the following.

Definition 7.4 Define

(i)

$$\underline{Gr}_{F_c}^\nu \mathrm{CH}^r(X/\mathbb{C}; \mathbb{Q}) \;\; := \;\; \frac{F_c^\nu \mathrm{CH}^r(X/\mathbb{C}; \mathbb{Q})}{\{F_c^\nu \mathrm{CH}^r(X/\mathbb{C}; \mathbb{Q})\} \cap \{F^{\nu+1}\mathrm{CH}^r(X/\mathbb{C}; \mathbb{Q})\}},$$

$$\subset \;\; Gr_F^\nu \mathrm{CH}^r(X/\mathbb{C}; \mathbb{Q}).$$

(ii)

$$\underline{Gr}_N^{r-\nu} H^{2r-\nu}(X, \mathbb{Q})$$

$$:= \frac{\left\{ [w]_* \big(H^1(C_1, \mathbb{Q}) \otimes \cdots \otimes H^1(C_\nu, \mathbb{Q}) \big) \,\middle|\, \begin{array}{c} \text{over all curves} \\ C_1, \ldots, C_\nu \,\&\, w \in \\ \mathrm{CH}^r(C_1 \times \cdots \times C_\nu \times X) \end{array} \right\}}{N^{r-\nu+1} H^{2r-\nu}(X, \mathbb{Q}) \,\cap\, \text{Numerator}}$$

$$\subset \quad Gr_N^{r-\nu} H^{2r-\nu}(X, \mathbb{Q}).$$

For our next result, the reader can consult [L4] for the statements of the Hodge conjecture (HC) in its classical form, as well as the (Grothendieck amended) general Hodge conjecture (GHC) for smooth complex projective varieties.

Theorem 7.5   (i) *Under the assumption of the HC,*

$$\underline{Gr}_N^{r-\nu} H^{2r-\nu}(X, \mathbb{Q}) = 0 \Leftrightarrow \underline{Gr}_{F_c}^{\nu} \mathrm{CH}^r(X/\mathbb{C}; \mathbb{Q}) = 0.$$

(ii) *Note that under the assumption of the GHC for X, the aforementioned composite map in Definition 7.2 yields an inclusion of the generating set:*

$$\underline{Gr}_N^{r-\nu} H^{2r-\nu}(X, \mathbb{Q}) \hookrightarrow H_{\overline{\mathbb{Q}}, c}^{r-\nu, r}(X/\overline{\mathbb{Q}}),$$

and therefore

$$H_{\overline{\mathbb{Q}}, c}^{r-\nu, r}(X/\overline{\mathbb{Q}}) = 0 \Leftrightarrow \underline{Gr}_{F_c}^{\nu} \mathrm{CH}^r(X/\mathbb{C}; \mathbb{Q}) = 0.$$

Using the same ideas as in §4, we arrive at an analog of the main result in [MSa], viz.,

Corollary 7.6 *Let us assume the GHC. If $\underline{Gr}_{F_c}^{\nu} \mathrm{CH}^r(X/\mathbb{C}; \mathbb{Q}) \neq 0$, then there exists a subfield $K \subset \mathbb{C}$ of transcendence degree $\nu - 1$ over $\mathbb{Q}$ such that $\underline{Gr}_{F_c}^{\nu} \mathrm{CH}^r(X_K; \mathbb{Q})$ has infinite rank.*

Proof of Theorem 7.5 Put $\overline{V} = C_1 \times \cdots \times C_\nu$. A class in $\underline{Gr}_N^{r-\nu} H^{2r-\nu}(X, \mathbb{Q})$ will be in the image of some w given as in Definition 7.4 (ii). Further, by working with the Chow-Künneth decomposition of the diagonal classes of the curves $\{C_j\}$ as in §5, we can assume that $w \in \mathrm{CH}^r((\overline{V} \times X)/\mathbb{C})$ is given such that $w_*\big(\mathrm{CH}^\nu(\overline{V}/\mathbb{C})\big) \subset F_c^\nu \mathrm{CH}^r(X/\mathbb{C})$. Next, by Remark 7.3 (i), we can assume that $\overline{V}$ and w are defined over $\overline{\mathbb{Q}}$. Let us first assume that $\underline{Gr}_{F_c}^{\nu} \mathrm{CH}^r(X/\mathbb{C}; \mathbb{Q}) = 0$. Thus we also have $w_*\big(\mathrm{CH}^\nu(\overline{V}/\mathbb{C}; \mathbb{Q})\big) \in F^{\nu+1} \mathrm{CH}^r(X/\mathbb{C}; \mathbb{Q})$. From the discussion in the first part of §3, it is clear that $[w] \mapsto 0 \in H^\nu(V, \mathbb{Q}) \otimes H^{2r-\nu}(X, \mathbb{Q})$, for some Zariski open subset $V \subset \overline{V}$. Thus $[w] \in \mathrm{Image}\big(H_D^\nu(\overline{V}, \mathbb{Q}) \otimes H^{2r-\nu}(X, \mathbb{Q}) \to H^\nu(\overline{V}, \mathbb{Q}) \otimes H^{2r-\nu}(X, \mathbb{Q})\big)$, for some divisor $D \subset \overline{V}$. Let $\tilde{D} \to D$ be a desingularization with corresponding induced map $j_0 : \tilde{D} \to \overline{V}$. Then by the HC, and working on the level of (co)homology, it follows that w is homologous to a cycle on $\overline{V} \times X$ arising from a $\tilde{w} \in \mathrm{CH}^{r-1}(\tilde{D} \times X; \mathbb{Q})$. Let $j_1 : \tilde{D}_H \hookrightarrow \tilde{D}$ be a smooth hyperplane section. Thus $\dim \tilde{D}_H = \nu - 2$, and let $\tilde{w}_H \in \mathrm{CH}^{r-1}(\tilde{D}_H \times X)$ be the restriction of $\tilde{w}$. Note the image of the projection $Pr_{2,*} : \mathrm{CH}^{r-1}(\tilde{D}_H \times X) \to \mathrm{CH}^{r-\nu+1}(X)$, and hence together the commutative diagram:

$$H^\nu(\overline{V}, \mathbb{Q}) \xrightarrow{\ [w]_*\ } H^{2r-\nu}(X, \mathbb{Q})$$

$$\downarrow j_0^* \qquad\qquad \nearrow {[\tilde{w}]_*}$$

$$H^\nu(\tilde{D}, \mathbb{Q})$$

$$\text{PD} \Vert \wr \qquad\qquad\qquad\qquad {[\tilde{w}_H]_*}$$

$$H_{\nu-2}(\tilde{D}, \mathbb{Q})$$

$$\uparrow j_{1,*}$$

$$H_{\nu-2}(\tilde{D}_H, \mathbb{Q}) \xrightarrow{\ \text{PD}\ \cong\ } H^{\nu-2}(\tilde{D}_H, \mathbb{Q})$$

we deduce that the image of $[w]_*$ in $\underline{Gr}_N^{r-\nu} H^{2r-\nu}(X, \mathbb{Q})$ is zero, and hence

$$\underline{Gr}_{F_c}^\nu \mathrm{CH}^r(X/\mathbb{C}; \mathbb{Q}) = 0 \Rightarrow \underline{Gr}_N^{r-\nu} H^{2r-\nu}(X, \mathbb{Q}) = 0.$$

Now suppose $\underline{Gr}_N^{r-\nu} H^{2r-\nu}(X, \mathbb{Q}) = 0$, and consider a class $\xi \in \underline{Gr}_{F_c}^\nu \mathrm{CH}^r(X/\mathbb{C}; \mathbb{Q})$ in the image of some w_*, for appropriate $w \in \mathrm{CH}^r(\overline{V} \times X)$. It suffices to show that on the graded level, we have $w_*(Gr_F^\nu \mathrm{CH}_0(\overline{V}/\mathbb{C}; \mathbb{Q})) = 0$ in $Gr_F^\nu \mathrm{CH}^r(X/\mathbb{C}; \mathbb{Q})$, since one can again argue, using the Chow-Künneth decomposition of the diagonals of the $\{C_j\}$ as in §5, that $\xi \in w_*(Gr_F^\nu \mathrm{CH}_0(\overline{V}/\mathbb{C}; \mathbb{Q}))$. Since we are dealing with graded terms, we need only work on the level of cohomology. Since $[w] = 0 \in 0 = \underline{Gr}_N^{r-\nu} H^{2r-\nu}(X, \mathbb{Q})$, it follows from the HC that we can assume w is supported on $\overline{V} \times Y$, where $\mathrm{codim}_X Y = r - \nu + 1$. Let $\tilde{Y} \to Y$ be a desingularization. Then w is induced from $\tilde{w} \in \mathrm{CH}^{\nu-1}(\overline{V} \times \tilde{Y}; \mathbb{Q})$. Note that $\dim \tilde{Y} = d - r + \nu - 1 = (\nu - 2) + (d + 1 - r) \geq \nu - 2$. Choose $\tilde{Y}_H \subset \tilde{Y}$, $\dim \tilde{Y}_H = \nu - 2$, where $\tilde{Y}_H$ is a complete intersection of $\tilde{Y}$ cut out by $(d + 1 - r)$ general hyperplane sections. We have

$$[w]_* : H^\nu(\overline{V}, \mathbb{Q}) \to H^{2r-\nu}(X, \mathbb{Q}),$$

factoring through

$$H^\nu(\overline{V}, \mathbb{Q}) \xrightarrow{[\tilde{w}]_*} H^{\nu-2}(\tilde{Y}, \mathbb{Q}) \to H^{2r-\nu}(X, \mathbb{Q});$$

moreover by the weak Lefschetz theorem, an injection

$$H^{\nu-2}(\tilde{Y}, \mathbb{Q}) \hookrightarrow H^{\nu-2}(\tilde{Y}_H, \mathbb{Q}),$$

with surjective left inverse

$$H^{\nu-2}(\tilde{Y}_H, \mathbb{Q}) \twoheadrightarrow H^{\nu-2}(\tilde{Y}, \mathbb{Q}),$$

which we can assume is cycle induced by the HC. Thus $[w]_* : H^\nu(\overline{V}, \mathbb{Q}) \to H^{2r-\nu}(X, \mathbb{Q})$ factors through

$$H^\nu(\overline{V}, \mathbb{Q}) \xrightarrow{[\tilde{\tilde{w}}_H]} H^{\nu-2}(\tilde{Y}_H, \mathbb{Q}) \twoheadrightarrow H^{\nu-2}(\tilde{Y}, \mathbb{Q}) \to H^{2r-\nu}(X, \mathbb{Q}),$$

where

$$\tilde{\tilde{w}}_H \in \mathrm{CH}^{\nu-1}(\overline{V} \times \tilde{Y}_H; \mathbb{Q}).$$

Finally we have the projection

$$\mathrm{CH}^{\nu-1}(\overline{V} \times \tilde{Y}_H) \to \mathrm{CH}^1(\overline{V}),$$

which implies that
$$w_*\big(Gr^\nu_F\mathrm{CH}^\nu(\overline{V}/\mathbb{C};\mathbb{Q})\big) = 0.$$
This implies that $\underline{Gr}^\nu_{F_c}\mathrm{CH}^r(X/\mathbb{C};\mathbb{Q}) = 0$, and we are done. $\square$

References

[Ar]   D. Arapura, *The Leray spectral sequence is motivic,* Invent. Math. **160** (3), 567-589 (2005).

[As]   M. Asakura, *Arithmetic Hodge structure and nonvanishing of the cycle class of 0-cycles,* K-Theory **27** (2002), no. 3, 273-280.

[F-G]   G. Frey and M. Jarden, *Approximation theory and the rank of Abelian varieties over large algebraic fields,* Proc. London Math. Soc. **28** (3),(1974), 112-128.

[G-G-P]   M. Green, P. A. Griffiths and K. Paranjape, *Cycles over fields of transcendence degree 1,* Michigan Math. J. **52**, (2004), 181-187.

[K]   M. Kerr, *A survey of transcendental methods in the study of Chow groups of zero-cycles,* Mirror Symmetry V, Proceedings of the BIRS workshop on Calabi-Yau Varieties and Mirror Symmetry, Noriko Yui, Shing-Tung Yau, James D. Lewis, Editors, AMS/IP Studies in Advanced Mathematics, Volume **38** (2006), 295-349.

[K-L]   M. Kerr and J. D. Lewis, *The Abel-Jacobi map for higher Chow groups II,* Inventiones Math. **170** (2), (2007), 355-420.

[L1]   J. D. Lewis, *A filtration on the Chow group of a complex projective variety,* Compositio Math. **128**, (2001), 299-322.

[L2]   J. D. Lewis, *Cylinder homomorphisms and Chow groups,* Math. Nachr. **160**, (1993), 205-221.

[L3]   J. D. Lewis, *Generalized Poincaré classes and cubic equivalences,* Math. Nachr. **178** (1996), 249-269.

[L4]   J. D. Lewis, *Introductory Lectures in Transcendental Algebraic Geometry: A Survey of the Hodge Conjecture,* **CRM** Monograph Series, Vol. **10**. American Mathematical Society, Providence, R.I. (1999).

[L-S]   J. D. Lewis and S. Saito, *Algebraic cycles and Mumford-Griffiths invariants,* Amer. J. Math. **129** (2007), no. 6, 1449-1499.

[M]   J. P. Murre, *Lectures on motives, in Transcendental Aspects of Algebraic Cycles,* Edited by S. Müller-Stach and C. Peters, London Math. Soc. Lecture Note Series, No. **133**, Cambridge University Press, (2004), 123-170.

[R-S]   A. Rosenschon and M. Saito, *Cycle map for strictly decomposable cycles,* Amer. J. Math **25**, no. 4, (2003), 773-790.

[HSa]   H. Saito, *The Hodge cohomology and cubic equivalences,* Nagoya Math. J., Vol. **94** (1984), 1-41.

[MSa]   M. Saito, *Filtrations on Chow groups and transcendence degree.* `math.AG/0212057`.

[SSa]   S. Saito, *Motives and filtrations on Chow groups,* Invent. Math. **125** (1996), 149-196.

[Sch1]   C. Schoen, *Varieties dominated by product varieties,* International Journal of Math., Vol. **7**, No. 4 (1996), 541-571.

[Sch2]   C. Schoen, *Zero cycles modulo rational equivalence for some varieties over fields of transcendence degree one,* Proc. Symp. Pure Math **46**, (1987), part 2, 463-473.

[T]   T. Terosoma, *Complete intersections with middle Picard number 1 defined over* $\mathbb{Q}$, Math. Z. **189** (1985), 289-296.

Fields Institute Communications
Volume **56**, 2009

Euler Characteristics and Special Values of Zeta-Functions

Stephen Lichtenbaum
Department of Mathematics
Brown University
Providence, RI 02912 USA
slicht@math.brown.edu

Abstract. This article gives evidence for the philosophy that all special values of arithmetic zeta and L-functions are given by Euler characteristics. We treat here the case of L-functions of 1-motives, where we indicate what the answer ought to be, and sketch an argument showing that this answer is correct up to a rational number modulo the conjecture of Birch and Swinnerton-Dyer.

We would like to propose a very general philosophy for computing the orders of zeroes and special values of arithmetic L-functions. We begin with the notion of Euler characteristic. We say that a collection $\mathcal{C}$ of cohomology groups $(H^i(M))$ is a "set of Euler data" if it satisfies the following conditions (we write $N_{\mathbb{C}}$ for $N \otimes \mathbb{C}$ throughout):

(1) The groups $H^i(M)$ are finitely generated abelian groups which are zero for all but finitely many i.

(2) There exist maps θ_i mapping $H^i(M)_{\mathbb{C}}$ to $H^{i+1}(M)_{\mathbb{C}}$ such that $\theta_{i+1} \circ \theta_i = 0$, and the complex $(H^*(M)_{\mathbb{C}}, \theta)$ is acyclic.

Given a finite exact sequence of finite-dimensional vector spaces $V_{\bullet}$ over a field F:

$$0 \to V_0 \to V_1 \to \cdots \to V_n \to 0$$

with given bases B_i for V_i, there is a standard way to define the determinant of (*) as an element of F^*: If V_i has dimension d_i let $\Lambda(V_i) = \Lambda^{d_i} V_i$,. Then there is a canonical identification of $\Lambda(V_{\bullet}) = \bigotimes \Lambda(V_i)^{(-1)^i}$ with F. If we let b_i be the exterior product of the elements of B_i the b_i are bases for the one-dimensional spaces $\Lambda(V_i)$ and their alternating product gives an element of F, which is the determinant of $V_{\bullet}$. If the basis of one of these vector spaces V is changed by an element A of $GL(V)$, then the determinant is changed by the determinant of A. If V is obtained by tensor product with F from a finitely generated abelian group M we may take

2000 *Mathematics Subject Classification.* Primary 14G10.

The author was partially supported by N.S.F. grant DMS-0501064 and also by the Institute for Advanced Study and the University of Tokyo.

a basis of V which comes from M. If we do this for all the vector spaces in a set of Euler data, we get a determinant in F^* which is well defined up to ± 1.

Now, given a set $\mathcal{C}$ of Euler data, we define the *Euler characteristic* of $\mathcal{C}$ to be the alternating product and quotient of the orders of the torsion subgroups of $H^i(M)$, divided by the determinant of $H^*(M)_\mathbb{C}$, which is a well-defined element of $\mathbb{C}^*/(\pm 1)$. We also define the *rank* of $\mathcal{C}$ to be $\sum_i (-1)^i i \operatorname{rank}(H^i(M))$. The basic idea is that, given any motive M, there should be sets $\mathcal{C}_j$ of Euler data such that the order of the zero of $L(M,s)$ at $s = 0$ is given by the sum of the ranks of the $\mathcal{C}_j$'s and the leading term $L^*(M,0)$ of $L(M,s)$ at $s = 0$ is given by the product of the Euler characteristics of the $\mathcal{C}_j$'s.

If M is a 1-motive, there are two sets of Euler data associated with M. One should be thought of as the cohomology groups of M itself, and the other the cohomology of the tangent space $t(M)$ of M. The cohomology groups should be Weil-étale cohomology with compact support of M. Unfortunately this cohomology theory has only been defined (see [L]) for discrete 1-motives, and so we have to make educated guesses about what it should be in other situations.

We begin with the simplest examples. First, let $M = \mathbb{Z}$ over a number field F. Then we put $H^i(\mathbb{Z})$ equal to the Weil-étale cohomology groups $H^i_c(Y, \mathbb{Z})$ as defined in [L], which we will recall below. (Here O_F is the ring of integers in F, $Y = \operatorname{Spec} O_F$, and H^i_c denotes the cohomology with compact support at the infinite primes.)

Define $\widetilde{Pic}(\bar{Y})$ to be the idèle class group of F divided by the image of the unit idèles, and $\widetilde{Pic}^1(\bar{Y})$ to be the kernel of the norm map to $\mathbb{R}^*$. (We think of $\bar{Y}$ as being the union of Y and the infinite primes of Y.) We define μ_F to be the group of roots of unity in F, and let D denote Pontriagin dual.

It is an easy exercise to verify that there exists an exact sequence

$$0 \to Pic(F)^D \to (\widetilde{Pic}(\bar{Y}))^D \to Hom(O_F^*, \mathbb{Z}) \to 0$$

where O_F^* is the group of units of O_F.

We recall the following computation from [L]: $H^0_c(Y, \mathbb{Z}) = 0$, $H^1_c(Y, \mathbb{Z}) = \coprod_v \mathbb{Z}/\mathbb{Z}$, where the sum is over the infinite places v, and $\mathbb{Z}$ is embedded in the sum by the diagonal map, $H^2_c(Y, \mathbb{Z}) = (\widetilde{Pic}^1(\bar{Y}))^D$, and $H^3_c(Y, \mathbb{Z}) = (\mu_F)^D$. We define the higher cohomology groups to be zero. Note that we say "define" rather than "compute" because Flach has shown ([Fl2] that the higher Weil-étale cohomology groups of [L] are in general enormous, and certainly not zero.

The map from $H^1_c(Y, \mathbb{Z}) \otimes \mathbb{R} = \coprod_v \mathbb{R}/\mathbb{R}$ to $H^2_c(Y, \mathbb{Z}) \otimes \mathbb{R} = Hom(O_F^*, \mathbb{R})$ is given by $1_v \mapsto (u \mapsto log|u|_v)$. In fact this is given by cup-product with the element $log(Norm)$ in $H^1(\bar{Y}, \mathbb{R})$, but this is not necessary at this point in the story. The determinant of this map is the classical regulator R_F, and the Euler characteristic is given by $h_F R_F / w_F$, where $h_F = |(Pic(F)|$ is the class number of F, and $w_F = |(\mu_F)|$ is the number of roots of unity in F. This is well known to be $-\zeta^*(F,0)$. The rank of this set of Euler data is given by $(-1)(r_1 + r_2 - 1) + 2(r_1 + r_2 - 1) = r_1 + r_2 - 1$ which is the order of the zero of $\zeta(F,s)$ at $s = 0$. In this case, the tangent space $t(\mathbb{Z})$ is equal to 0, and so its rank and Euler characteristic do not contribute to the formula. For the next example, let $M = \mathbb{Z}_F(1)$, so that $L(M,s) = \zeta(F, s+1)$. Here we cannot use the Weil-étale topology defined in [L], because we do not know how to define a sheaf corresponding to $\mathbb{Z}(1)$. Nonetheless, we describe what the cohomology groups should be (up to torsion):

Let σ denote an embedding of F into $\mathbb{C}$. We allow real embeddings and we count conjugate embeddings as distinct. Let F_σ be the completion of F at the valuation induced by σ. The embedding σ of F in $\mathbb{C}$ extends to F_σ. We have the following diagram, where the row is exact:

$$0 \longrightarrow \mathbb{Z}^{r_2} \longrightarrow \coprod_\sigma F_\sigma \xrightarrow{\exp} \coprod F_\sigma^*$$

$$\uparrow$$

$$O_F^*$$

The map from O_F^* to $\coprod F_\sigma^*$ is given by $u \mapsto (\sigma(u))$.

We define $H_c^0(Y, \mathbb{Z}(1))$ to be zero, and $H_c^1(Y, \mathbb{Z}(1))$ to be the fibered product $\coprod_\sigma F_\sigma \times_{\coprod F_\sigma^*} O_F^*$. We define $H_c^2(Y, \mathbb{Z}(1))$ to be the kernel of the sum map from $\coprod_\sigma \mathbb{Z}$ to $\mathbb{Z}$, and define the higher cohomology groups to be zero. Observe that we have the exact sequence

$$0 \to \mathbb{Z}^{r_2} \to H_c^1(Y, \mathbb{Z}(1)) \to O_F^*$$

where the last map has a finite 2-torsion cokernel. We see that $H_c^1(Y, \mathbb{Z}(1))$ and $H_c^2(Y, \mathbb{Z}(1))$ are both finitely generated abelian groups with rank $r_1 + 2r_2 - 1$. So the H_c^i's form a set of Euler data with rank equal to $r_1 + 2r_2 - 1$.

Define a map θ_c from $H_c^1(Y, \mathbb{Z}(1))_\mathbb{C}$ to $H_c^2(Y, \mathbb{Z}(1))_\mathbb{C}$ as follows: Let $x = ((a_\sigma,), u)$ be an element of H^1, so $\exp(a_\sigma) = \sigma(u)$. Map x to the collection $\sigma(a_\sigma)$, and verify, using the product formula, that the sum is equal to zero because u is a unit. An easy computation shows the Euler characteristic is given by $(2\pi i)^{r_2} R_F$, up to a rational number.

We now introduce another set of Euler data associated with the motive $\mathbb{Z}(1)_F$. Let $H_a^0(Y, \mathbb{Z}(1))$ be O_F, and $H_a^1(Y, \mathbb{Z}(1))$ be $\coprod_\sigma \mathbb{Z}$. Define a map θ_a from O_F to $(\coprod_\sigma \mathbb{Z})_\mathbb{C} = \coprod_\sigma \mathbb{C}$ by sending x in O_F to the collection $(\sigma(x))$. Extend θ_a by linearity to $(O_F)_\mathbb{C}$. It is well known that the determinant of θ_a is given by $\sqrt{(d_F)}$ where d_F is the discriminant of F. Hence the rank of $H_a^*(Y, \mathbb{Z}(1))$ is equal to $-(r_1 + 2r_2)$ and the Euler characteristic is equal to $(\sqrt{(d_F)})^{-1}$.

Notice that the sum of the ranks of these two sets of Euler data is $(r_1 + 2r_2 - 1) - (r_1 + 2r_2) = -1$ which is the order of the zero of $\zeta(F, s)$ at $s = 1$. Also the product of the two Euler characteristics is equal to $(2\pi i)^{r_2} R_F (\sqrt{d_F})^{-1}$ which is equal, up to a rational number, to the classical formula for the residue $\zeta^*(F, 1)$ of the zeta-function of F at $s = 1$, namely $2^{r_1}(2\pi)^{r_2} h_F R_F w_F^{-1} \sqrt{|d_F|}^{-1}$, in view of the fact that $i^{r_2} \sqrt{d_F} = \sqrt{|d_F|}$, up to sign.

We next consider the case of a general 1-motive over $\mathbb{Q}$, (in fact, over $\mathbb{Z}$, see below).

We need to begin by saying a few words about "1-motives with torsion". It will be convenient for our purposes to enlarge Deligne's category of 1-motives by allowing torsion. Let first K be an algebraically closed field. Given any discrete group X we can construct a group scheme over K whose points correspond exactly to the points of X and where the group law is given by the group law on X. If X is a finitely generated abelian group, the group scheme $T = Hom(X, G_m)$ is said to be a group of multiplicative type, and if X is free T is a torus. A group scheme G is a semi-abelian variety with torsion if it is an extension of an abelian variety by a group of multiplicative type.

Define $M = u : X \to G$ to be a 1-motive with torsion over F if X is a group scheme which is locally for the étale topology isomorphic to a discrete finitely generated abelian group, and G is a semi-abelian variety with torsion over F. It is possible to extend Deligne's definition of duality for 1-motives to the category of 1-motives with torsion, and this extension will be included in the Brown thesis of Donghoon Park. This construction will be necessary to give formulas valid up to sign and not just up to $\mathbb{Q}^*$.

We will be working essentially in the category of Deligne 1- $\mathbb{Q}$-motives over $\mathbb{Q}$. More precisely, the objects of our category are equivalence classes of triples (X, G, u), where X is a group scheme over $\mathbb{Q}$ locally in the étale topology isomorphic to a constant group scheme corresponding to a finitely generated abelian group, G is a semi-abelian variety with torsion over $\mathbb{Q}$, (so we write $0 \to T \to G \to A \to 0$, where T is a group of multiplicative type and A is an abelian variety) and u is a map of group schemes from X to G. We have the obvious notion of morphisms (f_X, f_G) of triples, and we take the equivalence relation generated by saying that $M_1 = (X_1, G_1, u_1)$ is equivalent to $M_2 = (X_2, G_2, u_2)$ if there exists a map (f_X, f_G) from M_1 to M_2 such that f_X and f_G have finite kernels and cokernels.

Note that any triple (X, G, u) is equivalent to one where X is torsion-free and G is semi-abelian, so an actual Deligne 1-motive. Let $M = (X, G, u)$ be a 1-motive over $\mathbb{Q}$, and choose an algebraic closure $\bar{\mathbb{Q}}$ of $\mathbb{Q}$. Let $M^\sharp$ be the 1-motive (X, A, v), where v is obtained by composing u with the projection from G to A.

Following [De2], [Fl] and [Fo] we associate with M various finite- dimensional vector spaces over $\mathbb{Q}$; the Betti cohomology $H_B(M) = H_B(M, \mathbb{Z}) \otimes \mathbb{Q}$, the motivic cohomology spaces $H_f^0(M)$ and $H_f^1(M)$, and the de Rham cohomology $H_{DR}(M)$. If $X = 0$ we identify $H_f^1(M)$ with $G_{\mathbb{Z}} \otimes \mathbb{Q}$, where we define $G_{\mathbb{Z}}$ to be the subset of $G(\mathbb{Q})$ consisting of those points whose image lies in the maximal compact subgroup of $G(\mathbb{Q}_p)$ for all finite primes p. The motivic cohomology groups may be described in the general case as follows: $H_f^0(M)$ is equal to $Ker(H_f^0(X) \to G_{\mathbb{Z}} \otimes \mathbb{Q})$, and $H_f^1(M)$ is equal to $Coker(H_f^0(X) \to G_{\mathbb{Z}} \otimes \mathbb{Q})$.

Now assume in addition that M is a 1-motive over $\mathbb{Z}$, by which we mean that the image of the natural map $H_f^0(X) \to G(\mathbb{Q}) \otimes \mathbb{Q}$ lies in $G_{\mathbb{Z}} \otimes \mathbb{Q}$. Let $\widetilde{M}$ be the 1-motive given by $G_{\mathbb{Z}} \times_{G(\mathbb{C})} X(\mathbb{C}) \to G$, where the map into G is given by $(x, y) \mapsto x - u(y)$. Complex conjugation c acts on the Betti cohomology. We define H_B^+ to be H_B^c (the fixed points of H_B under c). We have the period isomorphism γ_M which maps $H_B(M)_{\mathbb{C}}$ to $H_{DR}(M)_{\mathbb{C}}$. This induces Deligne's map α_M which maps $H_B^+(M)_{\mathbb{C}}$ to $(t_M)_{\mathbb{C}}$, which of course is not in general an isomorphism.

Let $M' = \widetilde{M^D}$, where M^D denotes Deligne's dual 1-motive ([De2]) to M (recall that M^D is actually $M^*(1)$, where, confusingly, M^* is the dual motive to M), and let M'' denote $(M')'$. We write $M' = X' \to G'$, $M'' = X'' \to G''$, and $0 \to T' \to G' \to A' \to 0$ writing the semi-abelian variety G' as an extension of the abelian variety A' by the torus T'. Note that if $M = A$ is an abelian variety, then M'^D is the semiabelian variety attached to A by Bloch in [B]. This construction is very closely related to one introduced by Scholl ([S]) for the full (but conjectural) category of motives over $\mathbb{Q}$.

Let $v_M = H_B(M^D)^\sharp / H_B^-(A^t)$.

Direct computation yields the following two exact sequences:

$$0 \;\rightarrow\; H_f^0(M) \rightarrow H_B^+(M) \rightarrow H_B^+(\widetilde{M}) \rightarrow H_f^1(M) \rightarrow 0, \tag{1}$$

$$0 \;\rightarrow\; H_f^0(M^D) \rightarrow v_M \rightarrow v_M'' \rightarrow H_f^1(M^D) \rightarrow 0. \tag{2}$$

Now define $H_c^i(M)$ to be $Hom(H_f^{2-i}(M^D), \mathbb{Q})$, for $i = 1, 2$. and let $u_M = Hom(v_M, \mathbb{Q})$. Taking the $\mathbb{Q}$-dual of (2), we obtain:

$$0 \rightarrow H_c^1(M) \rightarrow u_{M''} \rightarrow u_M \rightarrow H_c^2(M) \rightarrow 0. \tag{3}$$

It is easy to check that we may identify $(u_M)_{\mathbb{C}}$ with $(t_M)_{\mathbb{C}}$, so we have:

$$0 \rightarrow H_c^1(M)_{\mathbb{C}} \rightarrow (t_{M''})_{\mathbb{C}} \rightarrow (t_M)_{\mathbb{C}} \rightarrow H_c^2(M)_{\mathbb{C}} \rightarrow 0. \tag{4}$$

Further, we have:

Proposition 0.1 *For any 1-motive M over $\mathbb{Z}$, the canonical map α_M'' from $H_B^+(M'')_{\mathbb{C}}$ to the tangent space $(t_M'')_{\mathbb{C}}$ is an isomorphism.*

Proof This immediately reduces to the cases when the 1-motive M is equal to $X, T,$ or A. If $M = X$ standard techniques reduce to the case when $M = R_{F/\mathbb{Q}}(\mathbb{Z})$, and if $M = T$, the same techniques reduce to the case when $M = R_{F/\mathbb{Q}}(\mathbb{Z}(1))$, which were both essentially treated in the first part of this article. (The computation for $\mathbb{Z}_F$ is the same as for $R_{F/\mathbb{Q}}(\mathbb{Z})$, and the computation for $\mathbb{Z}(1)_F$ is the same as that for $R_{F/\mathbb{Q}}\mathbb{Z}(1)$. If $M = A$, the result follows from [Fo]). □

Warning. The obvious generalization of Proposition 0.1 to 1-motives over number fields is false. It can be generalized, but care must be taken.

We are now going to explain our conjectures about L-functions of 1-motives and explain how they relate to the Bloch-Kato conjectures. Actually, many people have made very substantial contributions to these conjectures, which ideally should be called (at least) the Deligne-Beilinson-Scholl-Bloch-Kato-Fontaine-Perrin-Riou conjectures, but I think Bloch-Kato is the most accepted term.

It is an easy exercise to verify that $H_B^+(\widetilde{M})$ is naturally isomorphic to $H_B^+(M'')$. (More precisely, there is a functorial map from M'' to $\widetilde{M}$ which induces this isomorphism). To define our first set of Euler data we put $H^1(M)$ equal to either of these, and $H^2(M) = u_{M'^D} = u_{M''}$. Technically speaking, this is not a set of Euler data, but since we are only working up to $\mathbb{Q}^*$, we may replace finitely-generated abelian groups by finite-dimensional vector spaces over $\mathbb{Q}$, and ranks by dimensions. Since we have seen that $H^2(M)_{\mathbb{C}}$ is naturally isomorphic to $(t_{M''})_{\mathbb{C}}$, Proposition 1.1 implies that the map $\alpha_{M''}$ induces an isomorphism from $H^1(M)_{\mathbb{C}}$ to $H^2(M)_{\mathbb{C}}$, thus giving us a set of Euler data.

We define our second set of Euler data by $H_a^0(M) = t_M$, and $H_a^1(M) = u_M$. The rank of the first set of Euler data is equal to the common dimension of $H^1(M)$ and $H^2(M)$. The rank of the second set is equal to the negative of the common dimension of u_M and t_M. Hence the rank of the sum of these two sets of Euler data is is equal to dimension $(H^2(M))$ - dimension (t_M), which by the second exact sequence is equal to dimension $(H_c^1(M))$ - dimension $(H_c^2(M))$, which is what the Bloch-Kato conjecture says should be equal to the order of the zero of the L-function $L(M, s)$ at $s = 0$, so the two conjectures are equivalent.

As for $L^*(M, 0)$, our conjecture is that this is equal, up to $\mathbb{Q}^*$, to the product of the Euler characteristics of our two sets of Euler data. On the other hand, in our language, the Bloch-Kato conjecture says that $L^*(M, 0)$ is equal to the product of the Euler characteristic of the sequence (6) with the Euler characteristic of the sequence:

$$0 \to Ker(\alpha_M) \to (H_B^+(M))_{\mathbb{C}} \to (t_M)_{\mathbb{C}} \to Coker(\alpha_M) \to 0. \tag{5}$$

(Note that all the vector spaces in these sequences have natural $\mathbb{Q}$-bases except for Ker (α_M) and Coker (α_M), but we may choose any rational bases here since they each occur twice with opposite signs.)

An immediate formal calculation shows the equivalence of these two conjectures. Of course, the actual Bloch-Kato conjecture predicts the value of $L^*(M, 0)$ up to sign, not just up to $\mathbb{Q}^*$. We hope to return to this point in a future paper.

Of course, we know the truth of these for $\mathbb{Z}$ and $\mathbb{Z}(1)$, hence for any discrete group X and any torus T, and the Bloch-Kato conjectures for A are equivalent to the conjecture of Birch and Swinnerton-Dyer for A ([Fo]).

If we apply our sequences to the case when M is an abelian variety A over $\mathbb{Q}$, we obtain the following diagram with exact rows:

$$
\begin{array}{ccccccccc}
0 & \longrightarrow & H_B^+(A)_{\mathbb{C}} & \xrightarrow{f} & H_B^+(\widetilde{A})_{\mathbb{C}} & \xrightarrow{g} & A(\mathbb{Q})_{\mathbb{C}} & \longrightarrow & 0 \\
 & & & & \alpha \downarrow & & & & \\
0 & \longrightarrow & Hom(B(\mathbb{Q}), \mathbb{C}) & \xrightarrow{\phi} & (t_{A''})_{\mathbb{C}} & \xrightarrow{\gamma} & (t_A)_{\mathbb{C}} & \longrightarrow & 0
\end{array}
$$

In view of the fact that the composite $\gamma \alpha f$ is equal to the isomorphism α_A, we see that we get a map from $A(\mathbb{Q})_{\mathbb{C}}$ to $Hom(B(\mathbb{Q}), \mathbb{C})$ by lifting an element x in $A(\mathbb{Q})_{\mathbb{C}}$ to an element y in $H_B^+(\widetilde{A})_{\mathbb{C}}$, changing y to a new element y' mapping to x such that $\gamma(\alpha(y')) = 0$ by subtracting an element of the form $f(z)$, and then taking the unique u in $Hom(B(\mathbb{Q}), \mathbb{C})$ such that $\phi(u) = \alpha(y')$. It is then not difficult to see, using the description of the height pairing given in [B], that this map too gives the height pairing.

As a final remark, we point out that if we tensor sequences (1) and (3) with $\mathbb{C}$ and make use of the isomorphism $\alpha_{M''}$ from $H_B^+(\widetilde{M})_{\mathbb{C}}$ to $(t_{M''})_{\mathbb{C}}$, we obtain by diagram-chasing the exact sequence

$$0 \to H_f^0(M)_{\mathbb{C}} \to Ker(\alpha_M) \to H_c^1(M)_{\mathbb{C}} \to H_f^1(M)_{\mathbb{C}}$$
$$\to Coker(\alpha_M) \to H_c^2(M)_{\mathbb{C}} \to 0. \tag{6}$$

This sequence is given (over $\mathbb{R}$) in [Fo] and [Fl1], p. 81.

In order to do the diagram-chasing, we recall that the isomorphism β_M from $H_B^+(M'')$ to $H_B^+(\widetilde{M})$ leads to a description of α_M as the map from $(H_B^+(M))_{\mathbb{C}}$ to $(H_B^+(\widetilde{M}))_{\mathbb{C}}$ followed by $(\beta_M^{-1})_{\mathbb{C}}$ followed by the isomorphism $\alpha_{M''}$ followed by the natural map from $(t_{M''})_{\mathbb{C}}$ to $(t_M)_{\mathbb{C}}$.

But notice that although all the maps in (5) are transcendental, the maps in (1) and (3) are all defined over $\mathbb{Q}$, making it clear that all the transcendentality comes from the isomorphism $\alpha_{M''}$.

References

[B]   Bloch, S. A note on height pairings, Tamagawa numbers, and the Birch and Swinnerton-Dyer conjecture, Inv. Math. 58 (1980) 65-76

[Del]   Deligne, P. Théorie de Hodge III, Inst. Hautes Études Sci. Publ. Math. 44, 5-77

[De2] Deligne, P. Valeurs de fonctions L et pèriodes d'intégrales. Proc. Sympos. Pure Math., XXXIII, Automorphic forms, representations and L-functions (Proc. Sympos. Pure Math., Oregon State Univ., Corvallis, Ore., 1977), Part 2, pp. 313–346, Amer. Math. Soc., Providence, R.I., 1979.

[Fl1] Flach, M. The equivariant Tamagawa number conjecture: a survey. With an appendix by C. Greither. Contemp. Math., 358, Stark's conjectures: recent work and new directions, 79–125, Amer. Math. Soc., Providence, RI, 2004.

[FL2] Flach, M. Cohomology of topological groups with applications to the Weil group , to appear, Comp. Math.

[Fo] Fontaine, J.-M. Valeurs spéciales des fonctions L des motifs. Séminaire Bourbaki, Vol. 1991/92. Astèrisque No. 206 (1992), Exp. No. 751, 4, 205–249.

[L] Lichtenbaum, S. The Weil-étale topology for number rings, to appear, Annals of Math.

[S] Scholl, A. J. Height pairings and special values of L-functions. Motives (Seattle, WA, 1991), 571–598, Proc. Sympos. Pure Math., 55, Part 1, Amer. Math. Soc., Providence, RI, 1994.

Fields Institute Communications
Volume **56**, 2009

Local Galois Symbols on $E \times E$

Jacob Murre

Department of Mathematics
University of Leiden
2300 RA Leiden
Leiden, Netherlands
murre@math.leidenuniv.nl

Dinakar Ramakrishnan

Department of Mathematics
California Institute of Technology
Pasadena, CA 91125 USA
dinakar@caltech.edu

To Spencer Bloch, with admiration

Abstract. This article studies the Albanese kernel $T_F(E \times E)$, for an elliptic curve E over a p-adic field F. The main result furnishes information, for any odd prime p, about the kernel and image of the Galois symbol map from $T_F(E \times E)/p$ to the Galois cohomology group $H^2(F, E[p] \otimes E[p])$, for E/F ordinary, without requiring that the p-torsion points are F-rational, or even that the Galois module $E[p]$ is semisimple. A key step is to show that the image is zero when the finite Galois module $E[p]$ is acted on non-trivially by the pro-p-inertia group I_p. Non-trivial classes in the image are also constructed when $E[p]$ is suitably unramified. A forthcoming sequel will deal with global questions.

Introduction

Let E be an elliptic curve over a field F, $\overline{F}$ a separable algebraic closure of F, and ℓ a prime different from the characteristic of F. Denote by $E[\ell]$ the group of ℓ-division points of E in $E(\overline{F})$. To any F-rational point P in $E(F)$ one associates by Kummer theory a class $[P]_\ell$ in the Galois cohomology group $H^1(F, E[\ell])$,

2000 *Mathematics Subject Classification.* Primary 11G07, 14C25, 11G10; Secondary 11S25.
Partially supported by the NSF.

represented by the 1-cocycle

$$\beta_\ell : \mathrm{Gal}(\overline{F}/F) \to E[\ell], \ \sigma \mapsto \sigma\left(\frac{P}{\ell}\right) - \frac{P}{\ell}.$$

Here $\frac{P}{\ell}$ denotes any point in $E(\overline{F})$ with $\ell\left(\frac{P}{\ell}\right) = P$. Given a pair (P,Q) of F-rational points, one then has the cup product class

$$[P,Q]_\ell := [P]_\ell \cup [Q]_\ell \in H^2(F, E[\ell]^{\otimes 2}).$$

Any such pair (P,Q) also defines a F-rational algebraic cycle on the surface $E \times E$ given by

$$\langle P,Q\rangle := [(P,Q) - (P,0) - (0,Q) + (0,0)],$$

where $[\cdots]$ denotes the class taken modulo *rational equivalence*. It is clear that this *zero cycle of degree zero* defines, by the parallelogram law, the trivial class in the Albanese variety $Alb(E \times E)$. So $\langle P,Q\rangle$ lies in the *Albanese kernel* $T_F(E \times E)$. It is a known, but non-obvious, fact that the association $(P,Q) \to [P,Q]_\ell$ depends only on $\langle P,Q\rangle$, and thus results in the *Galois symbol map*

$$s_\ell : T_F(E \times E)/\ell \to H^2(F, E[\ell]^{\otimes 2}).$$

For the precise technical definition of s_ℓ we refer to Definition 2.5.3. In essence, s_ℓ is the restriction to the Albanese kernel $T_F(E \times E)$ of the cycle map over F with target the continuous cohomology (in the sense of Jannsen [J]) with $\mathbb{Z}/\ell$-coefficients. Due to the availability of the norm map for finite extensions (see Section 1.7), it suffices to study this map s_ℓ on the subgroup $ST_F(E \times E)$ of $T_F(E \times E)$ generated by symbols, whence the christening of s_ℓ as the *Galois symbol map*. Moreover, it is possible to study this map via a Kummer sequence (see Sections 2.2 and 2.3).

It is a conjecture of Somekawa and Kato that this map is always injective ([So]). It is easy to verify this when F is $\mathbb{C}$ or $\mathbb{R}$. In the latter case, the image of s_ℓ is non-trivial iff $\ell = 2$ and all the 2-torsion points are $\mathbb{R}$-rational, which can be used to exhibit a non-trivial global 2-torsion class in $T_\mathbb{Q}(E \times E)$, whenever E is defined by $y^2 = (x-a)(x-b)(x-c)$ with $a,b,c \in \mathbb{Q}$. It should also be noted that injectivity of the analog of s_ℓ fails for certain surfaces occurring as quadric fibrations (cf. [ParS]). However, the general expectation is that such pathologies do not occur for *abelian* surfaces.

Let F denote a non-archimedean local field, with ring of integers $\mathcal{O}_F$ and residual characteristic p. Let E be a semistable elliptic curve over F, and $\mathcal{E}[\ell]$ the kernel of multiplication by ℓ on $\mathcal{E}$, which defines a finite flat groupscheme $\mathcal{E}[\ell]$ over $S = \mathrm{Spec}(\mathcal{O}_F)$. Let $F(E[\ell])$ denotes the smallest Galois extension of F over which all the ℓ-division points of E are rational. It is easy to see that the image of s_ℓ is zero if (i) E has good reduction *and* (ii) $\ell \neq p$, the reason being that the absolute Galois group G_F acts via its maximal unramified quotient $\mathrm{Gal}(F_{nr}/F) \simeq \hat{\mathbb{Z}}$, which has cohomological dimension 1. So we will concentrate on the more subtle $\ell = p$ case.

Theorem A   *Let F be a non-archimedean local field of characteristic zero with residue field $\mathbb{F}_q$, $q = p^r$, p odd. Suppose E/F is an elliptic curve over F, which has good, ordinary reduction. Then the following hold:*

   (a) s_p *is injective, with image of $\mathbb{F}_p$-dimension ≤ 1.*

   (b) *The following are equivalent:*

   (bi) $dim\, Im(s_p) = 1$,
   (bii) $F(E[p])$ *is unramified over* F, *with the prime-to-p part of* $[F(E[p]) : F]$
        being ≤ 2, *and* $\mu_p \subset F$.
 (c) *If* $E[p] \subset F$, *then* $T_F(E \times E)/p \simeq \mathbb{Z}/p$ *consists of symbols* $\langle P, Q \rangle$ *with* P, Q
    in $E(F)/p$.

Note that $[F(E[p]) : F]$ is prime to p iff the $\mathrm{Gal}(\overline{F}/F)$-representation ρ_p on $E[p]$ is semisimple. We obtain:

Corollary B $T_F(E \times E)$ *is p-divisible when* $E[p]$ *is non-semisimple and wildly ramified.*

When all the p-division points of E are F-rational, the injectivity part of part (a) has already been asserted, without proof, in [R-S], where the authors show the interesting result that $T_F(E \times E)/p$ is a quotient of $K_2(F)/p$. Our techniques are completely disjoint from theirs, and besides, the delicate part of our proof of injectivity is exactly when $[F(E[p]) : F]$ is divisible by p, which is equivalent to the Galois module $E[p]$ being non-semisimple. Our results prove in fact that when $T_F(A)/p$ is non-zero, i.e., when $F(E[p])/F$ is unramified of degree ≤ 2, it is isomorphic to the p-part $\mathrm{Br}_F[p]$ of the Brauer group of F, which is known ([Ta1]) to be isomorphic to $K_2(F)/p$.

A completely analogous result concerning s_ℓ holds when E/F has multiplicative reduction, for both $\ell = p$ and $\ell \neq p$, and this can be proved by arguments similar to the ones we use in the ordinary case. However, there is already an elegant paper of Yamazaki ([Y]) giving the analogous result (in the multiplicative case) by a different method, and we content ourselves to a very brief discussion in section 7 on how to deduce this analogue from [Y].

The relevant preliminary material for the paper is assembled in the first two sections and in the Appendix. We have, primarily (but not totally) for the convenience of the reader, supplied proofs of various statements for which we could not find published references, even if they are apparently known to experts or in the *folklore*. After this the proof of Theorem A is achieved in the following sections via a number of Propositions. In section 3, we first prove, via the key Proposition C, that the image of s_p is at most one and establish part of the implication $(bi) \implies (bii)$; the remaining part of this implication is proved in section 4 via Proposition D. The converse implication $(bii) \implies (bi)$ and part (c) are proved in section 5. Finally, part (a), giving the injectivity of s_p, is proved in section 6.

In a sequel we will use these results in conjunction with others (including a treatment of the case of supersingular reduction, p-adic approximation, and a local-global lemma) and prove two global theorems about the Galois symbols on $E \times E$ modulo p, for any odd prime p.

The second author would like to acknowledge support from the National Science Foundation through the grants DMS-0402044 and DMS-0701089, as well as the hospitality of the *DePrima Mathematics House* at Sea Ranch, CA. The first author would like to thank Caltech for the invitation to visit Pasadena, CA, for several visits during the past eight years. This paper was finalized in May 2008. We are grateful to various mathematicians who pointed out errors in a previous version, which we have fixed. Finally, it is a pleasure to thank the referee of this paper, and also Yamazaki, for their thorough reading and for pointing out the need for further explanations; one such comment led to our noticing a small gap, which we have luckily been able to take care of with a modified argument.

It gives us great pleasure to dedicate this article to Spencer Bloch, whose work has inspired us over the years, as it has so many others. The first author (J.M.) has been a friend and a fellow *cyclist* of Bloch for many years, while the second author (D.R.) was Spencer's post-doctoral mentee at the University of Chicago during 1980-82, a period which he remembers with pleasure.

1 Preliminaries on Symbols

1.1   Cycles

Let F be a field and X a smooth projective variety of dimension d and defined over F. Let $0 \leq i \leq d$ and $\mathcal{Z}_i(X)$ be the group of algebraic cycles on X defined over F and of dimension i, i.e. the free group generated by the subvarieties of X of dimension i and defined over F (see [F], section 1.2). A cycle of dimension i is called rationally equivalent to zero over F if there exists $T \in \mathcal{Z}_{i+1}(\mathbb{P}^1 \times X)$ and two points P and Q on $\mathbb{P}^1$, each rational over F, such that $T(P) = Z$ and $T(Q) = 0$, where for $R \in \mathbb{P}^1$ we put $T(R) = \mathrm{pr}_2(T.(R \times X))$ in the usual sense of calculation with cycles (see [F]). Let $\mathcal{Z}_i^{\mathrm{rat}}(X)$ be the subgroup of $Z_i(X)$ generated by the cycles rationally equivalent to zero and $\mathrm{CH}_i(X) = \mathcal{Z}_i(X)/\mathcal{Z}_i^{\mathrm{rat}}(X)$ the Chow group of rational equivalence classes of i-dimensional cycles on X defined on F.

If $K \supset F$ is an extension then we write $X_K = X \times_F K$, $Z_i(X_K)$ and $\mathrm{CH}_i(X_K)$ for the corresponding groups defined over K and if $\bar{F}$ is the algebraic closure of F then we write $\bar{X} = X_{\bar{F}}$, $\mathcal{Z}_i(\bar{X})$ and $\mathrm{CH}_i(\bar{X})$.

1.2   The Albanese kernel

In this paper we are only concerned with the case when X is of dimension 2, i.e. X is a *surface* (and in fact we shall only consider very special ones, see below) and with groups of 0-dimensional cycles contained in $\mathrm{CH}_0(X)$. In that case note that if $Z \in \mathcal{Z}_0(X)$ then we can write $Z = \sum n_i P_i$ when $P_i \in X(\bar{F})$, i.e. the P_i are points on X – but themselves only defined (in general) over an extension field K of F. Put $A_0(X) = \mathrm{Ker}(\mathrm{CH}_0(X) \xrightarrow{\deg} \mathbb{Z})$ where $\deg(Z) = \sum n_i \deg(P_i)$, i.e. $A_0(X)$ are the cycle classes of degree zero and put further $T(X) := \mathrm{Ker}(A_0(X) \to \mathrm{Alb}(X))$, where $\mathrm{Alb}(X)$ is the Albanese variety of X (see [Bl]). Again if $K \supset F$, write $A_0(X_K)$, $T(X_K)$ or also $T_K(X)$ for the corresponding groups over K.

It is known that both $A_0(\bar{X})$ and $T(\bar{X})$ are *divisible* groups ([Bl]). Moreover if the group of transcendental cycles $H^2(\bar{X})_{\mathrm{trans}}$ is non-zero, and if $\bar{F}$ is a universal domain (ex. $\bar{F} = \mathbb{C}$), the group $T(\bar{X})$ is "huge" (infinite dimensional) by Mumford (in characteristic zero) and Bloch (in general) ([Bl], 1.22).

1.3   Symbols

We shall only be concerned with surfaces X which are abelian, even of the form $A = E \times E$, where E is an *elliptic curve* defined over F. If $P, Q \in E(F)$, put

$$\langle P, Q \rangle := (P, Q) - (P, O) - (O, Q) + (O, O)$$

where O is the origin on E, and the addition is the *addition of cycles* (or better: cycle classes). Clearly $\langle P, Q \rangle \in T_F(A)$ and $\langle P, Q \rangle$ is called the *symbol* of P and Q.

Definition $ST_F(A)$ denotes the subgroup of $T_F(A)$ generated by symbols $\langle P, Q \rangle$ with $P \in E(F)$, $Q \in E(F)$. It is called the *symbol group* of $A = E \times E$.

1.4   Properties and remarks

1. The symbol is *bilinear* in P and Q. For instance
$$\langle P_1 + P_2, Q \rangle = \langle P_1, Q \rangle + \langle P_2, Q \rangle$$
(where now the first $+$ sign is addition on E!).
2. The symbol is the Pontryagin product
$$\langle P, Q \rangle = \{(P, 0) - (0, 0)\} * \{(0, Q) - (0, 0)\}.$$
3. If F is finitely generated over the prime field, then it follows from the theorem of Mordell-Weil that $ST_F(A)$ is finitely generated.
4. Note the similarity with the group $K_2(F)$ of a field. Also the symbol group $ST_F(A)$ is related to, but different from, the group $K_2(E \times E)$ defined by Somekawa ([So]); compare also with [R-S].

1.5   Restriction and norm/corestriction

Let $K \supset F$. Consider the morphism $\varphi_{K/F} \colon A_K \to A_F$.

a. This induces homomorphisms, called the *restriction* homomorphisms:
$$\mathrm{res}_{K/F} = \varphi_{K/F}^* \colon \mathrm{CH}_i(A_F) \to \mathrm{CH}_i(A_K), T_F(A) \to T_K(A)$$
and $ST_F(A) \to ST_K(A)$ (see [F], section 1.7).

b. Also, when $[K : F] < \infty$, we have the *norm* or *corestriction* homomorphisms
$$N_{K/F} = (\varphi_{K/F})_* \colon CH_i(A_K) \to CH_i(A_F) \text{ and } T(A_K) \to T_F(A)$$
(see [F], section 1.4, page 11 and 12).

Remarks   1. If $[K \colon F] = n$ we have $\varphi_* \circ \varphi^* = n$
2. It is *not* clear if $N_{K/F} = \varphi_*$ induces a homomorphism on the *symbol groups themselves*. Note however that there is such norm map for cohomology ([Se2], p. 127) and for K_2-theory ([M], p. 137).

1.6   Some preliminary lemmas

Lemma 1.6.1 *Let $Z \in T_F(A)$. Suppose we can write $Z = \sum_{i=1}^{N}(P_i, Q_i) - \sum_{i=1}^{N}(P_i', Q_i')$ with P_i, P_i', Q_i and Q_i' all in $E(F)$ for $i = 1, \ldots, N$. Then $Z \in ST_F(A)$.*

Proof Since $Z \in T_F(A)$ we have that $\sum_{i=0}^{N} P_i = \sum_{i=0}^{N} P_i'$ and $\sum_{i=1}^{N} Q_i = \sum_{i=1}^{N} Q_i'$ as *sum of points on E*. From this it follows immediately that we can rewrite
$$Z = \sum_{i=1}^{N}\{(P_i, Q_i) - (P_i, 0) - (0, Q_i) + (0, 0)\} - \sum_{i=1}^{N}\{(P_i', Q_i') - (P_i', 0) - (0, Q_i') + (0, 0)\} =$$
$$\sum_{i=1}^{N} \langle P_i, Q_i \rangle - \sum_{i=1}^{N} \langle P_i', Q_i' \rangle; \text{ hence } Z \in ST_F(A). \qquad \square$$

Corollary 1.6.2 *Over the algebraic closure we have*
$$T(\bar{A}) = \varinjlim_{K} \{\mathrm{res}_{\bar{F}/K} ST_K(A)\}$$
where the limit is over all finite extensions $K \supset F$.

Proof Immediate from Lemma 1.6.1. □

1.7   Norms of symbols

Next we turn our attention again to $T(A) = T_F(A)$ itself. If $K \supset F$ is a finite extension and $P, Q \in E(K)$ then consider $N_{K/F}(\langle P, Q \rangle) \in T_F(A)$. Let $ST_{K/F}(A) \subset T_F(A)$ be the subgroup of $T_F(A)$ generated by such elements (i.e., coming as norms of the symbols from finite field extensions $K \supset F$). Note that clearly $ST_{F/F}(A) = ST_F(A)$ and also that $ST_{K/F}(A)$ consists of the norms of elements of $ST_K(A)$.

Lemma 1.7.1 *With the above notations let $T'_F(A)$ be the subgroup of $T_F(A)$ generated by all the subgroups of type $ST_{K/F}(A)$ of $T_F(A)$ for $K \supset F$ finite (with $K \subset \bar{F}$). Then $T'_F(A) = T_F(A)$.*

Proof For simplicity we shall (first) assume $\mathrm{char}(F) = 0$. If $\mathrm{cl}(Z) \in T_F(A)$, then $\mathrm{cl}(Z)$ is the (rational equivalence) class of a cycle $Z \in \mathcal{Z}_0(A_F)$ and moreover $Z = Z' - Z''$ with Z' and Z'' positive (i.e. "effective"), of the same degree and both $Z' \in \mathcal{Z}_0(A_F)$ and $Z'' \in \mathcal{Z}_0(A_F)$. Fixing our attention on $Z' \in \mathcal{Z}_0(A_F)$ we can, by definition, write $Z' = \sum_\alpha Z'_\alpha$ where the Z'_α are (0-dimensional) subvarieties of A and *irreducible* over F (Remark: in the terminology of Weil's Foundations [W] the Z' is a "rational chain" over F and the Z'_α are the "prime rational" parts of it, see [W], p. 207). For each α we have $Z'_\alpha = \sum_{i=1}^{n_\alpha}(P'_{\alpha i}, Q'_{\alpha i})$ where the $(P'_{\alpha i}, Q'_{\alpha i}) \in A(\bar{F})$ is a set of points which form a *complete set of conjugates* over F (see [W], 207). Note that also $\sum_{i=1}^{n_\alpha}(P'_{\alpha i}, 0) \in \mathcal{Z}_0(A_F)$ and similarly $\sum_{i=1}^{n_\alpha}(0, Q'_{\alpha i}) \in \mathcal{Z}_0(A_F)$ (Note: these cycles are rational, but not necessarily prime rational.) For each α fix an arbitrary $i(\alpha)$ and put $K_{\alpha i(\alpha)} = F(P'_{\alpha i(\alpha)}, Q'_{\alpha i(\alpha)})$ and consider now the *cycle* $(P'_{\alpha i(\alpha)}, Q'_{\alpha i(\alpha)}) \in \mathcal{Z}_0(A_{K_{\alpha i(\alpha)}})$. We have by definition (see [F], section 1.4, p. 11), $Z'_\alpha = N_{K_{\alpha i(\alpha)}/F}(P'_{\alpha i(\alpha)}, Q'_{\alpha i(\alpha)})$. Furthermore if we put

$$Z^*_{\alpha i(\alpha)} = \left\langle P'_{\alpha i(\alpha)}, Q'_{\alpha i(\alpha)} \right\rangle \in ST_{K_{\alpha i(\alpha)}}(A)$$

then in $T_F(A)$ we have

$$N_{K_{\alpha i(\alpha)}/F}(Z^*_{\alpha i(\alpha)}) = \sum_{i=1}^{n_\alpha} \langle P'_{\alpha i}, Q'_{\alpha i} \rangle$$

(again by the definition of the $N_{-/F}$) and clearly the cycle is in $ST_{K_{\alpha i(\alpha)}/F}(A)$, i.e. in $T'_F(A)$. Now doing this for every α and treating similarly $Z'' = \sum_\beta Z''_\beta$, we have, since $Z \in T_F(A)$, that

$$Z = Z' - Z'' = \sum_\alpha Z'_\alpha - \sum_\beta Z'_\beta = \sum_\alpha \sum_i \langle P'_{\alpha i}, Q'_{\alpha i} \rangle - \sum_\beta \sum_j \langle P''_{\beta j}, Q''_{\beta j} \rangle .$$

Hence $Z \in T'_F(A)$, which completes the proof (in char 0). □

Remark If $\mathrm{char}(F) = p > 0$, then we have that $Z'_\alpha = p^{m_\alpha} \sum_i (P'_{\alpha i}, Q'_{\alpha i})$ where the p^{m_α} is the degree of inseparability of the field extension $K_{\alpha i(\alpha)}$ over F (see again [W], p. 207). Note that p^{m_α} does not depend on the choice of the index

$i(\alpha)$, because the field $K_{\alpha i(\alpha)}$ is determined by α up to *conjugation* over F. From that point onwards the proof is the same (note in particular that we shall have $Z'_\alpha = N_{K_{\alpha i(\alpha)}/F}(P'_{\alpha i}, Q'_{\alpha i}))$.

Lemma 1.7.2 *For* $P, Q \in E(F)$ *we have* $\langle Q, P \rangle = -\langle P, Q \rangle$, *i.e., the symbol is skew-symmetric.*

Proof Since the symbol is bilinear we have a well-defined homomorphism

$$\lambda\colon E(F) \otimes_{\mathbb{Z}} E(F) \to ST_F(A) \text{ with } \lambda(P \otimes Q) = \langle P, Q \rangle.$$

Then $\lambda((P+Q) \otimes (P+Q)) = (P+Q, P+Q) - (P+Q, 0) - (0, P+Q) + (0, 0)$. On the other hand by the bilinearity, it also equals $\langle P, P \rangle + \langle P, Q \rangle + \langle Q, P \rangle + \langle Q, Q \rangle$. On the diagonal we have $(P+Q, P+Q) + (0,0) = (P,P) + (Q,Q)$, on $E \times 0$ we have $(P+Q, 0) + (0,0) = (P, 0) + (Q, 0)$, and on $0 \times E$ we have $(0, P+Q) + (0,0) = (0, P) + (0, Q)$. Putting these facts together we get $\langle P, Q \rangle + \langle Q, P \rangle = 0$. $\square$

1.8 A useful lemma

We will often have occasion to use the following simple observation:

Lemma 1.8.1 *Let* F *be any field and* ℓ *a prime. Let* P, Q *be points in* $E(F)$, $Q' \in E(\bar{F})$ *s.t.* $\ell Q' = Q$, *and put* $K = F(Q')$. *Consider the statements*

(a) $P \in N_{K/F} E(K) \bmod \ell E(F)$,
(b) $\langle P, Q \rangle \in \ell T_F(A)$.

Then (a) implies (b).

Proof Let $P = N_{K/F}(P') + \ell P_1$, with $P' \in E(K)$, $P_1 \in E(F)$. Then $\langle P, Q \rangle - \ell \langle P_1, Q \rangle = \langle N_{K/F}(P'), Q \rangle$, which equals, by the projection formula,

$$N_{K/F}(\langle P', \mathrm{Res}_{K/F} Q \rangle) = N_{K/F}(\langle P', \ell Q' \rangle) = \ell N_{K/F}(\langle P', Q' \rangle) \in \ell T_F(A).$$

$\square$

2 Symbols, cup products, and $H_s^2(F, E^{\otimes 2})$

2.1 Degeneration of the spectral sequence

Notations and assumption are as before. Let ℓ be a prime number with $\ell \neq \mathrm{char}(F)$. We work here with $\mathbb{Z}_\ell$ coefficients, but the results are also true for $\mathbb{Z}/\ell^s$ coefficients (any $s \geq 1$) and $\mathbb{Q}_\ell$-coefficients.

Lemma 2.1.1 *The Hochschild-Serre spectral sequence*

$$E_2^{pq} = H^p(F, H_{\mathrm{et}}^q(\bar{A}, \mathbb{Z}_\ell(s))) \implies H_{\mathrm{et}}^{p+q}(A, \mathbb{Z}_\ell(s))$$

degenerates at d_2*-level (and at all* d_t*-levels,* $t \geq 2$*) for all* r.

Remark We could take here any abelian variety A instead of $E \times E$.

Proof This follows from "weight" considerations. Consider on A multiplication by n, i.e. $n\colon A \to A$ is the map $x \to nx$. Then we have a commutative diagram

$$
\begin{array}{ccc}
H^p(F, H_{\mathrm{et}}^q(\bar{A}, -)) & \xrightarrow{\ d_2\ } & H^{p+2}(F, H_{\mathrm{et}}^{q-1}(\bar{A}, -)) \\
\Big\downarrow{\scriptstyle n^*} & & \Big\downarrow{\scriptstyle n^*} \\
H^p(F, H_{\mathrm{et}}^q(\bar{A}, -)) & \xrightarrow{\ d_2\ } & H^{p+2}(F, H_{\mathrm{et}}^{q-1}(\bar{A}, -))
\end{array}
$$

On the left we have multiplication by n^q, on the right by n^{q-1}. this being true for any $n > 0$, we must have $d_2 = 0$.   □

2.2   Kummer sequence (and some notations)

If E is an elliptic curve defined over F, we write (by abuse of notation)

$$E[\ell^n] := \ker\{E(\bar{F}) \xrightarrow{\ell^n} E(\bar{F})\},$$

and we have the (elliptic) Kummer sequence

$$0 \longrightarrow E[\ell^n] \longrightarrow E(\bar{F}) \xrightarrow{\ell^n} E(\bar{F}) \longrightarrow 0.$$

This is an exact sequence of $\mathrm{Gal}(\bar{F}/F)$-modules and gives us a short exact sequence of (cohomology) groups

$$0 \longrightarrow E(F)/\ell^n \longrightarrow H^1(F, E[\ell^n]) \longrightarrow H^1(F, \bar{E})[\ell^n] \longrightarrow 0.$$

Taking the limit over n, one gets the homomorphism

$$\delta_\ell^{(1)}\colon\ E(F) \longrightarrow H^1(F, T_\ell(E))$$

where $T_\ell(E) = \varprojlim_n E[\ell^n]$ is the Tate group.

This allows us to define

$$[\cdot,\cdot]_\ell\colon\ E(F) \otimes E(F) \longrightarrow H^2(F, T_\ell(E)^{\otimes 2})$$

by

$$[P, Q]_\ell = \delta_\ell^{(1)}(P) \cup \delta_\ell^{(1)}(Q).$$

We have similar maps (and we use the same notations) if we take $E(F)/\ell^n$ and $E[\ell^n]^{\otimes 2}$.

Explicitly the map $\delta_\ell^{(1)}$ is given by the following: For $P \in E(F)$ the cohomology class $\delta_\ell^1(P)$ is represented by the 1-cocycle

$$\mathrm{Gal}(\bar{F}/F) \to T_\ell(E),$$

$$\sigma \mapsto \left(\sigma\left(\frac{1}{\ell}P\right) - \frac{1}{\ell}P, \sigma\left(\frac{1}{\ell^2}P\right) - \frac{1}{\ell^2}P, \dots\dots \right).$$

2.3   Comparison with the usual cycle class map

For every smooth projective variety X defined over F there is the cycle class map to continuous cohomology as defined by Jannsen ([J], lemma 6.14)

$$\mathrm{cl}_\ell^{(i)}\colon\ \mathrm{CH}^i(X) \longrightarrow H^{2i}_{\mathrm{cont}}(X, \mathbb{Z}_\ell(i)).$$

Taking now $X = E$, resp. $X = A$, and using the degeneration of the Hochschild-Serre spectral sequence we get

$$\mathrm{cl}_\ell^{(1)}\colon\ CH^1_{(0)}(E) \longrightarrow H^1(F, H^1_{\mathrm{et}}(\bar{E}, \mathbb{Z}_\ell(1)))$$

where $\mathrm{CH}^1_{(0)}(E)$ is the Chow group of 0-cycles on E of degree 0, resp.

$$\mathrm{cl}^{(2)}_\ell \colon T_F(A) \longrightarrow H^2(F, H^2_{\mathrm{et}}(\bar{A}, \mathbb{Z}_\ell(2))).$$

Lemma 2.3.1 *There is a commutative diagram*

$$
\begin{array}{ccc}
E(F) & \xrightarrow{\ \delta^{(1)}_\ell\ } & H^1(F, T_\ell(E)) \\
\downarrow & & \downarrow{\scriptstyle \cong} \\
\mathrm{CH}^1_{(0)}(E) & \xrightarrow{\ cl^{(1)}_\ell\ } & H^1(F, H^1_{\mathrm{et}}(\bar{E}, \mathbb{Z}_\ell(1)))
\end{array}
$$

where the vertical map on the left is $P \mapsto (P) - (0)$ and the one on the right comes from the well-known isomorphism $T_\ell(E) \xrightarrow{\sim} H^1_{\mathrm{et}}(\bar{E}, \mathbb{Z}_\ell(1))$.

Proof See [R], proof of the lemma in the appendix. $\qquad\square$

2.4 The symbolic part of cohomology

Let K/F be any finite extension. Given a pair of points $P, Q \in E(K)/\ell$, with associated classes $\delta^{(1)}_\ell(P), \delta^{(1)}_\ell(Q)$ in $H^1(K, E[\ell])$. We have seen in section 2.2 that by taking cup product in Galois cohomology, we get a class $[P, Q]_\ell$ in $H^2(K, E[\ell]^{\otimes 2})$. By taking the norm (corestriction) from $H^2(K, E[\ell]^{\otimes 2})$ to $H^2(F, E[\ell]^{\otimes 2})$, we then get a class

$$N_{K/F}([P, Q]_\ell) \in H^2(F, E[\ell]^{\otimes 2}).$$

We define the *symbolic part* $H^2_s(F, E[\ell]^{\otimes 2})$ of $H^2(F, E[\ell]^{\otimes 2})$ to be the $\mathbb{F}_\ell$-subspace generated by such *norms of symbols* $N_{K/F}([P, Q]_\ell)$, where K runs over all possible finite extensions of F and P, Q run over all pairs of points in $E(K)/\ell$.

Note the similarity of this definition with the description of $T_F(E \times E)/\ell$ via Lemma 1.7.1.

2.5 Summary

The remarks and maps of the previous sections can be subsumed in the following:

Proposition 2.5.1 *There exist maps and a commutative diagram*

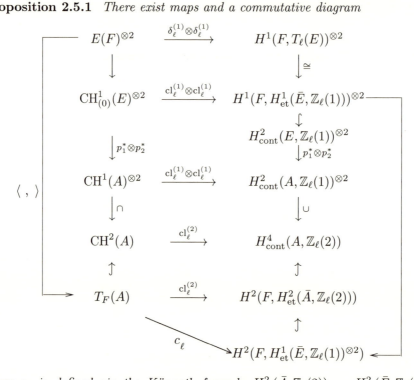

where c_ℓ is defined via the Künneth formula $H^2_{\mathrm{et}}(\bar{A}, \mathbb{Z}_\ell(2)) = H^2_{\mathrm{et}}(\bar{E}, \mathbb{Z}_\ell(2)) \oplus H^1_{\mathrm{et}}(\bar{E}, \mathbb{Z}_\ell(1))^{\otimes 2} \oplus H^2_{\mathrm{et}}(\bar{E}, \mathbb{Z}_\ell(2))$ and the injection on the relevant (= middle) term. In particular $[P,Q]_\ell := \delta^{(1)}_\ell(P) \cup \delta^{(1)}_\ell(Q) = c_\ell(\langle P, Q \rangle)$ for $P, Q \in E(F)$. Moreover the same holds for $\mathbb{Z}/\ell^n$-coefficients (instead of $\mathbb{Z}_\ell$) and also for $\mathbb{Q}_\ell$-coefficients.

Remark Note that the map $\langle , \rangle$ on the left from $E(F)^{\otimes 2}$ to $T_F(A)$ is the symbol map, and that the map from $H^1(F, H^1_{\mathrm{et}}(\overline{E}, \mathbb{Z}_\ell(1)))^{\otimes 2}$ to $H^2(F, H^1_{\mathrm{et}}(\bar{E}, \mathbb{Z}_\ell(1))^{\otimes 2})$ on the right is the one given by taking the cup product in group cohomology.

Proof and further explanation of the maps The map $p^*_1 \otimes p^*_2$ is induced by the projections $p_i \colon A = E \times E \to E$. The commutativity in upper rectangle comes from Lemma 2.3.1. The other rectangles are all "natural". $\qquad\square$

Corollary 2.5.2 *With respect to the decomposition*

$$H^1_{\mathrm{et}}(\bar{E}, \mathbb{Z}_\ell(1))^{\otimes 2} = T_\ell(E)^{\otimes 2} \simeq \mathrm{Sym}^2 T_\ell(E) \oplus \Lambda^2 T_\ell(E)$$

we have that $[P,Q]_\ell = c_\ell(\langle P, Q \rangle) \in H^2(F, \mathrm{Sym}^2 T_\ell(E))$ if $\ell \neq 2$ (and similarly for $\mathbb{Z}/\ell^s$-coefficients).

Proof Step 1. For the sake of clarity of the proof we shall first take two different elliptic curves E_1 and E_2 (or, if one prefers, two "different copies" E_1 and E_2 of E). Put $A_{12} = E_1 \times E_2$. There exist obvious analogs of the maps and the commutation diagram of Proposition 2.5.1; only – of course – one should now write $E_1(F) \times E_2(F)$, etc. In particular we have again the cup product on the right:

$$H^1(F, T_\ell(E_1)) \otimes H^1(F, T_\ell(E_2)) \xrightarrow{\cup} H^2(F, T_\ell(E_1) \otimes T_\ell(E_2))$$

where we write $T_\ell(E_1) = H^1_{\text{et}}(\bar{E}_i, \mathbb{Z}_\ell(1))$, $i = 1, 2$. We get $c_\ell(\langle P, Q \rangle) = \delta_\ell^{(1)}(P) \cup \delta_\ell^{(1)}(Q)$. Now consider also $A_{21} = E_2 \times E_1$ and the corresponding diagram for A_{21}. Consider the natural isomorphism $t \colon A_{12} \to A_{21}$ given by $t(x, y) = (y, x)$ for $x \in E_1$ and $y \in E_2$ and also the corresponding isomorphism

$$t_* \colon H^1(\bar{E}_1) \otimes H^1(\bar{E}_2) \longrightarrow H^1(\bar{E}_2) \otimes H^1(\bar{E}_1).$$

$\square$

Claim. If $P \in E_1(F)$ and $Q \in E_2(F)$ then

$$c_{\ell, A_{21}}(\langle Q, P \rangle) = -t_*(\delta_\ell^{(1)}(P) \cup \delta_\ell^{(1)}(Q)).$$

Here we have written – in order to avoid confusion – $c_{\ell, A_{21}}$ for c_ℓ in the diagram relative to A_{21}.

Proof of the claim

$$c_{\ell, A_{21}}(\langle Q, P \rangle) = \delta_\ell^{(1)}(Q) \cup \delta_\ell^{(1)}(P) = -t_* \left(\delta_\ell^{(1)}(P) \cup \delta_\ell^{(1)}(Q) \right)$$

where the first equality is from the diagram (for A_{21}) in Proposition 2.5.1 and the second equality is a well-known property in cohomology (see for instance [Br], p. 111, (3.6), or [Sp], chap.5, §6, p. 250).

Step 2. Returning to the case $E_1 = E_2 = E$, we have $\langle Q, P \rangle = -\langle P, Q \rangle$ in $T_F(A)$ by Lemma 1.7.2.

Step 3. Let us write $c_\ell(\langle P, Q \rangle) = \alpha + \beta$, with α in $H^2(F, \text{Sym}^2 T_\ell(E))$ and β in $H^2(F, \Lambda^2 T_\ell(E))$. We get $c_\ell(\langle P, Q \rangle) = -c_\ell \langle Q, P \rangle$ from Step 2, and next from Claim in Step 1 we have $-c_\ell(\langle Q, P \rangle) = t_*(\delta_\ell^{(1)}(P) \cup \delta_\ell^{(1)}(Q))$, hence $\alpha + \beta = t_*(\alpha + \beta) = \alpha - \beta$, hence $\beta = 0$ if $\ell \neq 2$. $\square$

Definition 2.5.3 *Let*

$$s_\ell \colon T_F(E \times E)/\ell \to H^2(F, E[\ell]^{\otimes 2})$$

be the homomorphism induced by the reduction of c_ℓ modulo ℓ and by using the isomorphism of $H^1(E_{\bar{F}}, \mathbb{Z}_\ell(1))$ with the ℓ-adic Tate module of E, which is the inverse limit of $E[\ell^n]$ over n.

Thanks to the discussion above, as well as the definition of the symbolic part of $H^2(F, E[\ell]^{\otimes 2})$ in section 2.4, and Lemma 1.7.1, we obtain the following:

Proposition 2.5.4 *The image of $T_F(E \times E)/\ell$ under s_ℓ is $H^2_s(F, E[\ell]^{\otimes 2})$.*

3 A key Proposition

Let E/F be an elliptic curve over a local field F as in Theorem A, i.e., non archimedean and with finite residue field of characteristic $p > 2$, and with E having good, ordinary reduction. We will henceforth take $\ell = p$.

A basic fact is that the representation ρ_p of G_F on $E[p]$ is reducible, so the matrix of this representation is triangular. Since the determinant is the mod p cyclotomic character χ_p, we may write

$$\rho_p = \begin{pmatrix} \chi_p \nu^{-1} & * \\ 0 & \nu \end{pmatrix}, \tag{3.1}$$

where ν is an unramified character of finite order, such that $E[p]$ is semisimple (as a G_F-module) iff $* = 0$. Note that ν is necessarily of order at most 2 when E has multiplicative reduction, with $\nu = 1$ iff E has *split* multiplicative reduction. On the other hand, ν can have arbitrary order (dividing $p - 1$) when E has good, ordinary reduction. In any case, there is a natural G_F-submodule C_F of $E[p]$ of dimension 1, such that we have a short exact sequence of G_F-modules:

$$0 \to C_F \to E[p] \to C'_F \to 0, \tag{3.2}$$

with G_F acting on C_F by $\chi_p \nu^{-1}$ and on C'_F by ν. Clearly, $E[p]$ is semisimple iff the sequence (3.2) splits.

The natural G_F-map $C_F^{\otimes 2} \to E[p]^{\otimes 2}$ induces a homomorphism

$$\gamma_F : H^2(F, C_F^{\otimes 2}) \to H^2 F, E[p]^{\otimes 2}). \tag{3.3}$$

The key result we prove in this section is the following:

Proposition C *Let F be a non-archimedean local field with odd residual characteristic p, and E an elliptic curve over F with good, ordinary reduction. Denote by $Im(s_p)$ the image of $T_F(E \times E)/p$ under the Galois symbol map s_p into $H^2 F, E[p]^{\otimes 2})$. Then we have*

(a) $Im(s_p) \subset Im(\gamma_F)$.
(b) *The dimension of $Im(s_p)$ is at most 1, and it is zero dimensional if either $\mu_p \not\subset F$ or $\nu^2 \neq 1$.*

Remark If E/F has multiplicative reduction, with ℓ an arbitrary odd prime (possibly equal to p), then again the Galois representation ρ_ℓ on $E[\ell]$ has a similar shape, and in fact, ν is at most quadratic, reflecting the fact that E attains split multiplicative reduction over at least a quadratic extension of F, over which the two tangent directions at the node are rational. An analogue of Proposition C holds in that case, thanks to a key result of W.McCallum ([Mac], Prop.3.1), once we assume that ℓ does not divide the order of the component group of the special fibre $\mathcal{E}_s$ of the Néron model $\mathcal{E}$. For the sake of brevity, we are not treating this case here.

The bulk of this section will be involved in proving the following result, which at first seems weaker than Proposition C:

Proposition 3.4 *Let F, E, p be as in Proposition C. Then, for all points P, Q in $E(F)/p$, we have*

$$s_p(\langle P, Q \rangle) \in Im(\gamma_F).$$

Claim 3.5 *Proposition 3.4 $\Longrightarrow$ Part (a) of Proposition C:*

Proof of this claim goes via some lemmas.

Lemma 3.6 (Behavior of γ under finite extensions) *Let K/F be finite. We then have two commutative diagrams, one for the norm map $N = N_{K/F}$ and the other for the restriction map $Res = Res_{K/F}$:*

$$
\begin{array}{ccc}
H^2(K, C_K^{\otimes 2}) & \xrightarrow{\gamma_K} & H^2(K, E[p]^{\otimes 2}) \\
N \downarrow \uparrow Res & & N \downarrow \uparrow Res \\
H^2(F, C_F^{\otimes 2}) & \xrightarrow{\gamma_F} & H^2(F, E[p]^{\otimes 2})
\end{array}
$$

This Lemma follows from the compatibility of the exact sequence (3.2) (of Galois modules) with base extension.

Lemma 3.7 *In order to prove that $Im(s_p) \subset Im(\gamma_F)$, it suffices to prove it for the image of symbols, i.e., that*

$$\mathrm{Im}\left(s_p(ST_{F,p}(A))\right) \subset \mathrm{Im}(\gamma_F),$$

where

$$ST_{F,p}(A) := \mathrm{Im}\left(ST_F(A) \to T_F(A)/p\right).$$

Proof Use the commutativity of the diagram in Lemma 3.6 for the norm. $\square$

This proves *Claim 3.5*. $\square$

Lemma 3.8 *In order to prove that $Im(s_p) \subset Im(\gamma_F)$, we may assume that $\mu_p \subset F$ and that $\nu = 1$, i.e., that we have the following exact sequence for groupschemes over S:*

$$0 \to \mu_{p,S} \to \mathcal{E}[p] \to (\mathbb{Z}/p)_S \to 0. \tag{*}$$

Proof There is a finite extension K/F such that $(*)$ holds over K, with $p \nmid [K : F]$. Now use the diagram(s) in Lemma 3.6 as follows: Let $P, Q \in E(F) \subset E(K)$, then $\mathrm{Res}(P) = P$, $\mathrm{Res}(Q) = Q$, and we have

$$[K : F]s_{F,p}(\langle P, Q \rangle) = N\{\mathrm{Res}(s_{F,p}(\langle P, Q \rangle)\},$$

which equals

$$N\{s_{K,p}\left(\langle \mathrm{Res}(P), \mathrm{Res}(Q) \rangle\right)\} = N\{s_{K,p}(\langle P, Q \rangle)\}.$$

Therefore, if $s_{K,p}(\langle P, Q \rangle) \subset \mathrm{Im}(\gamma_K)$, then (again by using Lemma 3.6 for norm) we see that $N\{s_{K,p}(\langle P, Q \rangle)\}$ is contained in $\mathrm{Im}(\gamma_F)$. Hence $[K : F]s_{F,p}(\langle P, Q \rangle)$ lies in $\mathrm{Im}(\gamma_F)$. Finally, since $p \nmid [K : F]$, $s_{F,p}(\langle P, Q \rangle)$ itself belongs to $\mathrm{Im}(\gamma_F)$. $\square$

3.9 Proof of Proposition 3.4. Since we may take $\mu_p \subset F$ and $\nu = 1$, the exact sequence $(*)$ in Lemma 3.8 holds, compatibly with the corresponding one over F of $\mathrm{Gal}(\overline{F}/F)$-modules. Taking cohomology, we get a commutative diagram of $\mathbb{F}_p$-vector spaces with exact rows:

$$
\begin{array}{ccccccc}
\mathcal{O}_F^*/p & \to & E(F)/p & \to & \mathbb{Z}/p & & \\
\downarrow & & \downarrow & & \downarrow & & \\
\overline{H}^1(F, \mu_p) & \to & \overline{H}^1(F, E[p]) & \to & \overline{H}^1(F, \mathbb{Z}/p) & \to & 0
\end{array}
$$

Here the vertical maps are isomorphisms (see the Appendix), which induce the horizontal maps on the top row.

Definition 3.10 *Put*

$$U_F := \mathrm{Im}\left(\mathcal{O}_F^*/p \to E(F)/p\right),$$

and choose a non-canonical decomposition

$$E(F)/p \simeq U_F \oplus W_F, \quad \text{with} \quad W_F \simeq \mathbb{Z}/p.$$

Notation 3.11 If S_1, S_2 are subsets of $E(F)/p$, we denotes by $\langle S_1, S_2 \rangle$ the subgroup of $ST_{F,p}(A)$ generated by the symbols $\langle s_1, s_2 \rangle$, with $s_1 \in S_1$ and $s_2 \in S_2$.

Lemma 3.12 $ST_{F,p}(A)$ *is generated by the two vector subspaces*

$$\Sigma_1 := \langle U_F, E(F)/p \rangle \quad \text{and} \quad \Sigma_2 := \langle E(F)/p, U_F \rangle.$$

Proof $ST_{F,p}(A)$ is clearly generated by Σ_1, Σ_2 and by $\langle W_F, W_F \rangle$. However, W_F is one-dimensional and the pairing $\langle \cdot, \cdot \rangle$ is skew-symmetric, so $\langle W_F, W_F \rangle = 0$. $\quad\square$

Now we have a commutative diagram

$$
\begin{array}{ccc}
\mathcal{O}_F^*/p \otimes E(F)/p & \xrightarrow{\alpha_1} & E(F)/p \otimes E(F)/p \\
\downarrow{\scriptstyle \sigma_1} & & \downarrow{\scriptstyle s_p} \\
H^2(F, \mu_p \otimes E[p]) & \xrightarrow{\beta_1} & H^2(F, E[p]^{\otimes 2}),
\end{array}
\qquad (3.13-i)
$$

where the top map α_1 factors as

$$
\mathcal{O}_F^*/p \otimes E(F)/p \to U_F \otimes E(F)/p \to E(F)/p \otimes E(F)/p.
$$

One can also obtain a similar diagram $(3.13 - ii)$, which is not given here, by replacing $\mathcal{O}_F^*/p \otimes (E(F)/p)$ (resp. $U_F \otimes (E(F)/p)$) by $(E(F)/p) \otimes \mathcal{O}_F^*/p$ (resp. $(E(F)/p) \otimes U_F$). The maps α_j and their factoring are obvious, s_p is the map constructed in section 2, and the vertical maps σ_j are defined entirely analogously.

Lemma 3.14 *The image of* $s_p : ST_{F,p}(A) \to H^2(F, E[p]^{\otimes 2})$ *is generated by* $\beta_1(Im(\sigma_1))$ *and* $\beta_2(Im(\sigma_2))$.

Proof Immediate by Lemma 3.12 together with the commutative diagrams $(3.13 - i)$ and $(3.13 - ii)$. $\quad\square$

Tensoring the exact sequence

$$
0 \to \mu_p \to E[p] \to \mathbb{Z}/p \to 0 \qquad (3.15)
$$

with μ_p from the left and the right, and taking Galois cohomology, we get two natural homomorphisms

$$
\gamma_1 : H^2(F, \mu_p^{\otimes 2}) \to H^2(F, \mu_p \otimes E[p]) \qquad (3.16 - i)
$$

and

$$
\gamma_2 : H^2(F, \mu_p^{\otimes 2}) \to H^2(F, E[p] \otimes \mu_p). \qquad (3.16 - ii)
$$

Lemma 3.17 $Im(\sigma_j) \subset Im(\gamma_j)$, *for* $j = 1, 2$.

Proof of Lemma 3.17 We give a proof for $j = 1$ and leave the other (entirely similar) case to the reader. By tensoring (3.15) by μ_p, we obtain the following exact sequence of G_F-modules:

$$
0 \to \mu_p^{\otimes 2} \to \mu_p \otimes E[p] \to \mu_p \to 0. \qquad (3.18)
$$

Consider now the following commutative diagram, in which the bottom row is exact:

$$
\begin{array}{ccccc}
& & U_F \otimes E(F)/p & & \\
& & \downarrow{\scriptstyle \tilde{\sigma}_1} & & \\
& & \overline{H}^2(F, \mu_p \otimes E[p]) & \xrightarrow{\varepsilon_0} & \overline{H}^2(F, \mu_p) \\
& & \downarrow{\scriptstyle i_1} & & \downarrow \\
H^2(F, \mu_p^{\otimes 2}) & \xrightarrow{\gamma_1} & H^2(F, \mu_p \otimes E[p]) & \xrightarrow{\varepsilon} & H^2(F, \mu_p)
\end{array}
\qquad (3.19)
$$

where $\sigma_1 = i_1 \circ \tilde{\sigma}_1$ is the map from $(3.13 - i)$.

To begin, the exactness of the bottom row follows immediately from the exact sequence (3.18). The factorization of σ_1 is the crucial point, and this holds because

U_F comes from $\mathcal{O}_F^*/p$ and is mapped to $H^1(F, \mu_p)$ via $\overline{H}^1(F, \mu_p)$, which is the image of $H_{\mathrm{fl}}^1(S, \mu_p)$ (see Lemma A.3.1 in the Appendix). Similarly, $E(F)/p \simeq \mathcal{E}(S)/p$ maps into $\overline{H}^1(F, E[p])$, the image of $H_{\mathrm{fl}}^1(S, \mathcal{E}[p])$; therefore the tensor product maps into $\overline{H}^2(F, \mu_p \otimes E[p])$, the image of $H_{\mathrm{fl}}^2(S, \mu_p \otimes \mathcal{E}[p])$.

To prove Lemma 3.17, it suffices, by the exactness of the bottom row, to see that $\mathrm{Im}(\sigma_1)$ is contained in $\mathrm{Ker}(\varepsilon)$, i.e., to see that $\varepsilon \circ \sigma_1 = 0$. Luckily for us, this composite map factors through $\varepsilon_0 \circ \tilde{\sigma}_1$, which vanishes because $H_{\mathrm{fl}}^2(S, \mu_{p,S})$, and hence $\overline{H}^2(F, \mu_p)$, is zero by [Mi2], part III, Lemma 1.1. $\qquad\square$

Putting these Lemmas together, we get the truth of Proposition 3.4.

3.20   Proof of Proposition C. As we saw earlier, Proposition 3.4, which has now been proved, implies (by Claim 3.5) part (a) of Proposition C. So we need to prove only part (b).

By part (a) of Prop. C, the dimension of the image of s_p is at most that of $H^2(F, C^{\otimes 2})$. Since $E[p]$ is selfdual, the short exact sequence (3.2) shows that the Cartier dual of C is C'. It follows easily that $(C^{\otimes 2})^D$ is $C'^{\otimes 2}(-1)$. By the local duality, we then get

$$H^2(F, C^{\otimes 2}) \simeq H^0(F, C'^{\otimes 2}(-1))^\vee,$$

As C' is a line over $\mathbb{F}_p$ with G_F-action, the dimension of the group on the right is less than or equal to 1, with equality holding iff $C'^{\otimes 2} \simeq \mu_p$.

Since G_F acts on C' by an unramified character ν, for $C'^{\otimes 2}$ to be μ_p as a G_F-module, it is necessary that $\mu_p \subset F$. So

$$\mu_p \not\subset F \implies \mathrm{Im}(s_p) = 0.$$

Now suppose $\mu_p \subset F$. Then for $\mathrm{Im}(s_p)$ to be non-zero, it is necessary that $C'^{\otimes 2} \simeq \mathbb{Z}/p$, implying that $\nu^2 = 1$. $\qquad\square$

4 Vanishing of s_p in the wild case

We start with a simple Lemma, doubtless known to experts.

Lemma 4.0 *Let E/F be ordinary. Then the Galois representation on $E[p]$ is one of the following (disjoint) types:*

(i) *The wild inertia group I_p acts non-trivially, making $E[p]$ non-semisimple;*
(ii) *$E[p]$ is semisimple, in which case $p \nmid F(E[p]) : F]$;*
(iii) *$E[p]$ is unramified and non-semisimple, in which case $[F(E[p]) : F] = p$.*

In case (i), we will say that we are in the *wild case*.

Proof Clearly, when I_p acts non-trivially, the triangular nature of the representation on $E[p]$ makes $E[p]$ non-semisimple. Suppose from here on that I_p acts trivially. If $E[p]$ is semisimple, $F(E[p])/F$ is necessarily a prime-to-p extension, again because the representation is solvable in the ordinary case. This leaves the final possibility when $E[p]$ is non-semisimple, but I_p acts trivially. We claim that the tame inertia group I_t also acts trivially. Since $I_t \simeq \lim_m \mathbb{F}_{p^m}^*$, its image in $\mathrm{GL}_2(\mathbb{F}_p)$ ($\simeq \mathrm{Aut}(E[p])$) will have prime-to-$p$ order, and so must be (conjugate to) a subgroup H of the diagonal group $D \simeq (\mathbb{F}_p^*)^2$. H cannot be central, as the semisimplification is a direct sum of two characters, at most one of which is ramified, and for the same reason, H cannot be all of D. It follows that, when I_t acts non-trivially, H must be conjugate to

$$\left\{ \begin{pmatrix} a & 0 \\ 0 & 1 \end{pmatrix} \mid a \in C \subset \mathbb{F}_p^* \right\},$$

for a subgroup $C \neq \{1\}$ of $\mathbb{F}_p^*$. An explicit calculation shows that any element of $\mathrm{GL}_2(\mathbb{F}_p)$ normalizing H must be contained in the group generated by D and the Weyl element $w = \begin{pmatrix} 0 & -1 \\ 1 & 0 \end{pmatrix}$. Again, since the representation is triangular, w cannot be in the image. Consequently, if I_t acts non-trivially, the image of $\mathcal{G}_k$ must also be in D, implying that $E[p]$ is semisimple, whence the claim. Thus in the third (and last) case, $E[p]$ must be unramified and non-semisimple. As $\mathcal{G}_k$ is pro-cyclic, the only way this can happen is for the image to be unipotent of order p.   □

Remark   An example of Rubin shows that for a suitable, ordinary elliptic curve E over a p-adic field, we can be in case (iii). To see this, start with an example E/F in case (i), with E[p] unipotent and non-trivial, which is easy to produce. Let L be the ramified extension of F of degree p obtained by trivializing $E[p]$. Let K be the unique unramified extension of F of degree p, and let M be an extension of F of degree p inside the compositum LK, different from both L and K. Then over M, $E[p]$ has the form we want. Since M does not contain L, $E[p]$ is a non-trivial $\mathcal{G}_M$-module. On the other hand, this representation (of $\mathcal{G}_M$) is unramified because all the p-torsion is defined over the unramified extension MK of M, which contains L.

The object of this section is to prove the following:

Proposition D   *Suppose E/F is an elliptic curve (with good ordinary reduction) over a non-archimedean local field F of odd residual characteristic p. Assume that we are in the wild case. Then we have*

$$s_p(T_F(E \times E)/p) = 0.$$

Combining this with Proposition C, we get the following

Corollary 4.1   *Let F be a non-archimedean local field with residual characteristic $p > 2$, and E an elliptic curve over F with good, ordinary reduction. If $\mathrm{Im}(s_p)$ is non-zero, then either $[F(E[p]) : F] \leq 2$ or $F(E[p])$ is an unramified p-extension.*

Indeed, if $\mathrm{Im}(s_p)$ is non-zero, Proposition D says that I_p acts trivially on $E[p]$. If the $\mathrm{Gal}(\overline{F}/F)$- action on $E[p]$ is semisimple, then by Proposition C, $[F(E[p]) : F] \leq 2$. Since the tame inertia group has order prime to p, the only other possibility, thanks to Lemma 4.0, is for $E[p]$ to be non-semisimple and unramified. As the residual Galois group is cyclic, this also forces $[F(E[p]) : F] = p$. Hence the Corollary (assuming the truth of Proposition D).

Proof of Proposition D   Let us first note a few basic things concerning base change to a finite extension K/F of degree m prime to p. To begin, since E/F has ordinary reduction, the Galois representation ρ_F on $E[p]$ is triangular, and it is semisimple iff the image does not contain any element of order p. It follows that ρ_K is semisimple iff ρ_F is semisimple. Moreover, the functoriality of the Galois symbol map relative to the respective norm and restriction homomorphisms, together with the fact that the composition of restriction with norm is multiplication by m, implies, as $p \nmid m$, that for any $\theta \in T_F(E \times E)/p$, we have

$$s_{p,K}(\mathrm{res}_{K/F}(\theta)) = 0 \implies s_p(\theta) = 0.$$

So it suffices to prove Proposition D, after possibly replacing F by a finite prime-to-p extension, under the assumption that F contains μ_p *and* $\nu = 1$, still with $E[p]$ non-semisimple and wild. Thus we have a non-split, short exact sequence of finite flat groupschemes over S:

$$0 \to \mu_{p,S} \to \mathcal{E}[p] \to (\mathbb{Z}/p)_S \to 0, \tag{4.2}$$

with the representation ρ_F of $G = \mathrm{Gal}(\overline{F}/F)$ on $E[p]$ having the form:

$$\begin{pmatrix} 1 & \alpha \\ 0 & 1 \end{pmatrix} \tag{4.3}$$

relative to a suitable basis. Here $\alpha : G \to \mathbb{Z}/p$ is a non-zero homomorphism, and $F(\alpha)$, the smallest extension of F over which α becomes trivial, is a ramified p-extension.

Using the exact sequence (4.2), both over S and over F, we get the following commutative diagram:

$$
\begin{array}{ccccccccc}
\mathcal{E}[p](S) & \xrightarrow{0} & \mathbb{Z}/p & \to & H^1_{\mathrm{fl}}(S, \mu_p) & \xrightarrow{\psi_S} & H^1_{\mathrm{fl}}(S, \mathcal{E}[p]) & \twoheadrightarrow & H^1_{\mathrm{fl}}(S, \mathbb{Z}/p) \\
\| & & \| & & \downarrow & & \downarrow & & \downarrow \\
E[p](F) & \xrightarrow{0} & \mathbb{Z}/p & \to & H^1(F, \mu_p) & \xrightarrow{\psi} & H^1(F, E[p]) & \to & H^1(F, \mathbb{Z}/p) \\
& & & & \| & & \cap & & \\
& & & & X + Y & & Z = \psi(Y) & &
\end{array}
$$

$$\tag{4.4}$$

where $X \subset \overline{H}^1(F, \mu_p) \subset H^1(F, \mu_p)$ is the subspace given by the image of $\mathbb{Z}/p$. Since μ_p is in F, we can identify $H^2(F, \mu_p^{\otimes 2})$ with $\mathrm{Br}_F[p] \simeq \mathbb{Z}/p$. By our assumption, $X \neq 0$ as the sequence (4.2) does *not* split. Now take $e \in X$, with $e \neq 0$. Then

$$e \in \mathcal{O}_F^*/p \subset F^*/F^{*p} = H^1(F, \mu_p).$$

Since $K = F(e^{1/p})$ is clearly the smallest extension of F over which (4.2) splits, K is also $F(\alpha)$ and hence a ramified p-extension of F, since the Galois representation on $E[p]$ is wild. Then, applying [Se2], Prop.5 (iii), we get a $v \in \overline{H}^1(F, \mu_p)$ which is not a norm from K, and so the cup product $\{e, v\}$ is non-zero in $\mathrm{Br}_F[p]$ (cf. [Ta1], Prop.4.3).

Now consider the commutative diagram

$$
\begin{array}{ccc}
H^1(F, \mu_p)^{\otimes 2} & \xrightarrow{\cup} & H^2(F, \mu_p^{\otimes 2}) \\
\psi^{\otimes 2} \downarrow & & \downarrow \gamma \\
H^1(F, E[p])^{\otimes 2} & \xrightarrow{\cup} & H^2(F, E[p]^{\otimes 2})
\end{array}
\tag{4.5}
$$

where ψ is the map defined in the previous diagram, and $\gamma = \gamma_F$ is the map defined in section 3 with $C_F = \mu_p$ and $C'_F = \mathbb{Z}/p$. By Proposition C, the image of s_p is contained in that of γ. On the other hand, $\psi(e) \cup \psi(v) = 0$ because $\psi(e) = 0$. Hence $\gamma(\{e, v\}) = 0$, and the image of s_p is zero as asserted. $\square$

5 Non-trivial classes in $\mathrm{Im}(s_p)$ in the non-wild case with $[F(E[p]) : F]' \leq 2$

For any finite extension K/F, we will write $[K : F]'$ for the prime-to-p degree of K/F.

Proposition E *Let E be an elliptic curve over a non-archimedean local field F of residual characteristic $p \neq 2$. Assume that E has good ordinary reduction, with trivial action on I_p on $E[p]$, such that*

  (a) $[F(E[p]) : F]' \leq 2$;
  (b) $\mu_p \subset F$.

Then $\mathrm{Im}(s_p) \neq 0$. Moreover, if $F(E[p]) = F$, i.e., if all the p-division points are rational over F, then there exist points P, Q of $E(F)/p$ such that

$$s_p(\langle P, Q \rangle) \neq 0.$$

In this case, up to replacing F by a finite unramified extension, we may choose P to be a p-power torsion point.

Proof First consider the case when $E[p]$ is semisimple, so that $[K : F]' = [K : F] \leq 2$, where $K := F(E[p])$.

Suppose K is quadratic over F. Recall that over F, C_F (resp. C_F') is given by the character $\chi \nu^{-1}$ (resp. ν), and since $\mu_p \subset F$ and ν quadratic, we have

$$H^2(F, C_F^{\otimes 2}) \simeq H^2(F, \mu_p) = Br_F[p] \simeq \mathbb{Z}/p.$$

Suppose we have proved the existence of a class θ_K in $T_K(E \times E)/p$ such that $s_{p,K}(\theta_K)$ is non-zero, and this image must be, thanks to Proposition C, in the image of a class t_K in $Br_K[p]$. Put $\theta := N_{K/F}(\theta_K) \in T_F(E \times E)/p$. Then $s_p(\theta)$ equals $N_{K/F}(s_{p,K}(\theta_K))$, which is in the image of $t := N_{K/F}(t_K) \in Br_F[p]$, which is non-zero because the norm map on the Brauer group is an isomorphism.

So we may, and we will, assume henceforth in the proof of this Proposition that $E[p]) \subset F$, i.e., $K = F$.

Now let us look at the basic setup carefully. Since E has good reduction, the Néron model is an elliptic curve over S. Moreover, since E has ordinary reduction with $E[p] \subset F$, we also have

$$\mathcal{E}[p] = (\mu_p)_S \oplus \mathbb{Z}/p$$

as group schemes over S (and as sheaves in S_{flat}). By the Appendix A.3.2,

$$E(F)/p \simeq \mathcal{E}(S)/p \xrightarrow{\ \partial_F\ } H^1_{\mathrm{fl}}(S, \mathcal{E}[p]) = H^1_{\mathrm{fl}}(S, \mu_p) \oplus H^1_{\mathrm{fl}}(S, \mathbb{Z}/p), \qquad (5.1)$$

where the boundary map ∂_F is an isomorphism. Moreover, the direct sum on the right of (5.1) is isomorphic to

$$H^1_{\mathrm{fl}}(S, \mu_p) \oplus H^1_{\mathrm{et}}(k, \mathbb{Z}/p) \simeq \mathcal{O}_F^*/p \oplus \mathbb{Z}/p,$$

thanks to the isomorphism $H^1_{\mathrm{fl}}(S, \mu_p) \simeq \mathcal{O}_F^*/p$ and the identification of $H^1_{\mathrm{fl}}(S, \mathbb{Z}/p)$ with $H^1_{\mathrm{et}}(k, \mathbb{Z}/p) \simeq \mathbb{Z}/p$ ([Mi1], p. 114, Thm. 3.9). Therefore we have a 1-1 correspondence

$$\bar{P} \longleftrightarrow (\bar{u}_p, \bar{n}_p) \qquad (5.2)$$

with $\bar{P} \in E(F)/p$, $\bar{u}_p \in \mathcal{O}_F^*/p$, $\bar{n}_p \in \mathbb{Z}/p$. The ordered pair on the right of (5.2) can be viewed as an element of $H^1_{\mathrm{fl}}(S, \mathcal{E}[p])$ or of its (isomorphic) image $\overline{H}^1(F, E[p])$ in $H^1(F, E[p])$.

In Galois cohomology, we have the decomposition

$$H^2(F, E[p]^{\otimes 2}) \simeq H^2(F, \mu_p^{\otimes 2}) \oplus H^2(F, \mu_p)^{\oplus 2} \oplus H^2(F, \mathbb{Z}/p). \qquad (5.3)$$

We have a similar one for $H^2_{\text{fl}}(S, \mathcal{E}[p]^{\otimes 2})$.

It is essential to note that by Proposition C,

$$Im(s_p) \hookrightarrow H^2(F, \mu_p^{\otimes 2}) \subset H^2(F, E[p]^{\otimes 2}), \qquad (5.4 - i)$$

and in addition,

$$H^2(F, \mu_p^{\otimes 2}) \simeq H^2(F, \mu_p) = \mathrm{Br}_F[p] \simeq \mathbb{Z}/p, \qquad (5.4 - ii)$$

which is implied by the fact that $\mu_p \subset F$ (since it is the determinant of $E[p]$).

In terms of the decomposition (5.2) from above, we get, for all $P, Q \in E(F)$,

$$s_p(\langle P, Q \rangle) = \bar{u}_P \cup \bar{u}_Q \in Br_F[p] \simeq \mathbb{Z}/p. \qquad (5.5)$$

$\square$

One knows that (cf. [L2], chapter II, sec. 3, Prop.6)

$$|\mathcal{O}_F^*/p| = |\mu_p(F)|p^{[F:\mathbb{Q}_p]}. \qquad (5.6)$$

Since we have assumed that F contains all the p-th roots of unity, the order of $\mathcal{O}_F^*/p$ is at least p^2.

Claim 5.7 *Let u, v be units in $\mathcal{O}_F$ which are linearly independent in the $\mathbb{F}_p$-vector space $V := \mathcal{O}_F^*/(\mathcal{O}_F^*)^p$. Then $F[u^{1/p}]$ and $F[v^{1/p}]$ are disjoint p-extensions of F.*

This is well known, but we give an argument for completeness. Pick p-th roots α, β of u, v respectively. If the extensions are not disjoint, we must have $\alpha = \sum_{j=0}^{p-1} c_j \beta^j$ in $K = F[v^{1/p}]$, with $\{c_j\} \subset F$. A generator σ of $\mathrm{Gal}(K/F$ must send β to $w\beta$ for some p-th root of unity $w \neq 1$ (assuming, as we can, that v is not a p-th power in F), and moreover, σ will send α to $w^i \alpha$ for some i. Thus α^σ can be computed in two different ways, resulting in the identity $\sum_j c_j w^i \beta^j = \sum_j c_j w^j \beta^j$, from which the Claim follows.

Consequently, since the dimension of V is at least 2 and since F has a unique unramified p-extension, we can find $u \in \mathcal{O}_F^*$ such that $K := F[u^{1/p}]$ is a ramified p-extension. Fix such a u and let $Q \in E(F)$ be given by $(\bar{u}, 0)$. By [Se2], Prop. 5 (iii) on page 72 (see also the Remark on page 95), there exists $v \in \mathcal{O}_F^*$ s.t. $v \notin N_{K/F}(\mathcal{O}_K^*)$. Then by [Ta1], Prop. 4.3, page 266, $\{v, u\} \neq 0$ in $\mathrm{Br}_F[p]$. Take $P \in E(F)$ such that $\bar{P} \leftrightarrow (\bar{v}, 0)$ then $s_p(\langle P, Q \rangle) = \{v, u\} \in H^2(F, E[p]^{\otimes 2}) \subset \mathrm{Br}_F[p]$ and $\{v, u\} \neq 0$.

It remains to show that we may choose P to be a p-power torsion point after possibly replacing F by a finite unramified extension. Since $E[p]$ is in F, μ_p is in F; recall that $\mu_p(F) \subset E[p](F)$. So we may pick a non-trivial p-th root of unity ζ in F. Let m be the largest integer such that $\zeta := w^{p^{m-1}}$ for some $w \in F^*$; in this case w is in $F^* - F^{*p}$. Let F'/F be the unramified extension of F such that over the corresponding residual extension, all the p^m-torsion points of $\mathcal{E}_s$ are rational; note however that $\mathcal{E}[p^m]$ need not be in F. This results in the following short exact sequence:

$$0 \to \mu_{p^m} \to \mathcal{E}[p^m] \to \mathbb{Z}/p^m \to 0,$$

leading to the inclusion

$$\mu_{p^m}(F') \subset E[p^m](F').$$

Since F'/F is unramified, w cannot belong to F'^{*p}, and the corresponding point P, say, in $E[p^m](F')$ is not in $pE(F')$. Put

$$L = F'(\frac{1}{p}P) = F'(w^{1/p}),$$

which is a ramified p-extension. So there exists a unit u in $\mathcal{O}_{F'}$ such that $\{w, u\}$ is not trivial in $\mathrm{Br}_{F'}[p]$. Now let Q be a point in $E(F')$ such that its class in $E(F')/p$ is given by $(\bar{u}, 0) \in \mathcal{O}_{F'}/p \oplus \mathbb{Z}/p$. It is clear that $\langle P, Q \rangle$ is non-zero in $T_{F'}(A)/p$.

Now suppose $E[p]$ is not semisimple. Since we are not in the wild case, we know by Lemma 4.0 that $K := F(E[p])$ is an unramified p-extension of F. By the discussion above, there are points $P, Q \in E(K)/p$ such that $s_{p,K}(\langle P, Q \rangle_K) \neq 0$. Put

$$\theta := N_{K/F}(\langle P, Q \rangle_K) \in T_F(A)/p.$$

Claim: $s_p(\theta)$ is non-zero.

Since we have

$$s_p(\theta) = N_{K/F}(s_{p,K}(\langle P, Q \rangle_K)),$$

the Claim is a consequence of the following Lemma, well known to experts.

Lemma 5.8 *If $K \supset F$ is any finite extension of p-adic fields, then $N_{K/F} \colon \mathrm{Br}_K \to \mathrm{Br}_F$ is an isomorphism.*

For completeness we give a

Proof The *invariant map* $\mathrm{inv}_F \colon \mathrm{Br}(F) \to \mathbb{Q}/\mathbb{Z}$ is an isomorphism and moreover (cf. [Se2], chap.XIII, Prop.7), if $n = [K : F]$, the following diagram commutes:

$$
\begin{array}{ccc}
\mathrm{Br}_F & \xrightarrow{\mathrm{Res}_{K/F}} & \mathrm{Br}_K \\
{\scriptstyle \mathrm{inv}_F} \downarrow & & \downarrow {\scriptstyle \mathrm{inv}_K} \\
\mathbb{Q}/\mathbb{Z} & \xrightarrow{\ n\ } & \mathbb{Q}/\mathbb{Z}
\end{array}
$$

As $\mathbb{Q}/\mathbb{Z}$ is divisible, every element in Br_K is the restriction of an element in Br_F. Now since $\mathrm{Res}_{K/F} \colon \mathrm{Br}(F) \to \mathrm{Br}(K)$ is already multiplication by n in $\mathbb{Q}/\mathbb{Z}$, the projection formula $N_{K/F} \circ \mathrm{Res}_{K/F} = [K : F]\mathrm{Id} = n\mathrm{Id}$ implies that the norm $N_{K/F}$ on Br_K must correspond to the identity on $\mathbb{Q}/\mathbb{Z}$.

We are now done with the proof of Proposition E.

$\square$

6 Injectivity of s_p

Proposition F *Let E be an elliptic curve over a non-archimedean local field F of odd residual characteristic p, such that E has good, ordinary reduction. Then s_p is injective on $T_F(E \times E)/p$.*

In view of Proposition E, we have the following

Corollary 6.1 *Let F, E, p be as in Proposition E. Then $T_F(E \times E)/p$ is a cyclic group of order p. Moreover, if $E[p] \subset F$, it even consists of symbols $\langle P, Q \rangle$, with $P, Q \in E(F)/p$.*

To prove Proposition F, we will need to consider separately the cases when $E[p]$ is semisimple and non-semisimple, the latter case being split further into the wild and unramified subcases.

Proof of Proposition F in the semisimple case. Again, to prove injectivity, we may replace F by any finite extension of prime-to-p degree. Since p does not divide $[F(E[p]) : F]$ when $E[p]$ is semisimple, we may assume (in this case) that all the p-torsion points of E are rational over F.

Remark The injectivity of s_p when $E[p] \subset F$ has been announced without proof, and in fact for a more general situation, by Raskind and Spiess [R-S], but the method of their paper is completely different from ours.

There are three steps in our proof of injectivity when $E[p] \subset F$:

Step I: Injectivity of s_p on symbols

Pick any pair of points P, Q in $E(F)$. We have to show that if $s_p(< P, Q >) = 0$, then the symbol $< P, Q >$ lies in $pT_F(A)$.

To achieve Step I, it suffices to prove that the condition (a) of lemma 1.8.1 holds.

In the correspondence (5.2), let $\bar{P} \leftrightarrow (\bar{u}_P, \bar{n}_P)$ and $\bar{Q} \leftrightarrow (\bar{u}_Q, \bar{n}_Q)$. Put $K_1 = F(\sqrt[p]{u_Q})$ and take K_2 to be the unique unramified extension of F of degree p if $\bar{n}_Q \neq 0$; otherwise take $K_2 = F$. Consider the compositum K_1K_2 of K_1, K_2, and $K := F\left(\frac{1}{p}Q\right)$, all the fields being viewed as subfields of $\bar{F}$.

From (5.1) we get the following commutative diagram:

$$
\begin{array}{ccccccc}
\mathcal{E}(\mathcal{O}_K)/p & \xrightarrow{\sim} & H^1_{\mathrm{fl}}(\mathcal{O}_K, \mathcal{E}[p]) & \simeq & \mathcal{O}^*_K/(\mathcal{O}^*_K)^p & \oplus & \mathbb{Z}/p \\
{\scriptstyle N_{K/F}}\downarrow\uparrow{\scriptstyle Res} & & \uparrow{\scriptstyle Res} & {\scriptstyle N_{K/F}}\downarrow\uparrow{\scriptstyle Res} & & \downarrow{\scriptstyle N_{K/F}=\mathrm{id}} & (6.2) \\
\mathcal{E}(\mathcal{O}_F)/p & \xrightarrow{\sim} & H^1_{\mathrm{fl}}(\mathcal{O}_F, \mathcal{E}[p]) & \simeq & \mathcal{O}^*_F/(\mathcal{O}^*_F)^p & \oplus & \mathbb{Z}/p.
\end{array}
$$

Here the map $Res = Res_{K/F}$ on $\mathcal{E}(\mathcal{O}_F)/p$ and $\mathcal{O}^*_F/p$ induced by the inclusion $F \hookrightarrow K$ is the obvious restriction map. However, on $\mathbb{Z}/p$, Res comes from the residue fields $\mathbb{F}$ of F and $\mathbb{F}'$ of K via the identifications

$$
H^1_{\mathrm{fl}}(S, \mathbb{Z}/p) \simeq H^1(\mathbb{F}, \mathbb{Z}/p) \simeq \mathrm{Hom}(\mathrm{Gal}(\bar{\mathbb{F}}/\mathbb{F}), \mathbb{Z}/p) \simeq \mathbb{Z}/p \tag{6.3}
$$

and the corresponding one for K and $\mathbb{F}'$. As $\mathrm{Gal}(\bar{\mathbb{F}}/\mathbb{F}) = \hat{\mathbb{Z}}$, we see by taking $\mathbb{F} \subset \mathbb{F}' \subset \bar{\mathbb{F}}$ with $[\mathbb{F}' : \mathbb{F}] = d$ dividing p, that $\mathrm{Gal}(\bar{\mathbb{F}}/\mathbb{F}')$ is the obvious subgroup of $\mathrm{Gal}(\bar{\mathbb{F}}/\mathbb{F})$ corresponding to $d\hat{\mathbb{Z}}$; consequently, the map Res on the $\mathbb{Z}/p$ summand is multiplication by d. Finally note that $N_{K/F} \circ Res_{K/F} = [K : F]\mathrm{id}$.

Lemma 6.4 *With K, K_1, K_2 as above, we have*

$$
K = K_1K_2.
$$

In other words we have the following diagram

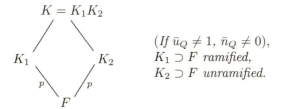

$$(\text{If } \bar{u}_Q \neq 1, \bar{n}_Q \neq 0),$$
$$K_1 \supset F \text{ ramified,}$$
$$K_2 \supset F \text{ unramified.}$$

Proof Put $L = K_1 K_2$. Let ∂_L be the boundary map given by the sequence (5.1).

a) $\mathbf{K} \subset \mathbf{L}$:   Apply the diagram (6.2) with L instead of K. We claim that $\partial_L(Res(Q)) = 0$ and hence $K \subset L$. If $\overline{n}_Q = 0$, so that $L = K_1$, we get $\partial_L(Res(Q)) = (Res(\overline{u}_Q), 0) = 0$. On the other hand, if $\overline{n}_Q \neq 0$, then K_2 is the unique unramified p-extension of F and the restriction map on $\mathbb{Z}/p$ is, by the remarks above, given by multiplication by p. So we still have $\partial_L(Res(Q)) = 0$, as claimed.

b) $\mathbf{K} \supset \mathbf{L}$:   In this case we have $\partial_K(Res(Q)) = 0$. On the other hand,

$$\partial_K(Res(Q)) = (Res(\overline{u}_Q), Res(\overline{n}_Q)) = (0, 0).$$

Hence $K \supset K_1$. If $\overline{n}_Q = 0$, $K_2 = F$ and we are done. So suppose $\overline{n}_Q \neq 0$. Then, since $Res(\overline{n}_Q)$ is zero, we see that the residual extension of K/F must be non-trivial and so must contain the unique unramified p-extension K_2 of F. Hence K contains $L = K_1 K_2$.   □

Lemma 6.5 *Let P, Q be in $E(F)/p$ with coordinates in the sense of (5.2), namely $P = (\overline{u}_P, \overline{n}_P)$ and $Q = (\overline{u}_Q, \overline{n}_Q)$. Let P_0, Q_0 be the points in $E(F)/p$ with coordinates $(\overline{u}_P, 0)$, $(\overline{u}_Q, 0)$ respectively. Then*

$$\langle P, Q \rangle = \langle P_0, Q_0 \rangle \in T_F(A)/p.$$

Proof of Lemma 6.5 Put $P_1 = (0, \overline{n}_P)$ and $Q_1 = (0, \overline{n}_Q)$. Then by linearity,

$$\langle P, Q \rangle = \langle P_0, Q_0 \rangle + \langle P_1, Q_0 \rangle + \langle P_0, Q_1 \rangle + \langle P_1, Q_1 \rangle.$$

First note that linearity,

$$\langle P_1, Q_1 \rangle = \overline{n}_P \overline{n}_Q \langle (0, 1), (0, 1) \rangle.$$

It is immediate, since $p \neq 2$, to see that $\langle (0, 1), (0, 1) \rangle$ is zero by the skew-symmetry of $\langle ., . \rangle$. Thus we have, by bi-additivity,

$$\langle P_1, Q_1 \rangle = 0.$$

Next we show that in $T_F(A)/p$,

$$\langle P_0, Q_1 \rangle = 0 = \langle P_1, Q_0 \rangle.$$

We will prove the triviality of $\langle P_0, Q_1 \rangle$; the triviality of $\langle P_1, Q_0 \rangle$ will then follow by the symmetry of the argument. There is nothing to prove if $\overline{n}_Q = 0$, so we may (and will) assume that $\overline{n}_Q \neq 0$. Then it corresponds to the unramified p-extension $M = F\left(\frac{1}{p} Q_1\right)$ of F. It is known that every unit in F^* is the norm of a unit in M^*/p. This proves that the point P_0 in $E(F)/p$ is a norm from M. Thus $\langle P_0, Q_1 \rangle$ is zero by Lemma 1.8.1. Putting everything together, we get

$$\langle P, Q \rangle = \langle P_0, Q_0 \rangle$$

as asserted in the Lemma.   □

Proof of Step I Let $P, Q \in E(F)/p$ be such that $s_p(\langle P, Q \rangle) = 0$. We have to show that $\langle P, Q \rangle = 0$ in $T_F(A)/p$. By Lemma 6.5 we may assume that $\overline{n}_P = \overline{n}_Q = 0$. So it follows that the element $\{\overline{u}_P, \overline{u}_Q\} := \overline{u}_P \cup \overline{u}_Q$ is zero in the Brauer group $\mathrm{Br}_F[p]$. Then putting $K_1 = K\left(u_Q^{1/p}\right)$, we have by [Ta1], Prop. 4.3, we have $u_P = N_{K_1/F}(u_1')$ with $u_1' \in K_1^*$ and where $u_P \in \mathcal{O}_F^*$ with image $\overline{u}_P$. It follows that u_1' must also be a unit in $\mathcal{O}_{K_1}$. Now with the notations introduced at the beginning of *Step I* we have, since $\overline{n}_Q = 0$, that $K = K_1$ and so $[K : F] = p$. In the diagram

(6.2), take in the upper right corner the element $(\overline{u'}, 0)$. Then we get an element $\overline{P'} \in E(K)/p$ such that $N_{K/F}\overline{P'} \equiv P \pmod{p}$. Hence condition (a) of Lemma 1.8.1 holds, yielding Step I. $\qquad \square$

Step II:

Put
$$\mathrm{ST}_{F,p}(A) \;=\; \mathrm{Im}\,(\mathrm{ST}_F(A) \to T_F(A)/p)\,.$$

Lemma 6.6 s_p *is injective on* $\mathrm{ST}_{F,p}(A)$.

Proof This follows from

Claim: If V is an $\mathbb{F}_p$-vector space with an alternating bilinear form $[\ ,\ ]\colon V \times V \to W$, with W a 1-dimensional $\mathbb{F}_p$-vector space, then the conditions (i) and (ii) of [Ta1], p. 266, are satisfied, which in our present setting read as follows :

  (i) Given a, b, c, d in V such that $[a,b] = [c,d]$, then there exist elements x and y in V such that
$$[a,b] = [x,b] = [x,y] = [c,y] = [c,d].$$

  (ii) Given a_1, a_2, b_1, b_2 in V, there exist c_1, c_2 and d in V such that
$$[a_1,b_1] = [c_1,d],\ \text{and}\ [a_2,b_2] = [c_2,d].$$

$\qquad \square$

Proof Recall that we are assuming in this paper that p is odd. The Claim is just Proposition 4.5 of [Ta1], p. 267. The proof goes through verbatim. In our case we apply this to $V = E(F)/p, W = \mathrm{Br}_F[p]$ and $[\bar{P}, \bar{Q}] = s_p(\langle P, Q \rangle) = \bar{u}_P \cup \bar{u}_Q$. By ([Ta1], Corollary on p. 266), the s_p is injective on $\mathrm{ST}_{F,p}(A)$. $\qquad \square$

Step III: Injectivity of s_p **in the general case** (but still assuming $E[p] \subset F$).

For this we shall appeal to Lemma 5.8.

Proof of Step III By Step II, we see from Lemma 1.7.1 that it suffices to prove the following result, which may be of independent interest.

Proposition 6.7 *Let* $K \supset F$ *be a finite extension, and* $E[p] \subset F$. *Then* $N_{K/F}(\mathrm{ST}_{K,p}(A))$ *is a subset of* $\mathrm{ST}_{F,p}(A)$ *and hence*
$$\mathrm{ST}_{F,p}(A) \;=\; T_F(A)/p.$$
In other words, $T_F(A)$ *is generated by symbols modulo* $pT_F(A)$.

It is an open question as to whether $T_F(A)$ is different from $\mathrm{ST}_F(A)$, though many expect it to be so.

Proof of Proposition 6.7 We have to prove that if $P', Q' \in E(K)$, then the norm to F of $\langle P', Q' \rangle$ is mapped into $\mathrm{ST}_{F,p}(A)$. Since we have
$$\mathrm{Im}\, s_p(\mathrm{ST}_F(A)) \;=\; \mathrm{Im}\, s_p(T_F(A)/p) \simeq \mathbb{Z}/p,$$
the assertion is a consequence of the following

Lemma 6.8 *If* $[K\colon F] = n$ *and* $P', Q' \in E(K)$, $P, Q \in E(F)$ *are such that*
$$s_{p,F}(N_{K/F}\langle P', Q' \rangle) \;=\; s_{p,F}(\langle P, Q \rangle),$$

then, assuming that all the p-torsion in $E(\overline{F})$ is F-rational,

$$N_{K/F}(\langle P', Q' \rangle) \equiv \langle P, Q \rangle \pmod{p\, T_F(A)}$$

Proof of Lemma 6.8. We start with a few simple sublemmas.

Sublemma 6.9 *If K/F is unramified, then Lemma 6.8 is true.*

Proof Indeed, in this case, every point in $E(F)$ is the norm of a point in $E(K)$ (cf. [Ma], Corollary 4.4). So we can write $P \equiv N_{K/F}(P_1) \pmod{pE(K)}$ with $P_1 \in E(K)$, from which it follows that

$$N_{K/F}(\langle P_1, Q \rangle) \equiv \langle P, Q \rangle \pmod{pT_F(A)}.$$

On the other hand, by the remark above (in the beginning of Step III) about the norm map on the Brauer group, we have

$$s_{p,K}(\langle P_1, Q \rangle) = s_{p,K}(\langle P', Q' \rangle)$$

and applying Step II to $\langle P_1, Q \rangle$ and $\langle P', Q' \rangle$ for K, we are done. □

Subemma 6.10 *If $K/F = n$ with $p \nmid n$, then Lemma 6.8 is true.*

Proof Indeed, as the cokernel of $N_{K/F}$ is annihilated by n which is prime to p, the norm map on $E(K)/p$ is surjective onto $E(F)/p$. Let $P_1 \in E(K)$ satisfy $P \equiv N_{K/F}(P_1) \pmod{pE(F)}$. Then by the projection formula,

$$\langle P, Q \rangle \equiv N_{K/F}(\langle P_1, Q \rangle) \pmod{pT_F(A)}. \qquad (*)$$

Since $s_{p,F} \circ N_{K/F} = N_{K/F} \circ s_{p,K}$ and since $N_{K/F}$ is non-trivial, and hence an isomorphism, on the one-dimensional $\mathbb{F}_p$-space $\mathrm{Br}_K[p]$, we see that

$$s_{p,K}(\langle P', Q' \rangle) = s_{p,K}(\langle P_1, Q \rangle).$$

Applying Lemma 6.6 for K we see that $\langle P', Q' \rangle)$ equals $\langle P_1, Q \rangle$ in $T_K(A)/p$. Now we are done by $(*)$. □

Sublemma 6.11 *If $K \supset F_1 \supset F$ and if 6.8 is true for both the pairs K/F_1 and F_1/F, then it is true for K/F.*

Proof Let $P, Q \in E(F)$ and $P', Q' \in E(K)$ be such that

$$s_{p,F}(N_{K/F}\langle P', Q' \rangle) = s_{p,F}(\langle P, Q \rangle). \qquad (a)$$

Applying Proposition E with F_1 in the place of F, we get points P_1, Q_1 in $E(F_1)$ such that

$$s_{p,F_1}(\langle P_1, Q_1 \rangle) = N_{K/F_1}(s_{p,K}(\langle P', Q' \rangle)). \qquad (b)$$

Since the right hand side is the same as $s_{p,F_1}(N_{K/F_1}(\langle P', Q' \rangle))$, we may apply 6.8 to the pair K/F_1 to conclude that

$$\langle P_1, Q_1 \rangle = N_{K/F_1}(\langle P', Q' \rangle). \qquad (c)$$

Applying $N_{F_1/F}$ to both sides of (b), using the facts that the norm map commutes with s_p and that $N_{K/F} = N_{K/F_1} \circ N_{F_1/F}$, and appealing to (a), we get

$$s_{p,F}(N_{F_1/F}\langle P_1, Q_1 \rangle) = s_{p,F}(\langle P, Q \rangle). \qquad (d)$$

Applying (6.8) to F_1/F, we then get

$$\langle P, Q \rangle = N_{F_1/F}(\langle P_1, Q_1 \rangle). \qquad (e)$$

The assertion of the Sublemma now follows by combining (c) and (e). $\square$

Sublemma 6.12 *It suffices to prove 6.8 for all finite Galois extensions K'/F' with $E[p] \subset F'$ and $[F' : \mathbb{Q}_p] < \infty$.*

Proof Assume 6.8 for all finite Galois extensions K'/F' as above. Let K/F be a finite, non-normal extension with Galois closure L, and let $E[p] \subset F$. Suppose $P, Q \in E(F)$ and $P', Q' \in E(K)$ satisfy the hypothesis of 6.8. We may use the surjectivity of $N_{L/K} : \mathrm{Br}_L \to \mathrm{Br}_K$ and Proposition E (over L) to deduce the existence of points $P'', Q'' \in E(L)$ such that

$$s_{p,K}(N_{L/K} \langle P'', Q'' \rangle) = s_{p,K}(\langle P', Q' \rangle).$$

As L/K is Galois, we have by hypothesis,

$$N_{L/K}(\langle P'', Q'' \rangle) = \langle P', Q' \rangle . \qquad (i)$$

By construction, we also have

$$s_{p,F}(N_{L/F} \langle P'', Q'' \rangle) = s_{p,F}(\langle P, Q \rangle).$$

As L/F is Galois, we have

$$N_{L/F}(\langle P'', Q'' \rangle) = \langle P, Q \rangle . \qquad (ii)$$

The assertion now follows by applying $N_{K/F}$ to both sides of (i) and comparing with (ii). $\square$

Thanks to this last sublemma, we may assume that K/F is Galois. Appealing to the previous three sublemmas, we may assume that we are in the following **key case**:

$(\mathcal{K})$ *K/F is a totally ramified, cyclic extension of degree p, with $E[p] \subset F$.*

So it suffices to prove the following

Lemma 6.13 *6.8 holds in the key case $(\mathcal{K})$.*

Proof Suppose K/F is a cyclic, ramified p-extension with $E[p] \subset F$, and let $P, Q \in E(F)$, $P', Q' \in E(K)$ satisfy the hypothesis of 6.8. Since $\mu_p = \det(E[p])$, the p-th roots of unity are in F and so

$$W := F^*/p = H^1(F, \mu_p) \simeq H^1(F, \mathbb{Z}/p) = \mathrm{Hom}(\mathrm{Gal}(\overline{F}/F), \mathbb{Z}/p).$$

In other words, lines in W correspond to cyclic p-extensions of F, and we can write $K = F(u^{1/p})$ for some $u \in \mathcal{O}_F^*$. For every $w \in \mathcal{O}_F^*$, let $\overline{w}$ denote its image in W. Put $m = \dim_{\mathbb{F}_p} W$. Then $m \geq 3$ by [L2], chapter II, sec. 3, Proposition 6. Let W_1 denote the line spanned in W by the unique unramified p-extension of F. Since $m > 2$, we can find some $v \in \mathcal{O}_F^*$ such that $\overline{v}$ is not in the linear span of W_1 and $\overline{u}$. Using the *Claim* stated in the proof of Proposition E, we see that $L := F(v^{1/p})$ is linearly disjoint from K over F. Both K and L are totally ramified p-extensions of F, and so is the (p, p)-extension KL. Then KL/K is a cyclic, ramified p-extension, and by local class field theory ([Se2], chap. V, sec 3), there exists $y \in \mathcal{O}_K^*$ not lying in $N_{KL/K}((KL)^*)$. Hence $\overline{v} \cup \overline{y}$ is non-zero in $\mathrm{Br}_K[p]$ (see [Ta1], Prop.4.3, p.266, for example). Let $P_1 \in E(F)$ and $Q_1 \in E(K)$ be such that in the sense of (5.2) we have

$$\overline{P}_1 \leftrightarrow (\overline{v}, 0), \quad \overline{Q}_1 \leftrightarrow (\overline{y}, 0).$$

Then
$$s_{p,K}(\langle P_1, Q_1 \rangle) = \overline{v} \cup \overline{y} \neq 0.$$
Since $\mathrm{Br}_K[p]$ is one-dimensional (over $\mathbb{F}_p$), we may replace Q_1 by a multiple and assume that
$$s_{p,K}(\langle P_1, Q_1 \rangle) = s_{p,K}(\langle P', Q' \rangle).$$
Hence
$$\langle P_1, Q_1 \rangle \equiv \langle P', Q' \rangle \bmod p T_K(A)$$
by *Step II* for K. So we get by the hypothesis,
$$s_{p,F}(N_{K/F} \langle P_1, Q_1 \rangle) = s_{p,F}(\langle P, Q \rangle).$$
But P_1 is by construction in $E(F)$, and so the projection formula says that
$$N_{K/F} \langle P_1, Q_1 \rangle = \langle P_1, N_{K/F}(Q_1) \rangle.$$
Now we are done by applying *Step II* to F.   □

Now we have completed the proof of Proposition F when $E[p]$ is semisimple.

Proof of Proposition F in the non-semisimple, unramified case. In this case $K := F(E[p])$ is an unramified p-extension of F. By Mazur ([Ma]), we know that as E/F is ordinary, the norm map from $E(K)$ to $E(F)$ is surjective. This shows the injectivity on symbols, giving Step I in this case. Indeed, if $s_p(\langle P, Q \rangle) = 0$ for some $P, Q \in E(F)/p$, we may pick an $R \in E(K)$ with norm Q and obtain
$$N_{K/F}(s_p(\langle P, R \rangle_K)) = s_p(\langle P, N_{K/F}(R) \rangle) = 0.$$
On the other hand, the norm map on the Brauer group is injective (cf. Lemma 5.8). This forces $s_p(\langle P, R \rangle_K)$ to be non-zero. As $E[p]$ is trivial over K, the injectivity in the semisimple situation gives the triviality of $\langle P, R \rangle_K = 0$.

The proof of injectivity on $ST_{F,p}(A)$ (Step II) is the same as in the semisimple case.

The proof of injectivity on $T_F(A)/p$ (Step III) is an immediate consequence of the following Lemma (and the proof in the semisimple case).

Lemma 6.14 *Let $\theta \in T_F(A)/p$. Then $\exists \tilde{\theta} \in T_K(A)/p$ such that $\theta = N_{K/F}(\tilde{\theta})$.*

Proof of Lemma In view of Lemma 1.7.1, it suffices to show, for any finite extension L/F and points $P, Q \in E(L)/p$, that $\beta := N_{L/F}(\langle P, Q \rangle_L)$ is a norm from K. This is obvious if L contains K. So let L not contain K, and hence is linearly disjoint from K over F (as K/F has prime degree). Consider the compositum $M = LK$, which is an unramified p-extension of L. By Mazur ([Ma]), we can write $Q = N_{M/L}(R)$, for some $R \in E(M)/p$. Put
$$\tilde{\beta} := N_{M/K}(\langle P, R \rangle_M) \in T_K(A)/p.$$
Then, using $N_{L/F} \circ N_{M/L} = N_{M/F} = N_{K/F} \circ N_{M/K}$, we obtain the desired conclusion
$$\beta = N_{L/F}\left(\langle P, N_{M/L}(R) \rangle_L\right) = N_{M/F}(\langle P, R \rangle_M) = N_{K/F}(\tilde{\theta}).$$
                                                                                 □

This finishes the proof when $E[p]$ is non-semisimple and unramified.

Proof of Proposition F in the non-semisimple, wild case. Here K/F is a wildly ramified extension, so we cannot reduce to the case $E[p] \subset F$. Still, we may, as in section 4, assume that $\mu_p \subset F$ and that $\nu = 1$, which implies that the p-torsion points of the special fibre $\mathcal{E}_s$ are rational over the residue field $\mathbb{F}_q$ of F. We have in effect (4.2) through (4.4).

Consider the commutative diagram (4.4). We can write

$$\mathcal{O}_F^*/p = H_{\mathrm{fl}}^1(S, \mu_p) \simeq \overline{H}^1(F, \mu_p) = X + Y_S,$$

with $X \simeq \mathbb{Z}/p$ and Y_S some complement. Furthermore, we have (cf. [Mi1], Theorem 3.9, p.114)

$$H_{\mathrm{fl}}^1(S, \mathbb{Z}/p) \simeq H_{\mathrm{et}}^1(k, \mathbb{Z}/p) = \mathbb{Z}/p.$$

Hence

$$H_{\mathrm{fl}}^1(S, \mathcal{E}[p]) = Z_S \oplus \mathbb{Z}/p, \quad \text{where} Z_S = \psi_S(Y_S). \tag{6.15}$$

(Note that the surjectivity of the right map on the top row of the diagram (4.4) comes from the fact that $H_{\mathrm{fl}}^2(S, \mu_p)$ is 0 ([Mi2], chapter III, Lemma 1.1). Recall (cf. (6.2)) that $E(F)/p$ is isomorphic to $H_{\mathrm{fl}}^1(S, \mathcal{E}[p])$. So by (6.15), we have a bijective correspondence

$$P \longleftrightarrow (\tilde{u}_P, \overline{n}_P) \tag{6.16}$$

where P runs over points in $E(F)/p$. Compare this with (5.2).

Lemma 6.17 *Fix any odd prime p. Let K be an arbitrary finite extension of F where $E[p]$ remains non-semisimple as a G_K-module. Suppose $\langle (u', 0), (u'', 0) \rangle = 0$ in $T_K(A)/p$ for all $u', u'' \in \mathcal{O}_K^*/p$. Then $\langle P, Q \rangle = 0$ for all $P, Q \in E(F)/p$.*

Proof of Lemma. Let $P = (u_P, n_P), Q = (u_Q, n_Q)$ be in $E(F)/p$. Then by the bilinearity and skew-symmetry of $\langle ., . \rangle$, the fact that $\langle 1, 1 \rangle = 0$, and also the hypothesis of the Lemma (applied to $K = F$), it suffices to show that

$$\langle (u_P, 0), (0, n_Q) \rangle = \langle (u_Q, 0), (0, n_P) \rangle = 0.$$

It suffices to prove the triviality of the first one, as the argument is identical for the second. Let L denote the unique unramified p-extension of F. Then there exists $u' \in \mathcal{O}_L^*/p$ such that $u_P = N_{L/F}(u')$. Since $\langle (u_P, 0), (0, n_Q) \rangle$ equals $N_{L/F}(\langle (u_P, 0), \mathrm{Res}_{L/F}((0, n_Q)) \rangle)$ by the projection formula, it suffices to show that

$$\langle (u', 0), \mathrm{Res}_{L/F}((0, n_Q)) \rangle = 0.$$

Now recall (cf. the discussion around (6.3)) that the restriction map $H^1(\mathcal{O}_F, \mathbb{Z}/p) \to H^1(\mathcal{O}_L, \mathbb{Z}/p)$ is zero. This implies that $\mathrm{Res}_{L/F}((0, n_Q)) = (u'', 0)$ for some $u'' \in \mathcal{O}_L^*/p$. (We cannot claim that this restriction is $(0, 0)$ in $E(L)/p$ because the splitting of the surjection $H^1(\mathcal{O}_K, \mathcal{E}[p]) \to H^1(\mathcal{O}_K, \mathbb{Z}/p)$ is not canonical when $E[p]$ is non-semisimple over K. This point is what makes this Lemma delicate.) Thus we have only to check that $\langle (u', 0), (u'', 0) \rangle = 0$ in $T_L(A)/p$. This is a consequence of the hypothesis of the Lemma, which we can apply to $K = L$ because $E[p]$ remains non-semisimple over any unramified extension of F. Done.

As in the semisimple case, there are three steps in the proof.

Step I Injectivity on symbols:

We shall use the following terminology. For $u \in \mathcal{O}_F^*$, write $\overline{u}$ and $\tilde{u}$ for its respective images in $\mathcal{O}_F^*/p$ and Y, seen as a quotient of $H_{\mathrm{fl}}^1(S, \mu_p)/X$. We will also denote by $\tilde{u}$ the corresponding element in $Z = \psi_S(Y_S) \subset H_{\mathrm{fl}}^1(S, \mathcal{E}[p])$, which should not cause any confusion as Y is isomorphic to Z. We use similar notation in

$\overline{H}^1(F,-)$. For $\overline{u}$ in $\mathcal{O}_F^*/p$, we denote by $P_{\overline{u}}$ the element in $E(F)/p$ corresponding to the pair $(\tilde{u},0)$ given by (6.16).

As in the proof of Proposition D under the diagram (4.4), pick a non-zero element $e \in X \subset \mathcal{O}_F^*/p$. By definition, $\tilde{e} = 0$, so that $P_e = (\tilde{e},0)$ is the zero element of $E(F)/p$. As we have seen in the proof of Proposition D, there exists a v in $\mathcal{O}_F^*/p$ such that $[\overline{e},\overline{v}] := \overline{e} \cup \overline{v}$ is non-zero in $\mathrm{Br}_F[p]$.

Now let $\overline{x}, \overline{y} \in \mathcal{O}_F^*/p$ be such that $s_p(\langle P_{\overline{x}}, P_{\overline{y}}\rangle) = 0$. We have to show that

$$\langle P_{\overline{x}}, P_{\overline{y}}\rangle = 0 \in T_F(A)/p. \tag{6.18}$$

There are two cases to consider.

Case (a) $[\overline{x},\overline{y}] = 0 \in \mathrm{Br}_F[p]$:

Then $\overline{x}$ is a norm from $K = F(y^{1/p})$. This implies, as in the proof in the semisimple case, that $P_{\overline{x}}$ is a norm from $E(K)/p$, and by Lemma 1.8.1, we then have (6.18).

Case (b) $[\overline{x},\overline{y}] \neq 0$:

Since $\mathrm{Br}_F[p] \simeq \mathbb{Z}/p$, we may, after modifying by a scalar, assume that $[\overline{x},\overline{y}] = [\overline{e},\overline{v}]$ in $\mathrm{Br}_F[p]$. Then by Tate ([Ta1], page 266, conditions (i), (ii)), there are elements $\overline{a}, \overline{b} \in \mathcal{O}_F^*/p$ such that

$$[\overline{x},\overline{y}] = [\overline{x},\overline{b}] = [\overline{a},\overline{b}] = [\overline{a},\overline{v}] = [\overline{e},\overline{v}].$$

Claim 6.19 *For each pair of neighbors in this sequence, the corresponding symbols are equal in $T_F(A)/p$.*

Proof of Claim 6.19 From $[\overline{x},\overline{y}] = [\overline{x},\overline{b}]$ we have, by linearity, that $[\overline{x},\overline{y}\overline{b}^{-1}] = 0$, hence x is a norm from $L := F((yb^{-1})^{1/p})$. Hence $P_{\overline{x}}$ is a norm from from $E(L)/p$, and so we get $\langle P_{\overline{x}}, P_{\overline{yb^{-1}}}\rangle = 0$ in $T_F(A)/p$. Consequently, now by the linearity of $\langle\cdot,\cdot\rangle$, $\langle P_{\overline{x}}, P_{\overline{y}}\rangle = \langle P_{\overline{x}}, P_{\overline{b}}\rangle$. The remaining assertions of the Claim are proved in exactly the same way. $\square$

This Claim finishes the proof of Step I since $\langle P_{\overline{e}}, P_{\overline{v}}\rangle = 0$, which holds because $P_{\overline{e}} = 0$ in $E(F)/p$. $\square$

Step II Injectivity on $\mathbf{ST_{F,p}(A)} \subset \mathbf{T_F(A)/p}$:

Proof We have just seen in Step I that each of the symbol in $T_F(A)/p$ is zero. Since $ST_{F,p}(A)$ is generated by such symbols, it is identically zero. Done in this case.

Step III Injectivity on $\mathbf{T_F(A)/p}$:

Proof Since $T_F(A)$ is generated by norms of symbols from finite extensions of F, it suffices to show the following for *every finite extension L/F and points $P, Q \in E(L)$*:

$$N_{L/F}(\langle P,Q\rangle) = 0 \in T_F(A)/p. \tag{6.20}$$

Fix an arbitrary finite extension L/F and put

$$F_1 = F(E[p]).$$

There are again two cases.

Case (a) $L \supset F_1$:

Here we are in the situation where $E[p] \subset L$. In particular, $s_{p,L}$ is injective.

In the correspondence of (5.2), let $P, Q \in E(L)/p$ correspond to $\overline{u}_P, \overline{u}_Q \in \mathcal{O}_L^*/p$ respectively. Put

$$t := s_{p,L}(\langle P, Q \rangle) = [\overline{u}_P, \overline{u}_Q] \in \mathrm{Br}_F[p].$$

If $t = 0$, we have $\langle P, Q \rangle = 0$ in $T_L(A)/p$ (as $E[p] \subset L$), and we are done.

So let $t \neq 0$. Now let e, as before, be the $\mathbb{F}_p$-generator of $X \subset \mathcal{O}_F^*/p$, where X is the image of $\mathbb{Z}/p$ encountered in the proof of Proposition D. The diagram (4.4) shows that $F_1 = F(e^{1/p})$. In particular, F_1/F is a ramified p-extension. Since $p \neq 2$, $\mathcal{O}_F^*/p$ has dimension at least 3, and so we can take $v \in \mathcal{O}_F^*/p$ such that the space spanned by $\overline{e}$ and $\overline{v}$ in $\mathcal{O}_F^*/p$ has dimension 2 and does not contain the line corresponding to the unique unramified p-extension M of F. Put

$$F_2 = F(v^{1/p}) \quad \text{and} \quad K = F_1 F_2 \subset \overline{F}.$$

Then K/F is totally ramified with $[K : F] = p^2$. Hence K/F_1 is still a ramified p-extension. By the local class field theory ([Se2], chapter V, section 3), we can choose $y \in \mathcal{O}_{F_1}^*/p$ such that $y \notin N_{K/F_1}(K^*)$, so that $[\overline{v}, \overline{y}] \neq 0$ in $\mathrm{Br}_F[p]$. By linearity, we can replace y by a power such that $[\overline{v}, \overline{y}] = t = [\overline{u}_P, \overline{u}_Q]$. Now we have

$$\langle P_{\overline{v}}, P_{\overline{y}} \rangle_L = \langle P, Q \rangle_L \in T_L(A)/p, \tag{6.21}$$

where $\langle \cdot, \cdot \rangle_L$ denotes the pairing over L. (This makes sense here because of the hypothesis $F_1 \subset L$.) Applying the projection formula to L/F_1 and F_1/F, using the facts that $P_{\overline{v}} \in E(F)/p$ and $P_{\overline{y}} \in E(F_1)/p$,

$$N_{L/F_1}(\langle P_{\overline{v}}, P_{\overline{y}} \rangle_L) = [L : F_1]\langle P_{\overline{v}}, P_{\overline{y}} \rangle_{F_1} \tag{6.22}$$

and

$$N_{F_1/F}(\langle P_{\overline{v}}, P_{\overline{y}} \rangle_{F_1}) = \langle P_{\overline{v}}, N_{F_1/F}(P_{\overline{y}}) \rangle.$$

Now since $N_{L/F} = N_{L/F_1} \circ N_{F_1/F}$ and since we have shown that the symbols in $T_F(A)/p$ are all trivial, we get the triviality of $N_{L/F}(\langle P, Q \rangle)$ by combining (6.21) and (6.22).

Case (b) $L \not\supset F_1$:

In this case $E[p]$ is not semisimple over L. Hence by Step II, we have $\langle P, Q \rangle_L = 0$ in $T_L(A)/p$. So (6.20) holds. Done.

This finishes the proof of Step III, and hence of Proposition F. $\square$

7 Remarks on the case of multiplicative reduction

Here one has the following version of Theorem A, which we will need in the sequel giving certain global applications:

Theorem G   *Let F be a non-archimedean local field of characteristic zero with residue field $\mathbb{F}_q$, $q = p^r$. Suppose E/F is an elliptic curve over F, which has multiplicative reduction. Then for any prime $\ell \neq 2$, possibly with $\ell = p$, the following hold:*

(a) *s_ℓ is injective, with image of $\mathbb{F}_\ell$-dimension ≤ 1.*

(b) *$Im(s_\ell) = 0$ if $\mu_\ell \not\subset F$.*

This theorem may be proved by the methods of the previous sections, but in the interest of brevity, we will just show how it can be derived, in effect, from the results of T. Yamazaki ([Y]).

The Galois representation ρ_F on $E[\ell]$ is reducible as in the ordinary case, giving the short exact sequence (cf. [Se1], Appendix)

$$0 \to C_F \to E[\ell] \to C_F' \to 0, \tag{7.1}$$

where C_F' is given by an unramified character ν of order dividing 2; E is split multiplicative iff $\nu = 1$. And C_F is given by $\chi\nu$ $(= \chi\nu^{-1}$ as $\nu^2 = 1)$, where χ is the mod ℓ cyclotomic character given by the Galois action on μ_ℓ.

Using (7.1), we get a homomorphism (as in the ordinary case)

$$\gamma_F : H^2(F, C_F^{\otimes 2}) \to H^2(F, E[\ell]^{\otimes 2}). \tag{7.2}$$

Here is the analogue of Proposition C:

Proposition 7.3 *We have*

$$Im(s_\ell) \subset Im(\gamma_F).$$

Furthermore, the image of s_ℓ is at most one-dimensional as asserted in Theorem G.

Proof of Proposition 7.3 As $\nu^2 = 1$ and $\ell \neq 2$, we may replace F by the (at most quadratic) extension $F(\nu)$ and assume that $\nu = 1$, which implies that $C_F = \mu_\ell$ and $C_F' = \mathbb{Z}/\ell$. We get the injective maps

$$\psi : H^1(F, \mu_\ell) \to H^1(F, E[\ell])$$

and

$$\delta : E(F)/\ell \to H^1(F, E[\ell]).$$

Claim 7.4 $Im(\delta) \subset Im(\psi)$.

Indeed, since $E(F)$ is the Tate curve $F^*/q^{\mathbb{Z}}$ for some $q \in F^*$ (cf. [Se1], Appendix), we have the uniformizing map

$$\varphi : F^* \to E(F).$$

Given $P \in E(F)$, pick an $x \in F^*$ such that $P = \varphi(x)$. Now $\delta(P)$ is represented by the 1-cocycle $\sigma \to \sigma(\frac{1}{\ell}P) - \frac{1}{\ell}P$, for all $\sigma \in G_F$. Let $y \in \overline{F}^*$ be such that $y^\ell = x$. Put $K = F(y)$. One knows that there is a natural uniformization map (over K) $\varphi_K : K^* \to E(K)$, which agrees with φ on F^*. Then we may take $\varphi_K(y)$ to be $\frac{1}{\ell}P$. It follows that the 1-cocycle above sends σ to $\frac{y^\sigma}{y} = \zeta_\sigma$, for an ℓ-th root of unity ζ_σ. In other words, the class of this cocycle is in the image of ψ in the long exact sequence in cohomology:

$$0 \to \mu_\ell(F) \to E[\ell](F) \to \mathbb{Z}/\ell \to H^1(F, \mu_\ell) \xrightarrow{\psi} H^1(F, E[\ell]) \to \mathbb{Z}/\ell \to \dots,$$

namely $\psi((\zeta_\sigma)) = \delta(P)(\sigma)$, with $(\zeta_\sigma) \in H^1(F, \mu_\ell)$. Hence the Claim.

Thanks to the Claim, the first part of the Proposition follows because $Im(s_\ell)$ is obtained via the cup product of classes in $Im(\delta)$, and $Im(\gamma_F)$ is obtained via cup product of classes in $Im(\psi)$.

The second part also follows because $H^2(F, \mu_\ell^{\otimes 2})$ is isomorphic to (the dual of) $H^0(F, \mathbb{Z}/\ell(-1))$, which is at most one-dimensional. $\qquad\square$

The injectivity of s_ℓ has been proved in Theorem 4.3 of [Y] in the split multiplicative case, and this saves a long argument we can give analogous to our proof in the ordinary situation. Injectivity also follows for the general case because we can, since $\ell \neq 2$, make a quadratic base change to $F(\nu)$ when $\nu \neq 1$. When combined with Proposition 7.3, we get part (a) of Theorem G.

To prove part (b) of Theorem G, just observe that, thanks to Proposition 7.3, it suffices to prove that when $\mu_\ell \notin F$, we have $H^2(F, \mu_\ell^{\otimes 2}) = 0$. This is clear because the one-dimensional Galois module $\mathbb{Z}/\ell(-1)$, which is the dual of $\mu_\ell^{\otimes 2}$, has G_F-invariants iff $\mathbb{Z}/\ell \simeq \mu_\ell$, i.e., iff $\mu_\ell \subset F$. Now we are done.

Appendix

In this Appendix, we will collect certain facts and results we use concerning the flat topology, locally of finite type (cf. [Mi1], chapter II).

A1. The setup

As in the main body of the paper, we assume that F is a non-archimedean *local field* of *characteristic zero*, with *residue field* $k = F_q$ of *characteristic* $p > 0$. Let $S = \operatorname{Spec} \mathcal{O}_F$, $j \colon U = \operatorname{Spec} F \to S$ and $i \colon s = \operatorname{Spec} k \to S$ where s is the closed point.

Let E be an *elliptic curve* defined on F, and let $A = E \times E$. Let $\mathcal{E}$ be the Néron model of E over S, so we have in particular $E(F) = \mathcal{E}(S)$ where $E(F)$ denotes the F-rational points of E and $\mathcal{E}(S)$ are the sections of $\mathcal{E}$ over S. We have $E = j^* \mathcal{E}$ and $i^* \mathcal{E}$ as pullbacks, and $i^* \mathcal{E}$ is the reduction of E. For any $n \geq 1$, let $\mathcal{E}[n]$ denote the kernel of multiplication by n on $\mathcal{E}$.

Let ℓ be a prime number > 2, possibly equal to p. Under these assumptions we can apply Proposition 2.5.1 and Corollary 2.5.2 to E and $A = E \times E$. Since $H^1_{\text{et}}(\bar{E}, \mathbb{Z}_\ell(1))$ is the Tate module $\mathcal{T}_\ell(E)$, we have the homomorphism $c_\ell \colon T_F(A) \to H^2(F, \mathcal{T}_\ell(E)^{\otimes 2})$. Reducing modulo ℓ, we obtain now a homomorphism

$$s_\ell \colon T_F(A)/\ell \longrightarrow H^2(F, E[\ell]^{\otimes 2}).$$

A2. Flat topology and the integral part

We shall need flat topology, or to be precise, *flat topology, locally of finite type* (cf. [Mi1]).

Let $\mathcal{F}$ be a sheaf on the *big flat site* S_{fl}. Consider the restriction $\mathcal{F}_{|U} = \mathcal{F}_U :=$ $j^* \mathcal{F}$ to U. Now assume that $\mathcal{F}_U$ is the pull back $\pi_U^*(\mathcal{G})$ of a sheaf $\mathcal{G}$ on the *étale site* U_{et} under the morphism of sites $\pi_U \colon U_{\text{fl}} \to U_{\text{et}}$. Then we have, in each degree i, a homomorphism

$$\pi_U^* \colon H^i_{\text{et}}(F, \mathcal{G}|_U) \to H^i_{\text{fl}}(F, \mathcal{F}|_U). \tag{A.2.1}$$

We will need the following well known fact (cf. [Mi1], Remark 3.11(b), pp. 116–117):

Lemma A.2.2 *This is an isomorphism when $\mathcal{G}$ is a locally constructible sheaf on U_{et}.*

Now let $\mathcal{F}$ and $\mathcal{G}$ be as above, with $\mathcal{G}$ locally constructible. Composing the natural map $H^i_{\text{fl}}(S, \mathcal{F}) \to H^i_{\text{fl}}(U, \mathcal{F})$ with the inverse of the isomorphism of Lemma A.2.2, we get a homomorphism

$$\beta \colon H^i(_{\text{fl}}(S, \mathcal{F}) \to H^i(F, \mathcal{G}_F),$$

where the group on the right is Galois cohomology in degree i of F, i.e., of $\mathrm{Gal}(\overline{F}/F)$, with coefficients in the Galois module $\mathcal{G}_F$ associated to the sheaf $\mathcal{G}$ on U_{et}.

Definition A.2.3 *In the above setup, define the integral part of $H^i(F, \mathcal{G}_F)$ to be*

$$\overline{H}^i(F, \mathcal{G}_F) := Im(\beta).$$

The sheaves of interest to us below will be even constructible, and in many cases, $\mathcal{G}$ will itself be the restriction to U of a sheaf on S_{et}. A key case will be when there is a finite, flat groupscheme C on S and $\mathcal{G} = C_U$, where C_U is the sheaf on U_{et} defined by $C_{|U}$. We will, by abuse of notation, use the same letter $\tilde{C}$ to indicate the corresponding sheaf on S_{fl} or S_{et} as long as the context is clear.

More generally, let $C_1, C_2, \ldots, C_r$ denote a collection of smooth, commutative, finite flat groupschemes over S, which are étale over U. (They need not be étale over S itself, and possibly, $C_i = C_j$ for $i \neq j$.) Let $\mathcal{F}_{\mathrm{et}}$, resp. $\mathcal{F}_{\mathrm{fl}}$, denote the sheaf for S_{et}, resp. S_{fl}, defined by the tensor product $\otimes_{j=1}^r \tilde{C}_j$. Since each $\mathcal{G}_i$ is étale over U, we have, for the restrictions to U via $\pi_U : U_{\mathrm{fl}} \to U_{\mathrm{et}}$, an isomorphism

$$\mathcal{F}_{\mathrm{fl}}|_U \simeq \mathcal{F}_{\mathrm{et}}|_U. \tag{A.2.4}$$

This puts us in the situation above with $\mathcal{F} = \mathcal{F}_{\mathrm{fl}}$ and $\mathcal{G} = \mathcal{F}_{\mathrm{fl}}|_U$.

In fact, for the $\tilde{C}_i$ themselves, (A.2.4) follows from [Mi1], p. 69, while for their tensor products, one proceeds via the tensor product of sections of presheaves. Note also that $j^*(\mathcal{F}_{\mathrm{et}})$ is constructible in U_{et} since it is locally constant. Ditto for the kernels and cokernels of homomorphisms between such sheaves on U_{et}.

A3.   Application: The image of symbols

We begin with a basic result:

Lemma A.3.1 *We have for any prime ℓ (possibly p),*

$$\mathcal{O}_F^*/\ell \xrightarrow{\sim} H_{\mathrm{fl}}^1(S, \mu_\ell) \xrightarrow{\sim} \overline{H}^1(F, \mu_\ell) \hookrightarrow H_{\mathrm{et}}^1(F, \mu_\ell) \xleftarrow{\sim} F^*/\ell$$

(with the natural inclusion).

Proof The first isomorphism follows from the exact sequence on S_{fl},

$$1 \longrightarrow \mu_\ell \longrightarrow \mathcal{G}_m \xrightarrow{\ell} \mathcal{G}_m \longrightarrow 1$$

and the fact that $H_{\mathrm{fl}}^1(S, \mathcal{G}_m) = \mathrm{Pic}(S) = 0$. (Of course, when $\ell \neq p$, this sequence is also exact over S_{et}. The second isomorphism then follows from the inclusion $\mathcal{O}_F^*/\ell \hookrightarrow F^*/\ell$. The last two statements follow from the definition or are well known.   □

Proposition A.3.2 *Let F be a local field of characteristic 0 with finite residue field k of characteristic p, E/F an elliptic curve with good reduction, and $\mathcal{E}$ the Néron model. Then for any odd prime ℓ, possibly equal to p, there is a short exact sequence*

$$0 \to E(F)/\ell \to H_{\mathrm{fl}}^1(S, \mathcal{E}[\ell]) \to H^1(k, \pi_0(\mathcal{E}_s))[\ell] \to 0. \tag{i}$$

In particular, since E has good reduction, we have isomorphisms

$$E(F)/\ell \simeq \mathcal{E}(S)/\ell \xrightarrow{\sim} H_{\mathrm{fl}}^1(S, \mathcal{E}[\ell]) \xrightarrow{\sim} \overline{H}^1(F, E[\ell]).$$

Proof Over $\mathcal{S}_{\text{fl}}$ we have the following *exact sequence of sheaves* associated to the group schemes (see [Mi2], p. 400, Corollary C9):

$$0 \longrightarrow \mathcal{E}[\ell] \longrightarrow \mathcal{E} \xrightarrow{\ell} \mathcal{E},$$

where the arrow on the right is surjective when E has good reduction. In any case, taking flat cohomology, we get an exact sequence

$$0 \to \mathcal{E}(S)/\ell = E(F)/\ell \to H^1_{\text{fl}}(S, \mathcal{E}[\ell]) \to H^1_{\text{fl}}(S, \mathcal{E})[\ell].$$

There is an isomorphism

$$H^1_{\text{fl}}(S, \mathcal{E}) \simeq H^1(k, \pi_0(\mathcal{E}_s)). \qquad (A.3.3)$$

Since E has good reduction, this is proved as follows: The natural map

$$H^1_{\text{fl}}(S, \mathcal{E}) \to H^1(s, \mathcal{E}_s), \ s = \text{spec} k,$$

is an isomorphism when $\mathcal{E}$ is smooth over S (see [Mi1], p. 116, Remark 3.11), and $H^1(s, \mathcal{E}_s)$ vanishes in this case by a theorem of Lang ([L1].

Clearly, we now have the exact sequence (i). $\qquad \square$

Proposition A.3.4 *Let E be an elliptic curve over F with good reduction, and let ℓ be an odd prime, possibly the residual characteristic p of F. Then the restriction s_ℓ^{symb} of s_ℓ to the symbol group $ST_{F,\ell}(A)$ factors as follows:*

$$s_\ell^{\text{symb}} : ST_{F,\ell}(A) \to \overline{H}^2(F, E[\ell]^{\otimes 2}) \hookrightarrow H^2(F, E[\ell]^{\otimes 2}).$$

Proof Thanks to Proposition A.3.2 and the definitions, this is a consequence of the following commutative diagram:

$$E(F)/\ell \times E(F)/\ell \to H^1_{\text{fl}}(S, \mathcal{E}[\ell]) \times H^1_{\text{fl}}(S, \mathcal{E}[\ell])$$

$$\downarrow \qquad\qquad\qquad \downarrow$$

$$ST_{F,\ell}(A) \quad \longrightarrow \quad H^2_{\text{fl}}(S, \mathcal{E}[\ell]^{\otimes 2}) \quad \longrightarrow \overline{H}^2(F, E[\ell]^{\otimes 2}) \hookrightarrow H^2(F, E[\ell]^{\otimes 2})$$

$\qquad\qquad\qquad\qquad\qquad\qquad\qquad\qquad\qquad\qquad\qquad\qquad\qquad\qquad\qquad\qquad \square$

A4. A lemma in the case of good ordinary reduction

Let E be an elliptic curve over a non-archimedean local field F of characteristic zero, whose residue field k has characteristic $p > 0$.

Lemma A.4.1 *Assume that E has good ordinary reduction, with Néron model $\mathcal{E}$ over $S = \text{Spec}(\mathcal{O}_F)$. Then the following hold:*

(a) *There exists an exact sequence of group schemes, finite and flat over S, and a corresponding exact sequence of sheaves on $\mathcal{S}_{\text{fl}}$, both compatible with finite field extensions K/F:*

$$0 \longrightarrow C \longrightarrow \mathcal{E}[p] \longrightarrow C' \longrightarrow 0, \qquad (A.4.2)$$

where $C = \mathcal{E}[p]^{\text{loc}}$ and $C' = \mathcal{E}[p]^{\text{et}}$.

(b) *If the points of $E_s^{\text{et}}[p]$ are all rational over k then the above exact sequence becomes*

$$0 \longrightarrow \mu_{p,S} \longrightarrow \mathcal{E}[p]_S \longrightarrow (\mathbb{Z}/p)_S \longrightarrow 0. \qquad (A.4.3)$$

(c) *If $E[p]$ is in addition semisimple as a $\mathrm{Gal}(\overline{F}/F)$-module, the exact sequence (A.4.3) splits as groupschemes (and à fortiori as sheaves on S_{fl}), i.e.,*

$$\mathcal{E}[p] \simeq \mu_{p,S} \oplus (\mathbb{Z}/p)_S. \qquad (A.4.4)$$

Proof (a) This sequence of finite, flat groupschemes is well known (see [Ta2], Thm. 3.4). The exactness of the sequence in S_{fl}, which is clear except on the right, follows from the fact that $\mathcal{E}[p] \to \mathcal{E}[p]^{\mathrm{et}}$ is itself flat (cf. [Ray], pp. 78–85). Moreover, the naturality of the construction furnishes compatibility with finite base change.

(b) Since C' is $\mathcal{E}[p]^{\mathrm{et}}$, it becomes $(\mathbb{Z}/p)_S$ when all the p-torsion points of $\mathcal{E}_s^{\mathrm{et}}$ become rational over k. On the other hand, since $\mathcal{E}[p]$ is selfdual, C is the Cartier dual of C', and so when the latter is $(\mathbb{Z}/p)_S$, the former has to be $\mu_{p,S}$.

(c) When $E[p]$ is semisimple, it splits as a direct sum $C_F \oplus C'_F$. In other words, the groupscheme $\mathcal{E}[p]$ splits over the generic point, and since we are in the Néron model, any section over the generic point furnishes a section over S. This leads to a decomposition $C \oplus C'$ over S as well. When we are in the situation of (b), the semisimplicity assumption yields (A.4.4). □

References

[Bl]   S. Bloch, *Lectures on Algebraic Cycles*, Duke University Press (1980).

[Br]   K. Brown, *Cohomology of groups*, Graduate Texts in Mathematics **87**, Springer-Verlag (1994).

[F]   W. Fulton, *Intersection theory* (second edition), Springer-Verlag (1998).

[J]   U. Jannsen, *Continuous étale cohomology*, Math. Annalen **280** (1988), no. 2, 207–245.

[L1]   S. Lang, Algebraic groups over finite fields. Amer. J. Math. **78** (1956), 555–563. (1956), 555–563.

[L2]   S. Lang, *Algebraic Number Theory*, Addison-Wesley (1970).

[Ma]   B. Mazur, Rational points of abelian varieties with values in towers of number fields, Inventiones Math. **18** (1972), 183–266.

[Mac]   W. McCallum, Duality theorems for Néron models, Duke Math. Journal **53** (1986), 1093–1124.

[Mi1]   J.S. Milne, *Étale cohomology*, Princeton University Press (1980), Princeton, NJ.

[Mi2]   J.S. Milne, *Arithmetic duality theorems*, Perspectives in Math. **1**, Academic Press (1986).

[M]   J. Milnor, *Algebraic K-theory*, Annals of Math. Studies **72**, Princeton University Press (1971).

[ParS]   R. Parimala and V. Suresh, Zero cycles on quadric fibrations: Finiteness theorems and the cycle map, Inventiones Math. **122** (1995), 83–117.

[R]   W. Raskind, Higher l-adic Abel-Jacobi mappings and filtrations on Chow groups. Duke Math. Journal **78** (1995), no. 1, 33–57.

[R-S]   W. Raskind and M. Spiess, Milnor K-groups and zero-cycles on products of curves over p-adic fields, Compositio Math. **121** (2000), no. 1, 1–33.

[Ray]   M. Raynaud, Passage au quotient par une rélation déquivalence plate, in *Proceedings of Local Fields*, Driebergen 1966, 78–85, Springer-Verlag.

[Se1]   J.-P. Serre, *Abelian ℓ-adic representations and elliptic curves* (second edition), Addison-Wesley (1989).

[Se2]   J.-P. Serre, *Local fields*, Graduate Texts in Mathematics **67**, Springer-Verlag (1979).

[So]   M. Somekawa, On Milnor K-groups attached to semi-abelian varieties, K-Theory **4** (1990), no. 2, 105–119.

[Sp]   E.H. Spanier, *Algebraic Topology*, McGraw-Hill (1966).

[Ta1]   J. Tate, K_2 and Galois cohomology, Inventiones Math. **36** (1976), 257–274.

[Ta2]   J. Tate, Finite flat group schemes, in *Modular forms and Fermat's last theorem* (Boston, MA, 1995), 121–154, Springer, NY (1997).

[W]   A. Weil, *Foundations of algebraic geometry* (second edition), American Mathematical Society, Providence, R.I. (1962).

[Y]   T. Yamazaki, On Chow and Brauer groups of a product of Mumford curves, Mathematische Annalen **333**, 549–567 (2007).

Fields Institute Communications
Volume **56**, 2009

Semiregularity and Abelian Varieties

V. Kumar Murty

Department of Mathematics
University of Toronto
Toronto, ON M5S 3G3, Canada
murty@math.toronto.edu

Abstract. Let Z be a local complete intersection subscheme of a smooth variety X over the complex numbers. Bloch defines a map

$$\pi : H^1(Z, N_{Z/X}) \longrightarrow H^{p+1}(X, \Omega^{p-1})$$

where $N_{Z/X}$ is the normal bundle of Z in X. He defines Z to be *semiregular* if this map is injective. He then shows that the variational Hodge conjecture holds if the Hodge class can be represented by a semiregular local complete intersection subscheme in one fibre. We discuss this result in outline and consider the case of X an Abelian variety.

1 Introduction

Let X be a smooth projective variety defined over $\mathbb{C}$. For each subvariety Z of X of codimension p, we have its cohomology class

$$[Z] \in H^{2p}(X, \mathbb{Q}) \cap H^{p,p}$$

where as usual $H^{a,b} = H^b(X, \Omega_X^a)$ is the b-th Dolbeault cohomology of the sheaf of a-forms. Denote by $C^p(X)$ the group of codimension p-cycles on X. By linearity, we have a cycle class map

$$cl_{X,p} : C^p(X) \otimes \mathbb{Q} \longrightarrow H^{2p}(X, \mathbb{Q}) \cap H^{p,p}.$$

The Hodge conjecture asserts that this map is surjective. We shall refer to elements of $H^{2p}(X, \mathbb{Q}) \cap H^{p,p}$ as Hodge classes of codimension p and elements in the image of the cycle class map $cl_{X,p}$ as algebraic Hodge classes.

Grothendieck proposed a variational version of the Hodge conjecture. This concerns families

$$f : X \longrightarrow S$$

where S is a smooth, connected, finite type scheme over $\mathbb{C}$ and f is a smooth projective morphism. Let ω be a horizontal class in $H^{2p}(X/S)$ of type (p, p). If

2000 *Mathematics Subject Classification.* Primary 14C25, 14K12; Secondary 14C30, 14K20.
Key words and phrases. semiregularity, algebraic cycle, Abelian variety.

for one $s_0 \in S$, the restriction ω_{s_0} of ω to the fibre over s_0 is an algebraic Hodge class, then Grothendieck conjectures that the restriction ω_s of ω to any fibre is an algebraic Hodge class.

Bloch's theorem on semiregularity gives one instance in which this can be proved. Let $Z \subset X$ be a local complete intersection subscheme of codimension p. Bloch defines what it means for Z to be *semiregular* and then shows that if ω_{s_0} is an algebraic Hodge class and is represented by a semiregular local complete intersection subscheme, then ω_s is an algebraic Hodge class for all $s \in S$.

In the first part of this note, we give a brief overview of Bloch's theorem. In the second part, we shall consider semiregular cycles in the context of Abelian varieties.

It is a pleasure to thank Ragnar Buchweitz, Spencer Bloch and Madhav Nori for some helpful and encouraging conversations on this topic and the referee for a careful reading of the paper and for many suggestions that improved the exposition.

2 Bloch's Theorem

2.1 Some remarks on deformation theory. A basic reference for this section is [5], Chapter 6. Let K be an algebraically closed field of characteristic zero. The local properties of a scheme Y near a point p are related to morphisms $f : Spec\, A \longrightarrow Y$ where A is a local Artin algebra over K and f maps the closed point to p.

With this in mind, we define a *deformation functor* D to be a covariant functor from local Artin K-algebras to sets with the property that $D(K)$ is a single point. A particular class of deformation functors is associated to local K-algebras R with residue field K and for which m_R/m_R^2 is a finite dimensional K-vector space. Here, m_R is the maximal ideal of R. We set

$$h_R(A) \; = \; \mathrm{Hom}_K(R, A)$$

for any local Artin K-algebra A. In case $R = \mathcal{O}_{X,p}$ for a scheme X and a point $p \in X$, $h_R(A)$ is the set of maps $Spec\, A \longrightarrow X$ sending the closed point to p. The *tangent space* of R is defined to be

$$T_R \; = \; (m_R/m_R^2)^*$$

which by assumption is a finite dimensional K-vector space. Its dimension is called the *embedding dimension* of R because in the case that $R = \mathcal{O}_{X,p}$, the smallest dimension of a smooth scheme containing an open neighbourhood of p in X as a closed subscheme is of dimension $\dim T_R$.

For a general deformation functor D, we might define the tangent space by

$$T_D \simeq D(K[\epsilon]/(\epsilon)^2).$$

Using the terminology of [5], Section 6.2, we say that a deformation functor D is *prorepresentable* if $D \simeq h_R$ for some local K-algebra R as above. A consequence of a result of Schlessinger ([5], Corollary 6.3.5) gives a criterion for a deformation functor to be pro-representable. It is stated in terms of the tangent space and the 'obstruction' space (see [5], §6.3).

2.2 The Hilbert scheme. Let X be a quasi-projective variety over K. For any locally Noetherian K-scheme T we define ([5], §5.1) $Hilb_{X/K}(T)$ to be the set of closed subschemes Y of $X \times_K T$ that are proper and flat over T. Grothendieck's main theorem is that this functor is representable by a locally Noetherian scheme over K.

Let Z be a closed subscheme of X. To study the behaviour of the Hilbert scheme near Z, we may consider the following deformation functor $H_{X,Z}$. For an Artin local-K algebra A, and with $S = Spec A$, set $H_{X,Z}(A)$ to be the set of subschemes $Z_S \subset X \times S$ whose base change to S_{red} is Z. Then $H_{X,Z}$ is pro-representable since the Hilbert scheme of X is representable.

2.3 Local complete intersections. Let X be a smooth variety over K, an algebraically closed field of characteristic zero. Let Z be a closed subscheme of X of codimension r. We say that Z is a *local complete intersection* subscheme in X if the ideal sheaf $\mathcal{I}_Z$ of Z in X can be locally generated by r elements at every point. If Z is smooth, it is a local complete intersection subscheme in any smooth X that contains it. (For this and other facts, the reader may consult Hartshorne [7], Chapter 2, Section 8.)

2.4 The normal sheaf. Let Z be a smooth subvariety of a smooth variety X. The locally free sheaf $\mathcal{I}_Z/\mathcal{I}_Z^2$ is called the *conormal sheaf* of Z in X. We have the exact sequence

$$0 \longrightarrow \mathcal{I}_Z/\mathcal{I}_Z^2 \overset{\epsilon}{\longrightarrow} \Omega_X \otimes \mathcal{O}_Z \longrightarrow \Omega_Z \longrightarrow 0. \qquad (2.1)$$

As Z is smooth, the sheaf Ω_Z is locally free of rank equal to $\dim Z$ and $\mathcal{I}_Z/\mathcal{I}_Z^2$ is locally free of rank r, the codimension of Z in X. The dual of this sheaf

$$N_{Z/X} = Hom_{\mathcal{O}_Z}(\mathcal{I}_Z/\mathcal{I}_Z^2, \mathcal{O}_Z)$$

is called the *normal sheaf* of Z in X. It is also locally free of rank r. Set

$$\omega_{Z/X} = \wedge^r N_{Z/X}.$$

Then, we have the factorization

$$\omega_Z \simeq \omega_X|_Z \otimes \omega_{Z/X}.$$

In fact, this formula holds for Z any local complete intersection subscheme (whether it is smooth or not) of an ambient smooth X if we take ω_Z to be the dualizing sheaf in Serre duality. (See Hartshorne [7], Chapter 3, Theorem 7.11 where the result is given in the case that the ambient space X equal to projective space. The general case is proved similarly as the argument is local.)

2.5 The tangent space of $H_{X,Z}$. The functor $H_{X,Z}$ has a tangent-obstruction theory with tangent space given by

$$T_Z = \mathrm{Hom}_{\mathcal{O}_X}(\mathcal{I}_Z, \mathcal{O}_Z).$$

Note that

$$\mathrm{Hom}_{\mathcal{O}_X}(\mathcal{I}_Z, \mathcal{O}_Z) \simeq \mathrm{Hom}_{\mathcal{O}_Z}(\mathcal{I}_Z/\mathcal{I}_Z^2, \mathcal{O}_Z) = H^0(Z, N_{Z/X}).$$

Thus, $\dim H^0(Z, N_{Z/X})$ is the embedding dimension at Z of the Hilbert scheme. In particular, this implies that the dimension of any component of the Hilbert scheme of X/K that contains Z has dimension $\leq \dim H^0(Z, N_{Z/X})$. Thus, if the dimension is equal to this number, then the Hilbert scheme is smooth at the point corresponding to Z. Moreover, if Z is a local complete intersection subscheme, then the 'obstruction space' is given by

$$\text{Ext}^1_{\mathcal{O}_X}(\mathcal{I}_Y, \mathcal{O}_Y) \simeq H^1(Z, N_{Z/X}).$$

2.6 Definition of the semiregularity map. Now let X be a smooth projective variety over $\mathbb{C}$ of dimension n and let $Z \subset X$ be a local complete intersection subscheme of codimension p. The map ϵ in (2.1) is defined for any local complete intersection, and so we have

$$\wedge^{p-1}\epsilon : \wedge^{p-1} N^*_{Z/X} \longrightarrow \Omega^{p-1}_X \otimes \mathcal{O}_Z. \tag{2.2}$$

This gives rise to a map

$$\Omega^{n-p+1}_X \times \wedge^{p-1} N^*_{Z/X} \longrightarrow \Omega^{n-p+1}_X \times \Omega^{p-1}_X \otimes \mathcal{O}_Z. \tag{2.3}$$

Now, we have a map given by exterior product

$$\Omega^{p-1}_X \times \left(\Omega^{n-p+1}_X\right) \longrightarrow \omega_X. \tag{2.4}$$

Inserting (2.4) into (2.3) gives a map

$$\Omega^{n-p+1}_X \longrightarrow \wedge^{p-1} N_{Z/X} \otimes \omega_X. \tag{2.5}$$

We have an isomorphism

$$\wedge^{p-1} N_{Z/X} \simeq N^*_{Z/X} \otimes \omega_{Z/X}. \tag{2.6}$$

Inserting (2.6) into (2.5), we get

$$\Omega^{n-p+1}_X \longrightarrow N^*_{Z/X} \otimes \omega_Z. \tag{2.7}$$

This gives rise to a map in cohomology

$$H^{n-p-1}(X, \Omega^{n-p+1}_X) \longrightarrow H^{n-p-1}(Z, N^*_{Z/X} \otimes \omega_Z). \tag{2.8}$$

By duality, this gives a map

$$H^{p+1}(X, \Omega^{p-1}_X) \longleftarrow H^1(Z, N_{Z/X}).$$

This is the semiregularity map $\pi = \pi_{Z,X}$.

2.7 Outline of Bloch's theorem. Let $Z \subset X$ be a local complete intersection subscheme of codimension p. Following Bloch and Kodaira-Spencer [9], we say that Z is *regular* if $H^1(Z, N_{Z/X}) = 0$ and *semiregular* if π is injective.

Bloch considers a family X over a smooth, connected base S over $\mathbb{C}$. More precisely, suppose given a smooth, projective morphism $f : X \longrightarrow S$ with S smooth, connected and of finite type over $\mathbb{C}$. Let ω be a horizontal deRham class $\omega \in \Gamma(S, R^{2p}f_*(\Omega^{\cdot}_{X/S}))$. His main theorem is that if the restriction $\omega_o \in H^{2p}(X_o, \mathbb{C})$ of ω to one fibre $o \in S$ is an algebraic Hodge class and is represented by a semiregular local complete intersection subscheme of X_o, then for all $s \in S$, the restriction ω_s to the fibre X_s over s is algebraic.

To prove this, Bloch first considers the infinitesimal case. Thus, consider $f : X \longrightarrow S$, a smooth projective morphism of schemes with $S = Spec(A)$ where A is an Artin local-$\mathbb{C}$ algebra. Let $Z \subset X$ be a local complete intersection subscheme of codimension p and let $I \subset \mathcal{O}_X$ be the ideal of Z. Grothendieck ([6], Exposé

149) defines a local cohomology theory $H_Z^*(X,F)$ for any coherent sheaf F of $\mathcal{O}_X$ modules, with the property that there is an isomorphism

$$H_Z^i(X,F) \simeq \varinjlim \operatorname{Ext}_{\mathcal{O}_X}^i(\mathcal{O}_X/I^k, F).$$

The local cohomology groups fit into a long exact sequence

$$H_Z^i(X,F) \longrightarrow H^i(X,F) \longrightarrow H^i(X-Z,F) \longrightarrow H_Z^{i+1}(X,F)$$

and there is a cycle class $\{Z\}' \in H_Z^p(X,\Omega_{X/S}^p)$ which maps to the class $[Z]$ of Z in $H^p(X,\Omega_{X/S}^p)$.

Bloch ([2], Proposition 6.2) shows that the local cohomology class of Z is related to the semiregularity map through the following commutative diagram:

$$
\begin{array}{ccc}
H^1(Z,N_{Z/X}) & \xrightarrow{\cdot\{Z\}'} & H_Z^{p+1}(X,\Omega_{X/S}^{p-1}) \\
{\scriptstyle\pi}\downarrow & & \downarrow \\
H^{p+1}(X,\Omega_{X/S}^{p-1}) & = & H^{p+1}(X,\Omega_{X/S}^{p-1}).
\end{array}
$$

Now let I be a square-zero ideal of A. Let $A_0 = A/I$ and let $S_0 = Spec(A_0)$ and $S = Spec(A)$. Let $f : X \longrightarrow S$ be as above and let $X_0 = X \times_S S_0$. Let Z_0 be a local complete intersection subscheme of X_0 with ideal J_0. Then ([2], Proposition (2.6)) the obstruction to extending Z_0 to a local complete intersection $Z \subset X$ is given by an element in $\operatorname{Ext}_{\mathcal{O}_{Z_0}}^1(J_0/J_0^2, I \otimes \mathcal{O}_{Z_0})$. Using the relation to local cohomology, Bloch then establishes a formula ([2], Proposition (6.8)) that relates this class to the obstruction to extending the cohomology class $[Z_0]$ to $F^p H_{dR}^{2p}(X/S)$. When Z_0 is semiregular, he deduces ([2], Corollary (6.10)) that this class vanishes if and only if the obstruction to extending $[Z_0]$ as a Hodge class also vanishes.

Thus, let $S = Spec(\mathbb{C}[[t_1,\cdots,t_r]])$ and X smooth and projective over S. Let X_o be the closed fibre and $Z_o \subset X_o$ be a semiregular local complete intersection subscheme of codimension p. Suppose that the topological cycle class $[Z_o] \in H^p(X_o, \Omega_{X_o}^p)$ lifts to a horizontal class $z \in F^p H_{dR}^{2p}(X/S)$. Let I be the ideal of $\mathbb{C}[[t_1,\cdots,t_r]]$ given by $I = (t_1,\cdots,t_r)$. Consider the thickenings

$$S_N = Spec(\mathbb{C}[[t_1,\cdots,t_r]]/I^{N+1})$$

of S. By applying the result of the previous paragraph, Bloch iteratively lifts Z_o to subschemes $[Z_N] \subset X \times_S S_N$ that are flat over S_N. These are, by construction, compatible in the sense that $Z_N \times_{S_N} S_{N-1} = Z_{N-1}$. Thus, this produces an S-point $\overline{Z} : S \longrightarrow H$ to the Hilbert scheme which extends Z_0.

Having $\bar{Z}$ as above, one appeals to a theorem of Artin ([1], Theorem 1.2) that allows one to deduce the existence of a complex neighbourhood $U \subset S$ of o and an analytic map $Z : U \longrightarrow H$ extending Z_0. Thus, Z_0 lifts to an analytic family over U and ω_s is algebraic for $s \in U$. On the other hand, the set $T = \{s \in S : \omega_s \text{ is algebraic}\}$ is contained in a countable union of closed sets. As S is connected, it follows that $T = S$.

Buchweitz and Flenner [4], [3] realized that some of the hypotheses of Bloch's theorem could be removed by invoking the cotangent complex of Illusie (and considerable technical prowess). Thus, they are able to associate a semiregularity map to any coherent sheaf on X, a smooth quasi-projective variety defined over K.

3 Semiregularity and Abelian varieties

In this section, we make some remarks about semiregularity in the context of Abelian varieties. In particular, we begin with the observation that an Abelian subvariety of an Abelian variety A is semiregular. We then look at subschemes Z that are a sum of two Abelian subvarieties of A and show that even though such subschemes are not local complete intersection subschemes, under some hypotheses, $Z \times \{1\}$ is homologically equivalent to a semiregular local complete intersection subscheme of $A \times E$ for any elliptic curve E.

3.1 Abelian subvarieties.

Proposition 3.1 *Let $X = Y \times Z$ and assume that Z and X are smooth projective varieties. Let $y_0 \in Y$. Then $\{y_0\} \times Z$ is semiregular.*

Proof. The exact sequence

$$ 0 \longrightarrow N^*_{Z/X} \overset{\epsilon}{\longrightarrow} \Omega^1_X \otimes \mathcal{O}_Z \longrightarrow \Omega^1_Z \longrightarrow 0 \qquad (3.1) $$

splits and we have an isomorphism

$$ \Omega^1_X \otimes \mathcal{O}_Z \simeq N^*_{Z/X} \oplus \Omega^1_Z. $$

Moreover, $N^*_{Z/X}$ is the trivial bundle on Z of rank $p = \dim Y$. We have

$$ \Omega^{p-1}_X \otimes \mathcal{O}_Z = \wedge^{p-1}(\Omega^1_X \otimes \mathcal{O}_Z) = \oplus_{i=0}^{p-1} (\Omega^i_Z \otimes \wedge^{p-1-i} N^*_{Z/X}). $$

The exterior product (2.4) pairs $\Omega^i_Z \otimes \wedge^{p-1-i} N^*_{Z/X}$ with $\Omega^{n-p-i}_Z \otimes \wedge^{i+1} N^*_{Z/X}$. Taking $i = 0$ and combining (2.3) and (2.4) shows that the map (2.7)

$$ \Omega^{n-p+1}_X \otimes \mathcal{O}_Z \longrightarrow N^*_{Z/X} \otimes \omega_Z $$

is the canonical projection onto a summand. Hence, the induced map on cohomology (2.8) is obviously surjective and π is injective.

Remark 3.2 Let us call a smooth subvariety $Z \subset X$ *split* if the sequence (3.1) splits. The above argument shows that a split subvariety is semiregular. It would be interesting to look for examples of split subvarieties when X is not an Abelian variety.

Proposition 3.3 *Let A be an Abelian variety and B an Abelian subvariety. Then B is semiregular in A.*

Proof. There is an Abelian subvariety C of A and an isogeny

$$ \phi : A \longrightarrow B \times C. $$

By Proposition 3.1, $B \times \{0\}$ is semiregular in $B \times C$. We have an isomorphism

$$ \Omega^1_A = \lambda^* \Omega^1_B \oplus \mu^* \Omega^1_C $$

where $\lambda = \mathrm{pr}_1 \circ \phi : A \longrightarrow B$ and $\mu = \mathrm{pr}_2 \circ \phi : A \longrightarrow C$ are the projections onto the two factors. We may therefore identify

$$ \mu^* \Omega^1_C \otimes \mathcal{O}_B \simeq N^*_{B/A}. $$

Now the argument proceeds as in Proposition 3.1.

Proposition 3.4 *Let E be an elliptic curve and $A = E^r$. Then any algebraic cycle on A is homologically equivalent to a linear combination of semiregular subvarieties.*

Proof. It is well-known (see, for example, [8], [12]) that modulo homological equivalence, a set of generators for the space of algebraic cycles on A is obtained as follows. For every partition I of $\{1, 2, \cdots, r\}$, choose $x_s \in E$ for each $s \in I$. Also, choose $\alpha = (\alpha_1, \cdots, \alpha_r) \in (\text{End } E)^r$. Set

$$\Delta_\alpha = \cup_{(x_s)}\{(\alpha_1 y_1, \cdots, \alpha_r y_r) \in E^r : y_i = x_s \text{ for } i \in s\}.$$

These cycles are Abelian subvarieties and hence are semiregular by Proposition 3.3. This completes the proof.

4 Cycles on Abelian varieties that are a sum of two Abelian subvarieties

The main result of this section is the following result.

Theorem 4.1 *Let A be an Abelian variety and A_1, A_2 two Abelian subvarieties of codimension p. Suppose that $p \geq 2$ and that A is generated by A_1 and A_2. Consider the cycle*

$$Z = A_1 + A_2$$

Let E be an elliptic curve and let $e \in E$. Then $Z \times \{e\}$ on $A \times E$ is homologically equivalent to a semiregular local complete intersection subscheme on $A \times E$.

Proof. The cycle $Z \times \{e\}$ is of codimension $p + 1$ in $A \times E$. Its cohomology class in $H^{2(p+1)}(A \times E, \mathbb{Q})$ is

$$[Z] \otimes \omega \in H^{2p}(A, \mathbb{Q}) \otimes H^2(E, \mathbb{Q})$$

where $[Z]$ (respectively, ω) is the fundamental class of Z (respectively, $\{e\}$) in A (respectively, E). Moreover, the reduced subscheme

$$\overline{Z} = A_1 \times \{x_1\} + A_2 \times \{x_2\}$$

where $x_1, x_2 \in E$ and $x_1 \neq x_2$ is homologically equivalent to $Z \times \{e\}$. Now $\overline{Z}$ is smooth and the semiregularity map

$$\pi : H^1(\overline{Z}, N_{\overline{Z}/A \times E}) \longrightarrow H^{p+2}(A \times E, \Omega^p_{A \times E})$$

is a sum of the semiregularity maps

$$\pi_i : H^1(A_i, N_{A_i/A \times E}) \longrightarrow H^{p+2}(A \times E, \Omega^p_{A \times E})$$

for $i = 1, 2$. We know from Proposition 3.3 that π_i is injective. Moreover, if we write

$$A_3 = A_1 \cap A_2$$

and

$$A_i \sim A_3 \times B_i$$

for $i = 1, 2$, then by the assumption that A_1 and A_2 together span A, we have

$$A \sim A_3 \times B_1 \times B_2.$$

Hence,

$$H^{p+2}(A \times E, \Omega^p_{A \times E})$$

is equal to

$$\bigoplus_{\substack{i+j+k+l=p+2 \\ a+b+c+d=p}} H^i(A_3, \Omega^a_{A_3}) \otimes H^j(B_1, \Omega^b_{B_1}) \otimes H^k(B_2, \Omega^c_{B_2}) \otimes H^l(E, \Omega^d_E).$$

Set

$$\overline{A_i} = A_i \times \{x_i\}.$$

As we may view the map

$$H^1(A \times E, T_{A \times E}) \longrightarrow H^1(\overline{A_i}, N_{A_i \times x_i})$$

as projection onto a direct summand, and as we may identify

$$N_{A_1 \times x_1/A \times E} \simeq T_{B_2 \times E} \otimes \mathcal{O}_{\overline{A_1}}$$

$$N_{A_2 \times x_2/A \times E} \simeq T_{B_1 \times E} \otimes \mathcal{O}_{\overline{A_2}}$$

and as $[\overline{A_1}]$ is the volume form on $B_2 \times E$, it follows that the image of

$$\pi_1 : H^1(\overline{A_1}, N_{\overline{A_1}/A \times E}) \longrightarrow H^{p+2}(A \times E, \Omega^p_{A \times E})$$

lies in the subspace

$$H^{p+2}(A \times E, \Omega^p_{B_2 \times E} \otimes \mathcal{O}_{A_1}).$$

This is equal to

$$\oplus_{\substack{i+j+k+l=p+2 \\ c+d=p}} H^i(A_3, \mathcal{O}_{A_3}) \otimes H^j(B_1, \mathcal{O}_{B_1}) \otimes H^k(B_2, \Omega^c_{B_2}) \otimes H^l(E, \Omega^d_E).$$

Similarly, the image of π_2 lies in

$$H^{p+2}(A \times E, \Omega^p_{B_1 \times E}).$$

Again, this is equal to

$$\oplus_{\substack{i+j+k+l=p+2 \\ b+d=p}} H^i(A_3, \mathcal{O}_{A_3}) \otimes H^j(B_1, \Omega^b_{B_1}) \otimes H^k(B_2, \mathcal{O}_{B_2}) \otimes H^l(E, \Omega^d_E).$$

Since $\overline{Z}$ is the disjoint union of $\overline{A_1}$ and $\overline{A_2}$, we have

$$H^1(\overline{Z}, N_{\overline{Z}/A \times E}) = H^1(\overline{A_1}, N_{\overline{A_1}/A \times E}) \oplus H^1(\overline{A_2}, N_{\overline{A_2}/A \times E}).$$

Now we see that $\mathrm{Im}\pi_1 \cap \mathrm{Im}\pi_2$ lies in the subspace

$$\oplus_{i+j+k+l=p+2} H^i(A_3, \mathcal{O}_{A_3}) \otimes H^j(B_1, \mathcal{O}_{B_1}) \otimes H^k(B_2, \mathcal{O}_{B_2}) \otimes H^l(E, \Omega^p_E).$$

But as $p \geq 2$, this is zero. Hence, the injectivity of π_1 and π_2 implies that $\pi = \pi_1 + \pi_2$ is also injective. This completes the proof.

Remark 4.2 We have actually proved more. Let $\mathcal{X}$ be a family of Abelian varieties of which A is a member. If the cohomology class of the cycle $A_1 + A_2$ is horizontal, then $A_1 + A_2$ moves in a family $\mathcal{Z} \times E$ and by topological considerations, the deformation Z continues to have two components. Moreover, the image of the semiregularity map associated to A_1 and A_2 is disjoint, and so Z must be a sum $Z_1 + Z_2$ where Z_i is a deformation of A_i. In particular, if A_i cannot be deformed, then $[A_1 + A_2]$ is not horizontal.

Remark 4.3 Let S be a connected locally Noetherian scheme and $X \longrightarrow S$ be a smooth projective morphism. Let $\epsilon : S \longrightarrow X$ be a section. Suppose that for one geometric point s of S, the fibre X_s is an Abelian variety with identity $\epsilon(s)$. Then by Mumford [11], Theorem 6.14), X is an Abelian scheme with identity ϵ. In other words, a deformation of an Abelian variety is the translate of an Abelian variety. The above result suggest that a similar result is true for the deformation of a cycle which is a sum of two Abelian varieties.

5 A generalization

The result of the previous section can be generalized as follows.

Theorem 5.1 *Let $A_1, \cdots, A_r$ be Abelian subvarieties of A of codimension p. Suppose that they span A but that no proper subset does. Suppose that one of the following conditions holds:*

1. *$r = 2$ and $p \geq 2$,*
2. *$r > 2$ and no A_i is contained in the span of $\{A_j, j \neq i\}$.*

Then the cycle $(A_1 + \cdots + A_r) \times \{e\}$ is homologically equivalent to a local complete intersection subscheme on $A \times E$ which is semiregular.

Proof. The case $r = 2, p \geq 2$ has been treated in Theorem 4.1, so we may assume that the second condition (2) above holds. We proceed as in the proof of Theorem 4.1. Write

$$B_i \ = \ < A_1, \cdots, \hat{A}_i, \cdots, A_r >$$

for the Abelian subvariety spanned by $A_1, \cdots, A_{i-1}, A_{i+1}, \cdots, A_r$. Set

$$C_i = B_i \cap A_i.$$

Write

$$A_i \sim C_i \times A_i'$$

for an Abelian subvariety A_i' of A_i. By condition (2), the dimension of A_i' is at least 1. Then

$$A_i' \cap < A_1', \cdots, \hat{A}_i', \cdots, A_r' >$$

is finite, as it is contained in

$$A_i' \cap B_i \ \subseteq A_i' \cap C_i.$$

Hence, there exists an Abelian subvariety C of A so that

$$A \sim C \times A_1' \times \cdots \times A_r'.$$

Note that if we write

$$C \sim C_i \times C_i'$$

then

$$p = \dim C_i' \ + \ \sum_{j \neq i} \dim A_j' \ \geq \ r - 1.$$

Now consider $A \times E$. Consider the cycle

$$A_1 + \cdots + A_r \times \{e\}$$

and then separate the components. Then, semiregularity of this cycle is the injectivity of the map

$$\oplus_{i=1}^r H^1(A_i, N_{A_i/A \times E}) \ \longrightarrow \ H^{p+2}(A \times E, \Omega^p_{A \times E}).$$

The image of

$$H^1(A_i, N_{A_i/A \times E})$$

is contained in

$$H^{p+2}(A \times E, \Omega^p_{A_1' \times \cdots \times A_{i-1}' \times C_i' \times A_{i+1}' \cdots A_r' \times E} \otimes \mathcal{O}_{A_i})$$

where

$$C \sim C_i \times C_i'.$$

By the Künneth formula, this is

$$\oplus H(C_i, \mathcal{O}_{C_i}) \otimes H(C_i', \Omega_{C_i'}) \otimes H(A_1', \Omega_{A_1'}) \otimes \cdots \otimes H(A_i', \mathcal{O}_{A_i'})$$
$$\otimes \cdots \otimes H(A_r', \Omega_{A_r'}) \otimes H(E, \Omega_E).$$

Consider $\operatorname{Im} \pi_i \cap \operatorname{Im} \pi_j$ for some $j \neq i$, say $i = 1, j = 2$. If it is nonzero, then it has a nonzero component in

$$H(C_1, \mathcal{O}_{C_1}) \otimes H(C_1', \Omega_{C_1'}^{a_1}) \otimes H(A_1', \mathcal{O}) \otimes \otimes_{m=2}^r H(A_m', \Omega_{A_m'}^{b_m}) \otimes H(E, \Omega_E^d)$$

and in

$$H(C_2, \mathcal{O}_{C_2}) \otimes H(C_2', \Omega_{C_2'}^{a_2}) \otimes H(A_1', \Omega^{b_1'}) \otimes H(A_2', \mathcal{O}) \otimes \otimes_{m=3}^r H(A_m', \Omega_{A_m'}^{b_m'}) \otimes H(E, \Omega_E^{d'}).$$

Thus, we must have $a_1 = a_2, b_1' = 0, b_2 = 0, b_m = b_m'$ for $m = 3, \cdots, r$ and $d = d'$. But then

$$a_1 + b_3 + \cdots b_r + d = p.$$

Now, $b_k \leq \dim A_k'$ and $a_1 \leq \dim C_1'$. Hence,

$$p - \dim A_1' - \dim A_2' + d \geq p.$$

Hence,

$$d \geq \dim A_1' + \dim A_2' \geq 2$$

which is a contradiction. Hence, $\operatorname{Im} \pi_i \cap \operatorname{Im} \pi_j = 0$ if $i \neq j$ and so

$$\operatorname{Ker} \pi = \operatorname{Ker}(\oplus \pi_i) = \oplus \operatorname{Ker} \pi_i = 0.$$

This proves the result.

6 Semiregularity on $A \times A$

In this section, we make some observations about the semiregularity of certain subschemes of $A \times A$. We begin with the following result.

Proposition 6.1 *Let X be a smooth projective variety. If $Z = Z_1 \times Z_2 \subset X \times X$ where Z_i is a local complete intersection subscheme of X and $\operatorname{codim}_X Z_i = p_i$ for $i = 1, 2$, then the semiregularity map*

$$\pi : H^1(Z, N_{Z/X \times X}) \longrightarrow H^{p_1 + p_2 + 1}(X \times X, \Omega_{X \times X}^{p_1 + 2 - 1})$$

factors into four maps

$$H^1(Z_1, N_{Z_1/X}) \otimes H^0(Z_2, \mathcal{O}_{Z_2}) \longrightarrow H^{p_1+1}(X, \Omega^{p_1-1}) \otimes H^{p_2}(X, \Omega^{p_2})$$

$$H^1(Z_1, \mathcal{O}_{Z_1}) \otimes H^0(Z_2, N_{Z_2/X}) \longrightarrow H^{p_1+1}(X, \Omega^{p_1}) \otimes H^{p_2}(X, \Omega^{p_2-1})$$

$$H^0(Z_1, N_{Z_1/X}) \otimes H^1(Z_2, \mathcal{O}_{Z_2}) \longrightarrow H^{p_1}(X, \Omega^{p_1-1}) \otimes H^{p_2+1}(X, \Omega^{p_2})$$

$$H^0(Z_1, \mathcal{O}_{Z_1}) \otimes H^1(Z_2, N_{Z_2/X}) \longrightarrow H^{p_1}(X, \Omega^{p_1}) \otimes H^{p_2+1}(X, \Omega^{p_2-1})$$

Proof. Let $p = p_1 + p_2$ and consider the local cycle class [2]

$$\{Z\}' \in H_Z^{2p}(X \times X, \Omega_{X \times X}^{2p}).$$

The class is constructed locally in terms of differential forms (see the description in Bloch [2], Section 5). We observe from the definition that

$$\{Z\}' = \{Z_1\}' \otimes \{Z_2\}'$$

where

$$\{Z_i\}' \in H_{Z_i}^{2p}(X, \Omega_X^p).$$

Then, from [2], Proposition 6.2, we have that the diagram

$$
\begin{array}{ccc}
H^1(Z, N_{Z/X\times X}) & \xrightarrow{\ \cdot\{Z_1\}'\otimes\{Z_2\}'\ } & H^{2p+1}_Z(X\times X, \Omega^{2p-1}_{X\times X}) \\
\pi\downarrow & & \downarrow \\
H^{2p+1}(X\times X, \Omega^{2p-1}_{X\times X}) & = & H^{2p+1}(X\times X, \Omega^{2p-1}_{X\times X})
\end{array}
$$

commutes. By the Künneth formula, $H^1(Z, N_{Z/X\times X})$ decomposes into the four summands listed in the statement of the Proposition. Contracting each of them with $\{Z_1\}'\otimes\{Z_2\}'$ shows that the images lie in the subspaces listed. This completes the proof.

Proposition 6.2 *Let D be a smooth ample divisor on an Abelian variety A. Then D is semiregular on A.*

Proof. If $D \subset A$ is a divisor, it is known [2], Proposition 1.1, that

$$\pi : H^1(D, N_{D/A}) \longrightarrow H^2(A, \mathcal{O}_A)$$

arises as the connecting homomorphism in the cohomology sequence of

$$0 \longrightarrow \mathcal{O}_A \longrightarrow \mathcal{O}_A(D) \longrightarrow N_{D/A} \longrightarrow 0.$$

As D is ample, $H^1(A, \mathcal{O}_A(D)) = 0$ ([10], Chapter 2, §16). Hence, π is injective. This completes the proof.

Proposition 6.3 *If A is an Abelian variety and D is a smooth ample divisor on A giving rise to a principal polarization, then $D \times A$ is semiregular on $A \times A$. If the polarization corresponding to D is not principal then $D \times A$ is not semiregular.*

Proof. By Proposition 6.1, we have to check the injectivity of four maps. The first

$$H^1(D, N_{D/A}) \otimes H^0(A, \mathcal{O}) \longrightarrow H^2(A, \mathcal{O}) \otimes H^0(A, \mathcal{O})$$

is injective as it is injective on the first factor and an isomorphism of one-dimensional vector spaces on the other. The second is injective as

$$H^1(D, \mathcal{O}_D) \otimes H^0(A, N_{A/A}) = 0.$$

The third is

$$H^0(D, N_{D/A}) \otimes H^1(A, \mathcal{O}_A) \longrightarrow H^1(A, \mathcal{O}_A) \otimes H^1(A, \mathcal{O}_A).$$

We have the exact sequence

$$0 \longrightarrow H^0(A, \mathcal{O}) \longrightarrow H^0(A, \mathcal{O}(D)) \longrightarrow H^0(D, N_{D/A})$$

$$\longrightarrow H^1(A, \mathcal{O}) \longrightarrow H^1(A, \mathcal{O}(D)).$$

As D is ample, $H^1(A, \mathcal{O}(D)) = 0$. As D gives rise to a principal polarization, $H^0(A, \mathcal{O}(D))$ is one dimensional (see Mumford [10], Chapter 2, §16) and

$$H^0(A, \mathcal{O}) \longrightarrow H^0(A, \mathcal{O}(D))$$

is an isomorphism. Hence the map

$$H^0(D, N_{D/A}) \longrightarrow H^1(A, \mathcal{O}_A) \tag{6.1}$$

is an isomorphism. If D does not give rise to a principal polarization, then $H^0(A, \mathcal{O}(D))$ has dimension greater than 1, and so (6.1) is not injective. Thus $D \times A$ is *not* semiregular. As for the second factor, we have an isomorphism of $H^1(A, \mathcal{O}_A)$ with its image as the map is given by cupping with the non-zero class in $H^0(A, \mathcal{O})$. For the fourth map, we again note that

$$H^0(D, \mathcal{O}_D) \otimes H^1(A, N_{A/A}) = 0.$$

This completes the proof.

Proposition 6.4 *Let A be an Abelian surface and D a smooth ample divisor of degree 1. Then $D \times D$ is semiregular on $A \times A$.*

Proof. Again, by Proposition 6.1, we have to check the injectivity of two maps, namely

$$H^1(D, N_{D/A}) \otimes H^0(D, \mathcal{O}) \longrightarrow H^2(A, \mathcal{O}) \otimes H^1(A, \Omega^1)$$

$$H^0(D, N_{D/A}) \otimes H^1(D, \mathcal{O}) \longrightarrow H^1(A, \mathcal{O}) \otimes H^2(A, \Omega^1).$$

The first is injective as D is semiregular (Proposition 6.2) and the map

$$\mathbb{C} \simeq H^0(D, \mathcal{O}) \longrightarrow H^1(A, \Omega^1)$$

sends

$$1 \mapsto [D].$$

As for the second, we have

$$H^1(D, \mathcal{O}) \simeq H^1(A, \mathcal{O})$$

as A is a surface. Moreover, the diagram

$$
\begin{array}{ccc}
H^1(A, \mathcal{O}) & \xrightarrow{\cdot [D]} & H^2(A, \Omega^1) \\
\downarrow & & \uparrow \\
H^1(D, \mathcal{O}) & = & H^1(D, \mathcal{O})
\end{array}
$$

commutes. The horizontal map is an isomorphism (by hard Lefschetz) and so

$$H^1(D, \mathcal{O}) \longrightarrow H^2(A, \Omega^1)$$

is an isomorphism. Moreover, the exact sequence

$$0 \longrightarrow H^0(A, \mathcal{O}) \longrightarrow H^0(A, \mathcal{O}(D)) \longrightarrow H^0(D, N_{D/A})$$

$$\longrightarrow H^1(A, \mathcal{O}) \longrightarrow H^1(A, \mathcal{O}(D))$$

shows that

$$H^0(D, N_{D/A}) \longrightarrow H^1(A, \mathcal{O})$$

is also an isomorhism. This proves the result.

Proposition 6.5 *Let $s_{\pm} : A \times A \longrightarrow A$ denote the sum or difference map, and let Z be semiregular on A. Then $s_{\pm}^* Z$ is semiregular on $A \times A$ if and only if $Z \times A$ is semiregular.*

Proof. The isomorphism

$$\psi : A \times A \longrightarrow A \times A$$

given by

$$(a, b) \mapsto (a + b, b)$$

gives an identification

$$\psi : s_{\pm}^* Z \simeq Z \times A.$$

This proves the result.

Proposition 6.6 *Let Z be a semiregular local complete intersection subscheme on A. Then $Z \times A$ is semiregular on $A \times A$ if the map*

$$H^0(Z, N_{Z/A}) \longrightarrow H^p(A, \Omega^{p-1})$$

is injective.

Proof. By Proposition 6.1, the semiregularity of $Z \times A$ is determined by the injectivity of the two maps

$$H^1(Z, N_{Z/A}) \otimes H^0(A, \mathcal{O}) \longrightarrow H^{p+1}(A, \Omega^{p-1}) \otimes H^0(A, \mathcal{O})$$

$$H^0(Z, N_{Z/A}) \otimes H^1(A, \mathcal{O}) \longrightarrow H^p(A, \Omega^{p-1}) \otimes H^1(A, \mathcal{O}).$$

This proves the result.

References

[1] M. Artin, On the solutions of analytic equations, *Invent. Math.*, **5**(1968), 277-291.

[2] S. Bloch, Semiregularity and de Rham cohomology, *Invent. Math.*, **17**(1972), 51-66.

[3] R. Buchweitz and H. Flenner, A semiregularity map for modules and applications to deformations, *Compositio Math.*, **137**(2003), 135-210.

[4] R. Buchweitz and H. Flenner, The Atiyah-Chern character yields the semiregularity map as well as the infinitesimal Abel-Jacobi map, in: CRM Proceedings and Lecture Notes, **24**(2000), 33-46.

[5] B. Fantechi, L. Göttsche, L. Illusie, S. Kleiman, N. Nitsure and A. Vistoli, *Fundamental Algebraic Geometry: Grothendieck's FGA Explained*, Mathemaical Surveys and Monographs Volume 123, American Mathematical Society, Providence, 2005.

[6] A. Grothendieck, *Fondements de la géométrie algébrique*, Secrétariat Mathématique, Paris, 1962.

[7] R. Hartshorne, *Algebraic Geometry*, Springer, New York, 1977.

[8] H. Imai, On the Hodge group of some Abelian varieties, *Kodai Math. Sem. Rep.*, **27**(1976), 367-372.

[9] K. Kodaira and D. Spencer, A theorem of completeness of characteristic systems of complete continuous systems, *Amer. J. Math.*, **81**(1959), 477-500.

[10] D. Mumford, *Abelian Varieties*, Oxford, Mumbai, 1970.

[11] D. Mumford, J. Fogarty and F. Kirwan, *Geometric Invariant Theory*, Springer Verlag, Berlin, 3rd Edition, 1994.

[12] V. Kumar Murty, Computing the Hodge group of an Abelian variety, in: Séminaire de Théorie des Nombres 1988-89, pp. 141-158, ed. C. Goldstein, *Progress in Mathematics*, volume 91, Birkhauser, Boston, 1990.

Fields Institute Communications
Volume **56**, 2009

Chern Classes, K-Theory and Landweber Exactness over Nonregular Base Schemes

Niko Naumann
Fakultät für Mathematik
Universität Regensburg
Regensburg, Germany
niko.naumann@mathematik.uni-regensburg.de

Markus Spitzweck
Fakultät für Mathematik
Universität Regensburg
Regensburg, Germany
Markus.Spitzweck@mathematik.uni-regensburg.de

Paul Arne Østvær
Department of Mathematics
University of Oslo
Oslo, Norway
paularne@math.uio.no

Abstract. The purpose of this paper is twofold. First, we use the motivic Landweber exact functor theorem to deduce that the Bott inverted infinite projective space is homotopy algebraic K-theory. The argument is considerably shorter than any other known proofs and serves well as an illustration of the effectiveness of Landweber exactness. Second, we dispense with the regularity assumption on the base scheme which is often implicitly required in the notion of oriented motivic ring spectra. The latter allows us to verify the motivic Landweber exact functor theorem and the universal property of the algebraic cobordism spectrum for every noetherian base scheme of finite Krull dimension.

1 Landweber exactness and K-theory

It has been known since time immemorial that every oriented ring spectrum acquires a formal group law [1]. Quillen [10] showed that the formal group law associated to the complex cobordism spectrum MU is isomorphic to Lazard's universal formal group law; in particular, the Lazard ring that corepresents formal group laws is isomorphic to the complex cobordism ring MU_*. This makes an important link between formal group laws and the algebraic aspects of stable homotopy theory.

2000 *Mathematics Subject Classification*. Primary 55N22, 55P42; Secondary 14A20, 14F43, 19E08.

Let KU denote the periodic complex K-theory spectrum. Bott periodicity shows the coefficient ring $\mathsf{KU}_* \cong \mathbf{Z}[\beta, \beta^{-1}]$. Under the isomorphism $\mathsf{KU}_2 \cong \widetilde{\mathsf{KU}}^0(S^2)$, the Bott element β maps to the difference $1 - \xi_{\infty|\mathbf{CP}^1}$ between the trivial line bundle and the restriction of the tautological line bundle ξ_∞ over $\mathbf{CP}^\infty$ to $\mathbf{CP}^1 \cong S^2$. Thus the class $\beta^{-1}(1 - \xi_\infty)$ in $\widetilde{\mathsf{KU}}^2(\mathbf{CP}^\infty)$ defines an orientation on KU. It is straightforward to show that the corresponding degree -2 multiplicative formal group law is $F_{\mathsf{KU}}(\mathsf{x}, \mathsf{y}) = \mathsf{x} + \mathsf{y} - \beta\mathsf{xy}$, and the ring map $\mathsf{MU}_* \to \mathsf{KU}_*$ classifying F_{KU} sends v_0 to p and v_1 to β^{p-1} for every prime number p and invariant prime ideal $(p, v_0, v_1, \dots)$ of the complex cobordism ring. This map turns $\mathbf{Z}[\beta, \beta^{-1}]$ into a Landweber exact graded MU_*-algebra because the multiplication by p map on the integral Laurent polynomial ring in the Bott element is injective with cokernel $\mathbf{Z}/p[\beta, \beta^{-1}]$, and the multiplication by β^{p-1} map on the latter is clearly an isomorphism. Having dealt with these well known topological preliminaries, we are now ready to turn to motivic homotopy theory over a noetherian base scheme S of finite Krull dimension.

Applying the considerations above to the motivic Landweber exact functor theorem announced in [8, Theorem 8.6] and extended to nonregular base schemes in Theorem 2.13, imply there exists a Tate object LK in the motivic stable homotopy category $\mathbf{SH}(S)$ and an isomorphism of motivic homology theories

$$\mathsf{LK}_{*,*}(-) \cong \mathsf{MGL}_{*,*}(-) \otimes_{\mathsf{MU}_*} \mathbf{Z}[\beta, \beta^{-1}].$$

Here MGL denotes the algebraic cobordism spectrum, and the category $\mathbf{SH}(S)_{\mathcal{T}}$ of Tate objects or cellular spectra refers to the smallest localizing triangulated subcategory of $\mathbf{SH}(S)$ containing all mixed motivic spheres [2], [8]. In addition, there exists a unique monoid structure on LK that represents the ring structure on the Landweber exact homology theory $\mathsf{LK}_{*,*}(-)$, cf. [8, Remark 9.8(ii)]. Recall from [13] the motivic spectrum KGL representing homotopy algebraic K-theory on $\mathbf{SH}(S)$. We show:

Proposition 1.1 *There is an isomorphism in $\mathbf{SH}(S)$ between* LK *and* KGL.

Proof It suffices to prove the result when $S = \mathsf{Spec}(\mathbf{Z})$ because LK and KGL pull back to general base schemes; the former by [8, Proposition 8.4] and the latter by the following argument, cf. [13, §6.2]: To compute the derived pullback one takes cofibrant replacements for a model structure such that the pullback and pushforward functors form a Quillen pair; this fails for the injective motivic model structure [7], but works for the projective motivic model structure [3]. Recall that $\mathbf{Z} \times \mathbf{BGL}$ comprises all the constituent motivic spaces in KGL. Let $f \colon Q\mathsf{KGL} \to \mathsf{KGL}$ be a cofibrant functorial replacement of KGL so that f is a trivial fibration. Now trivial fibrations in the stable model structure for motivic spectra coincide with the trivial fibrations in the projective model structure on motivic spectra, so it follows that this is a level weak equivalence. This model categorical argument shows that a levelwise derived pullback is a derived pullback of motivic spectra. Since the spaces $\mathbf{BGL}$ pull back this implies the pullback gives something level equivalent to the standard model for KGL_S.

The motivic Bott element turns KGL into an oriented motivic ring spectrum with multiplicative formal group law [11, Example 2.2]. It follows that there is a canonical transformation $\mathsf{LK}_{**}(-) \to \mathsf{KGL}_{**}(-)$ of $\mathsf{MGL}_{**}(-)$-module homology theories. Since $\mathbf{SH}(\mathbf{Z})_\mathcal{T}$ is a Brown category [8, Lemma 8.2], we conclude there is a map $\Phi\colon \mathsf{LK} \to \mathsf{KGL}$ of MGL-module spectra representing the transformation.

Now since LK and KGL are Tate objects; the former by construction and the latter by [2], Φ is an isomorphism provided the naturally induced map between motivic stable homotopy groups

$$\pi_{*,*}\Phi\colon \pi_{*,*}(\mathsf{LK} \longrightarrow \mathsf{KGL}) \tag{1.1}$$

is an isomorphism [2, Corollary 7.2]. The map in (1.1) is a retract of

$$\mathsf{MGL}_{*,*}\Phi\colon \mathsf{MGL}_{*,*}(\mathsf{LK} \longrightarrow \mathsf{KGL}). \tag{1.2}$$

To wit, there is a commutative diagram

$$
\begin{array}{ccccc}
\mathsf{LK} & \longrightarrow & \mathsf{MGL} \wedge \mathsf{LK} & \longrightarrow & \mathsf{LK} \\
\downarrow{\scriptstyle\Phi} & & \downarrow{\scriptstyle \mathsf{MGL}\wedge\Phi} & & \downarrow{\scriptstyle\Phi} \\
\mathsf{KGL} & \longrightarrow & \mathsf{MGL} \wedge \mathsf{KGL} & \longrightarrow & \mathsf{KGL}
\end{array}
$$

where the horizontal compositions are the respective identity maps. Thus it suffices to prove the map in (1.2) is an isomorphism.

By [8, Remark 9.2] there is an isomorphism

$$\mathsf{MGL}_{**}\mathsf{LK} \cong \mathsf{MGL}_{**} \otimes_{\mathsf{MU}_*} \mathsf{MU}_*\mathsf{KU}. \tag{1.3}$$

The latter identifies with the target of $\mathsf{MGL}_{*,*}\Phi$ via the isomorphisms

$$
\begin{aligned}
\mathsf{MGL}_{**}\mathsf{KGL} &\cong \mathrm{colim}_n\, \mathsf{MGL}_{*+2n,*+n}(\mathbf{Z} \times \mathbf{BGL}) \\
&\cong \mathsf{MGL}_{**} \otimes_{\mathsf{MU}_*} \mathrm{colim}_n\, \mathsf{MU}_{*+2n}(\mathbf{Z} \times \mathsf{BU}) \\
&\cong \mathsf{MGL}_{**} \otimes_{\mathsf{MU}_*} \mathsf{MU}_*\mathsf{KU}.
\end{aligned} \tag{1.4}
$$

In the third isomorphism we use compatibility of the structure maps of KGL and KU, and the computation $\mathsf{MGL}_{**}(\mathbf{BGL}) \cong \mathsf{MGL}_{**} \otimes_{\mathsf{MU}_*} \mathsf{MU}_*\mathsf{BU}$.

It remains to remark that $\mathsf{MGL}_{**}\Phi$ is the identity with respect to the isomorphisms in (1.3) and (1.4). $\qquad\square$

Remark 1.2 Proposition 1.1 reproves the general form of the Conner-Floyd theorem for homotopy algebraic K-theory [11]. We would like to point out that this type of result is more subtle than in topology: Recall that if E is a classical homotopy commutative ring spectrum which is complex orientable and the resulting formal group law is Landweber exact, then $E_*(-) = \mathsf{MU}_*(-)\otimes_{\mathsf{MU}_*} E_*$. Now suppose S has a complex point and let E be the image of HQ under the right adjoint of the realization functor. It is easy to see that E is an orientable motivic ring spectrum with additive formal group law and $\mathsf{E}_* \cong \mathsf{LQ}_*$, at least for fields. However, since $\mathsf{E}_{p,q} = \mathsf{E}_{p,q'}$ for all $p, q, q' \in \mathbf{Z}$, we get that $\mathsf{E} \not\cong \mathsf{LQ}$.

In [11] it is shown that the Bott inverted infinite projective space is the universal multiplicative oriented homology theory. Thus there exists an isomorphism between the representing spectra LK and $\Sigma^\infty \mathbf{P}^\infty_+[\beta^{-1}]$. By [8, Remark 9.8(ii)] there is a unique such isomorphism, and it respects the monoidal structure. This allows us to conclude there is an isomorphism in $\mathbf{SH}(S)$ between $\Sigma^\infty \mathbf{P}^\infty_+[\beta^{-1}]$ and KGL. An alternate computational proof of this isomorphism was given in [11] and another proof was announced in [4]. In the event of a complex point on S, the topological realization functor maps $\mathsf{LK}_{*,*}(-)$ to $\mathsf{KU}_*(-)$ and likewise for KGL. For fun, we note that running the exact same tape in stable homotopy theory yields Snaith's isomorphism $\Sigma^\infty \mathbf{CP}^\infty_+[\beta^{-1}] \cong \mathsf{KU}$. In the topological setting, the argument above could have been given over thirty years ago.

In the proof of Proposition 1.1 we used the universal property of the homology theory associated to MGL. For the proof this is only needed over $\mathbf{Z}$. In the following sections we shall justify the claim that this holds over general base schemes.

Theorem 1.3 *Suppose S is a noetherian scheme of finite Krull dimension and* E *a commutative ring spectrum in* $\mathbf{SH}(S)$. *Then there is a bijection between the sets of ring spectra maps* $\mathsf{MGL} \to \mathsf{E}$ *in* $\mathbf{SH}(S)$ *and the orientations on* E.

The proof also shows there is an analogous result for homology theories. For fields, this result is known thanks to the works of Vezzosi [12] and more recently of Panin-Pimenov-Röndigs [9]. The generalization of their results to regular base schemes is straightforward using, for example, the results in [8].

In order to prove Theorem 1.3 we compute the E-cohomology of MGL for E an oriented motivic ring spectrum; in turn, this makes use of the Thom isomorphism for universal vector bundles over Grassmannians. Usually the base scheme is tactically assumed to be regular for the construction of Chern classes of vector bundles. However, since the constituent spaces comprising the spectrum MGL are defined over the integers $\mathbf{Z}$ one gets Chern classes and the Thom isomorphism by pulling back classifying maps of line bundles. In any event, one purpose of what follows is to dispense with the regularity assumption for the construction of Chern classes. For the remaining part of the proof we note that the streamlined presentation in [9] following the route laid out in algebraic topology carries over verbatim.

2 Chern classes and oriented motivic ring spectra

The basic input required for the theory of Chern classes is that of the first Chern class of a line bundle. If the base scheme S is regular, there is an isomorphism

$$\mathsf{Pic}(X) \cong \mathsf{Hom}_{\mathbf{H}(S)}(X, \mathbf{P}^\infty)$$

between the Picard group of X and maps from X to the infinite projective space $\mathbf{P}^\infty$ in the unstable motivic homotopy category $\mathbf{H}(S)$ [7, Proposition 4.3.8]. If S is nonregular, there is no such isomorphism because the functor $\mathsf{Pic}(-)$ is not $\mathbf{A}^1$-invariant. The first goal of this section is to prove that the most naive guess as to what happens in full generality holds true, namely:

$$\mathbf{P}^\infty \text{ represents the } \mathbf{A}^1\text{-localization of } \mathsf{Pic}.$$

For τ some topology on the category of smooth S-schemes Sm/S, let $\mathbf{sPre}(\mathsf{Sm}/S)^\tau$ denote any one of the standard model structure on simplicial presheaves with τ-local weak equivalences and likewise for $\mathbf{sPre}(\mathsf{Sm}/S)^{\tau,\mathbf{A}^1}$ and $\tau - \mathbf{A}^1$-local weak equivalences. The topologies of interest in what follows are the Zariski and Nisnevich topologies, denoted by Zar and Nis respectively. Let $\mathbf{H}(S)$ denote the unstable motivic homotopy category of S associated to the model structure $\mathbf{sPre}(\mathsf{Sm}/S)^{\mathsf{Nis},\mathbf{A}^1}$. Throughout we will use the notation $\mathbf{G}_\mathsf{m}$ for the multiplicative group scheme over S, $\underline{\mathsf{Pic}}(X)$ for the Picard groupoid of a scheme X, and $\nu\underline{\mathsf{Pic}}$ for the simplicial presheaf obtained from (a strictification of) $\underline{\mathsf{Pic}}$ by applying the nerve functor.

Proposition 2.1   (i) *There exists a commutative diagram*

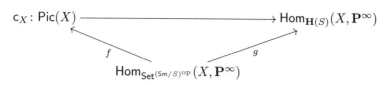

which is natural in $X \in \mathsf{Sm}/S$, where f is given by pulling back the tautological line bundle on $\mathbf{P}^\infty$ and g is the canonical map. The transformation c is an isomorphism when S is regular.
(ii) *There is a natural weak equivalence*

$$\mathbf{P}^\infty \simeq \nu\underline{\mathsf{Pic}}$$

in $\mathbf{sPre}(\mathsf{Sm}/S)^{\mathsf{Zar},\mathbf{A}^1}$.

To prepare for the proof of Proposition 2.1, suppose $\mathsf{G} \in \mathbf{sPre}(\mathsf{Sm}/S)$ is a group object acting on a simplicial presheaf X. Let $X/^h\mathsf{G}$ denote the associated homotopy quotient. It is defined as the total object (homotopy colimit) of the simplicial object:

$$X \Longleftarrow X \times \mathsf{G} \Lleftarrow X \times \mathsf{G}^2 \Lleftarrow \cdots$$

Pushing this out to any of the localizations yields local homotopy quotients since the localization is a left adjoint functor, and hence preserves (homotopy) colimits. If $X \to Y$ is a G-equivariant map with Y having trivial G-action, there is a naturally induced map $X/^h\mathsf{G} \to Y$ in the corresponding homotopy category.

Lemma 2.2 *The naturally induced map*

$$(\mathbf{A}^{n+1} \smallsetminus \{0\})/^h\mathbf{G}_\mathsf{m} \longrightarrow \mathbf{P}^n$$

is a Zariski local weak equivalence.

Proof Let $\pi : \mathbf{A}^{n+1} \smallsetminus \{0\} \to \mathbf{P}^n$ be the canonical map, fix $0 \leq i \leq n$ and consider the standard affine open $U_i := \{[x_0 : \ldots : x_n] \,|\, x_i \neq 0\} \subseteq \mathbf{P}^n$ and its pullback along π, i.e. $V_i := \pi^{-1}(U_i) \subseteq \mathbf{A}^{n+1} \smallsetminus \{0\}$. Working Zariski locally, it suffices to see that π induces a weak equivalence $V_i/^h\mathbf{G}_\mathsf{m} \to U_i$. There is an $\mathbf{G}_\mathsf{m}$-equivariant isomorphism $V_i \cong \mathbf{G}_\mathsf{m} \times \mathbf{A}^n$ with $\mathbf{G}_\mathsf{m}$ acting trivially on $\mathbf{A}^n$ and by multiplication on $\mathbf{G}_\mathsf{m}$. It remains to remark that $(\mathbf{G}_\mathsf{m} \times \mathbf{A}^n)/^h\mathbf{G}_\mathsf{m} \cong \mathbf{A}^n$ because the simplicial diagram defining $(\mathbf{G}_\mathsf{m} \times \mathbf{A}^n)/^h\mathbf{G}_\mathsf{m}$ admits an evident retraction. $\square$

By lemma 2.2 there is a weak equivalence $(\mathbf{A}^\infty \smallsetminus \{0\})/^h\mathbf{G}_\mathsf{m} \to \mathbf{P}^\infty$ in $\mathbf{sPre}(\mathsf{Sm}/S)^{\mathsf{Zar}}$. Let $\mathsf{B}\mathbf{G}_\mathsf{m}$ denote the quotient $\mathrm{pt}/^h\mathbf{G}_\mathsf{m}$ in any of the localizations of $\mathbf{sPre}(\mathsf{Sm}/S)$.

Lemma 2.3 *The map*

$$(\mathbf{A}^\infty \smallsetminus \{0\})/^h\mathbf{G}_\mathsf{m} \longrightarrow \mathsf{B}\mathbf{G}_\mathsf{m}$$

induced by $\mathbf{A}^\infty \smallsetminus \{0\} \to \mathrm{pt}$ is a weak equivalence in $\mathbf{sPre}(\mathsf{Sm}/S)^{\mathbf{A}^1}$.

Proof The map $\mathbf{A}^\infty \smallsetminus \{0\} \to \mathrm{pt}$ is an equivalence in $\mathbf{sPre}(\mathrm{Sm}/S)^{\mathbf{A}^1}$: In effect, the inclusions $\mathbf{A}^n \smallsetminus \{0\} \hookrightarrow \mathbf{A}^{n+1} \smallsetminus \{0\}$ are homotopic to a constant map by the elementary $\mathbf{A}^1$-homotopy

$$(t, (x_1, \ldots, x_n)) \longmapsto ((1-t)x_1, \ldots, (1-t)x_n, t).$$

A finite generation consideration finishes the proof. See [7, Proposition 4.2.3] for a generalization. $\qquad\square$

Remark 2.4 We thank the referee for suggesting that inserting t in the first coordinate gives a homotopy which is compatible with the inclusions i_n of $\mathbf{A}^n \smallsetminus \{0\}$ into $\mathbf{A}^{n+1} \smallsetminus \{0\}$ sending $(x_1, \ldots, x_n)$ to $(x_1, \ldots x_n, 0)$ and a contraction of $\mathbf{A}^\infty \smallsetminus \{0\}$ onto the point $(1, 0, \ldots)$. To wit, the homotopies

$$H_n : \mathbf{A}^1 \times \mathbf{A}^n - \{0\} \to \mathbf{A}^{n+1} - \{0\}, (t, x_1, \ldots, x_n) \mapsto (t, (1-t)x_1, \ldots, (1-t)x_n)$$

make the diagrams

$$
\begin{array}{ccc}
\mathbf{A}^1 \times \mathbf{A}^n - \{0\} & \xrightarrow{H_n} & \mathbf{A}^{n+1} - \{0\} \\
{\scriptstyle \mathrm{id} \times i_n} \downarrow & & \downarrow {\scriptstyle i_{n+1}} \\
\mathbf{A}^1 \times \mathbf{A}^{n+1} - \{0\} & \xrightarrow{H_{n+1}} & \mathbf{A}^{n+2} - \{0\}
\end{array}
$$

commute. Thus by considering $\mathbf{A}^\infty - \{0\}$ as a presheaf we get a homotopy

$$H := \mathrm{colim} H_n : \mathbf{A}^1 \times \mathbf{A}^\infty - \{0\} \to \mathbf{A}^\infty - \{0\}.$$

Note that $H(1, x_1, \ldots) = (1, 0, 0, \ldots)$ and $H(0, x_1, \ldots) = (0, x_1, x_2, \ldots)$, i.e. this is a homotopy from some constant map on $\mathbf{A}^\infty - \{0\}$ to the "shift map". The latter is homotopic to the identity map on $\mathbf{A}^\infty - \{0\}$ via the homotopy

$$H' : \mathbf{A}^1 \times \mathbf{A}^\infty - \{0\} \to \mathbf{A}^\infty - \{0\}; (t, x_1, \ldots)$$
$$\mapsto ((1-t)x_1, x_2 + t(x_1 - x_2), \ldots, x_n + t(x_{n-1} - x_n), \ldots).$$

Note that this map does not pass through the origin. Moreover, $H'(0, -)$ is the identity map and $H'(1, -)$ is the "shift map".

Corollary 2.5 *There is a natural weak equivalence*

$$\mathrm{BG_m} \simeq \mathbf{P}^\infty$$

in $\mathbf{sPre}(\mathrm{Sm}/S)^{\mathrm{Zar}, \mathbf{A}^1}$.

Proof Combine the equivalences

$$(\mathbf{A}^\infty \smallsetminus \{0\})/^h \mathrm{G_m} \to \mathbf{P}^\infty \text{ and } (\mathbf{A}^\infty \smallsetminus \{0\})/^h \mathrm{G_m} \to \mathrm{BG_m}$$

from Lemmas 2.2 and 2.3. $\qquad\square$

Lemma 2.6 *There is a natural weak equivalence*

$$\mathrm{BG_m} \simeq \nu \underline{\mathrm{Pic}}$$

in $\mathbf{sPre}(\mathrm{Sm}/S)^{\mathrm{Zar}}$.

Proof Denote by $\underline{\mathrm{Pic}}' \subset \underline{\mathrm{Pic}}$ the subpresheaf of groupoids consisting objectwise of the single object formed by the trivial line bundle. Then $\nu \underline{\mathrm{Pic}}'$ provides a model for $\mathrm{BG_m}$ in $\mathbf{sPre}(\mathrm{Sm}/S)$. By considering stalks it follows that $\nu \underline{\mathrm{Pic}}' \to \nu \underline{\mathrm{Pic}}$ is a Zariski local weak equivalence. $\qquad\square$

The fact that $\underline{\mathsf{Pic}}$ is a stack in the flat topology which is an instance of faithfully flat descent (for projective rank one modules) implies

$$\nu\underline{\mathsf{Pic}} \text{ satisfies Zariski descent.} \qquad (2.1)$$

Indeed, ν is a right Quillen functor for the local model structure on groupoid valued presheaves, so it preserves fibrant objects. Note that $\nu\underline{\mathsf{Pic}}$ satisfies descent in any topology which is coarser than the flat one. In general, $\nu\underline{\mathsf{Pic}}$ is not $\mathbf{A}^1$-local.

Proof (of Proposition 2.1) Combining Corollary 2.5 and Lemma 2.6 yields (ii). We define the natural transformation c in part (i) using the composition

$$\mathsf{Pic}(X) = \pi_0 \nu\underline{\mathsf{Pic}}(X)$$
$$\overset{(2.1)}{\cong} \mathsf{Hom}_{\mathsf{Ho}(\mathbf{sPre}(\mathsf{Sm}/S)^{\mathsf{Zar}})}(X, \mathsf{BG_m})$$
$$\to \mathsf{Hom}_{\mathbf{H}(S)}(X, \mathsf{BG_m})$$
$$\cong \mathsf{Hom}_{\mathbf{H}(S)}(X, \mathbf{P}^\infty).$$

If S is regular, then $\nu\underline{\mathsf{Pic}}$ is $\mathbf{A}^1$-local, and hence the composition is an isomorphism.

The triangle in (i) commutes since the map $\mathbf{P}^\infty \to \mathsf{BG_m}$ used to construct the transformation c classifies the tautological line bundle on $\mathbf{P}^\infty$. $\square$

Let $\mathsf{Th}(\mathcal{T}(1))$ denote the Thom space of the tautological vector bundle $\mathcal{T}(1) \equiv \mathcal{O}_{\mathbf{P}^\infty}(-1)$ with fiber $\mathbf{A}^1$ over the base point $S \hookrightarrow \mathbf{P}^\infty$. We recall the notion of oriented motivic ring spectra formulated in [9], cf. [5], [6] and [12]. The unit of a commutative, $\mathbf{P}^1$-ring spectrum E, i.e. a commutative monoid in $\mathbf{SH}(S)$, defines $1 \in \mathsf{E}^{0,0}(S_+)$. Applying the $\mathbf{P}^1$-suspension isomorphism to 1 yields the element $\Sigma_{\mathbf{P}^1}(1) \in \mathsf{E}^{2,1}(\mathbf{P}^1, \infty)$. The canonical covering of $\mathbf{P}^1$ defines motivic weak equivalences

$$\mathbf{P}^1 \overset{\sim}{\longrightarrow} \mathbf{P}^1/\mathbf{A}^1 \overset{\sim}{\longleftarrow} \mathbf{T} \equiv \mathbf{A}^1/\mathbf{A}^1 \smallsetminus \{0\}$$

of pointed motivic spaces inducing isomorphisms $\mathsf{E}^{**}(\mathbf{P}^1, \infty) \leftarrow \mathsf{E}^{**}(\mathbf{A}^1/\mathbf{A}^1 \smallsetminus \{0\}) \to \mathsf{E}^{**}(\mathbf{T})$. Let $\Sigma_{\mathbf{T}}(1)$ be the image of $\Sigma_{\mathbf{P}^1}(1)$ in $\mathsf{E}^{2,1}(\mathbf{T})$.

Definition 2.7 Let E be a commutative $\mathbf{P}^1$-ring spectrum.

(i) A Chern orientation on E is a class $\mathsf{ch} \in \mathsf{E}^{2,1}(\mathbf{P}^\infty)$ such that $\mathsf{ch}|_{\mathbf{P}^1} = -\Sigma_{\mathbf{P}^1}(1)$.

(ii) A Thom orientation on E is a class $\mathsf{th} \in \mathsf{E}^{2,1}(\mathsf{Th}(\mathcal{T}(1))$ such that its restriction to the Thom space of the fiber over the base point coincides with $\Sigma_{\mathbf{T}}(1) \in \mathsf{E}^{2,1}(\mathbf{T})$.

(iii) An orientation σ on E is either a Chern orientation or a Thom orientation. A Chern orientation ch coincides with a Thom orientation th if $\mathsf{ch} = z^*(\mathsf{th})$.

Example 2.8 Let $u_1 \colon \Sigma^\infty_{\mathbf{P}^1}(\mathsf{Th}(\mathcal{T}(1)))(-1) \to \mathsf{MGL}$ be the canonical map of $\mathbf{P}^1$-spectra. The -1 shift refers to $\mathbf{P}^1$-loops. Set $\mathsf{th}^{\mathsf{MGL}} \equiv u_1 \in \mathsf{MGL}^{2,1}(\mathsf{Th}(\mathcal{T}(1)))$. Since $\mathsf{th}^{\mathsf{MGL}}|_{\mathsf{Th}(1)} = \Sigma_{\mathbf{P}^1}(1) \in \mathsf{MGL}^{2,1}(\mathsf{Th}(\mathbf{1}))$, the class $\mathsf{th}^{\mathsf{MGL}}$ is an orientation on MGL.

Using Proposition 2.1 we are now ready to define the first Chern classes.

Definition 2.9 Suppose (E, σ) is an oriented motivic ring spectrum in $\mathbf{SH}(S)$, $X \in \mathsf{Sm}/S$ and $\mathcal{L}$ a line bundle on X. Then the first Chern class of $\mathcal{L}$ for the given orientation is defined as $\mathsf{c}_\mathcal{L} \equiv \mathsf{c}_X(\mathcal{L})^*(\sigma) \in \mathsf{E}^{2,1}(X)$.

It is clear that this definition recovers the previously considered construction in case S is regular.

The following key result and its proof due to Morel [6] are included for completeness.

Lemma 2.10 *For every $n \geq 1$ there is a canonical weak equivalence $\mathbf{P}^n/\mathbf{P}^{n-1} \cong (\mathbf{P}^1)^{\wedge n}$ such that the naturally induced diagram*

$$
\begin{array}{ccc}
\mathbf{P}^n & \xrightarrow{\Delta^{\wedge n}} & (\mathbf{P}^n)^{\wedge n} \\
\downarrow & & \uparrow \\
\mathbf{P}^n/\mathbf{P}^{n-1} & \xrightarrow{\cong} & (\mathbf{P}^1)^{\wedge n}
\end{array}
$$

commutes in $\mathbf{H}(S)$.

Proof For $0 \leq i \leq n$, let U_i be the open affine subscheme of $\mathbf{P}^n$ of points $[x_0, \ldots, x_n]$ such that $x_i \neq 0$, set $\Omega_i \equiv (\mathbf{P}^n)^{i-1} \times U_i \times (\mathbf{P}^n)^{n-i}$ and $\Omega \equiv \cup_{1 \leq i \leq n} \Omega_i \subseteq (\mathbf{P}^n)^n$. The projection $(\mathbf{P}^n)^n \to (\mathbf{P}^n)^n/\Omega$ induces a motivic weak equivalence $(\mathbf{P}^n)^{\wedge n} \to (\mathbf{P}^n/U_1) \wedge \cdots \wedge (\mathbf{P}^n/U_n)$. It allows to replace $(\mathbf{P}^n)^{\wedge n}$ by the weakly equivalent smash product of the motivic spaces $\mathbf{P}^n/U_i$ for $1 \leq i \leq n$. Note that $\mathbf{P}^n \to (\mathbf{P}^n/U_1) \wedge \cdots \wedge (\mathbf{P}^n/U_n)$ factors through $\mathbf{P}^n/\cup_{1 \leq i \leq n} U_i$ and the inclusion $\mathbf{P}^{n-1} \subseteq \cup_{1 \leq i \leq n} U_i$ is the zero section of the canonical line bundle over $\mathbf{P}^{n-1}$. Hence the latter map is a motivic weak equivalence and $\mathbf{P}^n \to \mathbf{P}^n/\cup_{1 \leq i \leq n} U_i$ induces a motivic weak equivalence $\mathbf{P}^n/\mathbf{P}^{n-1} \to \mathbf{P}^n/\cup_{1 \leq i \leq n} U_i$. The inclusion $U_0 \subseteq \mathbf{P}^n$ induces an isomorphism of pointed motivic spaces $\mathbf{A}^n/\mathbf{A}^n \smallsetminus \{0\} \cong \mathbf{P}^n/\cup_{1 \leq i \leq n} U_i$. Since there are canonical isomorphisms $\mathbf{A}^n/\mathbf{A}^n \smallsetminus \{0\} \cong (\mathbf{A}^1/\mathbf{A}^1 \smallsetminus \{0\})^{\wedge n}$ and $\mathbf{A}^1/\mathbf{A}^1 \smallsetminus \{0\} \cong \mathbf{P}^1/\mathbf{A}^1$ there is an induced map $(\mathbf{P}^1/\mathbf{A}^1)^{\wedge n} \to (\mathbf{P}^n/U_1) \wedge \cdots \wedge (\mathbf{P}^n/U_n)$ weakly equivalent to $(\mathbf{P}^1)^{\wedge n} \to (\mathbf{P}^n/U_1) \wedge \cdots \wedge (\mathbf{P}^n/U_n)$. $\qquad \square$

Next we state and prove the projective bundle theorem for oriented motivic ring spectra, cf. the computation for Grassmannians in [8, Proposition 6.1].

Theorem 2.11 *Suppose E is an oriented motivic ring spectrum, $\xi\colon Y \to X$ a rank $n + 1$ vector bundle over a smooth S-scheme X, $\mathbf{P}(\xi)\colon \mathbf{P}(Y) \to X$ its projectivization and $\mathsf{c}_\xi \in \mathsf{E}^{2,1}(\mathbf{P}(\xi))$ the first Chern class of the tautological line bundle.*

(i) *Then $\mathsf{E}^{*,*}(\mathbf{P}(\xi))$ is a free $\mathsf{E}^{*,*}(X)$-module on generators $1 = \mathsf{c}_\xi^0, \mathsf{c}_\xi, \ldots, \mathsf{c}_\xi^n$.*

(ii) *If ξ is trivial, then there is an isomorphism of $\mathsf{E}^{*,*}(X)$-algebras*

$$
\mathsf{E}^{*,*}(\mathbf{P}(\xi)) \cong \mathsf{E}^{*,*}(X)[\mathsf{c}_\xi]/(\mathsf{c}_\xi^{n+1}).
$$

Proof The claimed isomorphism is given explicitly by

$$
\bigoplus_{i=0}^n \mathsf{E}^{*-2i,*-i}(X) \xrightarrow{\cong} \mathsf{E}^{*,*}(\mathbf{P}(\xi));
$$

$$
(\mathsf{x}_0, \ldots, \mathsf{x}_n) \longmapsto \Sigma_{i=0}^n (\mathbf{P}(\xi))^*(\mathsf{x}_i)\mathsf{c}_\xi^i.
$$

Combining the choice of a trivialization of the vector bundle ξ with a Mayer-Vietoris induction argument shows that (i) follows from (ii). Part (ii) is proved by induction on the rank of ξ.

Assume the map is an isomorphism for the trivial vector bundle $Y \to X$ of rank n, and form $Y \oplus \mathcal{O}_X \to X$ of rank $n + 1$. Next form the cofiber sequence

$\mathbf{P}(Y) \to \mathbf{P}(Y \oplus \mathcal{O}_X) \to (\mathbf{T}_X \equiv \mathbf{A}_X^1 / \mathbf{A}_X^1 \smallsetminus \{0\})^{\wedge n} \wedge X_+$ and consider the induced diagram in E-cohomology:

$$
\begin{array}{ccccccc}
\longrightarrow & \mathsf{E}^{*-2n,*-n}(X_+) & \longrightarrow & \bigoplus_{i=0}^{n} \mathsf{E}^{*-2i,*-i}(X_+) & \longrightarrow & \bigoplus_{i=0}^{n-1} \mathsf{E}^{*-2i,*-i}(X_+) & \longrightarrow \\
& \downarrow & & \downarrow & & \downarrow & \\
\longrightarrow & \mathsf{E}^{*,*}((\mathbf{T}_X)^{\wedge n} \wedge X_+) & \longrightarrow & \mathsf{E}^{*,*}(\mathbf{P}(Y \oplus \mathcal{O}_X)) & \longrightarrow & \mathsf{E}^{*,*}(\mathbf{P}(Y)) & \longrightarrow
\end{array}
$$

The right hand square commutes using functoriality of first Chern classes and the fact that the restriction map is a ring map. The commutativity of the left hand square is exactly the explicit computation of the Gysin map and is where Lemma 2.10 enters the proof.

The left hand map is an isomorphism since smashing with $\mathbf{T}_X$ is an isomorphism in $\mathbf{SH}(S)$. Now use the induction hypothesis and apply the 5-lemma which proves part (i) in this special case. The relation $\mathsf{c}_\xi^{n+1} = 0$ follows using exactness of the lower row of the diagram. □

Theorem 2.11 allows to define the higher Chern class $\mathsf{c}_{i\xi} \in \mathsf{E}^{2i,i}(X)$ of ξ as the unique solution of the equation

$$
\mathsf{c}_\xi^n = \mathsf{c}_{1\xi}\mathsf{c}_\xi^{n-1} - \mathsf{c}_{2\xi}\mathsf{c}_\xi^{n-2} + \cdots + (-1)^{n-1}\mathsf{c}_{n\xi}.
$$

With the theory of Chern classes in hand, one constructs a theory of Thom classes by the usual script. In particular, as for regular base schemes one verifies the Thom isomorphism theorem:

Theorem 2.12 *If $\xi \colon Y \to X$ is a rank n vector bundle with zero section $z \colon X \to Y$ over a smooth S-scheme X, cup-product with the Thom class of ξ induces an isomorphism*

$$
(- \cup \mathsf{th}_\xi) \circ \xi^* \colon \mathsf{E}^{*,*}(X) \xrightarrow{\cong} \mathsf{E}^{*+2n,*+n}(Y/(Y \smallsetminus z(X))).
$$

Having dealt with these preparations, the E-cohomology computation of Grassmannians in [8] extends to nonregular base schemes. The other parts involved in the proof of the motivic Landweber exact functor theorem given in [8] do not require a regular base scheme. In conclusion, the following holds (for additional prerequisites we refer to [8]):

Theorem 2.13 *Let S be a noetherian base scheme of finite Krull dimension and M_* an Adams graded Landweber exact MU_*-module. Then there exists a motivic spectrum E in $\mathbf{SH}(S)_{\mathcal{T}}$ and a natural isomorphism*

$$
\mathsf{E}_{**}(-) \cong \mathsf{MGL}_{**}(-) \otimes_{\mathsf{MU}_*} \mathsf{M}_*
$$

of homology theories on $\mathbf{SH}(S)$.

Moreover, based on the results above, the following computations stated for fields in [9] hold for S. Let $\mathbf{Gr}$ denote the infinite Grassmannian over S.

Theorem 2.14 *Suppose E is an oriented motivic ring spectrum.*

(i) The canonical map

$$
\mathsf{E}^{*,*}(\mathsf{MGL}) \longrightarrow \varprojlim \mathsf{E}^{*+2n,*+n}(\mathsf{Th}(\mathcal{T}(n))) = \mathsf{E}^{*,*}[[\mathsf{c}_1, \mathsf{c}_2, \mathsf{c}_3, \dots]] \qquad (2.2)
$$

is an isomorphism. (Here c_i is the ith Chern class.)

(ii) The canonical map

$$\mathsf{E}^{*,*}(\mathsf{MGL} \wedge \mathsf{MGL}) \longrightarrow \varprojlim \mathsf{E}^{*+2n,*+n}(\mathsf{Th}(\mathcal{T}(n)) \wedge \mathsf{Th}(\mathcal{T}(n)))$$

$$= \mathsf{E}^{*,*}[[c_1', c_1'', c_2', c_2'', \dots]] \qquad (2.3)$$

is an isomorphism. Here c_i' is the ith Chern class obtained from projection on the first factor of $\mathbf{Gr} \times \mathbf{Gr}$, and likewise c_i'' is obtained from the second factor.

3 The universal property of MGL

In this section we observe, with no claims of originality other than for the cohomology computations in Theorem 2.14 and the Thom isomorphism Theorem 2.12, that the proof of the universal property of MGL given by Panin-Pimenov-Röndigs in [9] goes through for noetherian base schemes of finite Krull dimension.

Theorem 3.1 *Suppose E is a commutative $\mathbf{P}^1$-ring spectrum. Then the assignment $\varphi \mapsto \varphi(th^{\mathsf{MGL}}) \in \mathsf{E}^{2,1}(\mathsf{Th}(\mathcal{T}(1)))$ identifies the set of monoid maps $\varphi \colon \mathsf{MGL} \to \mathsf{E}$ in $\mathbf{SH}(S)$ with the set of orientations on E. Its inverse sends $th \in \mathsf{E}^{2,1}(\mathsf{Th}(\mathcal{T}(1)))$ to the unique map $\varphi \in \mathsf{E}^{0,0}(\mathsf{MGL}) = \mathrm{Hom}_{\mathbf{SH}(S)}(\mathsf{MGL}, \mathsf{E})$ such that $u_i^*(\varphi) = th(\mathcal{T}(i))$ in $\mathsf{E}^{2i,i}(\mathsf{Th}(\mathcal{T}(i)))$, where u_i denotes the canonical map $u_i \colon \Sigma_{\mathbf{P}^1}^\infty (\mathsf{Th}(\mathcal{T}(i)))(-i) \to \mathsf{MGL}$.*

Proof Note that $th := \varphi(th^{\mathsf{MGL}})$ is an orientation on E since

$$\varphi(th)|_{Th(1)} = \varphi(th|_{Th(1)}) = \varphi(\Sigma_{\mathbf{P}^1}(1)) = \Sigma_{\mathbf{P}^1}(\varphi(1)) = \Sigma_{\mathbf{P}^1}(1).$$

Conversely, note that th^{E} gives rise to a unique map $\varphi \colon \mathsf{MGL} \to \mathsf{E}$ in $\mathbf{SH}(S)$ as desired: In effect, the elements $th(\mathcal{T}(i))$ comprise an element in the target of the isomorphism $\mathsf{E}^{*,*}(\mathsf{MGL}) \to \varprojlim \mathsf{E}^{*+2i,*+i}(\mathsf{Th}(\mathcal{T}(i)))$ in (2.2). This shows uniqueness of $\varphi \in \mathsf{E}^{0,0}(\mathsf{MGL})$.

To check that φ respects the multiplicative structure, form the diagram:

$$
\begin{array}{ccc}
\Sigma_{\mathbf{P}^1}^\infty \mathsf{Th}(\mathcal{T}(i))(-i) \wedge \Sigma_{\mathbf{P}^1}^\infty \mathsf{Th}(\mathcal{T}(j))(-j) & \xrightarrow{\Sigma_{\mathbf{P}^1}^\infty(\mu_{i,j})(-i-j)} & \Sigma_{\mathbf{P}^1}^\infty \mathsf{Th}(\mathcal{T}(i+j))(-i-j) \\
\downarrow {\scriptstyle u_i \wedge u_j} & & \downarrow {\scriptstyle u_{i+j}} \\
\mathsf{MGL} \wedge \mathsf{MGL} & \xrightarrow{\mu_{\mathsf{MGL}}} & \mathsf{MGL} \\
\downarrow {\scriptstyle \varphi \wedge \varphi} & & \downarrow {\scriptstyle \varphi} \\
\mathsf{E} \wedge \mathsf{E} & \xrightarrow{\mu_{\mathsf{E}}} & \mathsf{E}
\end{array}
$$

The upper square homotopy commutes since

$$
\begin{aligned}
\varphi \circ u_{i+j} &\circ \Sigma_{\mathbf{P}^1}^\infty(\mu_{i,j})(-i-j) \\
&= \mu_{i,j}^*(th(\mathcal{T}(i+j))) = th(\mu_{i,j}^*(\mathcal{T}(i+j))) = th(\mathcal{T}(i) \times \mathcal{T}(j)) \\
&= th(\mathcal{T}(i)) \times th(\mathcal{T}(j)) = \mu_{\mathsf{E}}(th(\mathcal{T}(i)) \wedge th(\mathcal{T}(j))) \\
&= \mu_{\mathsf{E}} \circ ((\varphi \circ u_i) \wedge (\varphi \circ u_j)).
\end{aligned}
$$

Combining (2.3) with the equality $\varphi \circ u_{i+i} \circ \Sigma_{\mathbf{P}^1}^\infty(\mu_{i,i})(-2i) = \mu_{\mathsf{E}} \circ ((\varphi \circ u_i) \wedge (\varphi \circ u_i))$ shows that $\mu_{\mathsf{E}} \circ (\varphi \wedge \varphi) = \varphi \circ \mu_{\mathsf{MGL}}$ in $\mathbf{SH}(S)$.

It remains to show that the maps constructed above are inverses of each other. To begin with, if $th \in \mathsf{E}^{2,1}(\mathsf{Th}(\mathcal{O}(-1)))$, then $\varphi \circ u_i = th(\mathcal{T}_i)$ for all i and the induced orientation $th' := \varphi(th^{\mathsf{MGL}})$ coincides with th since $th' = \varphi(th^{\mathsf{MGL}}) = \varphi(u_1) = \varphi \circ u_1 = th(\mathcal{T}(1)) = th$. On the other hand, the monoid map φ' obtained from the orientation $\mathsf{th} := \varphi(\mathsf{th}^{\mathsf{MGL}})$ on E satisfies $u_i^*(\varphi') = \mathsf{th}(\mathcal{T}(i))$ for every $i \geq 0$. To check that $\varphi' = \varphi$, the isomorphism (2.2) shows that it suffices to prove that $u_i^*(\varphi') = u_i^*(\varphi)$ for all $i \geq 0$. Now since $u_i^*(\varphi') = \mathsf{th}(\mathcal{T}_i)$ it remains to prove that $u_i^*(\varphi) = \mathsf{th}(\mathcal{T}(i))$. In turn this follows if $u_i = \mathsf{th}^{\mathsf{MGL}}(\mathcal{T}(i))$ in $\mathsf{MGL}^{2i,i}(\mathsf{Th}(\mathcal{T}(i)))$, since then $u_i^*(\varphi) = \varphi \circ u_i = \varphi(u_i) = \varphi(\mathsf{th}^{\mathsf{MGL}}(\mathcal{T}(i))) = \mathsf{th}(\mathcal{T}(i))$. We note that the proof of the equality $u_i = \mathsf{th}^{\mathsf{MGL}}(\mathcal{T}(i))$ given in [9] carries over to our setting on account of the Thom isomorphism Theorem 2.12. $\qquad\square$

References

[1] J. F. Adams. *Stable homotopy and generalised homology*. Chicago Lectures in Mathematics. University of Chicago Press, Chicago, IL, 1995. Reprint of the 1974 original.

[2] D. Dugger and D. C. Isaksen. Motivic cell structures. *Algebr. Geom. Topol.*, 5:615–652 (electronic), 2005.

[3] B. I. Dundas, O. Röndigs, and P. A. Østvær. Motivic functors. *Doc. Math.*, 8:489–525 (electronic), 2003.

[4] D. Gepner and V. Snaith. On the motivic spectra representing algebraic cobordism and algebraic K-theory. Preprint, arXiv 0712.2817.

[5] P. Hu and I. Kriz. Some remarks on Real and algebraic cobordism. *K-Theory*, 22(4):335–366, 2001.

[6] F. Morel. Some basic properties of the stable homotopy category of schemes. Preprint, http://www.mathematik.uni-muenchen.de/~morel/Stable.pdf.

[7] F. Morel and V. Voevodsky. $\mathbf{A}^1$-homotopy theory of schemes. *Inst. Hautes Études Sci. Publ. Math.*, (90):45–143 (2001), 1999.

[8] N. Naumann, M. Spitzweck, and P. A. Østvær. Motivic Landweber exactness. Preprint, arXiv 0806.0274.

[9] I. Panin, K. Pimenov, and O. Röndigs. A universality theorem for Voevodsky's algebraic cobordism spectrum. *Homology Homotopy Appl.*, 10:211–226, 2008.

[10] D. Quillen. On the formal group laws of unoriented and complex cobordism theory. *Bull. Amer. Math. Soc.*, 75:1293–1298, 1969.

[11] M. Spitzweck and P. A. Østvær. The Bott inverted infinite projective space is homotopy K-theory. *Bull. London Math. Soc.*, 41:281–292, 2009.

[12] G. Vezzosi. Brown-Peterson spectra in stable $\mathbb{A}^1$-homotopy theory. *Rend. Sem. Mat. Univ. Padova*, 106:47–64, 2001.

[13] V. Voevodsky. $\mathbf{A}^1$-homotopy theory. In *Proceedings of the International Congress of Mathematicians, Vol. I (Berlin, 1998)*, number Extra Vol. I, pages 579–604 (electronic), 1998.

Fields Institute Communications
Volume **56**, 2009

Adams Operations and Motivic Reduced Powers

Victor Snaith
Department of Mathematics
University of Sheffield
Sheffield S3 7RH, England
V.Snaith@shef.ac.uk

To Spencer Bloch, with great admiration

Abstract. In a preface to a monograph by Spencer Bloch one finds
the quotation "Answer very simple but question very hard." This was
a motto of Charlie Chan, a 1940's movie detective.[1] In this note I
would like, in the spirit of Charlie Chan's motto, to draw attention
to phenomena in motivic stable homotopy in which Adams operations
should control reduced power operations in motivic cohomology.

1 Introduction

In [1] Michael Atiyah established (i) results about the behaviour of operations in
KU-theory with respect to the filtration which comes from the Atiyah-Hirzebruch
spectral sequence and (ii) results which relate operations in KU-theory to the Steenrod operations of [16] in mod p singular cohomology in the case of spaces whose
integral cohomology is torsion free. In the classical stable homotopy category the
series [2, 13, 14, 15] develops a systematic method (christened upper triangular technology (UTT) in [15]) to prove results similar to (ii) in the presence of 2-primary
torsion. The objective of this note is to describe the behaviour of Adams operations in algebraic K-theory and reduced power operations in motivic cohomology
which should result from the importation of UTT into the motivic stable homotopy
category.

Example 1.1 (The Arf-Kervaire invariant problem) In the classical stable
homotopy category 2-localized in the sense of Bousfield consider cofibrations of the
form

$$\Sigma^\infty S^{2t} \xrightarrow{\Theta} \Sigma^\infty \mathbb{R}P^{2t} \longrightarrow C(\Theta).$$

2000 *Mathematics Subject Classification.* Primary 19E15, 19E20; Secondary 19D99, 55S05,
55S25.

[1]A character created by Earl Derr Biggers in the 1920's and 1930's in a series of six mysteries.
Charlie Chan movies were a big hit in the 1940's with Charlie played (as I recall) by Werner Lowe.

Applying connective K-theory bu_* we obtain a short exact sequence

$$0 \longrightarrow bu_{2t+1}(\mathbb{R}P^{2t}) \longrightarrow bu_{2t+1}(C(\Theta)) \longrightarrow bu_{2t+1}(S^{2t+1}) \longrightarrow 0.$$

The 2-adic Adams operation ψ^3 satisfies

$$bu_{2t+1}(S^{2t+1}) = \mathbb{Z}_2\langle \iota \rangle, \ \psi^3(\iota) = \iota; \ bu_{2t+1}(\mathbb{R}P^{2t}) = \mathbb{Z}/2^t\langle \beta \rangle, \ \psi^3(\beta) = 3^{t+1}\beta.$$

Let $\tilde{\iota} \in bu_{2t+1}(C(\Theta))$ be any element which maps to ι. When Θ is the canonical map associated to the antipodal involution on the sphere the Adams operation satisfies

$$\psi^3(\tilde{\iota}) = \tilde{\iota} + (2s+1)\frac{(3^{t+1}-1)}{2}\beta$$

for some integer s. The following is the main UTT example from [15] (see also [14]).

Theorem 1.2 ([15] Chapter Eight, Theorem 1.2) *Let m be a positive integer and let $\Theta : \Sigma^\infty S^{8m-2} \longrightarrow \Sigma^\infty \mathbb{R}P^{8m-2}$ be as in Example 1.1. Then*

$$\psi^3(\tilde{\iota}) = \tilde{\iota} + (2s+1)\frac{(3^{4m}-1)}{4}\beta$$

is possible if and only if $m = 2^q$ and Θ is detected by the Steenrod operation $Sq^{2^{q+2}}$.

2 Motivic stable homotopy

Consider the analogue of Example 1.1 in the motivic category.[1]

Let $\overline{k}$ be an algebraically closed field of characteristic not equal to two. We shall work in the motivic stable homotopy category [19] 2-localized in the sense of Bousfield (see the Appendix of [12] for this localization) or, equivalently, in $\mathbf{MS}(\mathrm{Spec}(\overline{k}))$, the 2-localized stable homotopy category of motivic spectra modelled on $Sm/\overline{k}$ constructed in ([11] Appendix A).

Example 2.1 Let Q_t denote the quadric hypersurface in $\mathbb{P}^{t+1}$ given by

$$Q_t = \{(z_0, z_1, \ldots, z_{t+1}) \mid \sum_j z_j^2 = 0\}$$

and following ([4] p. 944) define the deleted quadric

$$DQ_t = \mathbb{P}^t - Q_{t-1}.$$

Recall the bi-indexed family of spheres in the motivic stable homotopy category. Let $S^n = S^{n,0}$ denote the n-fold smash product of the constant simplicial presheaf $\Delta^1/\partial\Delta^1$ with itself. The algebraic circle is the multiplicative group scheme $\mathbb{G}_m = \mathbb{A}^1 - \mathbb{A}^0$, base-pointed by the identity section $1 \to \mathbb{G}_m$; its n-fold smash product defines the algebraic sphere $S^{n,n}$. Putting the two together, we obtain the bi-indexed family of spheres

$$S^{p+q,q} = S^{p,0} \wedge S^{q,q}.$$

Suppose, for $i >> 0$, that we are given a morphism of the form

$$\Theta : S^{1-2^i+v,v} \wedge \frac{DQ_{2^i-2}}{DQ_{2^i-2t-2}} \longrightarrow S^{v-2t,v}$$

---

[1] I am very grateful to the referee for a number of useful and enlightening suggestions concerning this section.

where $\frac{DQ_{2^i-2}}{DQ_{2^i-2t-2}}$ is the quotient defined by an object in $Sm/\overline{k}$. We may form the cofibration

$$S^{v-1-2t,v} \longrightarrow C(\Theta) \longrightarrow S^{1-2^i+v,v} \wedge \frac{DQ_{2^i-2}}{DQ_{2^i-2t-2}} \longrightarrow S^{v-2t,v}.$$

Here i, t are positive integers , i being large, and v is an arbitrary integer. These dimensions are chosen in order to arrange, when $\overline{k}$ is the complex numbers, that the right-hand morphism corresponds, after passing to $\mathbb{C}$-valued points in the classical 2-localized stable homotopy category, to the S-dual of $\Sigma^\infty S^{2t} \longrightarrow \Sigma^\infty \mathbb{R}P^{2t}$.

Applying bigraded motivic algebraic K-theory, denoted by $BGL^{*,*}$ in [11], we obtain a long exact sequence of the form

$$\dots \xrightarrow{\Theta^*} BGL^{p,q}(S^{v+1-2^i,v} \wedge \frac{DQ_{2^i-2}}{DQ_{2^i-2t-2}}) \longrightarrow BGL^{p,q}(C(\Theta))$$

$$\longrightarrow BGL^{p,q}(S^{v-1-2t,v}) \longrightarrow \dots.$$

For every pointed motivic space A the cup-product map gives an bigraded isomorphism of the form ([11] §1.3.5)

$$BGL^{*,0}(A) \otimes \mathbb{Z}[\beta^{\pm 1}] \xrightarrow{\cong} BGL^{*,*}(A)$$

where $\beta \in BGL^{2,1}(\overline{k})$ is the motivic Bott element ([11] §1.3). In addition there is a graded ring isomorphism of cohomology theories ([11] §1.2.8)

$$Ad : K_*^{TT} \xrightarrow{\cong} BGL^{-*,0}$$

and K_*^{TT} is (Bousfield 2-localized) Thomason-Trobaugh K-theory [18], which coincides with Quillen K-theory for regular schemes.

Therefore we have isomorphisms

$$BGL^{p,q}(S^{v-1-2t,v}) \cong BGL^{p+v-1-2t,q+v}(\mathrm{Spec}(\overline{k}))$$

$$\cong K_{2q+v+2t+1-p}^{TT}(\mathrm{Spec}(\overline{k}))\langle \beta^{q+v} \rangle,$$

$$BGL^{p,q}(S^{1-2^i+v,v} \wedge \frac{DQ_{2^i-2}}{DQ_{2^i-2t-2}}) \cong BGL^{p+1-2^i+v,q+v}(\frac{DQ_{2^i-2}}{DQ_{2^i-2t-2}})$$

$$\cong K_{2q+v+2^i-1-p}^{TT}(\frac{DQ_{2^i-2}}{DQ_{2^i-2t-2}})\langle \beta^{q+v} \rangle.$$

Note that these are reduced cohomology groups so that, for example, before 2-localization, we would have $BGL^{2p,p}(S^1) \cong \overline{k}^*$ if S^1 denotes the simplicial circle but for the unreduced theory we would have $\mathbb{Z} \oplus \overline{k}^*$.

As remarked in ([4] p. 945), when $\overline{k} = \mathbb{C}$, as a topological space with the classical complex topology there is a homotopy equivalence ([10] §6.3) $DQ_t(\mathbb{C})^{top} \simeq \mathbb{R}P^t$. Using the motivic cohomology calculations of [4] together with [17] and the spectral sequence relating algebraic K-theory to motivic cohomology [5], which is available since $\frac{DQ_{2^i-2}}{DQ_{2^i-2t-2}}$ is represented in $Sm/\overline{k}$ by an object of finite type, (see also [5], [6], [9], [21]) we find that

$$K_{-m}^{TT}(\frac{DQ_{2^i-2}}{DQ_{2^i-2t-2}}) \cong bu^{-m}(\frac{DQ_{2^i-2}(\mathbb{C})^{top}}{DQ_{2^i-2t-2}(\mathbb{C})^{top}}) \cong bu_m(\underline{D}(\frac{DQ_{2^i-2}(\mathbb{C})^{top}}{DQ_{2^i-2t-2}(\mathbb{C})^{top}}))$$

where $\underline{D}(X)$ denotes the Spanier-Whitehead dual of X. The Spanier-Whitehead dual of $S^{1-2^i} \wedge \frac{DQ_{2^i-2}(\mathbb{C})^{top}}{DQ_{2^i-2t-2}(\mathbb{C})^{top}}$ is $\mathbb{R}P^{2t}$ [8] so that

$$K^{TT}_{-(2t+1)}\left(S^{1-2^i} \wedge \frac{DQ_{2^i-2}(\mathbb{C})}{DQ_{2^i-2t-2}(\mathbb{C})}\right) \cong bu_{2t+1}(\mathbb{R}P^{2t}) \cong \mathbb{Z}/2^t.$$

Therefore

$$BGL^{2t+1-v,-v}\left(S^{1-2^i+v,v} \wedge \frac{DQ_{2^i-2}}{DQ_{2^i-2t-2}}\right) \cong BGL^{2t+1,0}\left(S^{1-2^i,0} \wedge \frac{DQ_{2^i-2}}{DQ_{2^i-2t-2}}\right)$$

$$\cong K^{TT}_{-(2t+1)}\left(S^{1-2^i} \wedge \frac{DQ_{2^i-2}}{DQ_{2^i-2t-2}}\right)$$

$$\cong bu_{2t+1}(\mathbb{R}P^{2t})$$

$$\cong \mathbb{Z}/2^t$$

and

$$BGL^{2t+1-v,-v}(S^{-1-2t,v}) \cong BGL^{2t+1,0}(S^{-1-2t,0})$$

$$\cong K^{TT}_{-(2t+1)}(S^{-1-2t,0})$$

$$\cong bu_{2t+1}(S^{2t+1})$$

$$\cong \mathbb{Z}_2,$$

a copy of the 2-adic integers.

In this bigrading the cofibration yields a short exact sequence of the form ([15] Chapter 8, Example 3.2)

$$0 \longrightarrow \mathbb{Z}/2^t\langle\beta\rangle \longrightarrow BGL^{2t+1-v,-v}(C(\Theta)) \longrightarrow \mathbb{Z}_2\langle\iota\rangle \longrightarrow 0$$

on which the algebraic K-theory Adams operation ψ^3 acts via the formulae of Example 1.1. Let $\tilde{\iota} \in BGL^{2t+1-v,-v}(C(\Theta))$ be any element which maps to ι.

By the motivic rigidity results of [12], over any algebraically closed field $\bar{k}$ of characteristic different from 2, we still have

$$BGL^{2t+1-v,-v}(C(\Theta)) \cong \mathbb{Z}_2\langle\iota\rangle \oplus \mathbb{Z}/2^t\langle\beta\rangle.$$

Conjecture 2.2 Let $\bar{k}$ be an algebraically closed field of characteristic not equal to two. In the situation of Example 2.1 suppose that $t = 4m - 1$ for some integer $m \geq 1$. Then the relation

$$\psi^3(\tilde{\iota}) = \tilde{\iota} + (2s+1)\frac{(3^{t+1} - 1)}{4}\beta$$

is impossible unless $m = 2^q$ for some integer q and the morphism Θ is detected by reduced power operations [20] on the mod 2 motivic cohomology of $C(\Theta)$.

Remark 2.3 In Conjecture 2.2 I have omitted the converse result; namely, if Θ is detected by reduced power operations on the mod 2 motivic cohomology of $C(\Theta)$ then $m = 2^q$ and $\psi^3(\tilde{\iota})$ is given by the formula of Conjecture 2.2. This is because, in the classical case of Theorem 1.2, if Θ is detect by a reduced power operation then it is detected by $Sq^{2^{q+2}}$ in one and only one way and $m = 2^q$. This is the content of a famous result of Browder [3] (see also [15] Preface). This result may follow from the results of [20] (see also [7] Theorem 4) in the 2-localized motivic stable homotopy category but at the moment I do not know of a proof.

Without the converse, we have the following positive evidence for Conjecture 2.2.

Theorem 2.4 *Conjecture 2.2 is true when the characteristic of $\overline{k}$ is zero.*

Proof By 2-adic motivic rigidity [12] it suffices to consider the case $\overline{k} = \mathbb{C}$. In this case, as explained in Example 2.1, the map given by taking complex points is an isomorphism from algebraic K-theory to 2-adic *bu*-theory, in the dimension which concerns us. Therefore the formula

$$\psi^3(\tilde{\imath}) = \tilde{\imath} + (2s+1)\frac{(3^{t+1} - 1)}{4}\beta$$

holds motivically if and only if, by the discussion of Example 2.1, the same formula holds in *bu*-theory for the complex points. By Theorem 1.2 this implies that $m = 2^q$ and Θ is detected by the Steenrod operation $Sq^{2^{q+2}}$ on the mod 2 (co-)homology of the complex points of $C(\Theta)$. On the other hand reduced power operations are compatible with taking complex points [20] (see also [7]) which implies that the morphism Θ is detected by reduced power operations on the mod 2 motivic cohomology of $C(\Theta)$. □

Remark 2.5 Positive characteristic Let $\overline{k}$ be an algebraically closed field of positive characteristic not equal to two. In this case I do not know how to prove Conjecture 2.2, much less generalize the phenomenon. The UTT method ([14]; [15] Chapter Eight, Theorem 1.2) involves considerable calculational use of the classical mod 2 Adams spectral sequence. The recent construction of the 2-adic motivic Adams spectral sequence in $\mathbf{MS}(\mathrm{Spec}(\overline{k}))$ [7] and the calculations made with it suggest that at least part of the UTT calculations may be made to work motivically. Accordingly, it might be worthwhile to close with a sketch of the proof of Theorem 1.2.

In the 2-localized classical homotopy category theory there is a left $K\mathrm{Spec}(\overline{k})$-module equivalence of the form [13]

$$KA \wedge bo \simeq \vee_{j=0}^{\infty} KA \wedge (F_{4j}/F_{4j-1})$$

where *bo* is 2-adic connected orthogonal K-theory and $A \in Sm/\overline{k}$ with algebraic K-theory spectrum KA and (F_{4j}/F_{4j-1}) is the $4j$-th space in the Snaith splitting of $\Omega^2 S^3$. There is a corresponding left *bu*-module splitting when KA is replaced by *bu*

$$bu \wedge bo \simeq \vee_{j=0}^{\infty} bu \wedge (F_{4j}/F_{4j-1}).$$

This equivalence is proved by means of detailed calculations of the Adams spectral sequences of the summands $bu \wedge (F_{4j}/F_{4j-1})$ [13].

Incidentally from this one can derive related splittings, potentially more easy to use in algebraic K-theory, in which *bo* is replaced by $K\mathrm{Spec}(\overline{k}) \simeq bu$ and (F_{4j}/F_{4j-1}) by $(F_{4j}/F_{4j-1}) \wedge \Sigma^{-2}\mathbb{C}P^2$.

The UTT proof of Theorem 1.2 is based on the result that

$$1 \wedge \psi^3 : bu \wedge bo \longrightarrow bu \wedge bo$$

transports across to a map which is conjugate to the matrix [2] (with respect to the splitting) of the form

$$
\begin{pmatrix}
1 & 1 & 0 & 0 & 0 & \cdots \\
0 & 9 & 1 & 0 & 0 & \cdots \\
0 & 0 & 9^2 & 1 & 0 & \cdots \\
0 & 0 & 0 & 9^3 & 1 & \cdots \\
\vdots & \vdots & \vdots & \vdots & \vdots & \vdots
\end{pmatrix}.
$$

A similar result holds for

$$
1 \wedge \psi^3 : KA \wedge bo \longrightarrow KA \wedge bo.
$$

In the situation of Example 1.1 the maps representing the 1's on the superdiagonal

$$
bu \wedge (F_{4j}/F_{4j-1}) \wedge \mathbb{R}P^{2t} \longrightarrow bu \wedge (F_{4j-4}/F_{4j-5}) \wedge \mathbb{R}P^{2t}
$$

are virtually injective on homotopy groups in dimension $2t + 1$.

In the diagram of Fig. 1 the horizontal map composes with projection onto the $bu \wedge (F_0/F_{-1}) \simeq bu$ summand to give a map

$$
\pi_j(bo \wedge X) \longrightarrow \pi_j(bu \wedge X)
$$

for each j which differs from the complexification map by an isomorphism. Therefore chasing around Fig. 1 (carefully, since the right-hand trapezium does not commute) yields a series of linear equations starting in

$$
\pi_{4t+1}(bu \wedge (F_0/F_{-1}) \wedge C(\Theta))
$$

with a relation determined by the formula for $\psi^3(\tilde{\iota})$ and working inductively through increasing values of $j \geq 1$ of the groups

$$
\pi_{4t+1}(bu \wedge (F_{4j}/F_{4j-1}) \wedge C(\Theta)) \cong \pi_{4t+1}(bu \wedge (F_{4j}/F_{4j-1}) \wedge \mathbb{R}P^{2t}).
$$

The resulting equations are impossible unless $2t = 2^{q+3} - 2$ and in this case one finds that the component of $(\eta \wedge 1 \wedge 1)_*(\tilde{\iota})$ in $\pi_{4t+1}(bu \wedge (F_{2^{q+2}}/F_{2^{q+2}-1}) \wedge \mathbb{R}P^{2t})$ is detected by the Hurewicz map to mod 2 homology. The proof of Theorem 1.2 concludes by observing that the Hurewicz image of $(\eta \wedge 1 \wedge 1)_*(\tilde{\iota})$ in $H_*(bo \wedge \mathbb{R}P^{2t}; \mathbb{Z}/2)$, which has essentially only two terms in it, is annihilated by the dual of $Sq^{2^{q+2}}$.

There are other, non-Adams spectral sequence, calculations (for example, the calculation of $\pi_*(bu \wedge bo) \otimes \mathbb{Z}_2$ (modulo torsion) [2]) which go into the UTT method but, as a start, it should be possible to use the motivic Adams spectral sequence of [7] to establish a 2-adic left $K\mathrm{Spec}(\overline{k})$-module splitting of the form

$$
K\mathrm{Spec}(\overline{k}) \wedge K\mathrm{Spec}(\overline{k}) \simeq \vee_{j=0}^{\infty} K\mathrm{Spec}(\overline{k}) \wedge (F_{4j}/F_{4j-1}) \wedge \Sigma^{-2}\mathbb{C}P^2.
$$

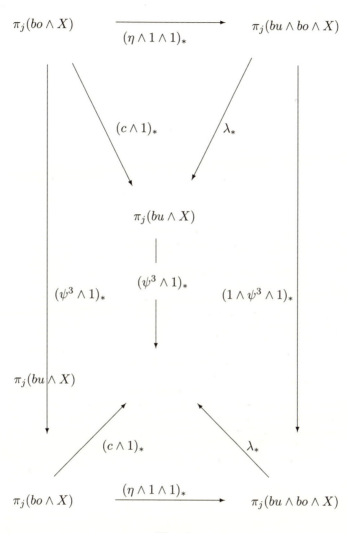

Fig. 1

References

[1] M.F. Atiyah: Power operations in K-theory, Quart. J. Math. (2) 17 (1966) 165-193.

[2] J. Barker and V.P. Snaith: ψ^3 as an upper triangular matrix; K-theory 36 (2005) 91-114.

[3] W. Browder: The Kervaire invariant of framed manifolds and its generalizations; Annals of Math. (2) 90 (1969) 157-186.

[4] D. Dugger and D.C. Isaksen: The Hopf condition for bilinear forms over arbitrary fields; Annals of Math. 165 (2007) 943-964.

[5] E.M. Friedlander and A.A. Suslin: The spectral sequence relating algebraic K-theory to motivic cohomology; Ann. Sci. École Norm. Sup. (4) 35 (6) (2002) 773-875.

[6] T. Geisser: Motivic cohomology, K-theory and topological cyclic homology; *Handbook of K-theory* vol I (eds. E.M. Friedlander and D.R. Grayson) Springer Verlag (2005) 193-234.

[7] P. Hu, I. Kriz and K. Ormsby: Some remarks on motivic homotopy theory of algebraically closed fields; K-theory archive (http://www.math.uiuc.edu/K-theory #904 August 14, 2008).

[8] D. Husemoller: *Fibre Bundles*; McGraw-Hill (1966).

[9] B. Kahn: Algebraic K-theory, algebraic cycles and arithmetic geometry; *Handbook of K-theory* vol I (eds. E.M. Friedlander and D.R. Grayson) Springer Verlag (2005) 351-428.

[10] P.S. Landweber: Fixed point free involutions on complex manifolds; Annals of Math. 86 (1967) 491-502.

[11] I. Panin, K. Pimenov and O. Röndigs: On Voevodsky's algebraic K-theory spectrum BGL; (April 2007) http://www.math.uiuc.edu/K-theory/0838.

[12] O. Röndigs and P.A. Østvaer: Rigidity in motivic homotopy theory; Math. Ann. 341 (2008) 651-675.

[13] V.P. Snaith: The upper triangular group and operations in algebraic K-theory; Topology 41 (6) 1259-1275 (2002).

[14] V.P. Snaith: Upper triangular technology and the Arf-Kervaire invariant; K-theory archive (http://www.math.uiuc.edu/K-theory #807 November 10, 2006).

[15] V.P. Snaith: *Stable Homotopy Around the Arf-Kervaire Invariant*; to appear in Birkhäuser Progress in Math. series (2009).

[16] N.E. Steenrod: *Cohomology operations*; Annals Math. Studies #50 (written and revised by D.B.A. Epstein) Princeton Univ. Press (1962).

[17] R.G. Swan: K-theory of quadric hypersurfaces; Annals of Math. 122 (1985) 113-153.

[18] R.W. Thomason and T. Trobaugh: Higher algebraic K-theory of schemes and of derived categories; Grothendieck Festschrift v.3 (1990) 247-436 Birkhäuser.

[19] V. Voevodsky: $\mathbb{A}^1$-homotopy theory; Doc. Math. Extra Vol. ICM I (1998) 579-604.

[20] V. Voevodsky: Reduced power operations in motivic cohomology; Pub. Math. I.H.E.S Paris 98 (2003) 1-57.

[21] V. Voevodsky, A. Suslin and E.M. Friedlander: *Cycles, Transfers and Motivic Homology Theories*; Annals of Math. Studies #143, Princeton Univ. Press (2000).

Fields Institute Communications
Volume **56**, 2009

Chow Forms, Chow Quotients and Quivers with Superpotential

Jan Stienstra

Mathematisch Instituut
Universiteit Utrecht
The Netherlands
J.Stienstra@uu.nl

Abstract. We recast the correspondence between 3-dimensional toric Calabi-Yau singularities and quivers with superpotential in the setting of an $(N-2)$-dimensional abelian algebraic group $\mathcal{G}_\mathbb{L}$ acting on a linear space $\mathbb{C}^N$. We show how the quiver with superpotential gives a simple explicit description of the Chow forms of the closures of the orbits in the projective space $\mathbb{P}^{N-1}$. The resulting model of the orbit space is known as the Chow quotient of $\mathbb{P}^{N-1}$ by $\mathcal{G}_\mathbb{L}$. As a by-product we prove a result about the relation between the quiver with superpotential and the principal $\mathcal{A}$-determinant in Gelfand-Kapranov-Zelevinsky's theory of discriminants.

1 Introduction

One of the important themes in the physics literature on AdS/CFT is a correspondence between complex 3-dimensional Calabi-Yau singularities and quiver gauge theories. A breakthrough in this field was the discovery of an algorithm that realizes the correspondence for toric CY3 singularities [5, 7]. Experience with, for instance, the Gelfand-Kapranov-Zelevinsky theory of hypergeometric systems or De Bruijn's construction of Penrose tilings shows that sometimes great simplifications and new insights can be achieved by passing to an equivariant setting in higher dimensions. In [8] we adapted the aforementioned algorithm to a higher dimensional viewpoint. In the present paper we clarify some equivariance aspects. We study an $(N-2)$-dimensional algebraic subgroup $\mathcal{G}_\mathbb{L}$ of $\mathbb{C}^{*N}$ acting by diagonal matrices on $\mathbb{C}^N$ and the induced action of a subgroup $\widetilde{\mathcal{G}}_\mathbb{L}$ of $\mathcal{G}_\mathbb{L}$ on $\mathbb{C}^N$ and of a quotientgroup $\overline{\mathcal{G}}_\mathbb{L}$ of $\mathcal{G}_\mathbb{L}$ on the projective space $\mathbb{P}^{N-1}$. Both groups $\widetilde{\mathcal{G}}_\mathbb{L}$ and $\overline{\mathcal{G}}_\mathbb{L}$ have dimension $N-3$. We make models of the orbit spaces by embedding $\mathbb{C}^{*N}/\widetilde{\mathcal{G}}_\mathbb{L}$ into some linear space and $\mathbb{C}^{*N}/\mathcal{G}_\mathbb{L}$ into some projective space and taking closures. We construct these embeddings directly from the quiver with superpotential P on the other side of the correspondence. We refer to Sections 2, 3, 4 for the precise details of the

2000 *Mathematics Subject Classification.* Primary 14M25, 14C25, 16G20.

relation between $\mathcal{G}_{\mathbb{L}}$, $\mathbb{L}$ and P. Here we only want to point out that by explicitly stating Condition 4.1 we make our results independent of any special realizations or algorithms for the correspondence.

The data of the quiver with superpotential P are recast in what we call the bi-adjacency matrix $\mathbb{K}_{\mathsf{P}}(\mathbf{z}, \mathbf{u})$; see Definition 3.2. Here $\mathbf{z}$ denotes a tuple of variables with for every quiver arrow e one variable z_e and $\mathbf{u}$ denotes a tuple of variables with for every quiver node i one variable u_i. The entries of $\mathbb{K}_{\mathsf{P}}(\mathbf{z}, \mathbf{u})$ are elements in the polynomial ring $\mathbb{Z}[\mathbf{z}, \mathbf{u}]$. Viewing $u_1, \ldots, u_N$ as coordinates on $\mathbb{C}^N$ we get an action of the group $\mathcal{G}_{\mathbb{L}}$ on the polynomial ring $\mathbb{C}[\mathbf{u}] = \mathbb{C}[u_1, \ldots, u_N]$. We show in Theorem 4.2 that under suitable conditions on P and $\mathcal{G}_{\mathbb{L}}$ there is a character χ of $\mathcal{G}_{\mathbb{L}}$ such that in the polynomial ring $\mathbb{C}[\mathbf{z}, \mathbf{u}]$

$$\det \mathbb{K}_{\mathsf{P}}(\mathbf{z}, \xi\mathbf{u}) = \chi(\xi) \det \mathbb{K}_{\mathsf{P}}(\mathbf{z}, \mathbf{u}), \qquad \forall \xi \in \mathcal{G}_{\mathbb{L}}.$$

The above mentioned group $\widetilde{\mathcal{G}}_{\mathbb{L}}$ is the kernel of χ. By evaluating $\mathbf{u} = (u_1, \ldots, u_N)$ at the points of $\mathbb{C}^{*N}$ we obtain well defined maps

$$\mathbb{C}^{*N}/\widetilde{\mathcal{G}}_{\mathbb{L}} \longrightarrow \mathbb{C}[\mathbf{z}]^{(\nu)}, \qquad \mathbf{u} \mapsto \det \mathbb{K}_{\mathsf{P}}(\mathbf{z}, \mathbf{u}), \tag{1.1}$$

$$\mathbb{C}^{*N}/\mathcal{G}_{\mathbb{L}} \longrightarrow \mathbb{P}(\mathbb{C}[\mathbf{z}]^{(\nu)}), \qquad \mathbf{u} \mapsto \det \mathbb{K}_{\mathsf{P}}(\mathbf{z}, \mathbf{u}) \bmod \mathbb{C}^*; \tag{1.2}$$

here $\mathbb{C}[\mathbf{z}]^{(\nu)}$ is the homogeneous part of degree $\nu = \deg_{\mathbf{z}} \det \mathbb{K}_{\mathsf{P}}(\mathbf{z}, \mathbf{u})$ in the polynomial ring $\mathbb{C}[\mathbf{z}]$ and $\mathbb{P}(\mathbb{C}[\mathbf{z}]^{(\nu)})$ is the corresponding projective space. Taking Zariski closures of the images of the maps (1.1) and (1.2) one obtains models for the orbit spaces of $\widetilde{\mathcal{G}}_{\mathbb{L}}$ acting on $\mathbb{C}^N$ and $\overline{\mathcal{G}}_{\mathbb{L}}$ acting on $\mathbb{P}^{N-1}$ together with embeddings into $\mathbb{C}[\mathbf{z}]^{(\nu)}$ and $\mathbb{P}(\mathbb{C}[\mathbf{z}]^{(\nu)})$, respectively.

In Definition 5.4 we define the *complementary bi-adjacency matrix* $\mathbb{K}_{\mathsf{P}}^c(\mathbf{z}, \mathbf{u})$ and in Formula (5.7) we define a ring homomorphism $\mathbf{y}$ from the polynomial ring $\mathbb{Z}[\mathbf{z}]$ to the ring $\mathcal{R}_{2,N}$ of homogeneous coordinates on the Grassmannian $\mathbb{G}(2, N)$. In Theorem 5.6 we show that for fixed $\mathbf{u} \in \mathbb{C}^{*N}$ the element $\det \mathbb{K}_{\mathsf{P}}^c(\mathbf{y}(\mathbf{z}), \mathbf{u})$ of $\mathcal{R}_{2,N}$ is a *Chow form* for the orbit closure $\overline{\mathcal{G}_{\mathbb{L}}[\mathbf{u}]}$ in $\mathbb{P}^{N-1}$. The maps (1.1) and (1.2) have obvious analogues:

$$\mathbb{C}^{*N}/\widetilde{\mathcal{G}}_{\mathbb{L}} \longrightarrow \mathcal{R}_{2,N}^{(\nu)}, \qquad \mathbf{u} \mapsto \det \mathbb{K}_{\mathsf{P}}^c(\mathbf{y}(\mathbf{z}), \mathbf{u}), \tag{1.3}$$

$$\mathbb{C}^{*N}/\mathcal{G}_{\mathbb{L}} \longrightarrow \mathbb{P}(\mathcal{R}_{2,N}^{(\nu)}), \qquad \mathbf{u} \mapsto \det \mathbb{K}_{\mathsf{P}}^c(\mathbf{y}(\mathbf{z}), \mathbf{u}) \bmod \mathbb{C}^*; \tag{1.4}$$

The closure of the image of the map (1.4) in $\mathbb{P}(\mathcal{R}_{2,N}^{(\nu)})$ is a model for the orbit space of $\overline{\mathcal{G}}_{\mathbb{L}}$ acting on $\mathbb{P}^{N-1}$ known as the *Chow quotient*; see [6]. In [6] one can also find a discussion of how this model compares with GIT quotients.

A priori $\det \mathbb{K}_{\mathsf{P}}^c(\mathbf{y}(\mathbf{z}), \mathbf{u})$ is an element in $\mathcal{R}_{2,N}[\mathbf{u}] = \mathcal{R}_{2,N}[u_1, \ldots, u_N]$, i.e. in the ring of homogeneous coordinates on $\mathbb{G}(2, N) \times \mathbb{P}^{N-1}$. In (1.4) this element is evaluated at points of $\mathbb{P}^{N-1}$. One can also evaluate it at points of $\mathbb{G}(2, N)$ and find polynomials in the ring $\mathbb{C}[\mathbf{u}] = \mathbb{C}[u_1, \ldots, u_N]$. A point of $\mathbb{G}(2, N)$ is a line in $\mathbb{P}^{N-1}$ and the corresponding polynomial is the equation for the image of this line in the Chow quotient for $\overline{\mathcal{G}}_{\mathbb{L}}$ acting on $\mathbb{P}^{N-1}$.

The $\mathbb{L}$ which appears in the above presentation as a subscript is in fact the character lattice of the group $\mathbb{C}^{*N}/\mathcal{G}_{\mathbb{L}}$. It is a rank 2 subgroup of $\mathbb{Z}^N$ and thus defines a point in the Grassmannian $\mathbb{G}(2, N)$. In Theorem 6.1 we show that evaluating $\det \mathbb{K}_{\mathsf{P}}^c(\mathbf{y}(\mathbf{z}), \mathbf{u})$ at this point yields the principal $\mathcal{A}$-determinant of Gelfand-Kapranov-Zelevinsky [4], if $\mathcal{G}_{\mathbb{L}}$ is connected. In [8] this result was observed in some examples and then conjectured to hold in general.

In Figures 1, 2, 3 and Section 7 we give one example of the theory presented in this paper. Some further examples can be found in [8].

2 The lattice $\mathbb{L}$ and associated groups

Throughout this paper $\mathbb{L}$ is a rank 2 subgroup of $\mathbb{Z}^N$ not contained in any of the standard coordinate hyperplanes and contained in the kernel of the map

$$h : \mathbb{Z}^N \longrightarrow \mathbb{Z}, \qquad h(z_1, \ldots, z_N) = z_1 + \ldots + z_N. \qquad (2.1)$$

Unlike what seems to be practice in the theory of Gelfand, Kapranov and Zelevinsky [4], we do allow torsion in the quotient group $\mathbb{Z}^N/\mathbb{L}$. Denoting by $\mathbf{e}_1, \ldots, \mathbf{e}_N$ the standard basis of $\mathbb{Z}^N$ we set

$$\mathcal{G}_{\mathbb{L}} := \mathrm{Hom}(\mathbb{Z}^N/\mathbb{L}, \mathbb{C}^*), \qquad \mathbf{a}_i := \mathbf{e}_i \bmod \mathbb{L} \text{ in } \mathbb{Z}^N/\mathbb{L}.$$

$\mathcal{G}_{\mathbb{L}}$ is an algebraic group for which the connected component containing the identity is isomorphic to the complex torus $(\mathbb{C}^*)^{N-2}$ and the group of connected components is isomorphic to the finite abelian group $(\mathbb{Z}^N/\mathbb{L})_{\mathrm{tors}}$. The inclusion $\mathbb{L} \hookrightarrow \mathbb{Z}^N$ induces the inclusion of groups

$$\mathcal{G}_{\mathbb{L}} \hookrightarrow \mathbb{C}^{*N}, \qquad \xi \mapsto (\xi(\mathbf{a}_1), \xi(\mathbf{a}_2), \ldots, \xi(\mathbf{a}_N))$$

and thus an action of $\mathcal{G}_{\mathbb{L}}$ on $\mathbb{C}^N$ and $\mathbb{P}^{N-1}$. Since $\mathbb{L}$ lies in the kernel of the map h there is a homomorphism

$$\overline{h} : \mathbb{Z}^N/\mathbb{L} \to \mathbb{Z} \qquad \text{such that} \qquad \overline{h}(\mathbf{a}_i) = 1 \quad \text{for all } i. \qquad (2.2)$$

As a consequence there is a subgroup $\mathcal{G}'_{\mathbb{L}}$ of $\mathcal{G}_{\mathbb{L}}$ consisting of those $\xi \in \mathcal{G}_{\mathbb{L}}$ for which $\xi(\mathbf{a}_1) = \ldots = \xi(\mathbf{a}_N)$. So, in fact $\mathcal{G}_{\mathbb{L}}$ acts on $\mathbb{P}^{N-1}$ via the quotient group $\overline{\mathcal{G}}_{\mathbb{L}} := \mathcal{G}_{\mathbb{L}}/\mathcal{G}'_{\mathbb{L}}$.

The subgroup $\widetilde{\mathcal{G}}_{\mathbb{L}}$ of $\mathcal{G}_{\mathbb{L}}$ is the kernel of some character χ of $\mathcal{G}_{\mathbb{L}}$. The construction is slightly involved. We choose a basis for $\mathbb{L}$ and represent the inclusion $\mathbb{L} \hookrightarrow \mathbb{Z}^N$ by a matrix $B = (b_{ij})_{i=1,2; j=1,\ldots,N}$ with entries in $\mathbb{Z}$. Let $\beta_1, \ldots, \beta_N \in \mathbb{Z}^2$ denote the columns of B. The complete fan in $\mathbb{R}^2$ with 1-dimensional rays $\mathbb{R}_{\geq 0}\beta_i$ ($i = 1, \ldots, N$) is the so-called *secondary fan* associated with $\mathbb{L}$; see Figure 1 for an example. Take $\mathbf{c} \in \mathbb{R}^2 \setminus \bigcup_{i=1}^N \mathbb{R}_{\geq 0}\beta_i$ and set $L_{\mathbf{c}} = \{\, \{i,j\} \subset \{1, \ldots, N\} \,|\, \mathbf{c} \in \mathbb{R}_{\geq 0}\beta_i + \mathbb{R}_{\geq 0}\beta_j \,\}$. We now define:

$$\mathbf{a}_0 := \sum_{\{i,j\} \in L_{\mathbf{c}}} |\det(\beta_i, \beta_j)|(\mathbf{a}_i + \mathbf{a}_j), \qquad (2.3)$$

$$\chi : \mathcal{G}_{\mathbb{L}} \to \mathbb{C}^*, \qquad \chi(\xi) := \xi(\mathbf{a}_0), \qquad (2.4)$$

$$\widetilde{\mathcal{G}}_{\mathbb{L}} := \ker \chi. \qquad (2.5)$$

It follows from [8] Eqs.(5), (7) that $\mathbf{a}_0$ does not depend on the particular choice of the vector $\mathbf{c}$. Changing the basis for $\mathbb{L}$ multiplies B with a 2×2-matrix with determinant ± 1. So, $\mathbf{a}_0$ does also not depend on the particular choice of the basis for $\mathbb{L}$.

3 The quiver with superpotential P

On the quiver side we look at a quiver with a special kind of superpotential $\mathsf{P} = (\mathsf{P}^0, \mathsf{P}^1, \mathsf{P}^{2\bullet}, \mathsf{P}^{2\circ}, s, t, \mathbf{b}, \mathbf{w})$. It consists of four finite sets $\mathsf{P}^0, \mathsf{P}^1, \mathsf{P}^{2\bullet}, \mathsf{P}^{2\circ}$ and four maps

$$s, t : \mathsf{P}^1 \to \mathsf{P}^0, \qquad \mathbf{b} : \mathsf{P}^1 \to \mathsf{P}^{2\bullet}, \qquad \mathbf{w} : \mathsf{P}^1 \to \mathsf{P}^{2\circ}. \qquad (3.1)$$

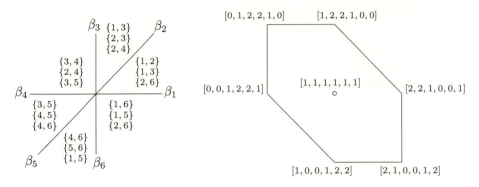

Figure 1 *Secondary fan with lists L_c (left) and secondary polygon with vectors $\sum_{\{i,j\}\in L_c}(\mathbf{e}_i + \mathbf{e}_j)$ (right) for $\mathbb{L} = \mathbb{Z}^2$* $\begin{bmatrix} 1 & 1 & 0 & -1 & -1 & 0 \\ 0 & 1 & 1 & 0 & -1 & -1 \end{bmatrix}$. *This shows* $\mathbf{a}_0 = (1,1,1,1,1,1) \bmod \mathbb{L}$.

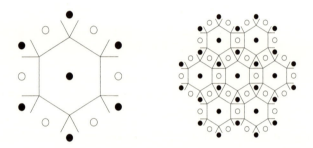

Figure 2 *On the left: the quiver with superpotential* P; *opposite sides of the hexagon must be identified;* P *actually lies on a torus. On the right: part of the lift of* P *on the universal cover of the torus.*

The elements of P^0 are called 0-cells or nodes or vertices. The elements of P^1 are called 1-cells or arrows or edges. The elements of $\mathsf{P}^{2\bullet}$ (resp. $\mathsf{P}^{2\circ}$) are called black (resp. white) 2-cells. An arrow e is oriented from $s(e)$ to $t(e)$.

Throughout this paper the following condition is assumed to be satisfied:

Condition 3.1   1. *The directed graph* $(\mathsf{P}^0, \mathsf{P}^1, s, t)$ *is connected and has no oriented cycles of length* ≤ 2 *and in every node* $v \in \mathsf{P}^0$ *there are as many incoming as outgoing arrows; i.e.* $\sharp\{e \in \mathsf{P}^1 | t(e) = v\} = \sharp\{e \in \mathsf{P}^1 | s(e) = v\}$.

2. *There are as many black as white cells; i.e.* $\sharp\mathsf{P}^{2\bullet} = \sharp\mathsf{P}^{2\circ}$.

3. *For every* $\mathbf{b} \in \mathsf{P}^{2\bullet}$ *and every* $\mathbf{w} \in \mathsf{P}^{2\circ}$ *the sets* $\{e \in \mathsf{P}^1 | \mathbf{b}(e) = \mathbf{b}\}$ *and* $\{e \in \mathsf{P}^1 | \mathbf{w}(e) = \mathbf{w}\}$ *are connected oriented cycles; by this we mean that the elements can be ordered* $(e_1, e_2, \ldots, e_r)$ *such that* $t(e_i) = s(e_{i+1})$ *for* $i = 1, \ldots, r-1$ *and* $t(e_r) = s(e_1)$.

One can realize P geometrically as an oriented surface without boundary by representing every black 2-cell $\mathbf{b}$ (resp. white 2-cell $\mathbf{w}$) as a convex polygon with sides labeled by the elements $e \in \mathsf{P}^1$ with $\mathbf{b}(e) = \mathbf{b}$ (resp. $\mathbf{w}(e) = \mathbf{w}$) and by gluing these polygons along sides with the same label. In quiver theory one views $(\mathsf{P}^0, \mathsf{P}^1, s, t)$ as a quiver (=directed graph) and the boundary cycles of the 2-cells as the terms of a superpotential; cf. [8] 8.6.

The structure of P described in (3.1) can be accurately captured in its bi-adjacency matrix:

Definition 3.2 The *bi-adjacency matrix* of P is the square matrix $\mathbb{K}_P(\mathbf{z}, \mathbf{u})$ with rows (resp. columns) corresponding with the black (resp. white) 2-cells, with entries in the polynomial ring $\mathbb{Z}[z_e, u_i \mid e \in P^1, i \in P^0]$ and $(\mathbf{b}, \mathbf{w})$-entry:

$$(\mathbb{K}_P(\mathbf{z}, \mathbf{u}))_{\mathbf{b}, \mathbf{w}} = \sum_{e \in P^1 : \, \mathbf{b}(e) = \mathbf{b}, \, \mathbf{w}(e) = \mathbf{w}} z_e \, u_{s(e)} u_{t(e)} . \tag{3.2}$$

This definition of the bi-adjacency matrix is the same as [8] Eq.(1), except that in loc. cit. we used a $\mathbb{C}$-valued function ϖ on P^1 instead of the formal variables z_e. The products $u_{s(e)} u_{t(e)}$ of commuting variables do not show the orientation of the edge e. As explained in [8] 8.10, the orientation can easily be reconstructed. So there is no loss of information in going from (3.1) to (3.2). The bi-adjacency matrix $\mathbb{K}_P(\mathbf{z}, \mathbf{u})$ of P is also the *Kasteleyn matrix of a twist* of P; see [8] Section 8.

4 Compatibility of $\mathbb{L}$ and P

In [8] §6 we described an algorithm, based on the *Fast Inverse Algorithm* of [5] and the *untwisting procedure* of [2], for constructing from a rank 2 lattice $\mathbb{L}$ as in Section 2 a quiver with superpotential P as in Section 3, such that also Condition 4.1 (below) is satisfied. We isolated Conditions 3.1 and 4.1 because this is exactly what we need in the present paper and because these conditions can be checked without worrying about the algorithm of [8].

Condition 4.1 *For every* $\mathbf{b} \in P^{2\bullet}$ *and* $\mathbf{w} \in P^{2\circ}$ *there are elements* $\varepsilon_{\mathbf{b}}$ *and* $\varepsilon_{\mathbf{w}}$ *in* $\mathbb{Z}^N / \mathbb{L}$ *such that for every* $e \in P^1$ *and with* $\mathbf{a}_0$ *as in (2.3) one has*

$$\mathbf{a}_{s(e)} + \mathbf{a}_{t(e)} = \varepsilon_{\mathbf{b}(e)} - \varepsilon_{\mathbf{w}(e)}, \qquad \mathbf{a}_0 = \sum_{\mathbf{b} \in P^{2\bullet}} \varepsilon_{\mathbf{b}} - \sum_{\mathbf{w} \in P^{2\circ}} \varepsilon_{\mathbf{w}} . \tag{4.1}$$

The ring $\mathbb{C}[u_i \mid i \in P^0] = \mathbb{C}[u_1, \ldots, u_N]$ is the coordinate ring of $\mathbb{C}^N$ and therefore carries an action of the group $\mathcal{G}_{\mathbb{L}}$:

$$\xi \, u_i := \xi(\mathbf{a}_i) u_i \qquad \text{for} \quad i = 1, \ldots, N, \ \xi \in \mathcal{G}_{\mathbb{L}} .$$

Theorem 4.2 *Assume* $\mathbb{L}$ *and* P *satisfy Conditions 3.1 and 4.1. Then we have for every* $\xi \in \mathcal{G}_{\mathbb{L}}$:

$$\mathbb{K}_P(\mathbf{z}, \xi \mathbf{u}) = \text{diag}(\xi(\varepsilon_{\mathbf{b}})) \, \mathbb{K}_P(\mathbf{z}, \mathbf{u}) \, \text{diag}(\xi(-\varepsilon_{\mathbf{w}})), \tag{4.2}$$

$$\det \mathbb{K}_P(\mathbf{z}, \xi \mathbf{u}) = \chi(\xi) \det \mathbb{K}_P(\mathbf{z}, \mathbf{u}), \tag{4.3}$$

$$\deg_{\mathbf{z}} \det \mathbb{K}_P(\mathbf{z}, \mathbf{u}) = \tfrac{1}{2} \deg_{\mathbf{u}} \det \mathbb{K}_P(\mathbf{z}, \mathbf{u}) = \sharp P^{2\bullet} = \sharp P^{2\circ} = \tfrac{1}{2} \overline{h}(\mathbf{a}_0) . \tag{4.4}$$

with character χ *as in (2.4), map* $\overline{h}$ *as in (2.2) and the entries of the diagonal matrices* $\text{diag}(\xi(\varepsilon_{\mathbf{b}}))$ *and* $\text{diag}(\xi(-\varepsilon_{\mathbf{w}}))$ *labeled with the black and white 2-cells in agreement with the labeling of the rows and columns of* $\mathbb{K}_P(\mathbf{z}, \mathbf{u})$.

Proof (4.2), (4.3) and the first three equalities in (4.4) are obvious. The first half of (4.1) implies that $\overline{h}(\varepsilon_{\mathbf{b}}) - \overline{h}(\varepsilon_{\mathbf{w}}) = 2$ for all $\mathbf{b} \in P^{2\bullet}$ and $\mathbf{w} \in P^{2\circ}$. The last equality in (4.4) now follows from the second half of (4.1). $\qquad \square$

$\varepsilon_\mathbf{b}\downarrow:\varepsilon_\mathbf{w}\rightarrow$	$(0,-1,-1,0,0,0)$	$(-1,-1,0,0,0,0)$	$(0,0,-1,-1,0,0)$
$(0,0,0,0,0,0)$	$z_1 u_2 u_3 + z_2 u_5 u_6$	$z_3 u_1 u_2 + z_4 u_4 u_5$	$z_5 u_3 u_4 + z_6 u_1 u_6$
$(0,0,-1,0,0,1)$	$z_7 u_2 u_6$	$z_8 u_2 u_4$	$z_9 u_4 u_6$
$(0,-1,0,0,1,0)$	$z_{10} u_3 u_5$	$z_{11} u_1 u_5$	$z_{12} u_1 u_3$

Figure 3 *The bi-adjacency matrix for* P *as in Figure 2 and on its sides vectors (to be read modulo the* L *of Figure 1) which verify Condition 4.1.*

5 Lines in projective space and Chow forms

Every line in $\mathbb{P}^{N-1}$ is of the form

$$\mathcal{L}_Y := \{[u_1 : \ldots : u_N] \in \mathbb{P}^{N-1} \mid u_j = t_1 y_{1j} + t_2 y_{2j}\,,\ [t_1 : t_2] \in \mathbb{P}^1\ \}$$

with $Y = (y_{ij})_{i=1,2;j=1,\ldots,N}$ a $2 \times N$ complex matrix of rank 2. Let $M(2,N)$ denote the set of $2 \times N$ complex matrices of rank 2. As $\mathcal{L}_{Y'} = \mathcal{L}_Y$ if and only if there is an invertible 2×2-matrix g such that $Y' = gY$, the lines in $\mathbb{P}^{N-1}$ correspond bijectively with the points of the Grassmannian

$$\mathbb{G}(2,N) := Gl(2)\backslash M(2,N)\,.$$

Natural homogeneous coordinates on $\mathbb{G}(2,N)$ are the *Plücker coordinates* Y_{km} of the matrix $Y = (y_{ij})_{i=1,2;j=1,\ldots,N}$:

$$Y_{km} := y_{1k}y_{2m} - y_{2k}y_{1m}\,, \qquad k,m = 1,\ldots,N\,. \tag{5.1}$$

In more algebraic terms (see [4] p.96), *the ring of homogeneous coordinates on* $\mathbb{G}(2,N)$ *in the Plücker embedding* is the ring

$$\mathcal{R}_{2,N} := \mathbb{C}[\Upsilon_{km} \mid k,m = 1,\ldots,N]/\mathcal{I}\,, \tag{5.2}$$

where $\mathcal{I}$ is the ideal in the polynomial ring $\mathbb{C}[\Upsilon_{km}]$ generated by the elements

$$\Upsilon_{km} + \Upsilon_{mk} \quad\text{and}\quad \Upsilon_{ij}\Upsilon_{km} + \Upsilon_{ik}\Upsilon_{mj} + \Upsilon_{im}\Upsilon_{jk}\,, \qquad \forall i,j,k,m\,. \tag{5.3}$$

Y_{km} is then the element $\Upsilon_{km} \bmod \mathcal{I}$ of $\mathcal{R}_{2,N}$.

Definition 5.1 [see [4] pp.99,100,123] Let $\mathcal{X}$ be an irreducible subvariety of $\mathbb{P}^{N-1}$ of codimension 2 and degree d. Then the set of lines $\mathcal{L}_Y$ which intersect $\mathcal{X}$ is an irreducible hypersurface $\mathcal{Z}(\mathcal{X})$ in the Grassmannian $\mathbb{G}(2,N)$, called the *associated hypersurface of* $\mathcal{X}$. The hypersurface $\mathcal{Z}(\mathcal{X})$ is given by the vanishing of a single, up to a multiplicative constant unique, irreducible homogeneous element of degree d in $\mathcal{R}_{2,N}$, the so-called *Chow form of* $\mathcal{X}$. The Chow form of a codimension 2 subvariety of $\mathbb{P}^{N-1}$ is the product of the Chow forms of its irreducible components.

Our aim is to explicitly compute the Chow form of the orbit closure $\overline{\mathcal{G}_\mathbb{L}[\mathbf{u}]}$ when $[\mathbf{u}] \in \mathbb{P}^{N-1}$ is a point with all homogeneous coordinates $\neq 0$.

Proposition 5.2 *For* $\mathbf{u} = (u_1,\ldots,u_N) \in \mathbb{C}^{*N}$ *we denote its inverse in* $\mathbb{C}^{*N}$ *by* $\mathbf{u}^{-1} = (u_1^{-1},\ldots,u_N^{-1})$ *and its image in* $\mathbb{P}^{N-1}$ *by* $[\mathbf{u}] = [u_1 : \ldots : u_N]$. *For* $Y \in M(2,N)$ *we denote by* $\mathbf{Y}_{st}$ *the tuple of complex numbers* $(Y_{s(e)t(e)})_{e\in\mathbb{P}^1}$, *where* Y_{km} *is the Plücker coordinate as in (5.1). Then*

$$\text{point } [\mathbf{u}] \text{ lies on line } \mathcal{L}_Y \quad\Rightarrow\quad \det\mathbb{K}_\mathsf{P}(\mathbf{Y}_{st},\mathbf{u}^{-1}) = 0\,. \tag{5.4}$$

Proof From the fact that a $3 \times N$-matrix has rank ≤ 2 if and only if all its 3×3-subdeterminants vanish, we see that *point* $[\mathbf{u}]$ *lies on line* $\mathcal{L}_Y$ if and only if

$$u_k^{-1} u_m^{-1} Y_{km} + u_m^{-1} u_j^{-1} Y_{mj} + u_j^{-1} u_k^{-1} Y_{jk} = 0 \tag{5.5}$$

for every triple of elements j, k, m in $\{1, \ldots, N\}$. Adding equations from the system (5.5) yields an equation

$$\sum_{i=1}^{r} Y_{k_i k_{i+1}} u_{k_i}^{-1} u_{k_{i+1}}^{-1} = 0 \tag{5.6}$$

for every cycle $(k_1, \ldots, k_r)$, $k_{r+1} = k_1$, of distinct elements of $\{1, \ldots, N\}$. In particular, there is such an equation (5.6) for the cycle of vertices of a black 2-cell of P and the terms of that equation match bijectively with the adjacent white 2-cells. This means that in every row of the matrix $\mathbb{K}_{\mathsf{P}}(\mathbf{Y}_{st}, \mathbf{u}^{-1})$ the sum of the entries is 0. Consequently $\det \mathbb{K}_{\mathsf{P}}(\mathbf{Y}_{st}, \mathbf{u}^{-1}) = 0$. $\quad\square$

An immediate consequence of Proposition 5.2 and Theorem 4.2 is:

Corollary 5.3 *Assume* $\mathbb{L}$ *and* P *satisfy Conditions 3.1 and 4.1. Let* $\mathbf{u}$ *and* Y *be as in Proposition 5.2. Then* $\det \mathbb{K}_{\mathsf{P}}(\mathbf{Y}_{st}, \mathbf{u}^{-1}) = 0$ *if the line* $\mathcal{L}_Y$ *intersects the closure* $\overline{\mathcal{G}_{\mathbb{L}}[\mathbf{u}]}$ *of the orbit* $\mathcal{G}_{\mathbb{L}}[\mathbf{u}]$ *in* $\mathbb{P}^{N-1}$.

Proof Indeed (4.3) and (5.4) imply $\det \mathbb{K}_{\mathsf{P}}(\mathbf{Y}_{st}, \mathbf{u}^{-1}) = 0$ if the line $\mathcal{L}_Y$ intersects the orbit $\mathcal{G}_{\mathbb{L}}[\mathbf{u}]$. The corollary follows because the set of lines which intersect $\overline{\mathcal{G}_{\mathbb{L}}[\mathbf{u}]}$ is the closure in the Grassmannian $\mathbb{G}(2, N)$ of the set of lines which intersect $\mathcal{G}_{\mathbb{L}}[\mathbf{u}]$. $\quad\square$

In order to remedy for the appearing inverses we now introduce the complementary bi-adjacency matrix:

Definition 5.4 The *complementary bi-adjacency matrix* $\mathbb{K}_{\mathsf{P}}^{c}(\mathbf{z}, \mathbf{u})$ is

$$\mathbb{K}_{\mathsf{P}}^{c}(\mathbf{z}, \mathbf{u}) := u_1 \cdot \ldots \cdot u_N \cdot \mathbb{K}_{\mathsf{P}}(\mathbf{z}, (u_1^{-1}, \ldots, u_N^{-1})).$$

with, as usual, $\mathbf{u} = (u_1, \ldots, u_N)$. The matrix $\mathbb{K}_{\mathsf{P}}^{c}(\mathbf{z}, \mathbf{u})$ has entries in the polynomial ring $\mathbb{Z}[\mathbf{z}, \mathbf{u}]$.

A further consequence of Proposition 5.2 and Theorem 4.2 is:

Corollary 5.5 *Assume that* $\mathbb{L}$ *and* P *satisfy Conditions 3.1 and 4.1. Let* $\mathbf{y}$ *denote the ringhomomorphism (cf. (5.2))*

$$\mathbf{y} : \mathbb{Z}[z_e \mid e \in \mathsf{P}^1] \longrightarrow \mathcal{R}_{2,N}, \qquad \mathbf{y}(z_e) = Y_{s(e)t(e)}. \tag{5.7}$$

Then for $\mathbf{u} \in \mathbb{C}^{*N}$ *the Chow form of* $\overline{\mathcal{G}_{\mathbb{L}}[\mathbf{u}]}$ *divides* $\det \mathbb{K}_{\mathsf{P}}^{c}(\mathbf{y}(\mathbf{z}), \mathbf{u})$ *in* $\mathcal{R}_{2,N}$.

Proof There is a slight subtlety because torsion in $\mathbb{Z}^N / \mathbb{L}$ makes $\overline{\mathcal{G}_{\mathbb{L}}[\mathbf{u}]}$ reducible. The Chow form of an irreducible component of $\overline{\mathcal{G}_{\mathbb{L}}[\mathbf{u}]}$ is an irreducible element of $\mathcal{R}_{2,N}$ (see [4] p.99 Prop.2.2), which according to Corollary 5.3 divides $\det \mathbb{K}_{\mathsf{P}}^{c}(\mathbf{y}(\mathbf{z}), \mathbf{u})$ in $\mathcal{R}_{2,N}$. Since an irreducible codimension 2 subvariety of $\mathbb{P}^{N-1}$ is uniquely determined by its Chow form (see [4] p.102 Prop.2.5), the Chow forms of different irreducible components of $\overline{\mathcal{G}_{\mathbb{L}}[\mathbf{u}]}$ are not equal. Moreover $\mathcal{R}_{2,N}$ is a factorial ring (see [4] p.98 Prop.2.1). It follows that $\det \mathbb{K}_{\mathsf{P}}^{c}(\mathbf{y}(\mathbf{z}), \mathbf{u})$ is divisible by the Chow form of $\overline{\mathcal{G}_{\mathbb{L}}[\mathbf{u}]}$. $\quad\square$

Theorem 5.6 *Assume that $\mathbb{L}$ and P satisfy Conditions 3.1 and 4.1. Let $\mathbf{u} \in \mathbb{C}^{*N}$.*
Then $\det \mathbb{K}_{\mathsf{P}}^{\mathsf{c}}(\mathbf{y}(\mathbf{z}), \mathbf{u})$ *is a Chow form for* $\overline{\mathcal{G}_{\mathbb{L}}[\mathbf{u}]}$.

N.B. We write here *a* Chow form to emphasize that it is only determined up to a non-zero multiplicative constant.

Proof In view of Corollary 5.5 we need only prove that the Chow form of $\overline{\mathcal{G}_{\mathbb{L}}[\mathbf{u}]}$ and $\det \mathbb{K}_{\mathsf{P}}^{\mathsf{c}}(\mathbf{y}(\mathbf{z}), \mathbf{u})$ are elements of $\mathcal{R}_{2,N}$ with the same degree.
Take $\mathbb{L}^0$ such that $\mathbb{L} \subset \mathbb{L}^0 \subset \mathbb{Z}^N$ and $\mathbb{L}^0/\mathbb{L} = (\mathbb{Z}^N/\mathbb{L})_{\mathrm{tors}}$. Set

$$\mathcal{G}_{\mathbb{L}}^0 := \mathrm{Hom}(\mathbb{Z}^N/\mathbb{L}^0, \mathbb{C}^*), \quad \mathbf{a}_i^0 := \mathbf{e}_i \bmod \mathbb{L}^0, \quad \mathcal{A}^0 := \{\mathbf{a}_1^0, \ldots, \mathbf{a}_N^0\}.$$

Then $\mathcal{G}_{\mathbb{L}}^0$ is the connected component containing the identity in $\mathcal{G}_{\mathbb{L}}$. Via the action of $\mathbb{C}^{*N}$, every irreducible component of $\overline{\mathcal{G}_{\mathbb{L}}[\mathbf{u}]}$ together with its embedding into $\mathbb{P}^{N-1}$ is isomorphic to $\overline{\mathcal{G}_{\mathbb{L}}^0 \mathbf{1}} \hookrightarrow \mathbb{P}^{N-1}$, where $\mathbf{1} \in \mathbb{P}^{N-1}$ has all its components equal to 1. In [4] p.166 the toric variety $\overline{\mathcal{G}_{\mathbb{L}}^0 \mathbf{1}}$ is denoted as $X_{\mathcal{A}^0}$. Since $\mathbb{L}$ is contained in the kernel of the map h (see (2.1)) the set $\mathcal{A}^0$ is contained in an affine hyperplane in $\mathbb{Z}^N/\mathbb{L}^0 = \mathbb{Z}^{N-2}$. By [4] p.203 Theorem 2.3 and p.99 Proposition 2.2 the degree of the toric variety $X_{\mathcal{A}^0} \subset \mathbb{P}^{N-1}$ and the degree of its Chow form in $\mathcal{R}_{2,N}$ are both equal to the volume $\mathrm{vol}_{\mathcal{A}^0}$ of the convex hull of $\mathcal{A}^0$, with the volume normalized so that the standard unit cube in $\mathbb{Z}^{N-3}$ has volume $(N-3)!$ (see [4] p.182). Let $\mathbf{a}_0^0 \in \mathbb{Z}^N/\mathbb{L}^0$ be the image of $\mathbf{a}_0 \in \mathbb{Z}^N/\mathbb{L}$. It is well-known and easy to prove (see e.g. [8] Eq.(19)) that $\mathrm{vol}_{\mathcal{A}^0} = \frac{1}{2} h(\mathbf{a}_0^0)$. On the other hand, it is obvious that $h(\mathbf{a}_0) h(\mathbf{a}_0^0)^{-1} = \sharp(\mathbb{L}^0/\mathbb{L})$ and that this equals the number of irreducible components of $\overline{\mathcal{G}_{\mathbb{L}}[\mathbf{u}]}$. Thus the degree of the Chow form of $\overline{\mathcal{G}_{\mathbb{L}}[\mathbf{u}]}$ is equal to $\frac{1}{2} h(\mathbf{a}_0)$. It follows from (3.2) and (4.4) that this is also the degree of $\det \mathbb{K}_{\mathsf{P}}^{\mathsf{c}}(\mathbf{y}(\mathbf{z}), \mathbf{u})$ as an element of $\mathcal{R}_{2,N}$. $\square$

6 The principal $\mathcal{A}$-determinant

In this section we prove Conjecture 10.5 of [8] by combining Theorem 5.6 and [1] Proposition 3.2. The result is:

Theorem 6.1 *Assume that Conditions 3.1 and 4.1 are satisfied and that $\mathbb{Z}^N/\mathbb{L}$ has no torsion. Let $B = (b_{ij})_{i=1,2;\, j=1,\ldots,N}$ be a $2 \times N$-matrix with entries in $\mathbb{Z}$ which gives the inclusion $\mathbb{L} \hookrightarrow \mathbb{Z}^N$ and let $\mathbf{B}_{st}$ denote the ring homomorphism*

$$\mathbf{B}_{st} : \mathbb{Z}[z_e \mid e \in \mathsf{P}^1] \longrightarrow \mathbb{Z}, \qquad \mathbf{B}_{st}(z_e) = b_{1s(e)} b_{2t(e)} - b_{1t(e)} b_{2s(e)}.$$

Set $\mathcal{A} = \{\mathbf{a}_1, \ldots, \mathbf{a}_N\}$ and let $E_{\mathcal{A}}(f)$ denote the principal $\mathcal{A}$-determinant defined in [4] p.297 Eq. (1.1) for the Laurent polynomial $f = \sum_{i=1}^N u_i \mathbf{x}^{\mathbf{a}_i}$.
Then the following equality holds in the polynomial ring $\mathbb{Z}[u_1, \ldots, u_N]$

$$E_{\mathcal{A}}(f) = \pm \mathbf{B}_{st}\left(\det \mathbb{K}_{\mathsf{P}}^{\mathsf{c}}(\mathbf{z}, \mathbf{u})\right). \tag{6.1}$$

Proof One can trace back the definition of the principal $\mathcal{A}$-determinant $E_{\mathcal{A}}(f)$ from [4] p.297 Eq.(1.1) to the definition of the Chow form of the subvariety $X_{\mathcal{A}} = \overline{\mathcal{G}_{\mathbb{L}} \mathbf{1}}$ of $\mathbb{P}^{N-1}$ on [4] p.100. That is also what Dickenstein and Sturmfels do before defining the principal $\mathcal{A}$-determinant, which they call the *full discriminant*, in [1] Definition 3.1. The result in [1] Eq. (3.2) is a more or less immediate consequence of this definition. The above Equation (6.1) now follows by substituting the expression for the Chow form given in Theorem 5.6 into [1] Eq. (3.2). $\square$

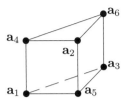

Figure 4 *The set* $\mathcal{A} = \{\mathbf{a}_1, \ldots, \mathbf{a}_6\}$ *for* $\mathbb{L} = \mathbb{Z}^2$ $\begin{bmatrix} 1 & 1 & 0 & -1 & -1 & 0 \\ 0 & 1 & 1 & 0 & -1 & -1 \end{bmatrix}$.

7 An example

Figures 1, 2 and 3 show $\mathbb{L}$, P, the bi-adjacency matrix and the verification of Condition 4.1 for the example which in the physics literature (e.g. [3]) is known as *model I of Del Pezzo* 3. The right-hand picture in Figure 2 is the dual of the *brane tiling* in Figure 1 of [3]. One should keep in mind that in [3] toric data for the singularity are extracted from the *Kasteleyn matrix* (see [3] Eqs. (3.4), (3.5)), whereas we work with the bi-adjacency matrix. In this particular example, the Kasteleyn matrix and the bi-adjacency matrix give the same toric data, but in general they lead to different toric data.

In the context of hypergeometric systems this example is Appell's F_1; see [8] §4. The group $\mathbb{Z}^6/\mathbb{L}$ is torsion free and can be identified with $\mathbb{Z}^4$. This puts $\mathbf{a}_1, \ldots, \mathbf{a}_6$ as points in a hyperplane in $\mathbb{Z}^4$. Figure 4 shows the points $\mathbf{a}_1, \ldots, \mathbf{a}_6$ and their convex hull (often called *the primary polytope*) situated in this 3-dimensional hyperplane. $\mathbb{L}$ is the lattice of affine relations between these six points.

From the bi-adjacency matrix in Figure 3 one easily gets the determinants

$\det \mathbb{K}_\mathsf{P}(\mathbf{z}, \mathbf{u})$

$$
\begin{aligned}
= & (z_2 z_8 z_{12} + z_3 z_9 z_{10} + z_5 z_7 z_{11} - z_6 z_8 z_{10} - z_1 z_9 z_{11} - z_4 z_7 z_{12}) \mathbf{u}^{[1,1,1,1,1,1]} \\
& + z_1 z_8 z_{12} \mathbf{u}^{[1,2,2,1,0,0]} + z_4 z_9 z_{10} \mathbf{u}^{[0,0,1,2,2,1]} + z_6 z_7 z_{11} \mathbf{u}^{[2,1,0,0,1,2]} \\
& - z_5 z_8 z_{10} \mathbf{u}^{[0,1,2,2,1,0]} - z_2 z_9 z_{11} \mathbf{u}^{[1,0,0,1,2,2]} - z_3 z_7 z_{12} \mathbf{u}^{[2,2,1,0,0,1]} .
\end{aligned}
$$

$\det \mathbb{K}_\mathsf{P}^c(\mathbf{y}(\mathbf{z}), \mathbf{u})$

$$
\begin{aligned}
= & (\mathbf{Y}_{56|24|13} + \mathbf{Y}_{12|46|35} + \mathbf{Y}_{34|62|51} - \mathbf{Y}_{61|24|35} - \mathbf{Y}_{23|46|51} - \mathbf{Y}_{45|62|13}) \mathbf{u}^{[2,2,2,2,2,2]} \\
& + \mathbf{Y}_{23|24|13} \mathbf{u}^{[2,1,1,2,3,3]} + \mathbf{Y}_{45|46|35} \mathbf{u}^{[3,3,2,1,1,2]} + \mathbf{Y}_{61|62|51} \mathbf{u}^{[1,2,3,3,2,1]} \\
& - \mathbf{Y}_{34|24|35} \mathbf{u}^{[3,2,1,1,2,3]} - \mathbf{Y}_{56|46|51} \mathbf{u}^{[2,3,3,2,1,1]} - \mathbf{Y}_{12|62|13} \mathbf{u}^{[1,1,2,3,3,2]} .
\end{aligned}
$$

with notations $\mathbf{u}^{[a,b,c,d,e,f]} := u_1^a u_2^b u_3^c u_4^d u_5^e u_6^f$ and $\mathbf{Y}_{ab|cd|ef} := Y_{ab} Y_{cd} Y_{ef}$; here the Y_{km} are the Plücker coordinates on $\mathbb{G}(2,6)$; see (5.1)-(5.3). These formulas yield for every $\mathbf{u} \in (\mathbb{C}^*)^6$ elements in $\mathbb{C}[\mathbf{z}] = \mathbb{C}[z_1, \ldots, z_{12}]$ and $\mathcal{R}_{2,6}$, respectively. Taking the closure of $\{\det \mathbb{K}_\mathsf{P}(\mathbf{z}, \mathbf{u}) \mid \mathbf{u} \in (\mathbb{C}^*)^6\}$ in $\mathbb{C}[\mathbf{z}]$ adds six lines which correspond to the six sides of the secondary polygon in Figure 1; the top side, for instance, gives the line $\{(z_1 z_8 z_{12} \, x_1 - z_5 z_8 z_{10} \, x_2) \mid [x_1 : x_2] \in \mathbb{P}^1\}$. A similar remark holds for the closure of $\{\det \mathbb{K}_\mathsf{P}^c(\mathbf{y}(\mathbf{z}), \mathbf{u}) \mid \mathbf{u} \in (\mathbb{C}^*)^6\}$.

$\mathbb{L}$ is a rank 2 subgroup of $\mathbb{Z}^6$ and therefore defines a point $\mathfrak{l}$ in the Grassmannian $\mathbb{G}(2,6)$. Evaluating $\det \mathbb{K}_\mathsf{P}^c(\mathbf{y}(\mathbf{z}), \mathbf{u})$ at the point $\mathfrak{l}$ amounts in this case to setting all $Y_{km} = 1$. This results in a polynomial in $\mathbb{C}[u_1, \ldots, u_6]$:

$$\det \mathbb{K}^c_{\mathsf{P}}(\mathbf{y}(\mathbf{z}), \mathbf{u})_{|\text{at } \mathfrak{l}}$$

$$= \mathbf{u}^{[2,1,1,2,3,3]} + \mathbf{u}^{[3,3,2,1,1,2]} + \mathbf{u}^{[1,2,3,3,2,1]} - \mathbf{u}^{[3,2,1,1,2,3]} - \mathbf{u}^{[2,3,3,2,1,1]} - \mathbf{u}^{[1,1,2,3,3,2]}$$

$$= u_1 u_2 u_3 u_4 u_5 u_6 (u_1 u_2 - u_4 u_5)(u_3 u_4 - u_1 u_6)(u_5 u_6 - u_2 u_3).$$

Note that in the first expression for $\det \mathbb{K}^c_{\mathsf{P}}(\mathbf{y}(\mathbf{z}), \mathbf{u})_{|\text{at } \mathfrak{l}}$ all monomials are invariant for the action of the group $\widetilde{\mathcal{G}}_{\mathbb{L}}$ on $\mathbb{C}[u_1, \ldots, u_6]$. This invariance does not hold for the factors in the second expression. On the other hand, the factors in this factorization correspond with the faces of the primary polytope in Figure 4, in agreement with the prime factorization of the principal $\mathcal{A}$-determinant in [4] p.299 Theorem 1.2.

There can not be a similar factorization of $\det \mathbb{K}^c_{\mathsf{P}}(\mathbf{y}(\mathbf{z}), \mathbf{u})$ in the ring $\mathcal{R}_{2,6}[u_1, \ldots, u_6]$, because that would upon specializing $u_1, \ldots, u_6 \in \mathbb{C}^*$ give a factorization in $\mathcal{R}_{2,6}$ of the Chow form of the toric variety $\overline{\mathcal{G}_{\mathbb{L}}[\mathbf{u}]} \subset \mathbb{P}^5$ and thus contradict the fact that Chow forms of irreducible subvarieties of $\mathbb{P}^5$ are irreducible.

Acknowledgement. I want to express special thanks to Alicia Dickenstein for an e-mail saying that some results in [8] reminded her of her paper with Bernd Sturmfels [1]. This made me aware of the interesting results on Chow forms in [1, 4] and triggered the present paper. Chow forms are functions on spaces of *algebraic cycles* in projective space. It is therefore a pleasure to present this work in a volume in honor of Spencer Bloch.

References

[1] Dickenstein, A., B. Sturmfels, *Elimination Theory in Codimension Two*, J. Symb. Comput. 34(2): 119-135 (2002); see also arXiv:math/0102204

[2] Feng, B., Y-H. He, K. Kennaway, C. Vafa, *Dimer Models from Mirror Symmetry and Quivering Amoebae*, Adv. Theor. Math. Phys. Volume 12, Number 3 (2008), 489-545.; see also: arXiv:hep-th/0511287

[3] Franco, S., A. Hanany, K. Kennaway, D. Vegh, B. Wecht, *Brane Dimers and Quiver Gauge Theories*, J. High Energy Phys. JHEP01(2006)096; see also: arXiv:hep-th/0504110

[4] Gelfand, I.M., M.M. Kapranov, A.V. Zelevinsky, *Discriminants, Resultants and Multidimensional Determinants*, Birkhäuser Boston, 1994

[5] Hanany, A., D. Vegh, *Quivers, Tilings, Branes and Rhombi*, J High Energy Phys JHEP10 (2007) 029; see also arXiv:hep-th/0511063

[6] Kapranov, M.M., B. Sturmfels, A.V. Zelevinsky, *Quotients of toric varieties*, Math. Ann. 290 (1991) 643-655

[7] Kennaway, K., *Brane Tilings*, Int. J. Mod. Phys. A vol.22, issue 18, 2977-3038 (2007); see also: arXiv:0706.1660

[8] Stienstra, J., *Hypergeometric Systems in two Variables, Quivers, Dimers and Dessins d'Enfants*, in: N. Yui, H. Verrill, C.F. Doran (eds.), *Modular Forms and String Duality*, Fields Institute Communications, vol. 54 (2008); see also: arXiv:0711.0464

Titles in This Series

TITLES IN THIS SERIES

For a complete list of titles in this series, visit the
AMS Bookstore at **www.ams.org/bookstore/**.